Network Models for Data Science

This text on the theory and applications of network science is aimed at beginning graduate students in statistics, data science, computer science, machine learning, and mathematics, as well as advanced students in business, computational biology, physics, social science, and engineering working with large, complex relational datasets. It provides an exciting array of analysis tools, including probability models, graph theory, and computational algorithms, exposing students to ways of thinking about types of data that are different from typical statistical data. Concepts are demonstrated in the context of real applications, such as relationships between financial institutions, between genes or proteins, between neurons in the brain, and between terrorist groups. Methods and models described in detail include random graph models, percolation processes, methods for sampling from huge networks, network partitioning, and community detection. In addition to static networks, the book introduces dynamic networks such as epidemics, where time is an important component.

Alan Julian Izenman is Professor of Statistics, Operations, and Data Science at Temple University. He received his Ph.D. from the University of California, Berkeley. He was a faculty member at Tel Aviv University and Colorado State University, and was a visiting faculty member at the University of Chicago, the University of Minnesota, Stanford University, and the University of Edinburgh. He was Program Director of Statistics and Probability at NSF (1992–94). A Fellow of the ASA, IMS, RSS, and ISI, he has served on the Editorial Boards of *JASA, Law, Probability, and Risk*, and *Statistical Analysis and Data Mining*. He is the author of *Modern Multivariate Statistical Techniques* (2013).

"Izenman offers readers an extensive set of descriptions of models for networks. Emphasis is on random networks, with a chapter devoted to parametric statistical models for dependent relational ties. The author is to be praised for pulling together different material into a single, very useful book."

– Stanley Wasserman, Indiana University

Network Models for Data Science
Theory, Algorithms, and Applications

Alan Julian Izenman

Temple University, Philadelphia

CAMBRIDGE
UNIVERSITY PRESS

CAMBRIDGE
UNIVERSITY PRESS

Shaftesbury Road, Cambridge CB2 8EA, United Kingdom

One Liberty Plaza, 20th Floor, New York, NY 10006, USA

477 Williamstown Road, Port Melbourne, VIC 3207, Australia

314–321, 3rd Floor, Plot 3, Splendor Forum, Jasola District Centre, New Delhi – 110025, India

103 Penang Road, #05–06/07, Visioncrest Commercial, Singapore 238467

Cambridge University Press is part of Cambridge University Press & Assessment, a department of the University of Cambridge.

We share the University's mission to contribute to society through the pursuit of education, learning and research at the highest international levels of excellence.

www.cambridge.org
Information on this title: www.cambridge.org/9781108835763

DOI: 10.1017/9781108886666

First published 2023

Printed in the United Kingdom by TJ Books Limited, Padstow Cornwall

A catalogue record for this publication is available from the British Library.

ISBN 978-1-108-83576-3 Hardback

Additional resources for this publication at www.cambridge.org/izenman

*This book is dedicated
to the loves of my life,
Betty-Ann and Kayla*

Contents

Preface

In recent years, we have witnessed many incredible technological breakthroughs, including huge reductions in the sizes of computational devices, major innovations in storage facilities that have allowed researchers to create enormous data repositories and databases at little cost, an explosion in the variety of types and sizes of data being collected, and vast improvements in computational speed to process such data. These amazing achievements have led, in turn, to the "big data revolution" and to the introduction of the discipline of *data science*.

As a rapidly growing new field, data science is generally viewed as a unification of the disciplines of probability, statistics, machine learning, data mining, database management systems, artificial intelligence, and algorithm development. Like statistics, data science deals with the acquisition, processing, analysis, and interpretation of a wide variety of types of digital data. Such data are obtained from multiple sources, where different data types require different techniques and computational tools. Today, the amount of digital information has grown exponentially, leading to the era of "big data," especially in the areas of healthcare, business, science, and government programs.

This book concentrates on an important data type encountered in machine learning, data mining, and data science, namely, *relational data*, which record connections, for example, between people who make up friendship networks. Relational data are visualized through a network graph and studied using probabilistic models and statistical analysis. The study of networks can help in understanding and visualizing complex information often buried in large relational datasets collected on financial, biological, physics, social, business, and technological applications. Understanding relational data is of great importance in characterizing the geometry and structure of complex data networks. The combination of probability models, statistical techniques, graph theory, and computational algorithms provides us with an exciting array of tools for the analysis of network data.

In a random network, the focus is on the interpretation and analysis of a graph consisting of a set of nodes and a set of edges. The *nodes* (or *vertices*) of a graph correspond to entities such as people, animals, websites, power stations, banks and other financial institutions, genes or proteins, and neurons in the brain. An *edge* (or a *link*) in a graph is a line joining a pair of nodes that indicates a connection (if any) between those entities. If no edge exists between a pair of nodes, then there is no connection between those entities. Because network models typically assume that the edges joining pairs of nodes are random variables (hence, the name "random networks"), the observed network can be viewed as a realization from a probability distribution. With such a probability model, researchers have tried to explain how these edge-generating mechanisms work, how networks grow, and what are the conditions that can be damaging to networks. Some of these networks, such as those encountered in social network analysis, tend to be relatively small and fairly easy to manipulate, whilst other networks, such as technological and information networks – think of the World Wide Web and the Internet – are huge and computationally challenging.

This book was written to bring together many of the wide-ranging contributions of the interdisciplinary study of networks, whether theoretical, computational, or applied. Although there have already been a large number of books published on random networks (e.g., Wasserman and Faust, 1994; Easley and Kleinberg, 2010; Newman, 2010; Barrat, Barthélemy, and Vespignani, 2013; Barabási, 2016; Kolaczyk, 2017; Crane, 2018), there does not appear to be any other book that treats all the topics covered in this book. Some of the novel items in this book include the development of percolation models on various types of graphs, extensive discussions on network partitioning, methods for sampling "hard-to-access" networks, descriptions on how to deal with very large networks, and examples of many important and unusual applications.

Although there are technical theorems and mathematical derivations in this book, I have tried to cross over the invisible boundary between the theoretical and applied parts of the same discipline. I have tried to make this book as readable as possible, with historical and other informational remarks and footnotes in the text. In discussing specific applications of network theory, I have tried to give as complete a description of each application as possible in the hope that this book will also be viewed as educational and informative to a general audience.

Overview of Chapters

This book is divided into 17 chapters, which can be viewed as being arranged into eight sections:

I. Chapters 1–3 can be considered as introductory material. Chapter 1 introduces the terminology, notation, and basic research issues of graphs and networks. Chapter 2 describes seven different types of networks: technological, information, financial, social, biological, ecological, and terrorist networks. The last is a novel feature of this book and shows how the myriad of worldwide terrorist organizations are related to each other. Chapter 3 sets up definitions of the basic tools used in networks, namely, the adjacency matrix, paths, connectivity, clusters, and hubs, and also describes the real-world networks of telephone call graphs and Facebook social graphs, and provides various algorithms for graph searching and minimum spanning tree.

II. Chapter 4 begins the study of random graph models, starting with the Erdős–Rényi random graphs. Some technical tools, such as first- and second-moment methods, Chernoff bounds, Cayley's formula, and a brief account of branching processes, are given, followed by graph properties and a discussion of the emergence of a giant component.

III. Chapters 5 and 6 introduce the topic of percolation, which is closely related to the random graph models of Chapter 4. Chapter 5 restricts the percolation process to a d-dimensional lattice and Chapter 6 allows percolation to live on more general state spaces. Several applications of percolation are then described: impurity doping of semiconductors, infectious diseases and epidemics, galactic structure and star formation, polymer gelation, and the modeling of microfabrication and cellular engineering in amorphous computing.

IV. Chapters 7 and 8 describe a variety of models for unstructured networks. Chapter 7 discusses the topology of networks, which includes small-world networks, the Watts–Strogatz model, degree distributions, the power law and scale-free networks. Chapter 8 describes models of network evolution and growth, which includes the configuration model, the expected-degree random graph model, the preferential attachment model, and the random copying model.

V. Chapter 9 describes how to sample from a huge network (by sampling nodes or sampling edges) and the various types of network sampling techniques (link-trace sampling, snowball sampling, respondent-driven sampling, random-walk sampling, and forest-fire sampling), illustrated by the problem of monitoring student health in an influenza study. Also described is what to do if sampling is to be carried out when the network is "difficult to access."

VI. Chapter 10 describes a variety of parametric statistical models and techniques of parameter estimation based upon the exponential family for unstructured networks. These models are the p_1, p_2, and p^* models, the last of which is also known as the exponential random graph model. There is also a description of latent space models. Examples described in this chapter involve a friendship network of lawyers and the important issue of bullying in schools.

VII. Chapters 11–14 deal with various methods of graph partitioning for networks of unknown structure. Chapter 11 describes the different types of graph-cutting algorithms, such as minimum cuts, ratio cuts, and normalized cuts for both binary and multiway cuts. An example of graph cuts discussed in this chapter is the difficult problem of legislative redistricting of a state to avoid gerrymandering. Chapter 12 discusses the notion of community detection, a type of cluster analysis for networks. Included are the concepts of stochastic equivalence and stochastic blockmodels, modularity, regularized stochastic blockmodels, and latent cluster random-effects models. Chapter 13 discusses spectral clustering by introducing unnormalized and normalized graph Laplacians, and the spectral clustering of graphs and its regularized version. These methods are illustrated by the problem of identifying different party affiliations in a political blogs network. Chapter 14 describes the problem of modeling the presence of overlapping communities, where nodes can be members of more than a single community.

VIII. Chapters 15–17 are more technical and discuss the difficult problem of how to work with large complex networks. Chapter 15 describes the nature of very large networks, similarity measures for comparing networks, exchangeable random structures (sequences and arrays), homomorphism densities, isomorphisms, the graph coloring problem, and property testing of networks. Chapter 16 describes discrete and continuous graphons, what networks look like in the limiting case, the process of generating networks by sampling graphons, how to estimate graphons, and how to compare graphon estimates. Chapter 17 deals with dynamic networks, networks that incorporate either continuous or discrete time as an important component. Examples illustrate longitudinal social networks, with a focus on monitoring social networks for change, and dynamic biological networks, which discuss finding and counting motifs, comparing networks using graph distances, and building network models for tracking and analyzing epidemics and the spread of infectious diseases.

Audience

This book is meant for anyone interested in the theory and application of network science. It is primarily directed towards graduate students in statistics, data science, computer science, machine learning, mathematics, business, computational biology, physics, social science, and engineering, although advanced undergraduate students may find much that is interesting in the material. This book should also be useful to researchers in many diverse fields.

Readers should have some background in statistical theory and methods, probability, a good understanding of matrix/linear algebra, and multivariable calculus. Much of the necessary background material is detailed throughout the book. The applications used in this book are taken from a wide range of disciplines and a great deal of effort has been expended on describing those applications so that the reader will find the network methodology interesting and important.

Software Packages

It is very difficult to study networks and their structures without access to specialized computer software, especially graphics software. Fortunately, great graphics software is publicly available in the R computer package.

We also highly recommend the 3D graphics package `Persistence of Vision Raytracer` that should be more well known; its website is `povray.org`.

A file entitled Software Packages, which describes the major computer packages and software routines (if they are publicly available and can be readily downloaded) for carrying out the network analyses of each chapter, will be provided on the book's website. Updates will also be listed as they become available.

The datasets used to describe and analyze networks for this book are taken from numerous sources and disciplines. If data are acquired through the Internet, the data repository is listed in a footnote at the place where the data are used as illustration. There are classic datasets and some new datasets used in this book. As is the case with network data, there are some small datasets and some huge datasets, and it is suggested that the reader become familiar with both types.

Exercises

Exercises are listed at the end of each chapter. Some challenge the reader to solve theoretical problems, some illustrate methods on specific real data, and some ask the reader to write software to implement an algorithm described in the text. There is no uniformity to the level of difficulty of the exercises, and the reader is urged to try as many of them as possible.

Book Website

The book's website is located at `http://sites.temple.edu/alan`, where additional materials and the latest information will be available.

Acknowledgments

I owe a special acknowledgment to my association with John M. Hammersley, who played a major role in developing the material on percolation (Chapters 5 and 6). Hammersley visited the Statistical Laboratory at the University of California, Berkeley, as a guest of Jerzy Neyman, during the summers of 1968, 1969, and 1970. During that time, I was fortunate to collaborate with Hammersley on two papers: Hammersley et al. (1969) and Hammersley (1972).

I have enjoyed every moment of this book's gestation. Some parts of this book were written in draft form whilst I was on sabbatical during the Fall semester 2012 at the Department of Applied Mathematics and Statistics, Johns Hopkins University. I would especially like to thank Carey Priebe and his graduate students, especially Daniel Sussman, for many helpful discussions and meetings on the general topic of network modeling and for listening to some of my wild ideas. I would also like to acknowledge Daniel Naiman and John Weirman, both of whom contributed much to a most pleasant visit.

I would like to thank all those who helped, in person or by e-mail, to clear up points of my confusion, including Edo Airoldi, Harry Crane, Keith Crank, Marijtje van Duijn, Rick Durrett, Paul Krapivsky, Subhadeep Mukhopadhyay, Mark E.J. Newman, Ben Recht, C. Seshadhri, Vincent A. Traag, René R. Veenstra, and John Wierman.

I would especially like to thank my publisher, Lauren Cowles, at Cambridge University Press in New York, and her editorial assistant, Johnathan Fuentes, both of whom provided excellent and professional support during the final stages of preparing this book. I would also like to thank those individuals (Karen De Angeles, Elizabeth Sandler, Autumn Moss, Paloma Hammond, Lauren Aileen Briskman, and Melissa LeBoeuf) who helped with permissions to reproduce certain figures in the book.

The material in this book formed the basis for a graduate statistics course on Random Networks at Temple University given during the Fall semesters of 2015 and 2017, in which students in various Master's and Ph.D. programs participated. I thank the following students, who were instrumental in providing helpful comments, suggestions, and graphics for the book: Chen Chen, Patrick Coyle, Nooreen Dabbish, Nairong Fan, Lu Fang, Doug Fletcher, Lucas Glass, David Jungreis, Emily Lynch, Rich Nair, Shinjini Nandi, Abdul-Naseh Soale, Kaijun Wang, Zhentian Wei, Xu Zhang, and Lili Zhu. Others who helped me with the graphics and figures in this book include Richard M. Heiberger and student assistants Benjamin Evans and Safaniya Paul.

Last, but certainly not least, thanks go to my family, my wife Betty-Ann and my daughter Kayla, for their understanding, patience, and support during the time I spent researching and writing this book.

CHAPTER ONE

Introduction and Preview

In recent years, the science of "networks" has become a very popular research topic and a growth area in many different disciplines. Two journals, *Social Networks* (first published in 1978) and *Network Science* (first published in 2013), have appeared that focus on network theory and applications. In its inaugural issue, the journal *Network Science* defined *network science* as the "study of the collection, management, analysis, interpretation, and presentation of relational data," and noted that the more one learns about networks, the more one sees networks everywhere. In this chapter, we describe why it is of interest to study relational data and networks, some background history, and the different types of network models that have been proposed.

1.1 What is a Network?

The focus of this book is on "relational data," which consist of relationships between pairs of entities (called *nodes* or *vertices*). These entities might refer to people, institutions, documents, webpages, genes, proteins, species of animals, transportation hubs, or even terrorist organizations. Interest usually focuses on understanding the nature of those relations (called *edges*) and formulating models that try to explain how those relations were created.

The Adjacency Matrix

How do we display relational data? If we have N nodes in the network, where the ith node is represented by the symbol v_i, $i = 1, 2, \ldots, N$,[1] the relational data are translated into a square $(N \times N)$-matrix, which we denote by

$$Y = (Y_{ij}), \tag{1.1}$$

called an *adjacency matrix* (or *sociomatrix* in the social network literature),[2] in which the rows and columns each refer to the N nodes, and, in the binary case, each entry is either a zero or a one. In this matrix, if $Y_{ij} = 1$, this means that there is a relationship

[1] We use N as the size of the population network and n as the size of a sample network. In some situations, we may have access to the entire network of N nodes, but if the network is very large, we may have access only to a sample of $n \ll N$ nodes.

[2] In many books and articles on networks, the adjacency matrix is denoted by A. Here, we prefer to use Y because later on we will use X to introduce covariates into the statistical model, and having Y as the elements to predict from X makes more pedagogical sense.

1

of a specified type between the pair of nodes v_i and v_j. On the other hand, if $Y_{ij} = 0$, this means that there is no such relationship between those two nodes. Usually, the diagonal entries, $\{Y_{ii}\}$, of the adjacency matrix are zero, indicating that v_i does not have a relation with itself; if $Y_{ii} = 1$, then we say that a *self-loop* is present at that node.

In certain instances, the elements of the adjacency matrix may be weights that are placed on the edges (rather than the binary 0 or 1). Such weights indicate the strength or importance of the relationships, or maybe the number of common nodes that a particular row and column have in common. For example, think of rows and columns as movies and the entries are the number of actors two different movies have in common.

Relationships between pairs of nodes can be either directed or undirected. With an *undirected network*, the adjacency matrix will be symmetric (the relation of node v_i to node v_j is the same as the relation of v_j to v_i), and so $Y_{ij} = Y_{ji}$. In the case of a *directed network*, the relation between two nodes will be directed (by an arrow $\rightarrow$) from one node to the other (i.e., $v_i \rightarrow v_j$), the adjacency matrix may not be symmetric (the relations between v_i and v_j may not be the same), and Y_{ij} may not be the same as Y_{ji}.

From the adjacency matrix, the pairwise relations are represented as a graph or network (called a *sociogram* in the social networks literature), which we denote as $\mathcal{G}$, and the connections (called *edges*) between pairs of nodes are modeled probabilistically. The set of nodes is represented as $\mathcal{V} = \{v_i\}$ and the set of edges is represented as $\mathcal{E} = \{E_{ij}\}$, where the edge joining nodes v_i and v_j is written as $E_{ij} = (v_i, v_j)$, so that the network is written as $\mathcal{G} = (\mathcal{V}, \mathcal{E})$. If there is no relation between nodes v_i and v_j, then that non-edge is not a member of the set $\mathcal{E}$. Some examples of networks are listed in **Table 1.1**.

1.2 Why Study Networks?

Different disciplines study networks for different reasons. Networks display interactions between various entities and it is of great interest to researchers to understand the nature of those interactions. The study of networks is truly a multidisciplinary topic with many research articles appearing in the journals of different fields. Here are a few of those motivations for studying networks.

A social network analyst would try to understand the edges of a network and how to interpret them. An edge may indicate that one individual "likes" another individual (i.e., a directed edge), that there is a "close friendship" between two individuals, that two individuals are members of the same organization, that an actor has appeared in the same movie as another actor, that two authors have collaborated in scientific research, that two nations trade with each other or have a strategic alliance, and so on. Social scientists are also interested in developing search algorithms for massive networks such as the World Wide Web that would identify sociological information and "cyber communities," where nodes are html pages and directed edges are hyperlinks between webpages.

A statistical physicist would seek to understand how networks are created, how they grow, and their topological properties. An engineer would be interested in simulating the failure of a national power grid, where the nodes represent generator substations (providing the source of electricity), transmission substations (providing connections), and distribution substations (providing power) that span the country

Table 1.1 Examples of random networks

Social networks: Examples include friendships between individuals (e.g., *Facebook*, *MySpace*, *Google Plus*, *LinkedIn*), or alliances between firms. The nodes represent people (e.g., students, company directors, film actors, or even just e-mail addresses), organizations, firms, or nations, and the edges reflect social relationships between pairs of nodes, such as e-mail messages. Social scientists refer to nodes as "actors" and edges as "ties." Social relationships do not have to be symmetric, and so the edges tend to be directed. Most social networks are sparse, but characterized by high local clustering of nodes and a low average diameter (exemplified by the "six degrees of separation" phenomenon).

Document networks: The nodes in a document network represent documents from some corpus of documents (e.g., scholarly manuscripts, webpages, text documents, medical records, or images) and the edges represent links from a given document to another document through some user interface.

Information networks: The most well-known example is the World Wide Web, with billions of webpages (nodes) and hyperlinks (directed edges) from one page to another. Hyperlinks are small bits of highlighted text or pushbuttons that, when clicked on, will take you to the address of another related webpage. There is usually no reciprocal arrangement here between webpages, and so edges are directed. A special type of information network is the citation network of academic papers.

Biological networks: There are many different kinds of biological networks. For example, in *gene regulatory networks*, the nodes represent genes, proteins, their corresponding mRNAs, and protein–protein complexes, and the edges represent individual molecular reactions, such as regulation, through which the products of one gene affect those of a target gene. Most gene regulatory networks are large and complicated. In *protein–protein interaction networks*, the nodes represent proteins and the edges represent pairs of proteins that have been shown to bind together. When proteins bind with each other, they form protein complexes that perform many of the functions in a cell. Large protein–protein interaction networks are especially of interest.

Ecological networks: Nodes represent species and edges are the number of shared interactions between different species. One type of interaction present in a forest ecosystem is a host–parasitic interaction. Another example is a *food web*, which represents the "who-eats-whom" system of hierarchical relationships between species; a directed edge between nodes represents members of one species consuming members of the other species.

Transportation networks: The nodes may be geographical locations and the edges joining pairs of nodes may represent roads, railways, or airport routes that join the locations.

and the edges represent high-voltage transmission lines. A computer scientist would be concerned about the spread of a computer virus, where the edges of a computer network represent the propagation of the virus.

A computational biologist would be interested in classifying functions of proteins and, therefore, studies protein–protein interaction networks to identify groupings of proteins. A neuroscientist would be interested in understanding how neural circuits and systems function in the human brain, how these circuits relate to one another, how they vary between individuals, and how the brain activates changes from memory to attention to movement; in this network, nodes represent individual brain regions (i.e., subnetworks of a large-scale brain network) and the directed edges represent interactions between these components. A health worker may be interested in understanding the spread of a certain disease (e.g., HIV, Ebola, Zika, SARS,

COVID-19), and the directed edges represent the transmission of the disease from one individual to another.

Lawyers would be interested in the e-mail correspondence between company executives (directed edges) following a bankruptcy filing and federal investigation of possible accounting fraud (e.g., the Enron case). A financial analyst would be interested in the links between financial institutions (e.g., banks, insurance companies, stock exchanges) that would explain periods of optimism and pessimism in financial markets. Law-enforcement personnel need to track terrorist cells, and the edges of a terrorist network represent contacts made between two members of such a cell.

1.3 A Little Bit of History

The study of network analysis originated from a number of different sources. Probably the earliest instance of a graph is the riddle called the *Seven Bridges of Königsberg* that the famous mathematician Leonard Euler studied in 1736 and published in 1841.[3] The problem involved the city of Königsberg, which was built on both sides of the Pregel River. The city included two large islands connected to each other and the mainland through seven bridges. See **Figure 1.1** The problem was to map out a path through the city that would cross each bridge only once, from end to end.

Euler reworded the problem in abstract terms, which later became the foundation of graph theory. He represented the four land masses as nodes of a graph and the seven bridges as edges. See **Figure 1.2**. He then showed that the key to the problem depended upon the degrees of the nodes.[4] Euler showed that for there to be a path

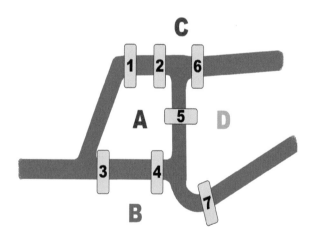

Figure 1.1 The Seven Bridges of Königsberg problem (freehand sketch). The four land masses are A: Big Island; B: Southern Bank; C: Northern Bank; and D: Small Island, and the edges are the seven bridges 1: Krämerbrückenfest (Merchant's Bridge); 2: Schmiedebrücke (Forge Bridge); 3: Grüne Brücke (Green Bridge); 4: Kottelbrücke (Connecting Bridge); 5: Honigbrücke (Honey Bridge); 6: Holzbrücke (Wooden Bridge); 7: Hohe Brücke (High Bridge).

[3]Königsberg, which was then in Prussia, was later renamed Kaliningrad, Russia.
[4]The "degree" of a node is the number of direct links (or edges) a node has to other nodes.

that satisfied the conditions of the problem, the graph would have to be connected,[5] and there would have to be zero or two nodes of odd degree. The map of the bridges showed that the latter condition did not hold (there were three nodes of degree 3 and one node of degree 5), and so no such solution exists to the problem.[6]

It was not until the 1930s that a systematic study of social networks (also called *sociometry*) was initiated by Jacob L. Moreno, a psychiatrist, and Helen Jennings, a social psychologist, who looked specifically at (1) the relations between prison inmates and (2) relations between residents in a girls' reform school (Moreno, 1932, 1934). At about the same time, W. Lloyd Warner and his business-school colleagues independently studied a social network component of the Western Electric research on industrial productivity. It was Moreno who introduced the *sociogram*, which was an early representation of a network with individuals as nodes and social relations between individuals as edges. Some development of these ideas using matrix representations of networks took place during the late 1940s and 1950s, and mathematicians introduced graph theory to social networks (Cartwright and Harary, 1958).

Interest in social networks was sporadic over the next decade, but some development of the field was carried out over the next 30 years, during which time 16 centers of social network research were set up in various countries. An additional such center was started in the early 1970s by Harrison C. White, who changed the way social networks were studied by developing mathematical models of social structure, including patterns of social relationships, and important applications to economics and sociology. During the 1970s, he and his students also introduced the concepts of "structural equivalence" and "blockmodel," which impacted the way that social network analysis was viewed by the social science community.

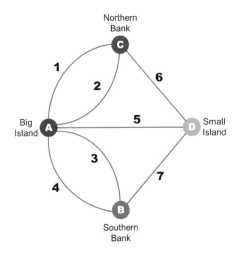

Figure 1.2 Euler's graph representation of the Seven Bridges of Königsberg problem. The nodes of the graph are the four land masses (*A, B, C, D*) and the edges are the seven bridges (1–7).

---

[5]A graph is *conected* if any two nodes can be linked by a sequence of nodes and edges.

[6]In 1875, the city of Königsberg built an eighth bridge between nodes *B* and *C*, which provided a simple solution to the problem. However, during the Second World War, most of the city was bombed and two of the bridges (Merchant's Bridge and Green Bridge) were destroyed.

During the 1950s, social network researchers started to consider the notion of "cohesive groups," which had been initiated in sociology many decades earlier, without actually defining what was meant by a "group." As social networks became a popular research area, the concept of a cohesive group became more formalized as a structural component of a network. In this scenario, social links within a group would be dense, whilst those between groups would be sparse. Since then, dozens of mathematical, probabilistic, statistical, and computational models for identifying cohesive groups have been proposed in the social networks literature.

During the late 1950s, mathematicians Paul Erdős, Alfréd Rényi, and Edgar N. Gilbert introduced probability models for the study of random graphs (Erdős and Rényi, 1959, 1960; Gilbert, 1959). The model studied by Erdős and Rényi had a fixed number of nodes and a given probability that a pair of nodes would be joined by an edge, and then they looked at what would happen when the number of edges increased. Although these probability models were much too simple, they did produce some astonishing theoretical results. One of the most surprising results was the discovery that a giant connected component would emerge when the edge probability exceeds a certain threshold.

Because of the state of mainframe computers and remote, batch-oriented computation during the 1960s, and the paucity of publicly available network data, the theoretical development of random graphs was pretty much the only game in town. The Internet was still in its infancy and could not yet play a role in social networking. E-mail was introduced in 1971 and, in 1979, *CompuServe* was the first Internet service that allowed e-mail exchanges. Compared with what we know today, theoretical models of social structure were not able to produce a reasonable representation of real-data networks.

In the 1980s, several important steps forward were made to help model social networks. First, parametric statistical models were proposed to model the structure of social relationships. The first such model was the p_1 model (Holland and Leinhardt, 1981). The p_1 model was proposed for the analysis of directed networks and was based upon the exponential family of distributions with unknown parameters to be estimated from network data called "dyads" (i.e., elements $D_{ij} = (Y_{ij}, Y_{ji})$, $i < j$, of the adjacency matrix Y). This was followed by the ERGM (exponential random graph model) or p^* model (Frank and Strauss, 1986) and later by the p_2 model that incorporated covariates into the p_1 model (van Duijn, 1995). Second, the introduction of *stochastic blockmodels* was proposed as a tool for community detection (Holland, Laskey, and Leinhardt, 1983). The theory underlying stochastic blockmodels involved a concept of *stochastic equivalence*, which was a generalized version of "structural equivalence" (by adding a probability component) that defined the notion of similarities amongst the nodes and edges.

Starting in the 1990s, the world began to shrink when the following dramatic developments occurred: the World Wide Web was introduced in 1990; the first major web browser *Mosaic* appeared in 1993, followed by *Netscape* in 1994 and Microsoft *Internet Explorer* in 1995; high-speed computational facilities became more readily available; high-density data storage became faster and more efficient, which led to the emergence of large-scale databases; more people were connected to the Internet, which could be used to transport all types of data for download; and computer software could be used for data manipulation, analysis, and graphics of networks with millions of nodes. Social networking sites, such as *Six Degrees* (1996, closed down in 2002), *Friendster* (2002, closed down in 2015), *LinkedIn* (2003), *MySpace* (2003), *Facebook* (2004), *Twitter* (2006), *Pinterest* (2010), and *Instagram* (2010), sprang up and changed the way people related to each other. With all these advances, more complicated network models were made possible. Researchers tried to make network models more realistic so that they could reproduce the complex structures of real networks. As a result, this desire to be more realistic led to research programs being focused on building statistical models to try to explain how networks were generated.

It was also during the late 1990s that physicists, who had not shown any previous interest in social networks, began to contribute significantly to the research work in this area. In efforts to build models for graph partitioning, they refocused attention from nodes to edges of the network. This development, however, was not taken kindly to at the time by those who had worked in the area for many decades and were irritated by these physicists, whom, they felt, had "crashed the world of social networks" (Bonacich, 2004). They claimed that physicists ignored the social networks literature, that they took the research topics of those working in social networks and claimed them as topics in physics, and that they encouraged many other physicists (viewed as "alien invaders") to get involved, who then "completely overwhelm[ed] the traditional social network analysts" (Freeman, 2011).

Fortunately, many of the models proposed by physicists, based upon ideas that had previously appeared in the social science literature, came to be accepted by social network analysts and, thus, helped to accelerate a revolution in social networks. These ideas included:

1. The model for the "small world" effect (Watts and Strogatz, 1998) and the "six-degrees-of-separation" phenomenon in which individuals are connected to each other by a succession of a few individuals who know each other.
2. A "power-law" model for the degree distributions of large networks (Barabási and Albert, 1999) described networks in which a few nodes each had a large degree and many nodes each had a small degree, so that the degree distributions are skewed and deviate significantly from the Poisson-distributed degree distributions of random graphs.

The small-world effect had a long history, reaching back to the late 1970s, and skewed node-degree distributions went back to the 1930s with the work of Moreno and Jennings (1938), and later de Solla Price (1976). Apparently, Barabási and Albert were not aware of either of those two articles. These physicists also succeeded in showing that networks were present in many different areas and disciplines (other than the social sciences) and, in effect, they actually broadened the study of social networks.

A series of articles by Mark Newman and Michelle Girvan (Girvan and Newman, 2002; Newman and Girvan, 2004) introduced physicists and computer scientists to the problem of cohesive groups, now called *communities*. This led to a focus on the algorithmic study of graph partitioning and community detection. Although the Girvan–Newman algorithm was shown to have its problems, it did alter the way researchers thought about modeling networks and that it was important to design efficient algorithms for doing so. The next step in the development process was focused on creating fast and efficient algorithms that would partition a very large network into communities. These concerns also succeeded in bringing computational biologists, who have networks with millions of nodes, and computer scientists into the world of network research.

So, major advances in network science since the late 1930s came from the fundamental contributions of social scientists, mathematicians, physicists, computer scientists, probabilists and statisticians, including Nobel Laureates who worked on network-related problems. By the mid-1980s, however, very little research on social networks had appeared in statistical journals. But then, perhaps because of the developments due to physicists and the various computational advances, probabilists and statisticians became interested in the analysis of random networks and network structure. Articles on networks, which had mostly been published in physics and social science journals, now also started to appear with some regularity in the major journals in statistics and probability.

The focus of the articles appearing in the statistical literature has been on studying statistical models for the analysis of networks, especially parameter estimation of ERGMs using maximum likelihood, pseudo-likelihood, and approaches using MCMC (Markov chain Monte Carlo) algorithms, analyses of stochastic blockmodels, and other related issues such as the quality of network data (primarily social network data) and what to do when confronted with bias in parameter estimates.

As network sizes grew larger and larger, it became important to consider the following question: If networks were now too large and complex to analyze in their entirety, how can one sample from a network? The answer is not as obvious as one might think. A large literature has grown up on the many ways of sampling from a very large network. Sampling can be carried out by nodes or by edges, and variations on these themes have been proposed. There is also the question of sampling from a "difficult-to-access" network (such as individuals with HIV/AIDS, homeless persons, illegal drug users, or undocumented aliens) if neither the members of the network are known (perhaps they want to be "hidden" from public attention) nor the size of the network is known. The social science literature has produced some remarkable advances related to sampling difficult-to-access populations.

The most recent addition to the theoretical literature on networks is the work on large networks and their limiting properties (Lovász, 2012). This topic has become very popular, and the literature, which includes studies on exchangeable random arrays, graph coloring, property testing in networks, and graphons and graphon estimation, has grown quite rapidly.

1.4 Building Network Models

During the last couple of decades, we saw the rapid development of probabilistic, statistical, and computational tools for modeling, analyzing, and graphing network data. As large amounts of network data became readily available on the Internet, new probabilistic methods and new statistical models were proposed to try to understand the structure of complex networks.

1.4.1 How are Networks Generated?

Given a set of nodes that make up a network, a major research goal is to discover how the edges that link pairs of nodes are formed. Many ideas have been put forward:

- *Random graph models*. Pairs of nodes are randomly joined according to a given probability.

——— **8** ———

- *Percolation models.* Bonds on a lattice are randomly opened or closed according to a given probability.
- *Small-world model.* Existing edges in a regular ring lattice are randomly rewired.
- *Configuration models.* Pairs of nodes are randomly joined subject to the condition that their degrees agree with a given degree distribution.
- *Expected-degree random graph models.* Edges are randomly selected so that the total number of edges is a Poisson random variable and the expected degree of a particular node is the weight attached to that node.

1.4.2 How do Networks Grow?

Attention has also been paid to the process of adding nodes and edges to an existing network with the goal of understanding how a network grows in size whilst still retaining its underlying properties:

- *Preferential attachment models.* Nodes and edges are randomly added in a sequential fashion to an existing network by linking, at each step, each new node to an existing node with probability proportional to its degree, so that the "rich get richer."
- *Random copying models.* A new node is added to an existing network, and then edges that emanate from that node are added by duplicating the edges of a randomly selected existing node.

Some of these ideas produce more realistic networks than others. When researchers were able to study real networks in detail, they found a number of common features that were not predicted by earlier models. They included power-law degree distributions, the presence of "hubs" (certain nodes that have very high degrees), and possible disjoint (or overlapping) communities in the network. So, any model worth its salt was expected to reproduce those features in its generated networks.

1.4.3 Statistical Models

One way of modeling network data is to use parametric statistical models, and several of these have been proposed for this purpose:

- The p_1 *model*, which uses the idea of independent "dyads" (pairs of entries in the adjacency matrix) for a binary network, has the form of a log-linear model whose likelihood is a member of an exponential family, where the model parameters are considered to be fixed effects, and maximum likelihood is used in a similar way as for contingency tables.
- The p_2 *model* is a logistic regression model that incorporates covariates into the p_1 model, where the network adjacency matrix is set as the dependent variable, and where the model parameters are considered to be random effects.
- The *exponential random graph model* or ERGM (also referred to as the p^* model in the social networks literature) is a member of an exponential family that incorporates variables based upon the number of edges (i.e., 1-stars), 2-stars, 3-stars, and triangles in the network, and a parameter that represents the average degree of the network.

- The *latent space model* is a conditional probability model that associates each node in the network with a point in some k-dimensional continuous latent space, where the presence of an edge between a pair of nodes is determined by the distance between the two points in the latent space (a small distance means likelihood of an edge, a large distance means likelihood of no edge).
- In the case of large networks, a theory of *graphons* has been proposed, and the problem of how to estimate a graphon is being considered and is now under serious development.

The p_1 and p_2 models have been used primarily within the social network community, although the ERGM has become generally more popular. Fitting an ERGM to network data by maximum likelihood is difficult, but workarounds have been proposed using logit models, pseudo-likelihood methods, and MCMC sampling methods.

1.5 Discovering Network Structure

Recognition of a number of common network features that have been observed in real-world networks has transformed the way in which network scientists now approach the study of network structure. Such features include power-law degree distributions, the presence of hubs, and disjoint or overlapping communities. These features led researchers to propose a variety of algorithms for discovering specific structures in networks. To accomplish these goals, they often adopted machine-learning techniques such as algorithms for graph partitioning and community detection.

1.5.1 Partitioning Algorithms

Any algorithm for partitioning a network into "communities" should be designed so that the communities each consist of groups of nodes that have the property that there are many edges between nodes in the same group, but only a few edges between nodes in different groups. The methods used include the following:

- *Graph cuts* employs a number of methods (binary cuts, normalized cuts, ratio cuts, multiway cuts) for dividing the nodes into two or more groups, where the objective function is different for each method.
- *Stochastic blockmodels* (SBMs) divide the nodes into disjoint communities or blocks where nodes are members of the same community if and only if they are stochastically equivalent, which means that nodes within the same community are exchangeable (with respect to some probability distribution).
- *Degree-corrected stochastic blockmodels* are stochastic blockmodels that include an additional parameter for dealing with heterogeneity in the degree of each node.
- *Spectral clustering* uses eigenvalues and eigenvectors of the graph Laplacians to divide the nodes of a network into two or more groups.
- *Overlapping-community models* relax the requirement of SBMs that networks be partitioned into disjoint communities, so that nodes can belong to more than one community.
- *Latent cluster models* extend the idea of latent space models, where communities in the network are formed, possibly using Bayesian model-based clustering of the points in a continuous latent space.

1.5.2 Estimating Network Models

A variety of approaches have been proposed to estimate the parameters that describe network models, especially the parameters of the stochastic blockmodel (or its degree-corrected version), and to determine the adequacy of a given partition of the network into a number of communities. These approaches have been used to determine the most likely number of communities and which nodes belong to which communities. They include:

- *Maximum likelihood* (ML) is a standard statistical tool that, in this instance, tries to find those community labels for the nodes that maximize the exponential-family likelihood (or log-likelihood) of the edge probabilities.
- *Maximum pseudo-likelihood*, which is equivalent to weighted least squares, can be used to analyze a large class of models (e.g., dyad-dependent models, which would not be possible using ML).
- *Bayesian approach*, which uses Gibbs's sampling and other MCMC algorithms to approximate the posterior distribution.
- *Maximum modularity* is a measure that tries to find a vector whose entries designate which nodes are members of which community. The goal, subject to certain restrictions, is to maximize a quadratic form whose compounding matrix is the adjacency matrix adjusted for the degrees of each node.

There are difficulties with some of these estimation techniques, ranging from algorithms that deliver biased estimates, to incorrect network structure determination.

1.6 Preliminaries

In this section, we give some of the technical background that the reader may find helpful in following arguments presented in this book. First, we introduce the basics of matrix notation; second, we set out Landau's notation of "big-O" and "little-o"; third, we describe the notion of computational complexity; fourth, we give definitions of random variables and moments, and then various abbreviations used in this book.

1.6.1 Matrix Notation

In this book, we often use random vectors and matrices to develop network theory. We give here an abbreviated account of the notation used so that the reader will be able to follow the developments as they are given. A more detailed description can be found in Chapter 3 of Izenman (2013).

Vectors will be denoted by uppercase or lowercase boldface letters (e.g., $\mathbf{X}$ or $\mathbf{x}$). Matrices will be uppercase boldface letters (e.g., $\mathbf{A}$, $\mathbf{\Sigma}$) or, in some cases, capital script letters (e.g., $\mathcal{X}$, $\mathcal{Y}$, $\mathcal{Z}$). All vectors will be column vectors and if the vector has J elements, it will be described, for convenience, as a J-vector. If a matrix $\mathbf{A}$ has J rows and K columns, it will be described as a $(J \times K)$-matrix. The transpose of a $(J \times K)$-matrix $\mathbf{A} = (A_{jk})$ will be a $(K \times J)$-matrix denoted by $\mathbf{A}^{\tau} = (A_{kj})$, and similarly the transpose of a column vector will be a row vector. The $(J \times J)$ identity matrix is $\mathbf{I}_J$, which has diagonal elements equal to 1s and all other entries 0s. The J-vector $\mathbf{1}_J$ is a vector of 1s. The null matrix $\mathbf{0}$ has all entries equal to zero.

If $\mathbf{A}$ is a square, $(J \times J)$-matrix, then the equation $|\mathbf{A} - \lambda \mathbf{I}_J| = 0$ has J roots, called *eigenvalues*, which are denoted by $\lambda_j = \lambda_j(\mathbf{A})$, $j = 1, 2, \ldots, J$. The set $\{\lambda_j\}$ of eigenvalues is called the *spectrum* of $\mathbf{A}$. Associated with λ_j is a J-vector $\mathbf{v}_j = \mathbf{v}_j(\mathbf{A})$ such that $(\mathbf{A} - \lambda_j \mathbf{I}_J)\mathbf{v}_j = \mathbf{0}$. The vector $\mathbf{v}_j$ is called the *eigenvector* associated with λ_j. Eigenvectors $\mathbf{v}_j$ and $\mathbf{v}_k$ associated with distinct eigenvalues ($\lambda_j \neq \lambda_k$) are orthogonal to each other (i.e., $\mathbf{v}_j^\tau \mathbf{v}_k = 0$).

The *spectral theorem* expresses a $(J \times J)$-matrix as $\mathbf{A} = \sum_{j=1}^{J} \lambda_j \mathbf{v}_j \mathbf{v}_j^\tau$, where $\sum_{j=1}^{J} \mathbf{v}_j \mathbf{v}_j^\tau = \mathbf{I}_J$. The *rank* of $\mathbf{A}$ is the number of nonzero eigenvalues of $\mathbf{A}$, the *trace* of $\mathbf{A}$ is the sum of the diagonal elements of $\mathbf{A}$ (which is the same as the sum of the eigenvalues of $\mathbf{A}$), and the *determinant* of $\mathbf{A}$ is the product of the eigenvalues of $\mathbf{A}$. If $\phi : \mathbb{R} \to \mathbb{R}$ is a function, then $\phi(\mathbf{A}) = \sum_{j=1}^{J} \phi(\lambda_j) \mathbf{v}_j \mathbf{v}_j^\tau$, where $(\lambda_j, \mathbf{v}_j)$ are the jth eigenvalue/eigenvector pair for $\mathbf{A}$. For example, $\mathbf{A}^{-1} = \sum_{j=1}^{J} \lambda_j^{-1} \mathbf{v}_j \mathbf{v}_j^\tau$ and $\mathbf{A}^{1/2} = \sum_{j=1}^{J} \lambda_j^{1/2} \mathbf{v}_j \mathbf{v}_j^\tau$.

The *singular value decomposition* of a $(J \times K)$-matrix $\mathbf{A}$, where $J \leq K$, is given by $\mathbf{A} = \mathbf{U}\Phi\mathbf{V}^\tau = \sum_{j=1}^{J} \lambda_j^{1/2} \mathbf{u}_j \mathbf{v}_j^\tau$, where $\mathbf{U} = (\mathbf{u}_1, \ldots, \mathbf{u}_J)$ is a $(J \times J)$-matrix, $\mathbf{u}_j$ is the jth eigenvector of $\mathbf{A}\mathbf{A}^\tau$, $j = 1, 2, \ldots, J$, $\mathbf{V} = (\mathbf{v}_1, \ldots, \mathbf{v}_K)$ is a $(K \times K)$-matrix, $\mathbf{v}_k$ is the kth eigenvector of $\mathbf{A}^\tau \mathbf{A}$, $k = 1, 2, \ldots, K$, and $\Phi = (\Phi_\sigma \vdots \mathbf{0})$ is a $(J \times K)$-matrix, where Φ_σ is a $(J \times J)$ diagonal matrix with the non-negative *singular values* $\sigma_1 \geq \sigma_2 \geq \cdots \geq \sigma_J \geq 0$ of $\mathbf{A}$ along the diagonal, $\sigma_j = \lambda_j^{1/2}$ is the square root of the jth largest eigenvalue of the matrix $\mathbf{A}\mathbf{A}^\tau$, $j = 1, 2, \ldots, J$.

1.6.2 Landau Notation

We need the following definitions:

Definition 1.1 A function $f(x)$ is said to be "big-O" of another function $g(x)$, written as $f(x) = \mathcal{O}(g(x))$, if, for some constants M and x_0, $|f(x)/g(x)| \leq M$ for all $x \geq x_0$.

Actually, $f(x) = \mathcal{O}(g(x))$ is a common abuse of notation because $\mathcal{O}(g(x))$ is a set. Although $f(x) \in \mathcal{O}(g(x))$ is a more correct notation, we use $f(x) = \mathcal{O}(g(x))$ with the understanding that "=" does not have the usual mathematical meaning of equality, but represents the idea that $f(x)$ "is" $\mathcal{O}(g(x))$.

Properties of the big-O notation include the following:

- The sum of two big-O functions is big-O of their sum; i.e., if $f_1(x) = \mathcal{O}(g_1(x))$ and $f_2(x) = \mathcal{O}(g_2(x))$, then, $f_1(x) + f_2(x) = \mathcal{O}(g_1(x) + g_2(x))$, assuming that g_1 and g_2 are positive functions.
- The product of two big-O functions is big-O of their product; i.e., $f_1(x)f_2(x) = \mathcal{O}(g_1(x)g_2(x))$.
- If $c \neq 0$ is a constant, then $\mathcal{O}(cf(x)) = \mathcal{O}(f(x))$.

Note that $f(x) = \mathcal{O}(1)$ implies that $f(x)$ is bounded above for all $x \geq x_0$.

Definition 1.2 A function $f(x)$ is said to be "little-o" of another function $g(x)$, written as $f(x) = o(g(x))$, if, for every choice of a constant k, there exists a constant a such that the inequality $f(x) < kg(x)$ holds, for every $x > a$.

This implies that $f(x)$ grows no faster than $g(x)$. Also, this is equivalent to saying that, if $g(x) \neq 0$, $f(x)/g(x) \to 0$ as $x \to \infty$. More correctly, $f(x) \in o(g(x))$ because $o(g(x))$, like $\mathcal{O}(g(x))$, is actually a set. For example, $1/x = o(1)$, $x = o(x^2)$, $x^2 = o(x^3)$.

Note that every f that is $o(g)$ is also $\mathcal{O}(g)$, but not every f that is $\mathcal{O}(g)$ is also $o(g)$.

1.6.3 Computational Complexity

Many of the techniques presented in this book will be described by algorithms. It is important for such algorithms to run as fast and efficiently as possible. Algorithms that run in polynomial time (i.e., $\mathcal{O}(n^k)$, for some constant k, where n is the size of the input) are called "tractable" and are considered to be "efficient"; they are much faster than the "intractable" problems that take exponential time (i.e., $\mathcal{O}(2^{n^k})$, for some constant k) to run.

The study of which kind of computational resources (e.g., time, storage space) it takes for an algorithm to solve a particular task is referred to as *computational complexity*, which has become a central part of computer science (Goldreich, 2008). Because several of the proposed techniques in this book are referred to as $\mathcal{NP}$-hard or $\mathcal{NP}$-complete, which necessitate finding approximations to the desired solutions, we explain briefly the major naming conventions in this field:

- *P*. This is a class of all deterministic[7] decision problems (problems whose output is either Yes or No) that can be solved quickly in polynomial time. Examples include linear search, binary search, minimum cut, shortest path, and minimum spanning tree. $\mathcal{P}$ stands for polynomial time.
- *NP*. This is a class of all nondeterministic decision problems such that if the answer to the problem is Yes, its proof can be *verified* quickly in polynomial time if we know the solution. $\mathcal{NP}$ stands for "nondeterministic polynomial time."
- *NP-hard*. This is a class of decision problems that are at least as hard as the hardest problems in $\mathcal{NP}$. Problems that are $\mathcal{NP}$-hard do not have to be members of the class $\mathcal{NP}$. In other words, a decision problem A is $\mathcal{NP}$-hard if, for any other problem B that is in $\mathcal{NP}$, B can be reduced to A in polynomial time. An example is the traveling salesman problem.
- *NP-complete*. This is a class of decision problems $A \in \mathcal{NP}$ for which any other problem $B \in \mathcal{NP}$ can be reduced to A in polynomial time. This is an incredibly strong requirement; $\mathcal{NP}$-complete problems are the most difficult problems in $\mathcal{NP}$. A problem is $\mathcal{NP}$-complete if it is in $\mathcal{NP}$ and is also $\mathcal{NP}$-hard. Thus, $\mathcal{NP}$-complete is a stronger concept than $\mathcal{NP}$-hard.

The computational-complexity classes of $\mathcal{P}$ and $\mathcal{NP}$ are formally defined in terms of languages and Turing machines. For more details, see, for example, Marion (1994).

P vs. NP

Every $\mathcal{P}$ problem is also an $\mathcal{NP}$ problem. So, $\mathcal{P} \subseteq \mathcal{NP}$. However, it is unknown whether $\mathcal{P} = \mathcal{NP}$ or $\mathcal{P} \subset \mathcal{NP}$. If $\mathcal{P} = \mathcal{NP}$, then a problem that can be verified

---

[7]Given a specific input, a deterministic algorithm will always yield the same output. On the other hand, given a specific input, a nondeterministic (or probabilistic) algorithm, because it relies on random choices, can yield different outputs each time the algorithm is run.

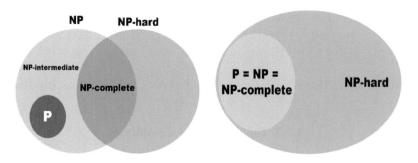

Figure 1.3 Left panel: $P \subset NP$. Right panel: $P = NP$.

quickly (i.e., in polynomial time) can also be solved quickly. In fact, there is a $1 million international prize (the first of the seven Millenium Prizes listed by the Clay Mathematics Institute) for the first person who can prove whether or not $\mathcal{P} = \mathcal{NP}$. This is a famous problem that has remained unsolved for decades. See **Figure 1.3** for diagrams of the $P \subset NP$ case and the $P = NP$ case.

If $\mathcal{P} \subset \mathcal{NP}$, then it has been shown that there are $\mathcal{NP}$ problems that are neither in $\mathcal{P}$ nor in $\mathcal{NP}$-complete; they are called $\mathcal{NP}$-*intermediate* problems. It is thought, for example, that the *graph isomorphism problem* (determining whether two finite graphs are isomorphic to each other; see Section 15.4.2) may be a member of $\mathcal{NP}$-intermediate; in fact, it is unknown whether the graph isomorphism problem is in $\mathcal{P}$, $\mathcal{NP}$-complete, or $\mathcal{NP}$-intermediate, although it is generally felt that it is not $\mathcal{NP}$-complete. Also, if $\mathcal{P} \subset \mathcal{NP}$, then $\mathcal{NP}$-complete problems cannot be solved in polynomial time.

1.6.4 Outcomes, Variables, and Moments

If A is a random event or outcome, then $P\{A\}$ represents the probability that A will occur. The conditional probability that an outcome A will occur given that outcome B has already occurred will be denoted by $P\{A|B\}$.

The expected value (or mean) of a random variable X is $\mu_X = E\{X\}$, and the variance of X is $\sigma_X^2 = \text{var}\{X\} = E\{(X - \mu_X)^2\} = E\{X^2\} - \mu_X^2$. The standard deviation of X is $\sigma_X = \text{sd}\{X\} = (\text{var}\{X\})^{1/2}$. If X and Y are two random variables, the covariance between them is $\text{cov}\{X, Y\} = E\{(X - \mu_X)(Y - \mu_Y)\}$. The correlation coefficient between X and Y is $r = \text{cov}\{X, Y\}/(\text{sd}\{X\} \cdot \text{sd}\{Y\})$, and $-1 \le r \le 1$.

If $\mathbf{X}$ is a random vector, then its expected value is $\boldsymbol{\mu}_X = E\{\mathbf{X}\}$ and its covariance matrix is $\Sigma_{XX} = \text{cov}\{\mathbf{X}\} = E\{(\mathbf{X} - \boldsymbol{\mu}_X)(\mathbf{X} - \boldsymbol{\mu}_X)^\tau\}$.

1.6.5 Abbreviations

Throughout this book, we have employed several common abbreviations: "aka" means *also known as*; "wrt" means *with respect to*; "iff" means *if and only if*; "iid" means *independently and identically distributed*; "wpr" means *with probability*; "a.s." means *almost surely*; "a.e." means *almost everywhere*; and "lhs" and "rhs" mean *left-hand side* and *right-hand side*, respectively.

1.6.6 The Nature of Infinity

In our study of networks, we come across many instances in which infinity plays a role: infinite networks, infinite arrays, infinite sequences of random variables, and number of nodes tending to infinity. The *Oxford English Dictionary, Vol. I* (1971) defines "infinite" as "unbounded, unlimited, . . . , immeasurably great in extent, duration, or other respect."

Our preference, however, and we hope that it will be amusing also for the reader, is the following definition of the concept of infinite and infinity:

Infinite: *Bigger than the biggest thing ever and then some. Much bigger than that in fact, really amazingly immense, a totally stunning size, real "wow, that's big," time. Infinity is just so big that by comparison, bigness itself looks really titchy. Gigantic multiplied by colossal multiplied by staggeringly huge is the sort of concept we're trying to get across here.*

Attributed to *The Hitchhiker's Guide to the Galaxy* in *The Restaurant at the End of the Universe*, by Douglas Adams, New York: Harmony Books, pp. 127–128 (1980).

1.7 Further Reading

There have been many excellent books written on networks and network modeling, including Kolaczyk (2009); Easley and Kleinberg (2010); Newman (2010); Kolaczyk (2017) and its accompanying volume, Kolaczyk and Csárdi (2014), that uses the computer language R to analyze networks; and Crane (2018). Other books include Palmer (1985), Wasserman and Faust (1994), Watts (1999), Janson, Luczak, and Rucinski (2000), Bollobás et al. (2001), Barabási (2002), Doreian, Batagelj, and Ferligoj (2005), Durrett (2007) and Barrat, Barthélemy, and Vespignani (2013). A very interesting introductory and nontechnical book on networks, particularly for the social sciences, is Watts (2004). The book edited by Newman (2006) contains all the main papers specific to the development of network theory and applications.

Special issues of statistics journals highlighting network modeling include *Statistical Analysis and Data Mining*, **4** (October 2011), *The Annals of Applied Statistics*, **4** (March and June 2010), and *Statistical Science*, **19** (August 2004). Journals dedicated to network research are *Social Networks* and *Network Science*.

Review articles include Albert and Barabási (2002), Dorogovstev and Mendes (2002), Newman (2003), Goldenberg et al. (2009), Fortunato (2010) and Salter-Townshend et al. (2012).

An excellent historical account of the development of social networks is given by Freeman (2004), who describes the various approaches, influences, and contributions to social network analysis by social scientists, physicists, biologists, and computer scientists. This historical account was updated by Freeman (2011).

CHAPTER TWO

Examples of Networks

In this book, we develop the theory and methodology for modeling networks. Before we do that, however, we will find it useful to describe examples of the types of real-world networks that exist today. We will be referring to these networks throughout the book. Indeed, there are many different kinds of networks, some small and some big, some simple and some, by the nature of their underlying structure, very complicated.

2.1 Introduction

In this chapter, a network $\mathcal{G}$ will be represented as a set of nodes (or vertices) $\mathcal{V}$ and a set of edges (or connections, links, or interactions) $\mathcal{E}$. We write this as $\mathcal{G} = (\mathcal{V}, \mathcal{E})$. In each of these examples, we describe the $\mathcal{V}$ and $\mathcal{E}$ that are associated with the specific network application.

2.2 Technological Networks

2.2.1 The Internet

The *Internet* is a worldwide system of computer networks that represents the physical links between computers and other telecommunication devices. It is usually confused in everyday speech with the World Wide Web.

The Internet originated sometime during the mid-1980s, although precursor networks had already appeared 20 years earlier. In the early 1960s, the main ideas behind packet-switching theory had been published (Kleinrock, 1964). By the end of the 1960s, ARPANET,[1] the first packet network, had been developed and the first four sites were UCLA (University of California, Los Angeles), SRI (Stanford Research Institute), University of Utah, and UCSB (University of California, Santa Barbara). In 1972, a basic version of electronic mail (e-mail) was demonstrated and a standardized *Internet Protocol Suite*, also known as TCP/IP,[2] was introduced. In 1973, the ideas behind the *Ethernet*, a networking system, were outlined in a memo by

[1]Developed at the Advanced Research Projects Agency (ARPA).

[2]Short for *Transmission Control Protocol/Internet Protocol*. The Internet Protocol provides addressing systems for computers on the Internet. An IP address consists of a unique 32-bit number, which is divided into four 8-bit fields; each field is a decimal number 0–255, and the four fields are separated by dots.

Robert Metcalfe at Xerox PARC[3] and followed up with published articles and patents over the next few years. During the 1980s, PCs, workstations, and LANs (local area networks) were developed and the fledgeling Internet was in widespread operation. By 1982, private, public, academic, business, and government network devices were linked to each other through TCP/IP.

The role of TCP is to establish a reliable connection between computers for transferring information (i.e., data) from one computer to another (e.g., sending an e-mail message or downloading a webpage). This process works by breaking up each data file into *packets*, where each packet is limited to at most 1460 bytes; the packets are each sent out by the *source computer* over a succession of transmission links (which could be copper cable, optical fiber, or wireless) and *routers* (as determined by a *routing table*) along a *network path*; when received by the *destination computer*, the packets are reassembled into the original file. Routers are highly specialized, high-speed computers that act as intermediaries through which data packets are forwarded between computer networks. A *routing table* is a database, usually stored in a router and constructed from information provided by routers using the *Border Gateway Protocol (BGP)*; the BGP is responsible for exchanging routing information between all of the major *Internet Service Providers (ISPs)* that connect users to their networks. The routing table lists the possible routes to the destination computer address, makes decisions about the best route to send the packets it receives, and also contains information about the topology of the surrounding network. The role of IP is to enable the packets to be routed across the Internet. To facilitate its journey through the Internet, each packet is imprinted with a 20-byte *TCP header* (containing information to control the connection and reassemble packets) and a 20-byte *IP header* (containing the IP addresses of the source and destination computers and the packet size).

As a network, the Internet can be viewed as a two-level hierarchy: the first level is the router level and the second level is the interdomain level. In the *Internet router (IR) level*, the nodes are the routers and the edges are the physical connections between them. **Figure 2.1** displays the Internet network of three billion users through *skitter* data, a tool for dynamically discovering Internet topology and connectivity, with selected backbone ISPs colored separately.

The *interdomain level* (more often referred to as the *Autonomous System (AS) level*) has many *domains* (consisting of hundreds of routers and computers), each of which is represented by a single node, and an edge joins two domains if there is at least one route that connects them. Each AS is identified by a globally unique 16-bit *Autonomous System Number (ASN)* that can have any number between 0 and 65535, as long as no conflicts occur. When routers have the same ASN, they are all sharing the same routing table. Most ISPs, large corporations, and universities have their own specific ASN. For example, Google uses AS 15169. Some of these numbers, however, are reserved for special purposes: the block of ASNs from 64512 through 65534 is reserved for private use, and the ASNs 0, 23456, and 65535 are also reserved. Currently, there is a proposal pending that would expand the size of the ASNs from 16 bits to 32 bits.

In 1990, the ARPANET was decommissioned and the U.S. National Science Foundation together with various sources of private funding were instrumental in

---

[3]PARC is short for Palo Alto Research Center.

Figure 2.1 The Internet is pictured here as a network of about 3 billion users. The nodes are routers and the edges are the physical connections between them. Source: Center for Applied Internet Data Analysis (CAIDA), located at UC-San Diego. Image created by Bill Cheswick and Hal Burch. Reproduced with permission from FireMon.

merging the many networks into the backbone of a global network. The Internet was originally an academic tool, but, by 1995, it quickly became a commercial success, and has now become an indispensible part of everyday life. At the end of 2014, it was estimated that there were over 3 billion Internet users worldwide, representing about 40% of the world's population.[4]

So far, there is no centralized governance in the Internet. Each constituent network sets its own policies. However, there is an organization, the *Internet Corporation for Assigned Names and Numbers (ICANN)*, that directs and maintains the two principal name spaces on the Internet, the *Internet Protocol address* space and the *Domain Name System*.

2.2.2 North American Power Grid

The electric power grid of North America is a giant network of high-voltage, power-transmission wires. It has been described as the "supreme engineering achievement of the 20th century." Instead of having a national grid, North America has a network consisting of three grids (called *interconnections*) that move electricity around the country, from hundreds of power plants to millions of consumers. The two largest are the *Eastern Interconnection System*, which operates in states east of the Rocky Mountains, and the *Western Interconnection System*, which covers the Pacific Ocean to the Rocky Mountain states. The Eastern and Western Systems are tied at six

[4]www.internetlivestats.com

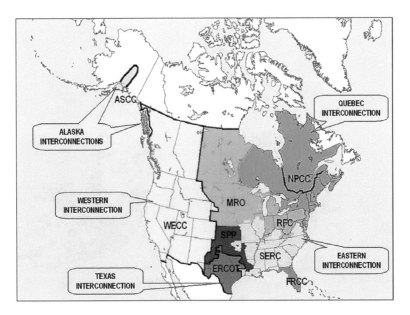

Figure 2.2 North America Electric Power Grid.

points. The third, and smallest, grid is the *Electric Reliability Council of Texas* (aka the *Texas Interconnection System* or ERCOT), which is an alternating-current (AC) electrical grid that covers most of Texas (El Paso, the upper Panhandle, and some of East Texas are each on other grids). **Figure 2.2** displays a geographical view of the North America power grid system.

For local control, the entire North American grid is subdivided into 150 smaller subsections so that power can be monitored locally. There are also two minor AC electrical grids in North America: the *Québec Interconnection System*, which covers all of the Province of Québec and is tied at two points to the Eastern Interconnection, and the *Alaska Interconnection System*, which covers a portion of the state of Alaska (South-Central Alaska and the Alaska Panhandle) and is not tied to either of the other interconnection systems. Power stations are usually located away from heavily populated areas, typically near a fuel source (e.g., at a dam site), and tend to be quite large.

Although falling trees can disrupt transmission lines, large-scale failures rarely occur on a modern power grid. There have been some huge failures of the power grid, most notably the August 2003 blackout in the Northeastern United States and Canada, which was triggered by a series of monitoring and communication breakdowns, and bad vegetation management, and the February 2021 historic snow-storm and single-digit temperatures in Texas that led to the collapse of much of ERCOT's power grid, resulting in lengthy blackouts due to wind turbines that froze and stopped spinning, thousands of burst pipes, and power plants that were not fully weatherized. The power grid is susceptible to natural disasters (e.g., hurricanes, floods, earthquakes, and even solar flares) and to physical attacks. Identifying and shielding the power grid at its most vulnerable sites is obviously a necessity. This raises questions regarding the robustness and resiliance of the network. Some have recommended adding a small number of longer transmission lines to provide shortcuts to different parts of the grid (Bashan et al., 2013).

Figure 2.3 The Western Interconnection System. There are 4,941 nodes and 6,594 edges. The nodes represent transformers, substations, and generators, and the edges represent high-voltage transmission lines of the grid. The nodes are colored by modularity class, which groups nodes by how connected they are and sized by degree.

Network studies of the North American power grid have typically focused on the Western Interconnection System (Watts and Strogatz, 1998). The topology of this large power grid is represented by a sparse, undirected network with 4941 nodes (representing transformers, substations, and generators, which convert high-voltage electricity into low-voltage power for homes and businesses) and 6594 edges (representing high-voltage transmission lines of the grid).[5] The network is displayed in **Figure 2.3**. The maximum degree for any node is about 19 edges. Many of the nodes are connected linearly in strings, each connected only to one or two other nodes. A number of assumptions have been made so that the network can be analyzed (Watts, 2004, Section 5.2). For example, all transmission lines are assumed to be identical (i.e., unweighted), which ignores the fact that voltages differ considerably and that different lines have significantly different properties. The ERCOT network has a connected graph with 5514 nodes and 6522 edges.

2.3 Information Networks

[5]The Western power-grid data are available at www-personal.umich.edu/~mejn/netdata/ or as an R dataset at toreopsahl.com/datasets/#uspowergrid. The data are also available through the UCI Network Repository sites.google.com/site/ucinetsoftware/datasets/power.

2.3.1 The World Wide Web

The *World Wide Web* (*WWW* or *Web*)[6] is a gigantic and complex network of information built on top of the Internet. The nodes of the network represent text documents (i.e., *webpages*) and the edges represent *hyperlinks* (or URLs) that allow one to navigate from one text document to another. Documents on the Web can be located anywhere in the world, but distance is not an impediment to reaching them; one just needs to click on the hyperlink to reach the target webpage. The Web operates as a central component in our lives, from online consumer purchasing to electronic banking, business conferences, educational opportunities, news gathering, book publishing, music, movies, entertainment, games, and sporting events. The Web has encouraged the development of new ways people interact with each other, such as e-mail, instant messaging, two-way interactive video calls, and social networking.

The Web was invented by the British computer scientist Tim Berners-Lee,[7] who worked as a software engineer at CERN[8] in Switzerland. In 1989, he circulated a proposal of a global hypertext[9] project for a CERN-distributed information system, but soon realized that his idea could be applied more generally. In 1990, he and Belgian computer scientist Robert Cailliau proposed using hypertext "to link and access information of various kinds as a web of nodes in which the user can browse at will." Berners-Lee completed the first website that December, he wrote the first web browser *WorldWideWeb* (later to be renamed *Nexus*), and reported his work to a newsgroup in August 1991. The first server located outside Europe was installed in 1991 at SLAC.[10] Berners-Lee then proposed that hypertext should be married to the Internet, but when no-one else took up his challenge, he completed the task himself, creating *HyperText Markup Language (HTML)*[11], the publishing format for the Web; *HyperText Transfer Protocol (HTTP)*, which allows all linked resources on the Web to be retrieved; and a system of globally unique identifiers, including the *Uniform Resource Locator (URL)* and *Uniform Resource Identifier (URI)*, which provides each resource on the Web with a unique address. In April 1993, CERN announced that the WWW would be placed in the public domain, free to everyone.

Also in 1993, the *Mosaic* web browser that was developed by a team at the University of Illinois at Urbana-Champaign became publicly available; *Mosaic*'s graphical user interface (GUI) was the first graphical web browser and, although it was very slow and did not handle downloading pictures well, it quickly became the most popular Internet protocol. In 1994, Marc Andreesen founded Netscape Corporation, whose *Netscape Navigator* soon became the primary web browser, Berners-Lee formed the World Wide Web Consortium to administer the development of WWW standards, and *Amazon* was founded, predicting that online shopping

---

[6]In a November 1999 issue of *The Independent on Sunday*, the author Douglas Adams made the comment that "The World Wide Web is the only thing I know of whose shortened form takes three times longer to say than what it's short for."

[7]Berners-Lee was knighted in 2004 for "services to the global development of the Internet."

[8]Short for *Conseil Européen pour la Recherche Nucléaire*. It is the European Particle Physics Laboratory, located in Geneva, Switzerland.

[9]The term "hypertext" was coined in 1963 by Ted Nelson.

[10]Short for Stanford Linear Accelerator Center, located in Palo Alto, California.

[11]The term "markup language" was coined in 1970 by Charles Goldfarb.

would be the way of the future. Over the next decade, the following web browsers appeared: Microsoft's *Internet Explorer* (1995), *Mozilla* (1998), Apple's *Safari* (2003), Mozilla's *Firefox* (2004), and Google's *Chrome* (2008).

Although tools for searching files on the Web have been available since 1990, no real search engine – as we would know it today – existed prior to mid-1993. Then several search engines appeared, including *Excite* (1993), *Lycos* (1994), *Yahoo! Search* (1994), *Infoseek* (1994), *WebCrawler* (1994), *AltaVista* (1994), and *Inktomi* (1996). In 1998, Larry Page and Sergey Brin applied their research to an information-retrieval problem, leading to their PageRank algorithm (Brin and Page, 1998), and, in the same year, they introduced the *Google* search engine. *Google* quickly achieved better results than any of the other search engines and has since become the world's most popular search engine, with a market share of 60–70%, and handles more than three billion queries every day.[12] Other prominent search engines include the Chinese *Baidu* (2000) and Microsoft's *Bing* (2009), which currently powers *Yahoo! Search*.

How do search engines know where to look for information when the Web has so many webpages? The answer is that they use *web crawlers* (also known as *spiders* or *bots*, short for *robots*), which are computer programs that are designed to explore the interconnections between webpages (Olston and Najork, 2010). For example, *Bing* has web crawler *bingbot* and *Google* has *Googlebot*. Essentially, web crawling is an application of breadth-first search on the Web. Starting with a carefully selected set of "seed" URLs to visit, the web crawler visits those URLs, downloads all the webpages addressed by those URLs into a *repository*, identifies and stores all hyperlinks that are encountered in those webpages, follows all links to other webpages, reads the documents on those webpages, and creates a searchable index based upon the words and their locations contained in each document.

Different web crawlers store the retrieved information in different ways. For example, *Google* stores all or part of the source page (or *cache*) as well as information about the webpages, whereas *AltaVista* stores every word of every page it finds. A cached page also retains the actual search text, which is useful if and when the page is updated and the search terms are no longer included. The web crawler has to keep track not only of the URLs that have yet to be downloaded (the *frontier*) but also of those that have been already downloaded (the *URL-seen*), so as not to keep downloading the same page over and over again. Nevertheless, it has been shown that there is a significant amount (estimated at about 11%) of exact duplication of webpages (Broder, Glassman, and Manasse, 1997).

Web crawlers are only able to access, download, and index a fraction of the Web's total content. There are several reasons for this. First, web crawling has to be carried out selectively and in a carefully controlled order so that websites with malicious content are avoided. Second, web crawlers are required to obey a convention called the *Robots Exclusion Protocol*, which allows a website administrator to forbid web crawlers from crawling their site, or some pages within the site. These restrictions are specified in a file /robots.txt, and, if the file exists, the web crawer is supposed to follow those rules. Third, many web crawlers do not go very "deep" into websites, where "deep" refers to the number of directories in a URL. In this way, they avoid the possibility of downloading an infinite number of dynamically generated webpages.

---

[12] In 2015, Google Inc. created a new public company, Alphabet Inc., that would be the parent to a collection of companies including Google.

Fourth, websites and webpages appear and disappear with rapid frequency: it is estimated that during the course of a year, 80% of webpages disappear; further, new links are created at the rate of 25% per week and old links are retired at a rate of 80% a year.

Web crawlers have been used in various ways, including being a main component of web search engines, web archiving, web data mining, and web monitoring.

2.3.2 Citation Networks

Citations are derived from the list of references to other papers that are found in the bibliography section of a published article. We can view citations as a network in which the nodes are papers published in academic journals and the edges are directed links (i.e., arrows); there is an arrow pointing from paper X to paper Y (i.e., $X \rightarrow Y$) if the earlier paper, Y, was cited in the bibliography (or in a footnote) in paper X. Then, the in-degree (i.e., the number of incoming arrows to a given node from other nodes in the network) for node Y is its citation count. Such a network of scientific papers was initiated by de Solla Price (de Solla Price, 1965), one of the founders of the field of bibliometrics, who counted a paper's number of direct citations as an assessment of scientific research. Citations to academic papers are now publicly available online at websites such as the *Science Citation Index*,[13] the *Social Science Citation Index*,[14] and the *Arts and Humanities Citation Index*.[15] All these citation indexes can be accessed online, for a fee, through the *Web of Science*. Other competing online citation databases include *Scopus*[16] and *Google Scholar*.

In **Figure 2.4**, we graph the high-energy physics-paper citation network from the Stanford University website on network data[17] that were originally made available for the 2003 KDD Cup competition (Gehrke, Ginsparg, and Kleinberg, 2003; Leskovec, Kleinberg, and Faloutsos, 2005).

There is also a very interesting website[18] that contains the citation network data of 30,288 majority opinions written by the U.S. Supreme Court and the cases they cite that were extracted from the *United States Reports*.

A recent article (Ji and Jin, 2016) reported on the collection of a network dataset[19] of citations from 3248 published papers in four of the top statistical journals, *Annals*

---

[13] ip-science.thomsonreuters.com/cgi-bin/jrnlst/jloptions.cgi?PC=K

[14] ip-science.thomsonreuters.com/cgi-bin/jrnlst/jloptions.cgi?PC=SS

[15] ip-science.thomsonreuters.com/cgi-bin/jrnlst/jlresults.cgi?PC=H

[16] www.elsevier.com/online-tools/scopus

[17] Data on citation networks can be found at the Stanford University website, snap.stanford.edu/data/#citnets, where there are two directed-network datasets on high-energy physics-paper citation networks and a directed-network dataset on the U.S. Patents citation network. Other websites that contain citation datasets include aminer.org/citation and the Microsoft Academic Graph website, www.microsoft.com/en-us/research/project/microsoft-academic-graph.

[18] jhfowler.ucsd.edu/judicial.htm

[19] The data can be found at any of three websites:
faculty.franklin.uga.edu/psji/sites/
faculty.franklin.uga.edu.psji/files/SCC2015.zip
www.stat.cmu.edu/~jiashun/StatNetwork/PhaseOne

Figure 2.4 High-energy physics-paper citation network from the e-print `arXiv`. Covers all citations within the dataset during the period January 1993 to April 2003 (124 months). There are 24,811 nodes and 348,826 edges in this graph. Nodes (and edges) that were connected only to one or two other nodes were removed. Source: Leskovec, Kleinberg, and Faloutsos (2005).

of Statistics, Biometrika, Journal of the American Statistical Association, and *Journal of the Royal Statistical Society, Series B*, during the period 2003–2012. A similar study of the above four journals plus *Statistical Science* during the period 1970–2015 was published by Anderlucci, Montanari, and Viroli (2019).

It is now recognized that just counting citations is a poor indicator of scientific contribution. For example, a paper's ranking depends to a large extent upon the size of the field of study; also, newer papers have fewer citations than older papers because they have not been around that long and so have not received as many citations. On the other hand, older, path-breaking papers may receive fewer citations as time proceeds because books and review articles that cite these papers themselves get the citations rather than the original papers.

Since that time, several measures have been proposed for assessing the impact of a published paper. These measures include the *h-index* (Hirsch, 2005), *CiteRank index* (Walker et al., 2007), *SARA* (Radicchi et al., 2009), and *Eigenfactor* (West, Bergstrom, and Bergstrom, 2010). *CiteRank* is a modification of Google's *PageRank* algorithm that reflects more appropriately the search activity of scientific researchers than would PageRank, which is designed for the Web. Whereas nodes and edges on the Web can appear and disappear over time, links between papers are permanent and cannot be removed or updated.

Another measure is the *citation wake* (Klosik and Bornholdt, 2014), which does not focus on the direct citation count of a paper. Instead, the citation wake focuses on whether a paper "started something" by estimating its "word-of-mouth dynamics" from the subsequent citation network. A paper's "wake" consists of all papers that have cited it, either directly or indirectly. Clearly, a paper can only receive citations from papers published at a later date. All papers in a paper's wake are then assigned to neighborhood layers (where the number of layers is up to the user) according to the length of the shortest path to the paper, where the "shortest path" is the minimal number of processing steps of an idea. The paper's *wake citation score (WCS)* is then computed as a weighted sum of the total number of papers in each layer. There is also a built-in adjustment factor for the effect of time: the earlier the paper was published, the more citations it is likely to have. The WCS yields a ranking of papers that is quite different from a list of papers ranked by number of citations. In a study of papers published in *Physical Review* during the last century, Klosik and Bornholdt show that WCS captures more papers coauthored by Nobel Prize laureates in physics amongst the top-ranked publications than any other citation measure.

Slightly different measures are the related concepts of co-citation and bibliographic coupling. *Co-citation* (Small, 1973) is the frequency with which two papers, A and B, say, are cited together by other papers. The more co-citations two papers receive, the higher their *co-citation strength*. Co-citations indicate that the two papers, A and B, deal with similar topics, and so is a type of similarity measure. Following a number of strange search engine results, it has been suggested in several popular blogs that Google may have introduced co-citation as a new way of ranking websites in search engine results. In the Web context, co-citation is taken to refer to a website that is cited or mentioned by two different sources without having a hyperlink involved. It is often referred to as "link building without links." The greater the number of websites that mention or refer to a given website, the higher that particular website is ranked. Co-citations may be an effort by Google to reduce manipulation of search engines (one cannot force anyone else to co-cite) and, at the same time, improve upon the PageRank algorithm. Another application of co-citation has been to the analysis of gene networks associated with honey bees, where pairwise gene interactions are inferred if they are mentioned within the same sentence of a written abstract published in *PubMed* (Mullen et al., 2014).

Bibliographic coupling (Kessler, 1963a,b) can be viewed as the opposite of co-citation. If two papers, A and B, say, each reference the same third paper, C, then A and B are said to be *bibliographically coupled*. It suggests that A and B are related using bibliographic coupling as a similarity measure. However, there is no reason to say that A and B are referencing the same information in C. The *coupling strength* of two papers is the number of common papers they both reference. For example, if papers A and B each reference papers C, D, and E, then their coupling strength is 3. The higher the number of common papers that are referenced by two papers, the greater their coupling strength. One way of thinking about the difference between co-citation and bibliographical coupling is that the former essentially looks forward in time, whilst the latter is a retrospective approach.

2.4 Financial Networks

The study of networks has played a major role in understanding financial markets. Financial dealings between traders can be viewed as directed graphs, where the nodes represent financial traders and the edges show which buy–sell transactions take place between pairs of traders during a given period of time. The stability of the financial network is a real concern for financial regulators, who have to keep an eye out for possibilities of turmoil in the markets due to price manipulation and distortion.

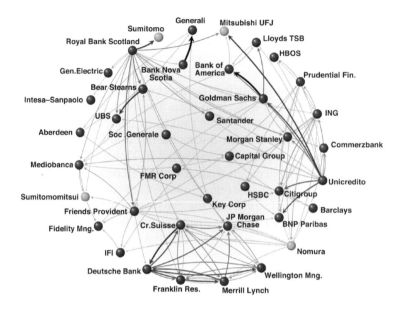

Figure 2.5 A portion of the international financial network as of end 2007. The nodes represent major financial institutions and the edges are directed and weighted and represent the strongest existing relations between them. Node colors show different geographical areas: red = European Union members, blue = North America, and green = other countries. Source: Schweitzer et al. (2009). Data provided by Orbis Database. Reprinted with permission from AAAS.

In this section, we first describe the network of banks and shadow banks, and the repercussions from the 2008 financial crisis. Then, we describe the transaction network of Bitcoin, the most well-known of the so-called cryptocurrencies.

2.4.1 Networks of Financial Institutions

Commercial banks, investment banks, insurance companies, and other financial institutions form an interconnected financial system that is concerned with the study of international capital flows in the presence of risk. The importance of the Internet, globalization, and the growth of electronic transactions has changed how financial institutions operate.

The worldwide system of financial institutions can be viewed as an extensive network (see **Figure 2.5**), where nodes of the network are "core" financial institutions (e.g., commercial banks, broker-dealers, insurance companies) and non-bank entities (also referred to as "peripheral" organizations or, more popularly, as "shadow banks"). Shadow banks are a system of diverse credit institutions, such as finance companies, investment banks, mortgage companies, money-market mutual funds, and credit hedge funds. They do not accept deposits as would a commercial bank, they do not own banking licenses, and they are subject to fewer regulations than commercial banks, but they do carry out certain traditional banking functions such as credit intermediation through maturity, credit, and liquidity transformation. While the assets of commercial banks are relatively safe and are guaranteed by the Federal Reserve Bank, the assets of shadow banks are risky and are not covered by federal deposit insurance.

The edges of the network are directed links through mutual exposures between banks obtained through the interbank market, where such links exist for the very-short-term transfer of funds. There are also directed links between financial institutions and the shadow banks representing total payments made from one to the other. Each edge of the financial network is usually assigned a weight that reflects the face value of the contact between the banks. Such a network uses a regulatory system to avoid a minor crisis turning into a major one that could test the stability of these financial institutions. However, this does not always protect the system from collapse. The 2008 subprime mortgage crisis is an example of what can go seriously wrong with the banking network.

The 2008 Subprime Mortgage Crisis

The turmoil in the U.S. subprime mortgage market that began in 2007 coincided with the deepest U.S. recession (2007–2009) since the end of World War II. This created a crisis said by former Federal Reserve Board (FRB) Chairman Ben Bernanke to be a "perfect storm" that took market analysts completely by surprise. In the United States, 9 million jobs were lost during 2008 and 2009, lost output was estimated at about 40% of 2007 gross domestic product, the stock market fell about 50% by early 2009, and banks, mortgage lenders, real estate investment trusts, and hedge funds incurred huge losses.

Two government-sponsored mortgage agencies, Fannie Mae (founded in 1938, privatized in 1968) and Freddie Mac (founded in 1970), suffered large losses and were taken over by the government in summer 2008. At the same time, the largest insurance company in the world, American International Group, Inc. (AIG), could not meet its payment obligations, including subprime mortgages, to several large banks; fortunately for AIG, the government provided funds in a bailout operation to prevent possible bank losses that would otherwise have wreaked havoc throughout the financial system. Again, in 2008, the bankruptcy of Lehman Brothers Holdings, an investment bank, led to the liquidation of the Reserve Primary Fund, a money market that held Lehman debt, triggering panic and a run on money market funds, leading to a credit market freeze; again, the government was prompted to step in and, this time, guarantee money market fund assets. The same year saw the disappearance of four other big Wall Street investment banks: Bank of America purchased Merrill Lynch, JPMorgan Chase purchased Bear Stearns, and Goldman Sachs and Morgan Stanley became financial holding companies, subject to various regulations. In 2010, the Dodd–Frank Act, amongst its actions, created the Financial Stability Oversight Committee, and permitted the Federal Reserve System to establish a "safety net" to regulate the shadow banking system.

Prior to the crisis, the FRB had drastically cut short-term interest rates to historically low levels in response to the crash of the dot-com bubble in 2000 and the following recession. These actions then led to the U.S. housing bubble. From 2004 to 2006, the FRB raised interest rates 17 times, house prices increased, and payment systems were functioning well at the time. During this period, however, there were many warnings of a potential housing crisis, but little or no attention was paid to them. When the crisis did hit, it spread from the housing market to other sectors of the economy and even to other countries. The stock markets in Germany, India, Britain, China, and Japan all reported significant declines on January 21, 2008 based upon fears that the U.S. crisis would spread to the world's financial system.

The U.S. crisis was triggered by a series of shocks to the system, which then precipitated a bursting of the housing bubble in 2006. The causes of these shocks have been debated at length. By 2006, the growth of "shadow banking" had surpassed in value that of the regular banks, and links between the two systems had became closely intertwined and inseparable, and, therefore, vulnerable to a banking crisis. Runs on the shadow banks led to a credit freeze, which started the crisis. Then, several major financial institutions collapsed in September 2008 and the flow of credit to consumers and businesses was severely disrupted; risky policies of mortgage lenders, of low introductory mortgage rates and no downpayment, encouraged home ownership for those who could not possibly repay; borrowers were suddenly unable to refinance their risky mortgages and many defaulted; residential overbuilding, based mostly on lax mortgage-lending standards and unrealistic assumptions that future home prices would increase, led to a surplus of unsold homes; the demand for homes became depressed and housing prices crashed, leading to the highest levels of foreclosure seen in the previous 10 years; bond funding of risky subprime mortgages collapsed; and record-high debt levels persisted, both for individuals and corporations.

The FRB, in its role as a federal regulator, has been accused of not anticipating the financial crisis until it was too late, and by its inaction that it actually escalated the effects of the crisis. Janet Yellen, then Vice-Chair of the FRB, in a 2013 speech (Yellen, 2013), suggested that crashes of financial markets could be due to "certain types of externalities, such as those arising from incomplete information or a lack of coordination among market participants." She added that "these externalities may do little harm or may even be irrelevant in normal times, but they can be devastating during a crisis."

What can be Learned from the Crisis?

From our experience of the 2008 financial crisis, we can ask several questions that focus on key aspects of financial networks. How are financial networks formed? How are connections made between financial institutions? How does network structure affect the financial system? In particular, what is the effect of a shock (or breakdown) to a single institution in the system and can the risk of contagion of the shock from one financial institution to another be controlled?

Two polar-opposite views of financial risk have been suggested. The first view is that although financial systems are inherently fragile, the risk of contagion can be controlled (or even reduced) by (a) increasing the amount of *integration* or *interconnectiveness* (i.e., number of connections) between banks and (b) increasing the amount of *diversification* of their holdings. As Elliott, Golub, and Jackson (2014) explained, "integration affects an organization's exposure to others compared to its exposure to its own assets, while diversification affects how many others one is (directly and indirectly) exposed to."

A perfectly diversified pattern of holdings provides optimal protection against moderate shocks, but it does not, by itself, offer adequate protection against very large shocks. Prior to 2007, for example, the financial system had exhibited greater complexity while greatly reducing financial diversity, and bank balance sheets had become increasingly homogeneous, so that banks looked like each other and responded like each other; the maxim of the day, which was that "complexity plus homogeneity equalled stability," would have set off alarm bells in nonfinancial disciplines (Haldane, 2009). Rather than "stability," Haldane argues that the equation should instead have read "fragility." Similarly, low levels of interconnectedness only affect peripheral organizations with maybe one core organization failing, moderate interconnectedness promotes widespread contagion, while highly interconnected core organizations tend to be more robust and resistant to failure.

The other view, which completely reverses the premise of the first, is that it is precisely the interconnectedness of the financial system that contributes to its fragility, because it makes it so much simpler to spread turmoil from one bank to the rest just like an epidemic does (see, e.g., Zhou, 2016).

The consequencies of a huge shock to the network depend upon when and where exactly in the network it hits, how large is the organization, and how connected is the network. For example, financial liquidity, which can be vulnerable to shocks due to uncertainty regarding where and when customers will withdraw funds, can be insured against by banks exchanging interbank deposits. One of the inadvertent effects of such connections is that inefficient banks would be protected against closure. An interesting sidenote is that banks often increase network connections where contagion rarely occurs.

Investment banks, which are part of the shadow banking system and whose role is to issue and underwrite securities, have found it important to develop two kinds of networks: an *information network*, in which the banks acquire information on how much investors (e.g., pension funds, insurance companies) are willing to pay for an issue, and a *liquidity network*, which is used to provide funds to purchase the issue (Morrison and Wilhelm, 2007). These networks do not function separately; there is a certain amount of overlap in the two networks as frequent traders in the liquidity network may also provide them with information regarding prices.

2.4.2 Bitcoin Transaction Network

Bitcoin appeared at the height of the 2008 financial crisis. It operates as an online peer-to-peer electronic cash system. Bitcoin has been described as "a radically new, decentralized system for managing the way societies exchange value. It is, quite simply, one of the most powerful innovations in finance in 500 years" (Casey and Vigna, 2015). The U.S. Internal Revenue Service calls Bitcoin an example of a "convertible virtual currency" in that it "can be digitally traded between users and can be purchased for, or exchanged into, U.S. dollars, Euros, and other real or virtual currencies" (Aqui, 2014).

Since the introduction of Bitcoin, many other so-called *cryptocurrencies* have been created (as of late May 2018, more than 1600). The growth of these cryptocurrencies has been attributed to speculative demand, technological developments, and a need for alternatives to official currencies that have not been adequately protected by their governments.

Bitcoin was proposed in 2008 by Satoshi Nakamoto[20] (Nakamoto, 2008) and it went online as open-source software in January 2009. Although the idea of digital money had been on people's minds since the introduction of the Internet, when Bitcoin appeared only a few people felt that it had any value in the real world.

---

[20] This name appears to be a pseudonym as the founder's true identity remains a mystery.

Then, in 2010, a Tokyo-based trading website, *Mt. Gox*,[21] started up and, as a byproduct, simplified the exchange of bitcoins. Financial activities on the website grew quickly and the Bitcoin system attracted widespread public interest. The world's largest (in terms of volume) cryptocurrency exchange currently is the Cayman Islands-registered *Binance*. The largest cryptocurrency exchange in the United States is *Coinbase*.[22]

The unit of currency is one bitcoin (commonly coded as BTC, which does not conform to ISO 4217, where the Bitcoin unofficial currency code is XBT), and the smallest transferable amount is 10^{-8} BTC, which is called a "satoshi" after the founder of bitcoins. To get an idea of a bitcoin's worth and how volatile it is, the exchange rate as of February 2, 2015 was 1 bitcoin = \$233.47; on December 16, 2017 it reached 1 bitcoin = \$20,089. Bitcoin then fell to \$3193.37 on December 16, 2018. It then rose to around \$12,000 in September 2020, then \$20,000 on December 16, 2020, \$30,000 on January 2, 2021, \$40,000 on January 7, 2021, \$50,000 on February 16, 2021, and passed \$60,000 on March 14, 2021. It subsequently dropped below \$18,000 on June 18, 2022.

Bitcoins are transacted through *digital keys*, *bitcoin addresses*, and *digital signatures*. A *digital signature* consists of a pair of *digital keys*, one public and the other private (and secret). Their relationship is given by the following process:

$$\text{Private key} \xrightarrow{\text{ECC}} \text{Public key} \xrightarrow{\text{SHA}} \text{Bitcoin address.} \tag{2.1}$$

The private key is used to generate a unique public key through an irreversible elliptic-curve cryptographic (ECC) algorithm. The public key then generates a *bitcoin address* (although that name can also refer to other items) through an irreversible secure hash algorithm (SHA). The bitcoin address, analogous to a bank account number, is associated with an owner's existing collection of bitcoins and can be handed out freely so that it can be used as a transaction destination for bitcoins.

The owner of the bitcoins signs an outgoing payment with his private key. The private key, analogous to a PIN, is a randomly generated 256-bit number[23] (but could be 128 or 512 bits long), which must always be kept secret and confidential; otherwise, anyone with access to the user's private key can control the user's collection and future purchases of bitcoins. As a means of protection, private keys will typically be encrypted by a password (or passphrase) and backed up regularly.

New payments are made public by broadcasting them to the network without revealing the identity of the owner, only the identity code of the owner's *digital wallet* (i.e., databases or structured files that manage Bitcoin holdings), which exists either on a user's computer or in the "cloud." The user's private keys are stored inside these wallets. Payments are validated by thousands of independently owned computers (referred to as *miners*) that check consistency with the entire transaction history.

Bitcoin users tend to acquire bitcoins in one of two ways: either buying them at an exchange or receiving them as compensation for goods or services. They can be

---

[21] Mt. Gox is now defunct, declaring bankruptcy due to a theft in early 2014 of more than \$470 million of bitcoins, as well as other scandals.

[22] See www.coinbase.com.

[23] In hexadecimal, 256 bits translates to 32 bytes (or 64 characters in the range 0–9 or A–F).

purchased with hard cash, credit/debit cards (usually not accepted by exchanges), or wire transfers. Every 10 minutes, details of transactions between Bitcoin addresses are arranged into *blocks*. These blocks, which contain different numbers of transactions, form a long list of entries in a general ledger known as a *blockchain*, each block referencing the previous block.[24] When a new block is added to the blockchain, a copy of that block (and each of its updates) is transmitted to every user who participates. Although Bitcoin transactions are kept anonymous, users may possess multiple Bitcoin addresses that are not linked to each other, thereby increasing their level of anonymity.

Tampering with a blockchain could be a serious problem, but this is guarded against through the following process. The miners compete with each other, using specialized software for mining blocks. They apply a mathematical formula to the transaction data in the block, which turns the data into a (unique) pseudo-random sequence of letters and numbers called a *hash*, which is 160 bits long. The hash has to have a certain format, including having a certain number of zeroes at the start. It is not easy to create a hash with the required format (because of another random piece of data, called the *nonce*, that has to be added to the block). Every time a miner successfully creates a hash, he/she gets rewarded with 6.25 bitcoins, and this news is transmitted to the network. The reward amount is cut in half every 210,000 blocks, usually around every 4 years. Although the data in a block can be transformed into a hash, the same data cannot be recovered from the hash. The hash is stored, together with the block and the hash of the previous block, at the end of the blockchain. Because of the uniqueness property of a hash, if a single character in a block is changed, this will radically alter the resulting hash, and so every user would know that someone had tampered with the blockchain and that the block was a fake.

The Bitcoin transaction network operates so that each node represents a Bitcoin address, and an edge is drawn between two nodes if there was at least one transaction between the addresses in question. It is also possible to record the time and amount of every payment so that the dynamic nature of the financial process (i.e., the flow and accumulation of bitcoins) may be analyzed.[25]

Problems with Bitcoin

The Bitcoin system has matured and is now considered to be a very secure and stable financial technology. Despite its popularity, however, the Bitcoin system has its problems.

Currently, the price of a bitcoin fluctuates a little too wildly for everyone's comfort and its anonymity encourages dealings in illegal substances. There is also the fact

---

[24] The origins of the blockchain technology are due to a physicist Scott Stornetta and a cryptographer Stuart Haber, who published a series of articles starting in 1991 on methods for building a cryptographically secure archive. They were named as co-inventors on Bellcore's patent for the blockchain technology. See Whitaker (2018).

[25] Two Bitcoin-related datasets, entitled `bitcoin-otc` and `bitcoin-alpha`, are currently available at the Stanford University website: `snap.stanford.edu/data/`. These two datasets are weighted, signed, directed networks that provide a record of Bitcoin users' reputations (i.e., "who-trusts-whom") so that "transactions with fraudulent and risky users" can be prevented. The weights range from -10 to $+10$ in steps of 1. An unweighted graph of `bitcoin-otc` is given in **Figure 2.6**. There are several websites that display charts of Bitcoin activities; see, for example, `Bitcoin.com`, `Bitcoin.org`, `coinbase.com/charts`, `coindesk.com`, `blockchain.info/charts`, `awebanalysis.com/en/coin-details/bitcoin`.

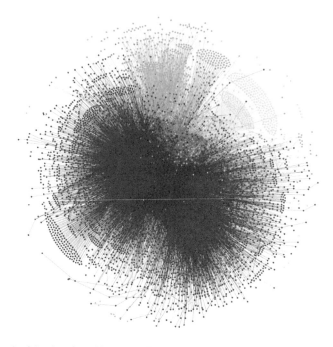

Figure 2.6 A graph of the data from bitcoin-otc. This is a weighted, signed, directed network that provides a record of Bitcoin users' reputations (i.e., "who-trusts-whom") so that "transactions with fraudulent and risky users" can be prevented. We have graphed it as an unweighted network with colors indicating the modularity of each node.

that wallets have proved to be vulnerable to cyberattacks and other computer hacking schemes. A popular cryptocurrency exchange, *Bithumb*, in South Korea, for example, recently reported a loss of over $30 million in Bitcoin and other cryptocurrencies in a cyberattack. Overall, to date, there have been 56 cyberattacks aimed at cryptocurrency exchanges, mostly in Asia, with reported losses of $1.63 billion.

Also, it is not a very practical way for an ordinary individual to purchase items: there have been too many cases reported of forgotten passwords, hard-drive crashes, and inadvertent discarding of computers, so that Bitcoin records are forever lost. Online discussions have considered how a single individual or a group could take control of over half of the Bitcoin network's *hashrate* (trillions of hashes per second). Experts believe, however, that the logistics required to pull it off would be too much of a challenge, and even if it could be carried out, it still could not crash the Bitcoin system.

2.5 Social Networks

Social networks are ubiquitous; they are everywhere. Typically, they consist of a group of individuals and the connections between them. If two individuals have a relationship (e.g., family or friends or people who share a common workspace), that connection is often referred to as a *tie*, and the individuals may be referred to as *actors*. The social networks literature also uses *ego* to refer to a network node of interest, and nodes that share ties with the ego are referred to as its *alters* or *contacts* (Ogburn and VanderWeele, 2017).

In some cases, friendship is a bidirectional relationship: A is a friend of B and B is a friend of A. In other cases, the direction is one-sided: A considers B to be a friend, while B does not consider A to be a friend. The question is whether such a one-sided relationship can really be viewed as friendship (Ball and Newman, 2013). One way out of this difficulty is to ignore the direction component of the relationship and consider friendship in all its forms, whether one-sided or two-sided; see, e.g., Airoldi, Choi, and Wolfe (2011), who studied friendship data from the National Longitudinal Study of Adolescent to Adult Health.

2.5.1 The Adolescent Health Survey

The National Longitudinal Study of Adolescent to Adult Health (usually abbreviated to *AddHealth*) is a longitudinal study of a nationally representative sample of 90,118 adolescents located in 144 schools in Grades 7–12 in the United States during the 1994–1995 school year. This study was designated as Wave I. From a sampling frame of 26,666 high schools, 80 high schools were chosen that were regarded as representative of U.S. high schools, with respect to region of the country, urbanicity, size, type, and ethnicity. The students answered in-school questionnaires on family background, school life, friendships, and health concerns. In addition, 20,745 adolescents responded to in-home interviews that consisted of more sensitive issues.

In order to provide sufficient data for network analysis, every enrolled student from each of 16 selected schools (two large, the rest small) had in-home interviews. This is referred to as the "core" in-home sample. Each student was presented with a roster that listed all students in the school by name and by a unique ID number, and was asked to name up to five of their best male friends and up to five best female friends, ranked in order of closeness, and to specify whether or not some sort of interaction occurred between them during the dates of the study. The resulting data constitute the friendship network data. The students were allowed to list friends outside school or not on the school roster. Students were stratified by grade level and gender. The questionnaire also requested information on race (Hispanic, Black, White, Asian, Native American, other), social and demographic characteristics of the students, education and occupation of parents, household structure, and the student's health status, with many other variables added in follow-up studies of the cohort of students.

The follow-up studies (Waves II (1996), III (2001–02), IV (2007–08), and V (2016–18)) were carried out on various subsets of the same individuals using in-home interviews, with Wave V having a target set of 19,828 adults aged 32–42. For further details, see Harris and Udry (2018) and Harris et al. (2019).

There are many missing data: students who did not complete the survey, students not listed on the roster, students whose parents told them not to participate in the study, and students who did not answer certain questions. The final dataset either did not include students with missing data or the missing data were imputed using statistical methods. Because of the size and complexity of this dataset, studies have focused on different subsets of the data.[26]

---

[26] The AddHealth data can be found at the website www.cpc.unc.edu/projects/addhealth. That website also provides more general information about the AddHealth survey. *Note:* Different subsets of these data have been used in many studies, with different numbers of nodes and different numbers of edges.

This friendship survey is usually examined for each school separately, which means that there is a separate friendship network for each school (see, e.g., Goodreau, Kitts, and Morris, 2009). The nodes of each network consist of the students in that school, and the edges represent the friendship information presented by those students. In the Goodreau et al. study, the number of nodes considered was 1681 and the number of edges was 1236. These networks are usually considered to have undirected edges, even though friendships may be either one-sided or two-sided. Important features of these networks include age (grade) distribution and race distribution of friendships.

2.5.2 E-mail Networks

E-mails are an electronic method of communication between people. To compose, deliver, and receive e-mails requires the use of a mail server that a person can connect to through the Web. An e-mail address consists of two parts joined by an "@" sign. The part before the @-sign is some personal identification, usually referred to as a *username*, and the part after the @-sign is called the *domain name*. We describe two large datasets extracted from organizations that provide the computer resources for e-mail correspondence between employees and the outside world.

A European Research Institute

These data (email-Eu) consist of all incoming and outgoing e-mail between members of a large European research institute for an 18-month period from October 2003 to May 2005 (Leskovec, Kleinberg, and Faloutsos, 2007; Yin et al., 2017). There are 1005 nodes and 25,571 edges in the network, each node of which represents a member of the institute. An edge exists in the network if a member of the institute sent another member at least one e-mail message. See **Figure 2.7**, which filtered out all singleton nodes.[27] The e-mails are only for institute members and do not include any e-mails between members and non-members. There are 42 departments in the institute. Each member can belong to only one department, and departments can be viewed as different communities. Colors represent those nodes that have different densities of edges (called "modularity" by the software package `Gephi`).

Enron Employees

The Enron Corporation was formed in 1985 through the merger of two local gas-supply companies, Houston Natural Gas and InterNorth of Omaha.[28] In 1995, Enron started a new business, EnronOnline, involving electronic trading on the Internet. In this venture, Enron was very successful and expanded its work to the sale of electricity and other brokering activities, including metals, weather derivatives, communication products, wood, paper, water, and so on. They also constructed the first nationwide natural-gas pipeline in the United States. By 2000, Enron had become a major player in international energy, with assets increasing from $4 billion in 1989

---

[27] The European Research Institute e-mail data can be found at the Stanford University website `snap.stanford.edu/data/email-Eu-core.html`.

[28] This description of the Enron case was adapted from Diesner, Frantz, and Carley (2005). See also Leskovec et al. (2009).

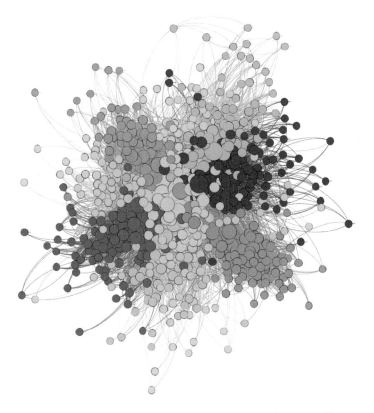

Figure 2.7 E-mail correspondence of members of a European research institute. There are 1005 nodes representing members of the institute and the 25,571 edges represent e-mail correspondence between pairs of members. Colors represent the modularity class of nodes.

to $100 billion in 2000, and had become the seventh-largest business organization in the United States (in terms of revenue).

Then, suddenly, in December 2001 Enron crashed. It found itself insolvent and was forced to declare Chapter 13 bankruptcy. Thousands of its employees were laid off. Enron had previously been accused, anonymously, of fraud and bad accounting practices (Enron's auditor since 1985 was Arthur Andersen, LLP). Public outcry and scandal followed. Independent enquiries into Enron's sudden collapse were carried out by the U.S. Securities and Exchange Commission (SEC) and the Federal Energy Regulatory Commission (FERC). The investigations disclosed that Enron's business was made possible through certain legal "loopholes" and decisions, which included exempting Enron from monitoring and control of its activities. There were also extensive fraudulent operations, including disguised loans having no interest, which operated under a culture that enabled personal enrichment and encouraged employees to "push the limits." The top executives at Enron were put on trial for their part in the Enron scandal and were convicted of securities and wire fraud. In 2002, Andersen was convicted of obstruction and received a probationary sentence, so that it could no longer audit public companies.

In May 2002, the FERC publicly released a corpus of 90% of Enron e-mails (in an unusable format) over a period of about 3.5 years. This was designed to enable the public to understand the various reasons for the Enron investigation. The Enron e-mail communication corpus originally consisted of 619,446 e-mail messages

Enron Executive Email Clusters

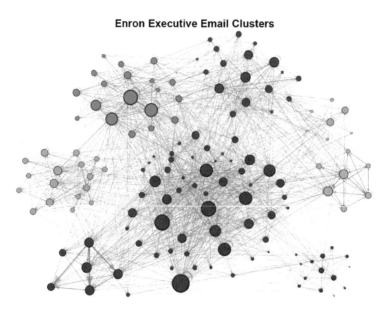

Figure 2.8 Enron e-mail network. There are 151 nodes (users) and 517,431 edges (distinct e-mail messages).

(reduced to 200,399 through "cleaning" efforts) organized into folders sent by 158 employees, mostly senior management at Enron. In the cleaning process, e-mail addresses were standardized and sensitive e-mails were deleted. These data were made publicly available and were uploaded to the Internet by the FERC during its investigation of Enron.[29]

Nodes of the network are e-mail addresses and an undirected edge corresponds to an e-mail sent from one address to another address. See **Figure 2.8**. There are 151 users (organized into 150 user folders, with numerous subfolders) and 517,431 distinct e-mail messages in the network.

2.5.3 Scientific Collaboration Networks

It is not unusual today for scientific journals to publish articles written by several coauthors. In fact, we would have to go back to 1665 to find the first academic paper having more than a single author, which was published in the journal *Philosophical Transactions of the Royal Society*. The trend now is to have many coauthors on scientific publications. Evidence shows that the average number of coauthors on papers in mathematics, high-energy, experimental, and condensed-matter physics, astrophysics, and computer science all have risen, some more substantially than others, in recent years (Newman, 2001a).

---

[29] There are many different versions of the Enron dataset with different numbers of users and different numbers of e-mail messages depending upon how the data were subsetted and cleaned. The Stanford version of the Enron e-mail data has 36,692 nodes and 183,831 edges and is available at snap .stanford.edu/data/email-Enron.html. The present version, which is similar to the FERC version, was made available in March 2004. See, for example, networkdata.ics.uci.edu/netdata/html/EnronMailUSC1.html or www.cs.cmu.edu/~enron/.

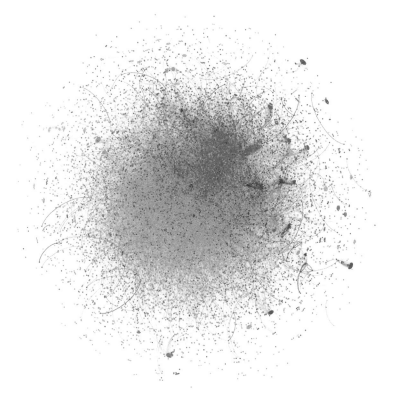

Figure 2.9 Astrophysics collaboration network between scientists posting preprints on the *Astrophysics E-Print Archive*, January 1993–April 2003 (124 months). There are 18,772 nodes and 198,110 edges in this graph.

Not so long ago, it was rare to see articles with more than 100 coauthors (Adams, 2012). This changed in the 1990s when there were more than 500 scientific articles with over 100 coauthors. In 2004, an article appeared with 1000 coauthors, and in 2011, there were 120 physics articles that had over 1000 coauthors. In 2010, a paper appeared in the journal *Physics Letters B* that had 3222 coauthors from 32 different countries. These highly multi-authored papers mostly reported on research at the Large Hadron Collider at CERN, Europe's particle-physics laboratory near Geneva, Switzerland. However, it is not only in physics that we see multi-authored publications, though maybe not on the same scale. We also see multi-authored papers appear in the fields of health, energy, climate, and social science, and we expect this trend to continue.

A scientific collaboration network is represented by a *bibliometric map*, in which the nodes are authors and an edge between two authors appears if they collaborated on one or more scientific papers. See, for example, **Figure 2.9**, which shows the collaboration network of astrophysicists. In some networks, the nodes represent institutions or countries, rather than authors, and the edges then display scientific collaborations between the institutions or countries. The display of bibliometric maps can be used to study an institution's past scientific research and identify future research directions and collaborative possibilities.

Scientific collaborations have a lot in common with social networks because they are both driven by the complexities of human interactions. In joint research between two or more individuals, one often has to consider each coauthor's contribution to the completed project. Some coauthors make valuable contributions, whilst others make only marginal contributions, and these types of contributions can change over the course of the project. Accordingly, the problem of giving appropriate credit to the amount of work contributed by a coauthor can be quite complicated. One reason for the increased level of scientific collaborations is the fact that science, in all its forms, has become highly specialized, and cross-disciplinary approaches to solving important problems have necessitated increasing collaboration between experts in each of the contributing disciplines[30] (Adams et al., 2004).

So-called "big science" has accomplished much because of international collaborations with institutions that possess large and expensive scientific equipment. International collaboration has been increasing, not only in basic research areas such as mathematics, biomedicine, and space sciences, but also in small research areas, where scientists are forced to seek international cooperation to overcome a dearth of researchers within their national boundaries.

Bibliometric research is based on large quantities of data that are provided by bibliographic databases. There are four prominent bibliographic databases in the sciences (Newman, 2001a,b):

Medline (biomedical research published in medical journals, 1961–present; medline.com);

LANL (the Los Alamos e-Print Archive, unrefereed preprints in experimental and theoretical physics, 1992–present; lanl.gov);

SPIRES (the Stanford Public Information Retrieval System, preprints and published papers in experimental and theoretical high-energy physics, 1974–present; slac.stanford.edu/spires);

NCSTRL (the Networked Computer Science Technical Reference Library, preprints in computer science, 1990–present; ncstrl.org).

2.5.4 Sampson's Monk Data

This example has become a classic in social network analysis. Over a period of a year, Sampson (1968) carried out an intensive study of the community structure of two cohorts of novice monks who were preparing to join St. Anthony's, a geographically isolated monastery in New England: first came the Cloisterville Group, and then the New-Wave Group.

The Cloisterville Group consisted of five cleric novices (Leo, Arsenius, Bruno, Thomas, and Bartholomew) and eight lay brother novices (Martin, Peter, Bonaventure, Berthold, Mark, Brocard, Victor, and Ambrose), who had been at St. Anthony's for nearly a year and many of them had previously attended the Cloisterville seminary together. Bartholomew was considered to be the leader of the group.

---

[30] The U.S. National Science Foundation has aggressively supported scientific collaboration through its funding policies and by expressing its belief that "Science knows no boundaries."

Table 2.1 Names, numbers, and group identifications of 18 novice monks as determined by Sampson (1968). YT = Young Turks, LO = Loyal Opposition, O = Outcasts, and I = Interstitials (Waverers)

1	John Bosco (YT)	7	Mark (YT)	13	Armand (I)
2	Gregory (YT)	8	Victor (I)	14	Hugh (YT)
3	Basil (O)	9	Ambrose (LO)	15	Boniface (YT)
4	Peter (LO)	10	Ramuald (I)	16	Albert (YT)
5	Bonaventure (LO)	11	Louis (LO)	17	Elias (O)
6	Berthold (LO)	12	Winifrid (YT)	18	Simplicius (O)

Seven of the Cloisterville Group graduated (Leo, Arsenius, Bruno, Thomas, Bartholomew, Martin, and Brocard) and were replaced by a new set of 12 novices (Gregory, John Bosco, Winifrid, Albert, Boniface, Hugh, Louis, Basil, Simplicius, Elias, Ramuald, and Amand), most of whom had not attended the Cloisterville seminary and had not known each other previously. Gregory was regarded as a "charismatic genius," whose strong views against orthodox dogma alienated some of the novices.

These 18 novices, termed the New-Wave Group, quickly fragmented into competing ideological factions. Two main groups were assessed by Sampson:

The Young Turks. Seven novices: John Bosco, Gregory (leader), Mark, Winifrid, Hugh, Albert, Boniface. This group had progressive tendencies that questioned monastery practices.

The Loyal Opposition. Five novices: Peter (leader), Bonaventure, Berthold, Ambrose, Louis. This group joined the monastery first. They were a more conservative group who defended the status quo and generally opposed the new arrivals.

Sampson also divided the remaining novices into two minor groups:

The Outcasts. Three novices: Basil, Elias, Simplicius. These novices were not accepted by the other groups (but were sympathetic to the Young Turks).

The Interstitials. Three novices: Victor, Ramuald, Amand. These novices vacillated between the Young Turks and the Loyal Opposition groups. Some sources refer to this group as the "waverers."

The list of novice monks and their group identifications is given in **Table 2.1**.[31]

After about two weeks, the regular staff of the monastery met and decided to expel all three members of the Outcasts (either because they had personality problems or were considered to be too immature) plus Gregory, the leader of the Young Turks, whom they felt considered himself above the rules. In quick order following this expulsion, five of the six remaining Young Turks departed voluntarily (John Bosco, Albert, Boniface, Hugh, and Mark), two of the interstitial group resigned (Amand and Victor), and then the remaining interstitial member resigned (Ramuald). Of the six novices remaining (Winifrid of the Young Turks, and Peter, Bonaventure,

---

[31] Articles and books that use these data have been inconsistent in using different numbering systems for the novices (see, e.g., Wang and Wong, 1987; Hoff, Raftery, and Handcock, 2002; Airoldi et al., 2008). We follow the ordering used by White, Boorman, and Breiger (1976). Also, Ramuald's name appears to have a number of different spellings.

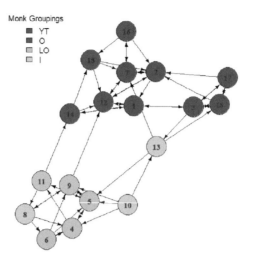

Figure 2.10 Sampson's monk network. The colors of the nodes reflect the classification of nodes to groups as given by Sampson: Young Turks (YT) are colored red, Loyal Opposition (LO) turquoise, Outcasts (O) purple, and Interstitials (I) green. The membership of each group is given in Table 2.1.

Berthold, Ambrose, and Louis of the Loyal Opposition), Winifrid and Peter left, leaving only Bonaventure, Berthold, Ambrose, and Louis.

In order to study the social structure of the novices within the monastery, Sampson carried out a survey of the novices by collecting 10 sets of responses on various sociometric criteria; for example, the novices were asked "whom do you personally like the most" (recorded as "like") and "whom do you personally like the least" (recorded as "dislike").

Most of these data were recorded retrospectively on the 18 novices, following the breakup. The relations were: like (recorded over three time points, labeled T_2, T_3, and T_4, during a 12-month period)/dislike; positive influence/negative influence; esteem/disesteem; praise/blame. Each of the 18 novices ranked the top three of their colleagues on each criterion (with 3 the highest choice and 1 the lowest choice), although there were some tied ranks.[32]

The network graph corresponding to the third time point (i.e., T_4) is displayed in **Figure 2.10**.

2.5.5 Zachary's Karate Club

This is another very popular dataset used to illustrate social network modeling. The data, which were collected by Wayne W. Zachary from the members of an American university karate club, consist of the presence or absence of a relationship between

---

[32] The UCINET version (Borgatti, Everett, and Freeman, 1999) of the complete Sampson dataset is available at the website vlado.fmf.uni-lj.si/pub/networks/data/ucinet/ucidata.htm in the form of 10 (18 × 18)-matrices labeled SAMPLK1, SAMPLK2, SAMPLK3; SANPDLK; SAMPES, SAMPDES; SAMPIN, SAMPNIN; SAMPPR, SAMPNPR. The monk data can also be found in the R package Monks in the LLN package. Part of the data (the three time points of the "like" relation) can be found in CRAN as sampson in the latentnet package at the website cran.r-project.org/web/packages/latentnet.

all pairs of members of the club (Zachary, 1977). The data were compiled over three years, 1970–1972, of direct observation of member interactions. During the observation period, the club maintained 50–100 members, and its activities included social affairs as well as regularly scheduled karate lessons taught by a part-time karate instructor, referred to as Mr. Hi. There was an informal political organization in the club with a constitution and four officers; however, most decisions were made by consensus at club meetings.

A conflict arose at the start of the study between Mr. Hi and the club president, John A.[33] The issue was the price of lessons: Mr. Hi, as the instructor, wished to raise prices, while John A. wanted to stabilize prices. The whole club became divided over this issue. Mr. Hi's supporters viewed him as a fatherly figure who was their spiritual and physical mentor, and who was only trying to meet his own physical needs after seeing to theirs. John A.'s supporters, on the other hand, saw Mr. Hi as a paid employee who was trying to raise his salary. After a number of confrontations, in which Mr. Hi tried unilaterally to raise prices, the officers, led by John A., fired Mr. Hi. Those members who supported Mr. Hi then resigned and started a new organization led by Mr. Hi.

Prior to the split of the club, while confrontations were taking place, decision-making still happened at club meetings. Whichever faction held a majority at a meeting would then pass resolutions reflecting its own ideological leanings. The other faction would then retaliate at a future meeting when it held the majority by repealing the unfavorable decisions and substituting ones favorable to itself. The two factions were really ideological groups and did not adhere to any political division; furthermore, no attempt was made by anyone to prevent any interaction between members of opposing factions. During this crisis within the club, friendship bonds within the ideological factions were strengthened, while bonds between members of opposing factions were weakened.

The data consisted of eight contexts that determined the strengths of the edges in the network:

1. Association in and between academic classes at the university.
2. Membership in Mr. Hi's private karate studio on the east side of the city, where Mr. Hi taught nights as a part-time instructor.
3. Membership in Mr. Hi's private karate studio on the east side of the city, where many of his supporters worked out on weekends.
4. Student teaching at Mr. Hi's east-side karate studio. Student teachers interacted with each other, but were prohibited from interacting with their students.
5. Interaction at the university rathskeller, which was located in the same basement as the karate club's workout area.
6. Interaction at a student-oriented bar located across the street from the university campus.
7. Attendance at open karate tournaments held through the area at private karate studios.
8. Attendance at intercollegiate karate tournaments held at local universities. Because both open and intercollegiate tournaments were held on Saturdays, attendance at both was impossible.

---

[33] Both names, Mr. Hi and John A., are pseudonyms.

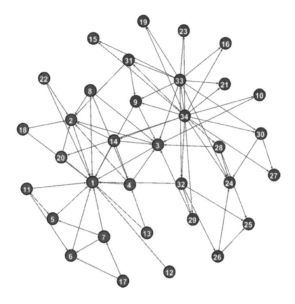

Figure 2.11 Zachary's karate club network. The nodes represent the 34 members of the club. The two colors, blue and red, represent a clustering of the data into two groups corresponding to the faction of 16 red nodes supporting Mr. Hi (node 1) and 18 blue nodes supporting John A. (node 34), respectively.

There are 34 nodes (club members) and 78 undirected edges (interactions between members outside the club) in the network. Club membership was near 60 at the time of the study, but 26 of them had no social contact with other club members outside of meetings and classes, and they belonged to neither faction; so, they were dropped by Zachary from the dataset.[34] Each edge in the network between a pair of club members showed that they interacted in at least one of the eight contexts.

The network graph corresponding to the adjacency matrix is displayed in **Figure 2.11**. Zachary ordered the nodes so that the first row and column of the adjacency matrix is Mr. Hi (node 1, with 16 edges) and the last row and column is John A. (node 34, with 17 edges).

2.6 Biological Networks

A momentous event in the history of science occurred when the competitive race to sequence the human genome led to two simultaneously published articles. The Human Genome Project, a government-sponsored consortium of universities and laboratories, and Celera, a private company, simultaneously published draft

[34] The karate club data can be found at the website www-personal.umich.edu/~mejn/netdata. UCINET has a version of the Zachary karate club data (Williams, 2011) that is available at the website sites.google.com/site/ucinetsoftware/datasets/zacharykarateclub and consists of two (34 × 34) symmetric matrices: ZACHE (binary) and ZACHC (valued). The ZACHE matrix (called the "existence" matrix by Zachary) displays the presence (1) or absence (0) of an edge between each pair of members, and is the usual adjacency matrix. The ZACHC matrix (called the "capacity" matrix by Zachary) displays the relative strength of the edges in the network (number, 0–7, of contexts inside and outside the club in which interactions occurred). The UCI Network Data Repository also contains the karate club data at the website networkdata.ics.uci.edu/data.php?id=105. There are many other websites that have the karate club data.

accounts of the human genome in *Nature* and *Science* on February 15 and 16, 2001, respectively; see Dennis and Gallagher (2001).

That event, combined with advances in computation and sequencing technology, was instrumental in creating the discipline of *bioinformatics*, which combined computer science, information technology, and biological science. Products emerging from this field include: a growing number of databases to store, manage, and manipulate huge quantities of biological data; efficient algorithms to analyze DNA, RNA, and protein sequence data; microarray technology that produced gene expression data; and the development by the pharmaceutical industry of new drugs for treating diseases. These advances in biotechnology led, first, to *genomics* (the study of gene activity and expression), and then *proteomics* (the study of proteins). In the future, we expect the field of *metabolomics* (the study of metabolic pathways) to increase in importance.

2.6.1 Gene Regulatory Networks

Although individual genes and proteins are fundamental parts of biology, studying them as isolated molecules does not enhance our understanding of most biological processes. More important are the interactions between genes or between proteins, or more complex interactions between genes, proteins, RNA, and DNA because they form the building blocks of biological systems. There are various kinds of biological networks involving these types of complex interactions. We describe two types of regulatory networks, namely, gene regulatory networks and, as a special type of subnetwork, transcription regulatory networks.

Gene regulatory networks (GRNs) are an important aspect of systems biology because they provide insight into the mechanisms of differential gene expression. Although this viewpoint employs more abstract concepts than, for example, transcriptional regulatory networks, it does take a more system-wide approach, which proves to be helpful in understanding the flow of genomic information over the entire biological system.

GRNs are usually represented by graphs, where the nodes represent specific proteins called *gene products*, which could include genes, transcription factors, signalling molecules, and regulatory RNAs, and directed edges represent regulatory connections between a gene product and its target genes. These connections may take the form of protein–DNA interactions, protein–protein interactions, or other, possibly indirect, relationships between genes. The topology of a GRN is "sparse" in the sense that the number of edges is much smaller than the number of nodes, so that each gene is regulated by only a small number of other genes.

The directed edges can be of two types: *activator* (represented by a "→" sign, as in $A \rightarrow B$) or *inhibitor* (represented by a "⊣" sign, as in $A \dashv B$). Sometimes, an activator, whose regulatory effect is positive (i.e., increases the expression level of another gene), is assigned a value of $+1$ to an edge, while an inhibitor, whose regulatory effect is negative (i.e., decreases the expression level of another gene), is assigned a value of -1. This hierarchical activity by which one set of genes regulates another set of genes and, in turn, these latter genes regulate still other genes, and so on, is the key that enables complicated organisms to be constructed.

Much of the work on GRNs has dealt with the regulatory relationships in the sea urchin (Davidson, 2009, 2010), the fruit fly, *Drosophila melanogaster*

(Biemar et al., 2006), the nematode worm, *Caenhorhabditis elegans* (Maduro, 2006), and the mouse, *Mus musculus* (Parfitt and Shen, 2014).

A specific subnetwork of a GRN is a *transcriptional regulatory network (TRN)*. TRNs are directed graphs that display the interactions between large numbers of genes and their "regulators." Most *regulators* constitute a class of proteins called *transcription factors (TFs)* that are special types of gene product. TFs bind directly to the DNA through their sequence-specific DNA binding sites. TFs control (or "regulate") exactly where, when, and at what rate gene expression takes place. TFs can either increase the rate of transcription of genes into mRNAs (i.e., *upregulate* gene expression), thereby becoming a gene *activator*, or decrease the rate of transcription (i.e., *downregulate* gene expression), thereby becoming a gene *inhibitor* or *repressor*. In addition to transcriptional factors, other gene regulators are small molecules, such as microRNA, other RNA, and various types of protein molecules.

The larger the genome size of an organism, the greater the number of TFs that can be found within that organism. Approximately 8–10% of genes in the human genome encode TFs, which makes TFs the largest class of human proteins. In fact, there are about 1500 different TFs in human cells. Many TFs help to regulate the cell cycle and they also provide a process by which cell size can be determined.

Mutations in TFs can produce bizarre results. In one example, the antennae of a fruit fly is replaced by a fully formed pair of legs; in another example, a thoracic segment towards the rear of a fruit fly, which normally produces a pair of small stabilizing structures called *halteres*, instead grows an extra pair of fully formed wings. Mutations in TFs have been linked to a number of human diseases such as diabetes, autoimmune diseases, and cancer, with the result that these TFs have become targets of many pharmaceutical drugs.

Amongst eukaryotic (i.e., unicellular) organisms, the first TRN to be studied in detail was that of *Saccharomyces cerevisiae* or baker's yeast (Lee et al., 2002), with its 6270 genes and 106 TFs, where the large quantity of gene expression data available on this organism enabled extensive research to be carried out on this TRN. The first prokaryotic organism to have its TRN studied was *Escherichia coli* (Huerta et al., 1998; Thieffry et al., 1998), which has 271 TFs that are assigned a DNA-binding domain (Babu and Teichmann, 2003).

A TRN consists of two types of nodes, representing genes and transcription factors, so that a TRN is *bipartite*. Thus, TRNs cannot be represented by the usual type of network graph in which every node is of the same type. The nodes are linked by directional edges (representing physical and/or regulatory interactions), where transcription factors regulate target genes, but not vice versa. As in GRNs, edges can display either an activator (a "→" sign, so that $A \rightarrow B$ means that TF A increases the transcription rate of gene B) or an inhibitor (a "⊣" sign, so that $A \dashv B$ means that TF A decreases the transcription rate of gene B). Transcription factors often interact with other transcription factors, and are themselves extensively regulated by other TFs.

When the edges of a network are directed, there are two types of *degrees*: the *in-degree* of a node is the number of incoming edges to that node and the *out-degree* is the number of outgoing edges from that node. In the case of TRNs, the in-degree is the number of transcription factors that bind a gene, while the out-degree is the number of genes that are bound by a transcription factor. In certain circumstances, nodes can have both an in-degree and an out-degree. Most nodes in a TRN tend to have relatively low degree (i.e., small number of edges per node). Consequently, TRNs are often referred to as *sparse* regulatory networks (see, e.g., James et al., 2010).

However, certain nodes are especially well-connected and are described as "TF-hubs" or "gene-hubs," where *TF-hubs* bind a huge number of target genes and *gene-hubs* are bound by a large number of TFs. A study of the ovarian network (Madhamshettiwar et al., 2012), for example, based upon microarray data from 12 normal and 12 unmatched cancerous ovarian epithelial tissues, predicted seven TFs and 171 TF-target genes; of the 282 differentially expressed genes, one TF (E2F1) regulated 134 genes, including five of the other six TFs, a second TF (SP3) regulated 51 genes, and a third TF (NFκB1) regulated 18 genes. In general, there exist a few TFs (called *global regulators*) that regulate a huge number of target genes, while a great number of TFs regulate a small number of target genes. There is evidence that these global regulators ensure that the TRN is "robust," where by *robust* we mean that the TRN still retains its functionality in the event that its structure is severely disrupted.

2.6.2 Protein–Protein Interaction Networks

Protein–protein interaction (PPI) networks are very important biological processes, and they have a special relevance for the study of systems biology, and especially the *interactome*, which is the collection of all molecular interactions occurring within a particular cell. It has been estimated that over 80% of proteins do not function alone but interact with other proteins in complexes (Rao et al., 2014). PPIs originate through a biochemical event and/or an electrostatic force, which, in turn, creates physical contacts between two or more proteins.[35]

Protein interactions are classified as either *permanent/stable* (i.e., protein complexes that are strong and irreversible) or *transient* (i.e., protein complexes that form and break down easily). Transient interactions can be either *strong* or *weak*. Most PPIs are expected to be transient. The definition of what actually constitutes a "real" interaction between proteins is controversial: should an interaction be regarded as "real" if it occurs only in a cell or if it occurs in a test tube, or does the interaction need to have a biological function? (Bonetta, 2010).

The human genome, which was once thought to contain 2 million protein-coding genes, is now estimated to have only approximately 20,000 genes.[36] This means that there are approximately 180 million protein pairs to test to generate a PPI map. Despite the huge quantities of molecular interaction data obtained from studies of PPIs, knowledge of the interactome still remains incomplete. Because of the limitations of current technology, data collected on the human interactome is estimated to account for only about 5–10% of the approximately 200,000–600,000 PPIs (Rolland et al., 2014), while the literature on the interactome of *S. cerevisiae* accounts for only about 20% of the interactions. PPIs have also been cited as one of the major areas of *proteomics*, the large-scale study of proteins, their structures, and their functions.

---

[35] Information on proteins and their crystal structures can be found at the Protein Data Bank's website www.rcsb.org/pdb/home/home.do or at the MODBASE website modbase.compbio.ucsf.edu/modbase-cgi/index.cgi if the protein's structure is unknown.

[36] The most recent estimate is 19,000, but, for convenience, we use 20,000 as our approximation. This number is smaller than the approximately 20,470 genes in the genome of the nematode worm and the 31,000 in the water flea.

Figure 2.12 The PPI network map of *S. cerevisiae* (baker's yeast). There are 1870 nodes and 2240 edges in the network. Source: Jeong et al. (2001). © SpringerNature. Reproduced by permission.

The map of all PPIs for a given organism constitutes the PPI network for that organism with single proteins as nodes and binding interactions between pairs of proteins as undirected edges. PPI maps have been constructed for many different kinds of organisms, including viruses (e.g., *human herpesvirus 8*, the virus responsible for the AIDS-linked cancer known as Kaposi's sarcoma), prokaryotes (e.g., the bacteria *Helicobacter pylori* and *E. coli*), and eukaryotes (e.g., baker's yeast *S. cerevisiae*, the roundworm *C. elegans*, and the fruit fly *D. melanogaster*). The PPI network of baker's yeast is given in **Figure 2.12**. A list of a few databases that contain information or catalogues on PPIs is given in **Table 2.2**.

PPI identification methods can be divided into three categories: *in-vitro* methods, in which a given procedure is performed in a controlled environment outside a living organism; *in-vivo* methods, in which a given procedure is performed on the entire living organism itself; and *in silico*, in which computers are used for simulation or prediction of PPIs. There are three popular methods of identifying PPIs, namely, co-immunoprecipitation (Co-IP, an *in-vitro* method), yeast two-hybrid screen system (Y2HS, an *in-vivo* method), and tandem affinity purification (TAP, an *in-vitro* method).

2.6.3 Neural Networks in the Brain

It has long been known that a brain can be described as a network of connections between nodes. The nodes can be individual neurons (i.e., nerve cells) or, more likely, anatomically segregated *brain regions* (e.g., primary visual cortex, dorsolateral prefrontal cortex, fusiform gyrus) linked together through *pathways* (edges).

Table 2.2   Databases that contain information or catalogues on PPIs

Name	Abbr.	Website (URL)
Database of Interacting Proteins	DIP	dip.doe-mbi.ucla.edu
Biological General Repository for Interaction Datasets	BioGRID	thebiogrid.org
Human Protein Interaction Database	HPID	wilab.inha.ac.kr/hpid
HitPredict	HitPredict	hintdb.hgc.jp/htp
Agile Protein Interaction DataAnalyzer	APID	bioinfow.dep.usal.es/apid

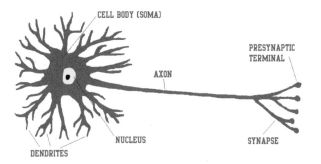

Figure 2.13 Schematic view of a biological neuron. Source: Izenman (2013, Figure 10.1). Reprinted by permission from *Springer*.

A schematic diagram of a biological neuron is given in **Figure 2.13**. We see the *cell body* (or *soma*) of a neuron, which contains a nucleus; outside the soma, there are *dendrites*, operating as input devices, and a single *axon*, which is a long fiber that operates as an output device. The dendrites convey signals from other neurons to the neuron. The axon branches out into a number of strands, each terminating in a *synapse*. Ion concentrations inside and outside the cell provide an electrical charge to the neuron, which, in turn, through an electrochemical process, fires an electrical pulse (called an *action potential* or *spike*) of fixed amplitude and duration to other neurons. The action potential travels down the axon to its endings, which are swollen to form a *synaptic knob* containing *neurotransmitters*. Neurons do not physically join with other neurons, although they may be connected. There is a tiny gap (the *synaptic cleft*) between the axon of the sending (or *presynaptic*) neuron and a dendrite of the receiving (or *postsynaptic*) neuron.

So, how do neurons communicate with each other? The neuron sends a signal to another neuron by releasing neurotransmitters from the presynaptic neuron across the synaptic cleft to the *receptor molecules* situated on the dendrites of the postsynaptic neuron. This action opens up a channel into the postsynaptic neuron and floods it with electrically charged sodium ions, which then produces a local electrical disturbance; this is followed by neighboring channels opening up, and an action potential shoots across the surface of the postsynaptic neuron towards the next neuron. A pulse received by a postsynaptic neuron may not always induce it to fire.

The axon may shut down (a *refractory period*) before it can fire again. Firing occurs at random times, but the firing rate depends upon a number of factors.

A different type of postsynaptic channel (an NMDA glutamic acid receptor) is unusual in that it will not open unless it receives two simultaneous signals, one an electrical discharge from the postsynaptic neuron and the other emitted by the axon from a presynaptic meuron. The two signals provide a mechanism for changes in the synapse.

Choices of nodes and edges in a brain network are often difficult to define and depend upon the type of network being studied (Bassett and Lynall, 2013). The simplest type of network would be individual neurons linked to each other through synaptic connections. Yet, such a network would be almost impossible to record because of the difficulty of observing such large numbers of individual neurons (Sporns, 2011, Chapter 3). This has led to the *parcellation* of the brain into distinct regions, and *connectivity* refers to the connections between different brain regions.

Brain networks are classified as "functional," "structural," or "effective," and the three types provide different views of brain structure and function (Friston, 1994):

- *Structural brain networks* are formed from diffusion-based neuroimaging data (e.g., diffusion tensor imaging (DTI), diffusion spectrum imaging) and network edges (referred to as *structural connectivity* or *connectome*) that represent white-matter connections linking gray-matter brain regions (e.g., synaptic links, fiber pathways). In essence, the connectome is a complete map of the brain's structural connections, which are taken to be undirected (absent information on directionality).
- *Functional brain networks* are formed from functional neuroimaging data (e.g., time-series data obtained from fMRI, DTI recordings) and neurophysiological recordings (e.g., EEG, MEG),[37] where nodes represent neurons, recording sites, or voxels, and undirected network edges represent *functional connectivity* links between neurons and brain regions.
- *Effective brain networks* are formed, like functional brain networks, from neuron interactions. However, in this type of network, the directed network edges represent time-dependent connections that define patterns of potential causal influences of one brain region towards another, and the causal connections are inferred from time-series analysis, statistical modeling, or experimental perturbations.

Time-series data that arise from functional or effective brain networks are obtained from fMRI, EEG, MEG, or other recordings; such data are highly time-dependent and can be highly nonstationary. Techniques used to analyze such data usually involve the frequency analysis of time series. Because of the statistical nature of the data and their methods of analysis, it is not possible to ascribe causal relationships between neurons when dealing with functional brain networks even if one shows that statistical dependencies exist.[38] No simple relationship appears to exist between the functional and structural networks of an individual's brain. However, a lot of research is being carried out trying to relate these two types of brain networks to each other.

---

[37] EEG = electroencephalography, MEG = magnetoencephalography, fMRI = functional magnetic resonance imaging, DTI = diffusion tensor imaging.

[38] This is a consequence of the admonition that "correlation does not imply causation" in observational studies.

The number of neurons in the brain varies remarkably from species to species. The nematode or roundworm, *C. elegans*, whose nervous system has been completely mapped, has 302 neurons, the fruit fly, *D. melanogaster*, has about 100,000 neurons, and the honey bee has about one million neurons in its brain. A small mouse brain, which was completely mapped in 2014 to have 75 million neurons, was found to possess a connectome of more than 1.8 petabytes of computer data. By comparison, it is estimated that the human brain, with about 10^{11} (= 100 billion) neurons, has a connectome consisting of 98,000 petabytes of data.

Recent advances in neuroscience research have led to an enormous increase in work relating brain research to network science. U.S. governmental funding agencies, such as the National Science Foundation, the National Institutes of Health, as well as recent government announcements, have embraced initiatives of brain-mapping research, including mapping the entire human connectome, possibly with a view to discovering important insights into the science of robotics.

The introduction of non-invasive techniques for measuring brain activity (e.g., MRI, fMRI, DTI in neuroimaging, and EEG and MEG for neurophysiological recordings) has enabled huge amounts of data to become readily available.[39] Network theory and methods were first applied to data from fMRI recordings in 2002, EEG recordings in 2007, and MEG recordings in 2004 (Stam, 2007).

2.7 Ecological Networks

Ecologists have long been interested in *ecosystems* and the interactions between living organisms, especially on the topic of which species eats which other species. Every ecosystem contains a *food web*, which is the collection of all interconnected food chains, and where several food chains overlap with each other. A *food chain* is a specific type of path that energy and nutrients may take as they move through an ecosystem. The study of food chains (and food cycles and food size) was pioneered and developed by Charles Elton in his book *Animal Ecology* (Elton, 1927). His use of the term "food cycle" was later renamed as "food web."

2.7.1 Food Webs

There are three different categories, or *trophic levels*, of organism that exist within a food web: producers, consumers, and decomposers. All organisms that occupy the same position within a food chain are considered to be members of the same trophic level within the food web. The trophic levels are:

1. *Primary producers* (or *autotrophs*) make their own food, using photosynthesis and sunlight to transform carbon dioxide and water into simple carbohydrates, which are then used to build other more complex organic molecules (e.g., proteins, lipids, starches). Autotrophs do not depend upon any other organism for nutrition. Examples of producers include plants, free-floating algae, phytoplankton, and bacteria.

---

[39] Network data of the brain recorded from various projects and appropriate Matlab software can be found at the website `brain-connectivity-toolbox.net`.

2. *Consumers* (or *heterotrophs*) eat the autotrophs. Consumers are herbivores, carnivores, or omnivores. *Herbivores* eat only plants, *carnivores* eat other animals, and *omnivores* (e.g., humans, bears) eat anything, including plants and other animals. There are different levels of consumer:

   (a) *Primary consumers* include herbivores such as deer, mice, and elephants in a grassland ecosystem, and fish, turtles, and sea urchins in the ocean ecosystem.

   (b) *Secondary consumers* eat the herbivores. For example, in a desert ecosystem, a snake (a secondary consumer) will eat a mouse.

   (c) *Tertiary consumers* eat the secondary consumers. In a desert ecosystem, an eagle (a tertiary consumer) will eat a snake.

   And so on, until we reach . . .

   (d) *Top predators*, such as lions, great white shark, bobcats, and mountain lions, who eat all other consumers, or they may be animals with few or no natural enemies, such as alligators, hawks, or polar bears.

3. *Decomposers* are small organisms, including invertebrates, fungi, and bacteria, that break down decaying matter such as nonliving plants and animal remains from all the other trophic levels. They turn organic wastes (e.g., decaying plants) into inorganic materials (e.g., nutrient-rich soil), thereby completing the food chain. *Detrivors* are organisms (e.g., vultures, crabs) that feed on detritus, dead and decomposing organisms. What they leave behind is used by decomposers.

Biomass decreases from the base of the chain to the top. Primary producers at the bottom are the most numerous and are the smallest organisms in the food web, whilst the higher trophic levels have the fewest individuals.

Ecosystems and their food webs have been affected by many types of human activity (McCann, 2011), such as hunting, habitat destruction, introduction of alien species, and climate change. An important example is given by Rachel Carson's book *Silent Spring* (Carson, 1962), which described the effect that widespread application of DDT had on the environment. DDT made birds' eggs less viable, which, in turn, removed a group of highly visible species. It is possible that the resulting reduction in birds and other vulnerable species could have drastically altered the ecosystem.

Another example of an intervention occurred in Yellowstone National Park, Wyoming. In the 1930s, hunters completely eliminated the wolves there. As wolves were top preditors, their loss cascaded through the ecosystem, and influenced the diversity within the park. The number of elk now grew to dominate herbivores in the absence of wolves, and as they no longer feared predators, the elk began to move into areas of the park they would not have visited had wolves been present. The increased numbers of elk had a negative effect on the growth of vegetation. Recently, Yellowstone reintroduced wolves into the park, which proved to be successful; the elk moved towards the higher elevations, leaving the vegetation to flourish again.

A food web can be viewed as a directed network, in which the nodes represents "trophic species" (species or groups of species at each trophic level), which interact with each other by preying on species in other trophic levels (i.e., preditor–prey interactions, where the larger predatory consumers are at the top and the smaller-sized, more diverse, prey are at the bottom because energy is lost between each trophic

Figure 2.14 Food web network illustrating the three tropic levels that consist of algae (red dots, lowest level), aquatic insects (orange dots, middle level), and fish (yellow dots, highest level) species and the trophic interactions between them. The data were collected by Thompson and Townsend (1999) in two locations, Dempsters and Sutton, both in Otago, New Zealand, during the autumn, summer, and spring. This figure displays the Dempsters network recorded during the spring. The data were organized in a binary, nonsymmetric matrix in which cells with a "1" indicate a trophic interaction between the preys (row names) and those with a "0" indicate no interaction.

level).[40] See **Figure 2.14**, which shows a food web network consisting of algae, aquatic insects, and fish. In the network representation of a food web, it is customary for ecologists to draw the directed edges so that they show the transference of energy from each node to other nodes (i.e., from prey to preditor), rather than from preditor to prey.

2.8 Terrorist Networks

The world is quickly becoming a more dangerous place because of the rapid rise in terrorist activity. Between 1970 and 2015, there were over 150,000 terrorist attacks worldwide. Over half of all terrorist attacks were nonlethal, and approximately 2% involved 20 or more fatalities. There were over 2000 named organizations plus nearly 700 other generic groups that carried out terrorist attacks during this period. The bulk of the action belonged to only 20 groups, which accounted for over 50% of all attacks, while over 70% of the groups were active only for a year or less and carried out fewer than four total attacks. Most terrorist attacks, including those attributed

---

[40] Data on many food webs can be found in the Pajek datasets at the website `vlado.fmf.uni-lj.si/pub/networks/data/bio/foodweb/foodweb.htm`.

Data on coral reef food webs in three Greater Antillean regions of the Caribbean (the Cayman Islands, Cuba, and Jamaica) can be found at the website `datadryad.org/resource/doi:10.5061/dryad.c213h` The UC-Irvine Network Data Repository contains datasets on the four different ecosystems of South Florida. There are two different networks, one for the dry season and one for the wet season. A node in each network represents a major component of the respective environment. Nodes are classified into five different categories: living/producing compartment; other compartment; input; output; respiration. The website is `networkdata.ics.uci.edu/netdata/html/Florida.html`.

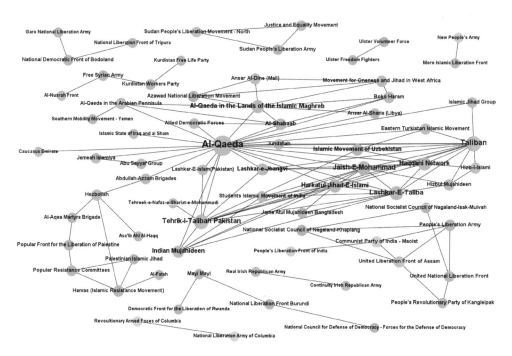

Figure 2.15 Worldwide terrorist network as it existed in 2012. The nodes are colored as follows: orange = religious, purple = ethnic, pink = separatist, blue = leftist, green = unknown.

to foreign organizations that seriously threaten the United States, attacked domestic targets in their own countries. The worldwide terrorist network as it existed in 2012 (the latest year for such data) is displayed in **Figure 2.15**.

During the period 1995–2015, there were 555 attacks in the United States, with a total of 3322 fatalities, including 2997 on September 11, 2001 alone, when two commercial jets, hijacked by al-Qaeda terrorists, crashed into the twin towers of the World Trade Center in New York City, a third jet, similarly hijacked, crashed into the Pentagon in Arlington County, Virginia, and a fourth jet, whose passengers tried to stop the hijacking, crashed into a field in Shanksville, Pennsylvania.

Initially affiliated with al-Qaeda and designated as al-Qaeda in Iraq, what is now known as *Islamic State* broke away from al-Qaeda and rebranded itself, first, as Islamic State in Iraq (ISI) in 2006, and then, in 2013, as Islamic State in Iraq and Syria (ISIS) or, alternatively, as Islamic State in Iraq and the Levant (ISIL), reflecting the different translations from the Arabic.

The organization had three leaders: the Jordanian Abu Musab al-Zarqawi (who founded the group in 1999), the Egyptian Abu Ayyub al-Masri, and the Iraqi Abu Bakr al-Baghdadi. In 2006, al-Zarqawi was killed by a U.S. airstrike and al-Masri was killed in 2010 in a joint Iraqi–U.S. raid on his Iraqi safehouse. In 2014, al-Baghdadi declared his intention to impose a worldwide caliphate (an Islamic institution that would incorporate all temporal and spiritual authorities) with himself as the caliph, and the group began to refer to itself as the Islamic State (IS).

During 2002–2015, ISI, ISIS/ISIL, and Islamic State and affiliated groups made over 4900 terrorist attacks (13% of all terrorist attacks), which led to 33,000 deaths and 41,000 injuries (Miller, 2016). In 2015, 74% of all deaths by terrorism were accounted for by ISIS/ISIL, the Taliban, Boko Haram, and al-Qaeda. Although Boko Haram (which means "Western Education is Forbidden") pledged allegiance to ISIS/ISIL in 2015, it is having financial and military difficulties due to its being hounded by Nigerian forces and African-Union troops. In 2016, over 13,400 terrorist attacks, with 34,000 total deaths, took place in 108 countries, but were mostly concentrated in the Middle East and North Africa (45% of the total), South Asia (27%, mostly in the Philippines), and Sub-Saharan Africa (15%, mostly in Nigeria with Boko Haram and Somalia with al-Shabaab). IS was able to carry out terrorist attacks in Europe (e.g., 2015 suicide attacks in Paris, 2016 attack in Manchester, England, and 2017 attack in Barcelona). IS carried out more than 1400 attacks in 2016, resulting in more than 11,700 total deaths, including more than 4400 of its own group. By mid-2019, however, IS had lost almost all of its territories due to major defeats by the various armies that were allied against it. Al-Baghdadi was killed in October 2019 in a U.S. military operation in northwest Syria.

This is not to say that al-Qaeda has relinquished its prominent role as a terrorist organization. The name "al-Qaeda," which means "The Base" in Arabic, was said by its leader Osama bin Laden in 2001 to have originated from the training camps used in Afghanistan to resist the Soviet occupation during 1979–89. Following the 9/11 attacks, al-Qaeda carried out (or were associated with) the Madrid train bombings in 2004 that killed 192 people, and London bombings in 2005, killing 52 people. Meanwhile, al-Qaeda moved into Yemen and Somalia because of these countries' poor economies and strategic locations. In 2009, those parts of al-Qaeda that operated in Saudi Arabia and Yemen merged to form al-Qaeda in the Arabian Peninsula, and in 2012, al-Shabaab (meaning "The Youth") joined al-Qaeda to form an official branch of al-Qaeda in Somalia. Osama bin Laden was killed in 2011 by U.S. Special Forces, where he was hidden out in Abbottabad, Pakistan.

2.9 Further Reading

Section 2.2.1. There are many accounts of the history of the Internet. Books on the Internet include Pastor-Satorras and Vespignani (2004). An excellent history of the Internet can be found at the website `internetsociety.org/internet/what-internet/history-internet/brief-history-internet`

Section 2.4.1. There have been many good articles written about the U.S. subprime mortgage crisis. The brief account described in Section 2.4.1 was adapted from several sources, including Allen and Babus (2008), Bianco (2008), Pozsar et al. (2012), Glasserman and Young (2015) and Zhou (2016).

Section 2.4.2. Excellent descriptions of the Bitcoin system are given by Kondor et al. (2014) and Antonopoulos (2015, Chapter 4).

Section 2.5. Books on social networks include Wasserman and Faust (1994), Scott (2000), and Freeman (2004).

Section 2.6.4. An excellent textbook on brain networks is Sporns (2011). See also Sporns (2013). Some of Section 2.6.4 on neural networks in the brain was adapted from Izenman (2013, Section 10.2).

Section 2.7. Books on ecological networks and food webs include Pascual and Dunne (2006) and McCann (2011).

Section 2.8. Information on terrorism can be found from many different sources, including Wikipedia. In particular, we mention the *National Consortium for the Study of Terrorism and Responses to Terrorism* (known as START), a Department of Homeland Security Center of Excellence headquartered at the University of Maryland.[41] START is also home to the *Global Terrorist Database* (GTD), which contains information on up to 120 variables that specify the details on each of the attacks worldwide up to 2015, including when and where the attack took place, the perpetrators, the victims, tactics, weapons, and the outcome of the attack. The data for Figure 2.9 were transcribed from the BAAD (Big, Allied, and Dangerous) website `start.umd.edu/baad/network/2012`.

Another source of information on terrorism is provided by the *Foundation for Defense of Democracies*,[42] where details on the main terrorist groups al-Qaeda, Boko Haram, Islamic State, and al-Shabaab, and their affiliations, are publicly available, including a database of entities and individuals designated by the U.S. Treasury's Office of Foreign Assets Control (OFAC) for sanctions, and a *Terror Finance Briefing Book* (Fanusie amd Entz, 2017) that spells out how the major terrorist groups fund their operations. See also Hoffman (2006).

2.10 Exercises

Exercise 2.1 Try out a software networks package that has good graphics capability (e.g., `igraph`, `Gephi`). Formulate which graphics commands would be best for certain types of network data. Which commands or package draws network graphs that looks the most pleasing visually?

Exercise 2.2 Find a network dataset of your choice and draw the graph corresponding to those data. Draw a histogram of the "degree distribution," where "degree" is the number of edges emanating from each node (degree number on the x-axis and number of nodes with that degree on the y-axis). What do you think the degree distribution tells you about the network?

Exercise 2.3 Find a large network dataset of your choice. Draw a graph of the network. Draw the histogram of the degree distribution of the data. Do you see a "hub," which is a node with a very high degree? What would happen to the network graph if you removed that hub?

Exercise 2.4 Download the data on Florentine family marriages that occurred during 1282–1500. The data on 16 families were collected from historical records by Padgett (1994). See also Padgett and Ansell (1993). The data are contained in the file `flomarriage` and are available at the UCINET website `networkdata.ics.uci.edu/netdata/html/florentine.html`.

They are also available at the website `rdrr.io/cran/network/man/flo.html`.

The data also include information on families who were locked in a struggle for political control of the city of Florence around the year 1430. The two most

---

[41] See the website `start.umd.edu`.
[42] See the website `fdd.org`.

powerful families were the Medicis (not wealthy, but married into power) and the Strozzis. Draw the network and the degree distribution of the marriage data.

Exercise 2.5 Download the Pennsylvania road network data `roadNet-PA` from the Stanford Large Network Dataset Collection at the website `snap.stanford.edu/data/roadNet-PA.html`.

There are 1,088,092 nodes and 1,541,898 edges that make up the network. The data are from Leskovec et al. (2009). Draw the network and the degree distribution of the data.

Exercise 2.6 Download any one of the food web datasets from the Pajek website (see footnote 40) and draw its network graph.

Exercise 2.7 Download data on a small, closed bottlenose dolphin community of Doubtful Sound, the second largest fjord of the 14 fjords that compose Fiordland in New Zealand (Lusseau et al., 2003). The population consisted of 62 dolphins. The data can be found at the website `www-personal.umich.edu/~mejn/netdata/`. Draw its network graph.

Exercise 2.8 Download data on a directed network of hyperlinks between weblogs on U.S. politics (Adamic and Glance, 2005). The data can be found at `www-personal.umich.edu/~mejn/netdata/`. Draw its network graph.

Exercise 2.9 Download data on Hollywood film music (Faulkner, 1983). There are 102 nodes (40 composers of film scores and the 62 producers who produced at least five movies in Hollywood during 1964–1976) and 192 weighted edges showing cooperation of producer and composer whose weights on the edges represent the number of films they cooperated on. The data can be found in the file `Movies.net` on the website `sites.google.com/site/ucinetsoftware/datasets/hollywoodfilmmusic`. Draw its network graph.

Exercise 2.10 Download data on the Marvel Universe, a series of comic books about superheroes such as Spiderman, the Hulk, the Fantastic Four, and the X-Men, plus Captain America. The dataset, which contains information about the superheroes and the comics, can be found at `www.kaggle.com/csanhueza/the-marvel-universe-social-network`. Draw its network graph.

CHAPTER THREE

Graphs and Networks

The aim of this chapter is to provide readers with an introduction to the basic ideas of networks and their representation by graphs. We will be using ideas, definitions, terminology, and notation from graph theory throughout this book.

3.1 Introduction

We first list many definitions of graph objects and terminology that we will need as we learn about random networks. We next describe the *adjacency matrix*, which formalizes the relationships found in networks. Then, we define the concept of a *path* through a network and the idea of the *connectivity* of a network. To illustrate some of these concepts in real-world networks, we describe the analyses of two huge networks, a telephone call network and the Facebook social network. Next, we define *clusters* and *hubs* that are commonly found in real networks. We then describe two graph-searching algorithms, *breadth-first search* and *depth-first search*, and the algorithm for coloring the nodes of a network. Finally, we describe the concept of a *minimal spanning tree* and the various algorithms that are used to compute it.

3.2 Definitions

A network is represented visually by a graph having nodes and edges.

Definition 3.1 A graph is denoted by

$$\mathcal{G} = (\mathcal{V}, \mathcal{E}), \tag{3.1}$$

where its *nodes* (or *vertices*, *sites*, *actors*) are given by the set $\mathcal{V}$, and its *edges* (or *links*, *bonds*, *ties*) by the set $\mathcal{E}$, where $\mathcal{E} \subseteq \mathcal{V} \times \mathcal{V}$.

Much of the notation, terminology, and theoretical results now common in the study of random networks originated in the mathematical discipline known as *graph theory*.

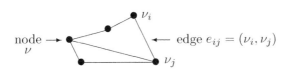

Figure 3.1 A network graph with $N = 5$ nodes and $M = 6$ edges.

3.2.1 Nodes and Edges

A generic *node* will be represented by $v \in \mathcal{V}$ with identifying subscripts as appropriate, and the *edge* from node $v_i \in \mathcal{V}$ to $v_j \in \mathcal{V}$ is represented as $e_{ij} = (v_i, v_j) \in \mathcal{E}$ or, for convenience, by just (i, j). See **Figure 3.1**. The number of nodes in $\mathcal{G}$ is denoted by $N = |\mathcal{V}|$ and the number of edges by $M = |\mathcal{E}|$. The maximum number of edges in $\mathcal{G}$ is $\binom{N}{2}$. A graph in which every pair of nodes is connected is called a *complete* graph.

Nodes and edges may have different names depending upon the discipline. In the literature of social networks, nodes are referred to as "actors" and edges as "ties," whilst in the physical sciences, nodes are "sites" and edges are "bonds."

Definition 3.2 If more than one edge links the same two nodes together, we refer to $\mathcal{G}$ as a *multi-edge* graph or a *multigraph*.

Definition 3.3 An edge (v, v) that links a node to itself is called a *self-loop*.

Definition 3.4 Two nodes that are connected by an edge to each other are said to be *neighbors* or *nearest neighbors* of each other. Neighboring nodes are also said to be *adjacent* to each other.

Definition 3.5 A *cycle* is a sequence of at least three (adjacent) nodes and edges such that the beginning node and the ending node are the same, whilst the intermediate nodes and edges of the sequence are different.

3.2.2 Graph Orientation

Graphs are described as "undirected" or "directed" depending upon whether their edges are undirected or directed, respectively.

Definition 3.6 An edge of an *undirected graph* is a line joining two nodes without any suggestion of direction of that line. An undirected edge is symmetrically related to both nodes in that the relationship netween node v_i and v_j is the same as that between v_j and v_i.

Definition 3.7 A *directed graph* (or *digraph*) has edges that are represented by arrows, where an arrow ($\rightarrow$) points away from one node, v_i, say, towards another node, v_j, say, in the sense that v_i causes a change in v_j to happen. There is no implication in a directed graph that the reverse arrow ($\leftarrow$) from v_j to v_i exists.

There are many examples of directed graphs, ranging from individuals who list their best friends, to website hyperlinks on the World Wide Web, to predators and prey in a food web. A directed graph without cycles is described as a *directed acyclic graph* or DAG.

3.2.3 Edge Weights

Graphs can have weighted edges. The default is that the weights are 1 or 0, depending upon whether the edge exists or does not exist, respectively. In certain types of networks, weights are different from 1 or 0, and are typically non-negative, although networks with signed (positive and negative) weights are known. In a *weighted signed network* (WSN), a negative weight may indicate the extent of one person's dislike or lack of trust in another person, whilst a positive weight would indicate the opposite.

An example of a weighted signed network is the Bitcoin OTC trust network with 5881 nodes and 35,592 edges.[1] The edge weights in this network are set to give Bitcoin traders, who are anonymous, some amount of confidence as to how well their potential trading partners can be trusted. These weights reflect a trader's reputation based upon his/her trading record, and are designed to prevent trading with "fraudulent and risky users." The weight that a trader uses to rate others ranges from -10 to $+10$ (excluding 0) in steps of 1.

3.2.4 Node Degree

Definition 3.8 The *degree* of node v in an undirected graph is the number of edges emanating from v to other nodes in $\mathcal{G}$. The histogram of node degrees ordered by degree size is called the *degree distribution*.

As we will see, the degree distribution is an important characteristic of a network. The total number of all node degrees is equal to twice the number of edges in the graph.

For a directed graph, we need a slightly different definition of degree because of the orientations of the links between pairs of nodes.

Definition 3.9 In a directed graph, the number of edges going out from node v is called the *out-degree* of v, whilst the number of edges incoming to node v is called the *in-degree* of v. There is no reason that the in-degree and the out-degree of node v should be equal. The *degree* for node v in a directed graph is the sum of the in-degree and the out-degree for that node.

3.2.5 Tree Graphs

Definition 3.10 A *tree graph* is a connected graph in which every node (considered as a "child" node) has only one other node (the "parent" node) from which

---

[1]See snap.stanford.edu/data//soc-sign-bitcoinotc.html. These and other WSN data are described and analyzed in Kumar et al. (2016).

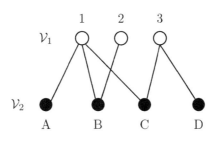

Figure 3.2 A bipartite network graph with two set of nodes: $\mathcal{V}_1$ with three nodes (1, 2, and 3) and $\mathcal{V}_2$ with four nodes (A, B, C, and D).

it is descended in a hierarchical graph structure. Tree graphs contain no cycles or self-loops. In a tree graph, $|\mathcal{V}| = |\mathcal{E}| + 1$. So, if $|\mathcal{V}| = N$, then for a tree graph, $|\mathcal{E}| = N - 1$.

Definition 3.11 The *root node* of a tree graph is the unique node that is connected by a path to every other node.

3.2.6 Bipartite Graphs

Definition 3.12 A *bipartite graph* is a graph $\mathcal{G} = (\mathcal{V}, \mathcal{E})$ whose set of nodes is partitioned into two disjoint sets, $\mathcal{V} = \mathcal{V}_1 \cup \mathcal{V}_2$, of nodes, where $\mathcal{V}_1 \cap \mathcal{V}_2 = \emptyset$, and every edge in $\mathcal{E}$ links a node in $\mathcal{V}_1$ to a node in $\mathcal{V}_2$.

See **Figure 3.2**. There is no edge from a node in $\mathcal{V}_1$ to another node in $\mathcal{V}_1$, nor any edge from a node in $\mathcal{V}_2$ to another node in $\mathcal{V}_2$.

Examples of bipartite networks include which actors appear in which movies, judges ranking ice skaters in tournaments, movie-goers rating movies, people who are members of certain clubs or organizations, and transportation networks (e.g., airline, train, or bus), where the starting locations are in one set and the destination locations (or possible routes) are in the other set. In some of these examples, the edges of the bipartite network would be weighted.

Some other examples of bipartite graphs:

- Bipartite graphs have been applied to human disease networks where the nodes divide into all known (1284) genetic disorders and all known (1777) disease genes in the human genome, and an edge exists between a disorder and a gene if mutations in that gene are thought to result in that disorder (Goh et al., 2007).
- A popular application of bipartite graphs is in modern coding theory (Arunkumar and Komala, 2015), where one group of nodes represents digits of a codeword whilst the other group of nodes represents combinations of digits that are expected to sum to zero in a codeword without errors.
- In a recent newspaper article (Dawson and Boston, 2018), a bipartite network was drawn (see **Figure 3.3**) that connected vehicles assembled in North America by 12 different automobile companies (BMW, Daimler, GM, FCA, Ford, Honda, KIA, Mazda, Nissan, Subaru, Toyota, and VW) that use engines or transmissions from 16

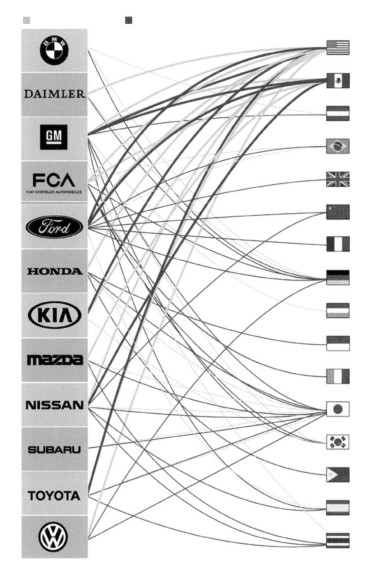

Figure 3.3 A bipartite network showing 12 different automobile companies (BMW, Daimler, GM, FCA, Ford, Honda, KIA, Mazda, Nissan, Subaru, Toyota, and VW) that use engines or transmissions from 16 countries (United States, Mexico, Austria, Brazil, Britain, China, France, Germany, Hungary, Indonesia, Italy, Japan, Korea, Phillippines, Spain, and Thailand). Reproduced from Dawson and Boston (2018). © *The Wall Street Journal.* Reproduced by permission.

countries outside the region (United States, Mexico, Austria, Brazil, Britain, China, France, Germany, Hungary, Indonesia, Italy, Japan, Korea, Phillippines, Spain, and Thailand).

3.2.7 Subgraphs

Definition 3.13 A *subgraph* $\mathcal{G}' = (\mathcal{V}', \mathcal{E}')$ of a graph $\mathcal{G} = (\mathcal{V}, \mathcal{E})$ is a graph in which $\mathcal{V}'$ is a subset of the nodes of $\mathcal{G}$, and $\mathcal{E}'$ is a subset of the edges in $\mathcal{E}$, so that $\mathcal{V}' \subseteq \mathcal{V}$ and $\mathcal{E}' \subseteq \mathcal{E}$.

Table 3.1 Adjacency matrix for Sampson's monk data on the "like" responses at the third time point (T_4), where # refers to the novice number listed in Table 2.1. A "1" indicates that an edge is present and a "0" indicates that no edge is present

#	1	2	3	4	5	6	7	8	9	10	11	12	13	14	15	16	17	18
1	0	0	1	0	0	0	0	0	0	0	0	1	0	1	0	0	0	0
2	1	0	0	0	0	0	1	0	0	0	0	1	0	0	0	0	0	0
3	1	0	0	0	0	0	0	0	0	0	0	0	1	0	0	0	1	1
4	0	0	0	0	1	1	0	0	0	0	1	0	0	0	0	0	0	0
5	0	0	0	1	0	0	0	0	1	0	1	0	0	0	0	0	0	0
6	0	0	0	1	1	0	0	0	1	0	0	0	0	0	0	0	0	0
7	0	1	0	0	0	0	0	0	0	0	0	1	0	0	0	1	0	0
8	0	0	0	1	0	1	0	0	1	0	0	0	0	0	0	0	0	0
9	0	0	0	0	1	0	0	1	0	0	0	1	0	0	0	0	0	0
10	0	0	0	1	1	0	0	0	1	0	0	0	1	0	0	0	0	0
11	0	0	0	0	1	0	0	1	0	0	0	0	0	1	0	0	0	0
12	1	1	0	0	0	0	1	0	0	0	0	0	0	0	0	0	0	0
13	0	0	0	0	1	0	1	0	0	0	0	0	0	0	0	0	0	1
14	1	0	0	0	0	0	0	0	0	0	0	1	0	0	1	0	0	0
15	0	1	0	0	0	0	1	0	0	0	0	1	0	0	0	0	0	0
16	0	1	0	0	0	0	1	0	0	0	0	0	0	0	1	0	0	0
17	0	1	1	0	0	0	0	0	0	0	0	0	0	0	0	0	0	1
18	0	1	1	0	0	0	0	0	0	0	0	0	0	0	0	0	1	0

Definition 3.14 A special subgraph is a *clique*, which is a complete subgraph of $\mathcal{G}$ in which all pairs of nodes in that subgraph are linked by edges.

Definition 3.15 A *maximal clique* is a clique that is not contained in a larger clique. So, every clique is contained in a maximal clique.

A graph with N nodes has at most $3^{N/3}$ maximal cliques (Moon and Moser, 1965) and these maximal cliques can be listed using the Bron–Kerbosch algorithm (Bron and Kerbosch, 1973).

3.3 The Adjacency Matrix

The usual way to specify an unweighted graph $\mathcal{G}$ and its nodes and edges is through its adjacency matrix.

Definition 3.16 The ($N \times N$) *adjacency matrix* $\mathbf{Y} = (Y_{ij})$ corresponding to $\mathcal{G}$ has elements

$$Y_{ij} = \begin{cases} 1, & \text{if } (v_i, v_j) \in \mathcal{E} \\ 0, & \text{otherwise} \end{cases}, \quad i, j = 1, 2, \ldots, N, \tag{3.2}$$

and $|\mathcal{V}| = N$.

In the case of an undirected graph, $Y_{ij} = Y_{ji}$, so that $\mathbf{Y}$ is a symmetric matrix. For example, **Table 3.1** displays the adjacency matrix for Sampson's monk network. For a directed graph, typically $Y_{ij} \neq Y_{ji}$, in which case $\mathbf{Y}$ is asymmetric.

Figure 3.4 Undirected graph for Worked Example 3.1.

For an undirected graph, the row sum $Y_{i+} = \sum_{j=1}^{N} Y_{ij}$ is equal to the degree of node v_i, the column sum $Y_{+j} = \sum_{i=1}^{N} Y_{ij}$ is equal to the degree of node v_j, and, by symmetry, $Y_{i+} = Y_{+i}$ for $i, j = 1, 2, \ldots, N$.

For a directed graph, Y_{i+} is the *out-degree* for node v_i and Y_{+j} is the *in-degree* for node v_j, and the *degree* of a directed graph is their sum, $Y_{i+} + Y_{+i}$.

Worked Example 3.1

Consider the following simple example. Suppose the undirected graph with five nodes and five edges is given by **Figure 3.4**. The corresponding adjacency matrix **Y** is given by

$$\mathbf{Y} = \begin{pmatrix} 0 & 1 & 0 & 0 & 0 \\ 1 & 0 & 1 & 1 & 0 \\ 0 & 1 & 0 & 1 & 0 \\ 0 & 1 & 1 & 0 & 1 \\ 0 & 0 & 0 & 1 & 0 \end{pmatrix}. \tag{3.3}$$

In this example, node v_1 has degree $Y_{1+} = Y_{+1} = 1$, node v_2 has degree $Y_{2+} = Y_{+2} = 3$, node v_3 has degree $Y_{3+} = Y_{+3} = 2$, node v_4 has degree $Y_{4+} = Y_{+4} = 3$, and node v_5 has degree $Y_{5+} = Y_{+5} = 1$.

3.4 The Incidence Matrix

A bipartite graph (or network) differs from an undirected graph in that the set of nodes $\mathcal{V}$ is divided into two disjoint sets, $\mathcal{V} = \mathcal{V}_1 \cup \mathcal{V}_2$, where $\mathcal{V}_1 \cap \mathcal{V}_2 = \emptyset$. There is no reason for the size of $\mathcal{V}_1$ to be the same as that of $\mathcal{V}_2$. There are no edges between nodes from $\mathcal{V}_1$ to nodes also in $\mathcal{V}_1$, and the same with $\mathcal{V}_2$. The matrix of edges from $\mathcal{V}_1$ to $\mathcal{V}_2$ is called an *incidence matrix* and has dimensions $|\mathcal{V}_1| \times |\mathcal{V}_2|$.

Definition 3.17 The $|\mathcal{V}_1| \times |\mathcal{V}_2|$ *incidence matrix* is denoted by $\mathbf{M} = (M_{ij})$, where the $|\mathcal{V}_1|$ rows represent the nodes of the network and the $|\mathcal{V}_2|$ columns represent groups of which the nodes could be members. Its elements are given by

$$M_{ij} = \begin{cases} 1, & \text{if node } v_i \text{ is a member of group } j \\ 0, & \text{otherwise} \end{cases}, \tag{3.4}$$

where $i = 1, 2, \ldots, |\mathcal{V}_1|$, $j = 1, 2, \ldots, |\mathcal{V}_2|$.

An example of an incidence matrix is given in **Table 3.2**. There are four rows (i.e., $|\mathcal{V}_1| = 4$) and six columns (i.e., $|\mathcal{V}_2| = 6$).

Table 3.2 Example of an incidence matrix. There are four nodes (a, b, c, d) in $\mathcal{V}_1$ and six nodes (A, B, C, D, E, F) in $\mathcal{V}_2$

	A	B	C	D	E	F
a	1	0	0	1	0	0
b	0	1	0	1	0	0
c	1	0	0	0	1	1
d	0	0	1	0	1	0

3.5 Paths and Connectivity

3.5.1 Paths and Walks

Definition 3.18 A *path* $\mathcal{P}$ in a graph $\mathcal{G} = (\mathcal{V}, \mathcal{E})$ is defined as an ordered collection of alternating nodes and edges that enables one to connect a pair of nodes in $\mathcal{G}$. Specifically, $\mathcal{P}_{ij}$ is a path from node v_i to node v_j if it can be written as a sequence of the form

$$\mathcal{P}_{ij} = \{v_i, (v_i, v_{i+1}), v_{i+1}, (v_{i+1}, v_{i+2}), v_{i+2}, \dots, v_{j-1}, (v_{j-1}, v_j), v_j\}. \quad (3.5)$$

Thus, a path runs from v_i to v_j, passing through a number of adjacent nodes that are connected pairwise through edges in $\mathcal{E}$. If the graph is directed, the path must follow the direction specified by each of the edges between nodes. In an undirected graph, the path can follow edges in either direction. A special type of path is a self-avoiding path.

Definition 3.19 A *self-avoiding path* is a path that does not cross itself and does not include a node more than a single time. The length of such a path is the number of edges in the path.

Definition 3.20 If a path can visit the same edges more than once in either direction, we call it a *walk*. The number of walks from node v_i to node v_j of length ℓ is given by the ijth entry in the matrix $\mathbf{Y}^\ell$, the ℓth power of $\mathbf{Y}$, where $\mathbf{Y}$ is the adjacency matrix associated with $\mathcal{G}$.

Definition 3.21 A *self-avoiding walk* is a walk that is not allowed to visit any node more than once.[2]

There may be several different paths that join two nodes. To be more specific, consider only those paths that are independent (or disjoint) of each other. There are two kinds of independent paths: *node-independent* paths and *edge-independent* paths. The former shares only the starting and ending nodes, whilst the latter shares no edges. Node-independent paths are also edge-independent paths, but the converse is false.

Consider again the adjacency matrix (3.3). Computing $\mathbf{Y}^2$ yields the number of walks of length 2 between pairs of nodes:

---

[2]The term "self-avoiding walk," which originated in the context of percolation on a lattice, was attributed by Cyril Domb to John Hammersley; see Hammersley and Morton (1954). It has been used as a model of linear polymers, which are molecules that form long chains of monomers. See Slade (1994).

$$\mathbf{Y}^2 = \begin{pmatrix} 1 & 0 & 1 & 1 & 0 \\ 0 & 3 & 1 & 1 & 1 \\ 1 & 1 & 2 & 1 & 1 \\ 1 & 1 & 1 & 3 & 0 \\ 0 & 1 & 1 & 0 & 1 \end{pmatrix}, \tag{3.6}$$

where the ith diagonal entry also shows the degree of node v_i, $i = 1, 2, \ldots, 5$. To find the number of walks of length 3 between pairs of nodes, we compute

$$\mathbf{Y}^3 = \begin{pmatrix} 0 & 3 & 1 & 1 & 1 \\ 3 & 2 & 4 & 5 & 1 \\ 1 & 4 & 2 & 4 & 1 \\ 1 & 5 & 4 & 2 & 3 \\ 1 & 1 & 1 & 3 & 0 \end{pmatrix}. \tag{3.7}$$

For example, to get from node v_2 to v_4 in a walk of length 3, we have the following five walks:

$$2 \rightarrow 4 \rightarrow 2 \rightarrow 4,$$
$$2 \rightarrow 3 \rightarrow 2 \rightarrow 4,$$
$$2 \rightarrow 1 \rightarrow 2 \rightarrow 4,$$
$$2 \rightarrow 4 \rightarrow 5 \rightarrow 4,$$
$$2 \rightarrow 4 \rightarrow 3 \rightarrow 4,$$

where, for simplicity, we have abbreviated v_i to just i.

3.5.2 Connectivity

Definition 3.22 A graph is called *connected* if any node in the graph can be reached (by a path) from any other node.

Definition 3.23 The strength of *connectivity* between two nodes is given by the number of (node- or edge-)independent paths between those nodes.

Definition 3.24 In a directed network, two nodes are said to be *weakly connected* if the undirected version of the network is connected. The two nodes are said to be *strongly connected* if the path between them has all directed edges oriented in the direction from one node to the other.

3.5.3 Components

Definition 3.25 A subgraph of a graph $\mathcal{G}$ is said to be a *component* of $\mathcal{G}$ if it is a maximally connected subgraph. In other words, if an extra node is added to the subset, the connectedness property of the subgraph will be broken.

Definition 3.26 Two components, A and B, say, of $\mathcal{G}$ are said to be *disconnected* from each other if no path exists from any node in component A to any node in component B. Conversely, if a path exists from every node in component A to every node in component B, the two components are said to be *connected*.

If the graph $\mathcal{G}$ is composed of several components, then the rows and columns of the adjacency matrix $\mathbf{Y}$ can be arranged in such a way that $\mathbf{Y}$ becomes a block-diagonal matrix, with the blocks corresponding to the different components. In practice, it may be difficult to identify the components if the rows and columns of **Y** are not arranged by components. Identifying the components of an undirected graph can, however, be accomplished by using the *breadth-first search* algorithm (see Section 3.7.2).

3.5.4 Geodesic Paths

There may be many paths that permit one to get from node v_i to node v_j. Which one is best? Obviously, the shortest path.

Definition 3.27 The distance between a pair of nodes, v_i and v_j, is taken to be the total number of edges that have to be visited along the shortest path connecting v_i to v_j. This distance measure is called the *shortest path length* or *geodesic distance*.

Note that, in the case of a directed graph, the geodesic distance from v_i to v_j may be different from the geodesic distance from v_j to v_i. The path whose length is the geodesic distance is called the *geodesic path*. If each of two nodes exists in a disconnected component of the graph, then we say that the geodesic distance between the two nodes is infinite.

Because a pair of nodes can be linked by two or more paths of equal length, geodesic paths may not be unique.

3.5.5 Diameter

Of special interest is the *diameter* of a graph.

Definition 3.28 The *diameter* of a graph is defined as the length of the longest geodesic path between any pair of nodes, assuming that such a path exists and not including infinite-length paths of nodes in disconnected components.

3.6 Two Huge Real-World Networks

In this section, we briefly describe the basic information obtained from two huge real-world networks that have hundreds of millions of nodes and between hundreds of millions and billions of edges. There are many computational challenges to analyzing networks of such size, and the authors of these studies have focused on how to solve these challenges.

3.6.1 Telephone Call Network

Abello, Pardalos, and Resende (1999) studied an example of a huge network of telecommunications traffic using telephone billing records. The study was carried out by members of AT&T Shannon Laboratories in Floral Park, New Jersey.

The *call graph* had 53,767,087 nodes (telephone numbers) and over 170 million edges (calls made from one number to another). The call graph is a *directed multigraph*, directed because there was an originator and receiver of each call and a multigraph because more than one call can be made between the same pair of telephones.

Although this particular call graph was not connected, there were 3,667,448 separate connected components (i.e., subsets of nodes that are connected), but just over 8% (302,468) of those components had size greater than 3. The vast majority (over 75%) of the components were pairs of telephones that only called each other. (A strange finding was that there were also 255 "self-loops," calls originating and received by the same person!)

A difficult computational problem was the identification of cliques (and the maximal clique) in this huge network. A probabilistic search strategy based upon the depth-first search algorithm (see Section 3.8.3) yielded many large cliques without showing any of them to be maximal. However, the largest clique (i.e., a subgraph in which every node is connected by an edge to every other node) that was found had 30 nodes. Over 14,000 of these cliques were identified as 30-node cliques.

3.6.2 Facebook Social Network

Ugander et al. (2011) carried out a huge structural analysis of *Facebook*, the largest existing social network. As in all social networks, the nodes represent individuals and the edges represent relationships between pairs of those individuals.[3]

At the time of this study, Facebook consisted of 721 million worldwide active users (nodes),[4] which is about 10% of the world population. An active user was defined as one who logged into the site during the last 28 days from the date of the study (May 2011) and had at least one Facebook friend.[5] There were 68.7 billion Facebook friends, which accounted for the edges in the network.

The degree of a Facebook user shows how many friends that user has. The degree distribution, p_k, of Facebook users shows what fraction of users in the network have exactly k friends. In this study, the authors showed that the degree distribution is skewed to the right (i.e., has a long right-hand tail) and degrees have a high variance with substantial curvature on a log–log scale. Such curvature was surprising because the literature often claimed that degree distributions generally followed a power-law shape, which would be a straight line on a log–log scale (see Section 7.5). The authors conclude that the power-law model is not appropriate for the Facebook network.

The Facebook social network was found to be almost fully connected. A connected component in the network consists of a collection of users for which each pair of users is connected by at least one path through the network. Although there are many connected components, most of them are tiny. In fact, the second-largest component consists of just over 2000 users. The largest connected component, however, is gigantic and accounts for 99.91% of the users in the network who have at least one friend.

Some of the most interesting findings from this study were that although there are about 30 million fewer active female users on Facebook than males, the average female degree of 198 is larger than the average male degree of 172, a random neighbor is more likely to be female than male and is more likely to be of the same age, and users have more friends from within their own country of origin than from outside.

---

[3]An anonymized sample of Facebook data (with 4039 nodes and 88,234 edges) can be downloaded from the website snap.stanford.edu/data/egonets-Facebook.html.

[4]Those eligible to have Facebook accounts have to be over 13 years old with access to the Internet. Facebook also limits a user to 5000 friends.

[5]The authors recognize that this is not Facebook's standard definition of "active user."

3.7 Clusters and Hubs

Many real networks tend to exhibit a "clustering" effect. What this means is that subcollections of the nodes tend to form groups (or clusters) with dense linkage patterns within clusters and sparse linkage patterns between clusters. Members of the same cluster are more likely to be connected than if they are members of different clusters. Clusters are often referred to as *communities* or *modules*, and it is of especial interest to identify such communities in a given network. In Section 12.5, we will assess the adequacy of a particular division of a network into modules, where we define a measure of "modularity."

It is common to assume that these communities are disjoint, that each node can be a member of one and only one community. Such an assumption, if false, can lead to the identification of non-existent communities. Most networks are more complicated than that, in that they may contain overlapping communities, meaning that nodes can be members of more than one community, which makes the identification problem more difficult.

The building blocks for identifying communities of nodes consist of cliques. Nodes form cliques in the neighborhoods around certain nodes of the graph. Several measures of clustering have been proposed, the most popular being the *clustering coefficient* (see Section 7.3.4), which quantifies the total number of triangles (i.e., three-node cliques) in a network, given that two neighbors of a node are neighbors of themselves. In other words, if node v_i is a neighbor of node v_j, and node v_j is a neighbor of node v_k, then there is a high probability that v_i is a neighbor of v_k. The clustering coefficient has a value that lies between 0 and 1, where 0 reflects no clustering and 1 perfect clustering (i.e., the network breaks down into a set of disjoint cliques). The ability to detect network communities depends upon context and field of application, and the complexity of the network.

Another important feature of a network is that it may contain at least one "hub." In the case of an undirected graph, a *network hub* is a high-degree node that occupies a central position in the network.

We next describe two types of network hubs, transportation hubs and brain hubs.

3.7.1 Transportation Hubs

An important type of network hub is a transportation hub, which acts as a bridge that is strategically located so that passengers and air cargo can switch from one mode of transportation (e.g., planes, trains, ships, buses) to another. For example, an airport hub is a very large central airport that has a huge number of flight connections to other airports and may also act as a main hub for one or more airline companies. Some of the larger airline companies have hubs in different locations.

Even though most studies of transportation networks tend to center on the behavior of specific networks and hubs, research shows that transportation networks do not operate as isolated entities; they are dependent upon one another through their interconnections. Thus, major airports tend to have close relationships to rail transportation systems and also to major seaports, where the common geographical location of each makes it worthwhile to maintain connecting services for their

Figure 3.5 The 150-foot-tall Oculus, the main station house of the World Trade Center Transportation Hub, New York City, which contains the Westfield World Trade Center plaza and shopping mall. It was designed by Spanish architect Santiago Calatrava in remembrance of 9/11. ©Alan Karchmer/OTTO. Reproduced by permission.

passengers. The recently completed (in 2016) World Trade Center Transportation Hub in New York City (see **Figure 3.5**), which incorporates retail and dining facilities, connects 250,000 daily commuters and millions of annual visitors to 11 different subway lines, a rail system, a ferry terminal, museums, and financial centers.

In an airline transportation system, the routes of flights are considered to be the "spokes" of a "hub-and-spokes" system, which became very popular following the U.S. federal government's deregularization of the airlines during the late 1970s. Many different types of transportation networks employ the hub-and-spoke system for daily operations to reduce costs, increase speed of deliveries, and provide improved shipment tracking. Hub-and-spoke models have drawbacks, however, due to their vulnerability to attacks, breakdowns, and disruptions. Such vulnerabilities, therefore, tend to encourage transportation hubs to cooperate as a global network. A *hub dependence index* (Ducruet and Lugo, 2013), which has been proposed as a measure of such vulnerability, corresponds to the percentage of the strongest traffic edge in the total traffic of each node in the network.

The increased volume of daily passenger traffic and goods, flight delays, and long lines at security checkpoints has led to passengers spending larger amounts of time in airports (as well as at train stations and central bus terminals) than ever before. As a result, many transportation hubs are now being redesigned and appear as impressive architectural structures. Many of these types of hubs each tend to operate as a self-contained city having many different "neighborhoods" with information centers, cell-phone charging stations, different food vendors, restaurants, and boutique clothes and jewelry stores.

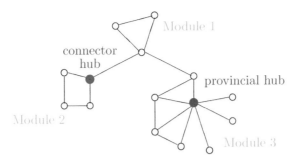

Figure 3.6 Types of brain hubs. Provincial hubs are high-degree nodes that connect mainly to nodes in the same module, while connector hubs are high-degree nodes that connect mainly to nodes in other modules. Adapted from van den Heuvel and Sporns (2013, Figure 1).

3.7.2 Brain Hubs

Another important type of hub occurs in the brain. Studies of networks in the human brain examine the central role of hubs in neural communication and brain function and how they differ in healthy and diseased (or injured) brains. For brain networks, nodes correspond to neurons and/or brain regions and edges correspond to synaptic connections. Network hubs are nodes that enjoy a very high degree of connectivity to other nodes.

As we noted in Section 2.6.4, the human brain contains different types of networks, including structural and functional networks. Structural networks deal with anatomical connectivity, which is stable for short periods of time (seconds to minutes), but not so much so in the long run (hours to days). Functional networks, on the other hand, are highly time-dependent and tend to exhibit nonstationary behavior. Although both structural and functional hubs play a major part in brain communications, it is more meaningful when discussing hubs in the brain to focus on structural brain networks because functional networks do not provide such a straightforward interpretation of brain hubs.

Hubs in the brain that are primarily connected to nodes within the same *module*[6] are known as *provincial hubs*, whereas hubs whose main connections are to nodes in other modules are known as *connector hubs* (van den Heuvel and Sporns, 2013). See **Figure 3.6**. In a directed graph, there could be in-degree ("incoming") hubs and out-degree ("outgoing") hubs. Hubs emerge in structural brain networks at a relatively early stage in brain development. These regions remain relatively stable from childhood to adolescence, even while their interactions with other network regions are transformed by developmental changes (e.g., gender-related differences in hormone levels).

Abnormal disease-related features in brain hub regions can lead to disruptions in brain activity. Such disturbances in network connectivity can cause behavioral and

---

[6]A module is a group of nodes having a large number of mutual connections but only a few connections to nodes outside that group.

cognitive impairment in brain disorders such as schizophrenia, autism, Alzheimer's disease, dementia, and neurodegenerative disease. In cases of traumatic brain injury and resulting decline in cognitive awareness, studies have shown connections between such injuries and consequent disruptions of network function. In some situations, when the individual is in a coma, there is evidence of a decrease in metabolic activity in certain hub regions, especially in those individuals who exist in a totally nonresponsive vegetative state.

3.8 Graph-Searching Algorithms

In almost every exploratory study of a network and its properties, there are many questions that attract our attention. For example, we might want to know whether an "end" node can be reached from some other given "start" node, and if it can, then what is the shortest path that exists between the two nodes? These issues crop up in contexts as varied as air-traffic safety problems (Rippel, Bar-Gill, and Shimkin, 2005), VLSI (Very Large-Scale Integrated) design problems (Sherwani, 1993, Chapter 3), and computer-aided music composition (Le Bel, 2017). Most of the algorithms that have been devised to answer these questions are classified as *graph-search algorithms*. In this section, we describe two basic types of search algorithms: "breadth-first", search and "depth-first" search.

3.8.1 Coloring the Nodes

Assume the graph $\mathcal{G} = (\mathcal{V}, \mathcal{E})$, with $N = |\mathcal{V}|$ nodes and $M = |\mathcal{E}|$ edges, is undirected. We describe two exhaustive search algorithms that are used to visit – in a systematic order – all nodes in a graph that form a single connected component.

The two algorithms are called "breadth-first search" and "depth-first search." Both types of graph searches move from node to node until a "goal" node (i.e., a node that satisfies certain conditions of the problem) is reached or until all reachable nodes are visited. The searches are "blind" (or "uninformed") searches in the sense that nothing is known about the topology of the graph or where the goal node is located. Each search process is defined only by its program, which does not adapt if the wrong path is taken. In the following, only if a node is directly connected to another node by an edge are they called "neighbors" of each other.

Graph coloring is used to keep track of the state of each node as the search proceeds. We start with all nodes "unvisited" and colored white. When the search visits a node, the color of that node changes from white to gray if there are neighboring nodes that have not yet been visited. The node remains gray until none of its neighboring nodes is colored white, at which time it changes its color to black.

Define the following three sets of nodes whose members are visually identified by their associated colors:

- W is the set of (white) *unvisited* nodes;
- $G = \mathcal{V} \backslash (W \cup B)$ is the set of (gray) nodes that are *visited but not examined*;
- B is the set of (black) nodes that have been *visited and examined*.

A node in B has been "examined" if it has been visited and all of its neighbors are, or have been, in G. Operations on G have their own terminology: we "push" a node v

---

Algorithm 3.1 *Breadth-first search*

---

1. Set $W = \mathcal{V}$, $G = \emptyset$, and $B = \emptyset$.
2. Choose a starting node $s \in W$ in the graph, mark it as "visited" by deleting it from W, moving it to G, and coloring it `gray`. The node s becomes the top node in G according to σ.
3. Examine the top node v in G by visiting all neighbors of v.
4. Pop the top node v from G to B, color it `black`, and examine it.

   - If the intended goal is found within v, stop the search and return the corresponding solution.
   - Otherwise, find all neighbors of v that are in W, mark them as visited, push them to G, and color them `gray`. Nodes are listed in G according to σ.

5. If G is empty, every node in the graph has been visited and examined. Stop the search and, if appropriate, return "goal not found."
6. If G is not empty, repeat from Step 3.

---

by moving v from W to G and we "pop" a node v by moving v from G to B. Visually, the process is as follows:

$$W \overset{\text{push}}{\to} G \overset{\text{pop}}{\to} B. \tag{3.8}$$

We assume that both algorithms are given an ordering σ of the nodes in G, and the algorithms prioritize the nodes according to σ.

3.8.2 Breadth-First Search

The *breadth-first search (BFS)* algorithm, which is given in **Algorithm 3.1**, proceeds level-by-level in the graph: it first visits all the children of the parent node, then all the grandchildren, all the great-grandchildren, and so on. It can find the geodesic distance from any given node v to every other node that is a member of the same component as v. Because the geodesic path may not be unique, it can find every such geodesic path. BFS employs a *queue* data structure using a FIFO (first in, first out) queue (i.e., nodes are removed from the queue in the same order that they were added) to manage the set of visited nodes.

The set G with an ordering given by σ plays the role of the queue (which is usually denoted by Q). The ordering σ allows us to define a "top" node in G. For example, σ may order nodes by increasing distance from v, in which case the top node in G is the node closest to v.

A common type of output is the *BFS search tree*, which is the set B of `black` nodes and corresponding edges.

3.8.3 Depth-First Search

The *depth-first search (DFS)* algorithm, given in **Algorithm 3.2**, is used to search along a tree or graph by selecting a starting node and then visiting its neighbors as

---

Algorithm 3.2 *Depth-first search*

---

1. Set $W = V$, $G = \emptyset$, and $B = \emptyset$.
2. Choose a starting node $s \in W$, mark it as "visited" by deleting it from W and moving it to G.
3. Examine the first node in G according to σ:

   - If the intended goal is found within this node, stop the search and return the corresponding solution.
   - Otherwise, query W for the neighbors of the last node v that was added to G, using σ to order neighbors of v in W. If v has a neighbor v' in W, mark v' as "visited," delete it from W, and move it to G. If v has no neighbor in W, then pop v out of G and into B.

4. Continue to visit neighbors sequentially, moving them one-by-one from W to G until a goal node is found, or until we arrive at a "dead end" (i.e., all nodes that are reachable from the first node in G have been visited, and there are no further unvisited neighbors in W of the last visited node). Pop the last visited node out of G and move it to B.
5. Backtrack to the most recent node that was moved from W to G. If that node has no unvisited neighbor in W, pop the node from G to B.
6. If G is empty, choose the first node of W according to σ, delete it from W, and push it into G. If W is also empty (i.e., $W \cup G = \emptyset$), conclude that every node in the graph has been visited and examined (i.e., $B = V$). Stop the search and, if appropriate, return "goal not found."
7. If G is not empty, repeat from Step 3.

---

deep as possible along each branch; when we encounter a goal node or we reach a node that has no unvisited neighbors, backtracking takes place to the most recent node that has been visited but not examined. DFS uses a LIFO (last in, first out) stack data structure (i.e., nodes are removed from the stack in the reverse order that they were added), and neighbors are considered before sibling nodes (as BFS does). The set G with an ordering σ plays the role of the *stack*. The ordering σ allows us to define a "first" node in G.

Although DFS visits all nodes, it does not traverse all edges. The nodes and edges that DFS has visited constitute a *graph-spanning tree*.

3.8.4 Algorithm Running Time

In terms of computational analysis, the total running time of both the BFS and DFS algorithms is $\mathcal{O}(N + M)$, where N is the number of nodes and M is the number of edges. Although both types of searches are described here for undirected graphs, where an edge can be traversed in either direction, they also work for directed graphs, but edges have to be traversed only in the intended direction, from the tail to the head of the arrow.

3.9 Minimum Spanning Trees

A *spanning tree* is a connected acyclic subgraph of a connected, undirected graph $\mathcal{G} = (\mathcal{V}, \mathcal{E})$, where $\mathcal{V}$ is the set of nodes of the graph, $\mathcal{E} \subset \mathcal{V} \times \mathcal{V}$ is the set of edges, and each edge is assigned a real number as a weight. A weight may be the distance from one node to another. A *minimal spanning tree* (MST) is a spanning tree of $\mathcal{G}$ with the property that the sum of the edge weights of the spanning tree is a minimum.

Thus, if there are N nodes in the graph $\mathcal{G}$, the MST connects all of those nodes by $N - 1$ edges, and connected edges in the MST join nodes that are closest to each other. The resulting MST will be unique as long as there are no ties in the pairwise distances. The construction of an MST is made easier if a dissimilarity matrix is provided that gives all $N(N - 1)/2$ pairwise distances between the nodes of $\mathcal{G}$.

There are two popular and fast algorithms that are used to create an MST. Let $\mathcal{A}$ start out as the empty set $\emptyset$.

Prim's algorithm (Prim, 1957) begins at an arbitrary node and no edges, and then applies the following greedy rule: add an edge to $\mathcal{A}$ from that node to another node not yet visited so that the edge has smallest weight; repeat the process so that at each step an edge is added to $\mathcal{A}$ that has the smallest weight from among all previously visited nodes to a node that has not yet been visited; stop when all nodes have been visited. The running time of Prim's algorithm is $\mathcal{O}(|\mathcal{E}| \log |\mathcal{V}|)$, where $|\mathcal{V}|$ and $|\mathcal{E}|$ are the number of elements in $\mathcal{V}$ and $\mathcal{E}$, respectively.

Kruskal's algorithm (Kruskal, 1956) is also a greedy algorithm that begins with all nodes and no edges. The algorithm first picks the shortest edge in the graph; then, at each subsequent step, it adds the next shortest edge from anywhere in the graph as long as it does not form a loop; the process stops when all nodes have been visited. Kruskal's algorithm has running time of $\mathcal{O}(|\mathcal{E}| \log |\mathcal{V}|)$.

3.10 Further Reading

The material in this chapter has also been covered in several books, including Barrat, Barthélemy, and Vespignani (2013, Chapter 1), Kolaczyk (2009, Chapter 2), Newman (2010, Chapter 6), and Easley and Kleinberg (2010, Chapter 2).

Section 3.6.1. An excellent collection of studies on large-scale random networks is given in the edited volume by Bollobás, Kozma, and Miklós (2008), which has a chapter (Chapter 12) on the telephone call network of landline telephone users (over two million nodes and over 50 million edges) within Hungary.

Section 3.7.2. This section on brain hubs is adapted from van den Heuvel and Sporns (2013).

3.11 Exercises

Exercise 3.1 Show that a node-independent path is also an edge-independent path.

Exercise 3.2 Show, by a counterexample, that an edge-independent path may not be a node-independent path.

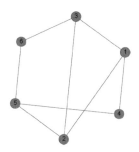

Figure 3.7 Graph for Exercise 3.10. There are six nodes and eight edges.

Exercise 3.3 Let $\mathbf{Y}$ denote the adjacency matrix for the undirected network $\mathcal{G}$. Show that $Y_{i+} = \sum_{j=1}^{N} Y_{ij}$ is the degree of node v_i.

Exercise 3.4 Let $\mathcal{G}$ be a graph and let x and y be any two nodes in $\mathcal{G}$. Let $d(x, y)$ denote the distance between x and y, where "distance" is geodesic distance (i.e., the number of edges that defines the shortest path from x to y). Define

$$\mathrm{ecc}(y) = \max_{z \in \mathcal{V}} d(x, z)$$

to be the *eccentricity* of y (i.e., the largest distance from y to any other node). If $\mathcal{G}$ is connected, the *radius* of $\mathcal{G}$ is the value of the smallest eccentricity, and the *diameter* of $\mathcal{G}$ is the value of the largest eccentricity. Suppose x and y are adjacent nodes in $\mathcal{G}$. Show that their eccentricities differ by at most one.

Exercise 3.5 Using the definitions in Exercise 3.4, draw a graph with eight nodes that has radius 3 and diameter 5.

Exercise 3.6 Show that the sum of the degrees of every node in $\mathcal{G}$ is twice the number of edges.

Exercise 3.7 Make up an undirected network with $N = 15$ nodes and $M = 20$ edges. Find the diameter and draw the degree distribution of your network.

Exercise 3.8 Using **Table 3.1** or from **Figure 2.10**, find the diameter and draw the degree distribution of Sampson's monk data.

Exercise 3.9 In Worked Example 3.1, find the diameter and draw the degree distribution of the network.

Exercise 3.10 Consider the graph in **Figure 3.7** that has six nodes and eight edges. Write down the adjacency matrix of the graph.

Exercise 3.11 Let $\mathbf{Y}$ denote the adjacency matrix for the undirected network $\mathcal{G}$. Show that the number of triangles in $\mathcal{G}$ is $(1/6)\mathrm{tr}(\mathbf{Y}^3)$, where $\mathrm{tr}(\mathbf{A})$ is the sum of the diagonal entries of $\mathbf{A}$.

Exercise 3.12 Let G be a labeled tree graph with N nodes. For $N = 50$, draw a tree graph whose node degrees are either 1 or 3. How many such trees have all N nodes with degree 1 or 3? Write a computer program to randomly generate such trees.

Exercise 3.13 Write down the incidence matrix of the bipartite network in **Figure 3.1**.

Exercise 3.14 Draw the graph corresponding to the incidence matrix in **Table 3.2**.

Exercise 3.15 If a graph has no self-loops, show that the trace of its adjacency matrix **Y** is zero.

Exercise 3.16 Suppose a graph has two connected components. Can you make any statement about the largest eigenvalue of its adjacency matrix **Y**?

Exercise 3.17 Let G be a graph having N nodes and more than $N - 1$ edges. Show that G contains a cycle.

Exercise 3.18 Show that if a matrix **M** is symmetric (i.e., $\mathbf{M}^\tau = \mathbf{M}$), then all integer powers of **M** (i.e., $\mathbf{M}^k$, for all $k \geq 1$) are each symmetric.

CHAPTER FOUR

Random Graph Models

Random graphs were introduced by the Hungarian mathematicians Erdős and Rényi (1959, 1960), who imposed a probabilistic framework on classical combinatorial graph theory. At the same time, Edgar N. Gilbert (1959) also studied the theoretical properties of random graphs. In this chapter, we describe the Erdős–Rényi–Gilbert work on defining and analyzing random graphs, and we define threshold functions and phase transitions. Then, various properties of random graphs, such as the emergence of trees, the emergence of cycles, the existence of a connected graph, and the emergence of a giant component are described and developed using probabilistic arguments.

4.1 Introduction

The basic question that Erdős and Rényi (and Gilbert) set out to answer was the following: Given a set of nodes in a graph, how would one arrange the edges of that graph assuming no prior knowledge or information regarding how graphs or networks are created in the real world? After all, there was little or no information available at that time on how real networks were generated. So, a theoretical approach to network formation was the only approach that was then possible. Their solution was that edges would be formed independently by choosing pairs of nodes at random with a certain probability $p \in (0, 1)$ and then joining them up. As a result, the Erdős–Rényi–Gilbert random graphs are now considered to be the simplest blueprint for a network graph model to which more complicated models can be compared. The real question today is whether real graphs are actually as random as the models that Erdős, Rényi, and Gilbert proposed.

The biggest discovery made by Erdős and Rényi was that, for a certain range of values of p, a "giant component" would emerge in the random graph. The emergence of a giant component refers to the fact that, as p increases past a threshold value, a very high proportion of the nodes of the graph (over 50%, but often more like 90%) would form a connected subgraph, while the remaining nodes form small groups that would be disconnected from the giant component. We explore the conditions for the existence of such a "giant component" in the random graph.

4.2 Erdős–Rényi Random Graphs

4.2.1 Graph Construction

Erdős and Rényi constructed random graphs in two equivalent ways, similar to how simple random sampling from a homogeneous population is defined:

- A random graph $G \in \mathcal{G}_{N,M}$ can be constructed by specifying the number M of edges from the $T = \binom{N}{2}$ possible edges, where N is the number of nodes, thereby implicitly specifying the marginal edge probability as M/T. Thus, there will be $\binom{T}{M}$ equally likely random graphs each having N nodes and M edges.
- Alternatively, a random graph $G \in \mathcal{G}_{N,p}$ can be constructed in sequential fashion by starting with N unconnected nodes and adding one edge at a time, where each pair of nodes is connected with probability p (or not, with probability $q = 1 - p$), independently of the other edges. Thus, the binomial probability of a graph with N nodes and M edges (out of the $T = \binom{N}{2}$ possible edges) is

$$\mathrm{P}_p\{G \in \mathcal{G}_{N,p} \text{ has } M \text{ edges}\} = \binom{T}{M} p^M q^{T-M}, \tag{4.1}$$

and the expected number of edges is $\mathrm{E}\{M\} = Tp$. When $p = \frac{1}{2}$, all 2^T possible graphs are equally likely to be chosen. If, for example, there are $N = 30$ nodes, this yields 2^{435} possible graphs, a very large number indeed.

Both $\mathcal{G}_{N,M}$ and $\mathcal{G}_{N,p}$ models are referred to as *Erdős–Rényi random graphs* (and sometimes *Erdős–Rényi–Gilbert random graphs*). Because $\mathcal{G}_{N,M}$ introduces an element of dependence into its construction, the more popular of these two constructions has been $\mathcal{G}_{N,p}$. When $p \approx M/T$, the two graph types are closely related.

4.2.2 Evolution of a Random Graph

Most large networks are dynamic in the sense that the number of nodes (and, hence, also edges) increases with the lifetime of the network. Adding nodes to a graph, which in turn adds random edges to the graph, has been referred to as *evolution* of the random graph. Examples include the explosive growth of webpages in the World Wide Web, the rapid growth of the research literature and its subsequent citation network, and the growth of the number of actors and actresses in movies. Thus, we are interested in properties of such graphs as we allow the number of nodes to increase without limit (i.e., $N \to \infty$).

4.2.3 Threshold Functions and Phase Transitions

Let Q be a particular property of a random graph. So, Q could be, for example, that a graph is connected (i.e., every pair of nodes is connected by a sequence of successive edges).

Definition 4.1 (Erdős and Rényi, 1960) We say that a property Q holds for "almost all" graphs (or "almost every" graph) if the probability of a random graph possessing that property converges to one as more nodes are added to the network; that is

77

Table 4.1 Examples of threshold functions (or critical probabilities), $p_c(N)$, for phase transitions of property Q for a random graph $G \in \mathcal{G}_{N,p}$, where $c > 0$. For these properties, as $N \to \infty$, if $c < 1$, $P\{Q\} \overset{a.s.}{\to} 0$, while if $c > 1$, $P\{Q\} \overset{a.s.}{\to} 1$

Property Q	Threshold $p_c(N)$
Tree of order k	$N^{-1} + c/N^{4/3}$
Cycle of order k	c/N
Complete subgraph of order k	$c/N^{2/(k-1)}$
Giant component	c/N
Connectivity	$c \log(N)/N$

$$P\{Q\} = P\{G \in \mathcal{G}_{N, p(N)} \text{ has property } Q\} \to 1 \text{ as } N \to \infty, \tag{4.2}$$

where we state explicitly that $p = p(N)$ depends upon N. If the probability in (4.2) converges instead to zero, we say that "almost no" graph has the property Q.

Definition 4.2 We say that $p_c(N)$ is a *threshold function* (or *critical probability*) for a graph property Q if the following holds for all $p = p(N)$ and $G \in \mathcal{G}_{N,p}$: If $p(N)$ grows slower than $p_c(N)$, as $N \to \infty$, then almost every graph will fail to have property Q, whilst if $p(N)$ grows faster than $p_c(N)$, then almost every graph will have property Q. In other words,

$$\lim_{N \to \infty} P\{G \in \mathcal{G}_{N, p(N)} \text{ has property } Q\} = \begin{cases} 0, & \text{if } \frac{p(N)}{p_c(N)} \to 0 \\ 1, & \text{if } \frac{p(N)}{p_c(N)} \to \infty \end{cases}. \tag{4.3}$$

Because any positive multiple of $p_c(N)$ is also a threshold function for Q, it follows that threshold functions are unique up to a multiplicative constant. If such a threshold function $p_c(N)$ exists for a particular property, we say that a *phase transition* has occurred for that property at that threshold function.

To derive a threshold function for a particular Q, it is necessary to demonstrate both limiting results in (4.3). We will show in Sections 4.4 and 4.5 the following: first, when $N \to \infty$ and $p(N) < p_c(N)$, almost no graph $\mathcal{G}$ has property Q and, second, that almost every $\mathcal{G}$ has Q when $p(N) > p_c(N)$.

Erdős and Rényi discovered a surprising fact, that, for many graphs, a property Q will suddenly appear due to the existence of a threshold function, $p_c(N)$, that governs the structure of the network. Several examples of properties Q and their respective threshold functions $p_c(N)$ are given in **Table 4.1** and are studied in this chapter.

4.3 Technical Tools

In order to study the properties of random graphs, we will need the following technical tools. For the properties we discuss here, X is often taken as any property Q that the graph $G \in \mathcal{G}_{N,p}$ is conjectured to possess (such as the existence in G of cycles or trees of a given order). As a result, trying to calculate the probability that G has property Q can be restated as a probability about the value of X. Thus, for example, if we wish to show that G has property Q, where Q is a count of something, then it is enough to show that $P\{X = 0\} \to 0$ if $N \to \infty$.

4.3.1 First-Moment Method

Theorem 4.1 *If X is a non-negative, integer-valued, random variable, then*

$$P\{X \neq 0\} = P\{X \geq 1\} \leq E\{X\}. \tag{4.4}$$

In particular, if $E\{X\} \to 0$, then $P\{X = 0\} \to 1$.

Proof Let X be a non-negative random variable. Markov's inequality states that $P\{X \geq a\} \leq \frac{1}{a}E\{X\}$ for all $a > 0$. Setting $a = 1$ yields the inequality $P\{X \geq 1\} \leq E\{X\}$. $\qquad\qquad\square$

4.3.2 Second-Moment Method

The following shortcut, which can be used to show that $P\{X = 0\} \to 0$, involves the second moment of X and can be used to demonstrate the existence of certain properties for random graphs.

Theorem 4.2 *If X is a random variable with $E\{X\} \neq 0$, then*

$$P\{X = 0\} \leq \frac{\text{var}\{X\}}{[E\{X\}]^2}. \tag{4.5}$$

In particular, if $E\{X^2\}/[E\{X\}]^2 \to 1$, then $P\{X = 0\} \to 0$.

Proof Consider the random variable $|X - E\{X\}|$. Chebyshev's inequality states that, for any $a > 0$:

$$P\{|X - E\{X\}| \geq a\} \leq \frac{\text{var}\{X\}}{a^2}. \tag{4.6}$$

By setting $a = E\{X\}$, we see that the event $\{|X - E\{X\}| \geq E\{X\}\}$ contains $\{X = 0\}$, and so, assuming that $E\{X\} \neq 0$:

$$P\{X = 0\} \leq P\{|X - E\{X\}| \geq E\{X\}\} \leq \frac{\text{var}\{X\}}{[E\{X\}]^2}. \tag{4.7}$$

Now, use the fact that $\text{var}\{X\} = E\{X^2\} - [E\{X\}]^2$, so that

$$P\{X = 0\} \leq \frac{E\{X^2\} - [E\{X\}]^2}{[E\{X\}]^2} = \frac{E\{X^2\}}{[E\{X\}]^2} - 1. \tag{4.8}$$

Thus, to show that $P\{X = 0\} \to 0$, it suffices to show that $E\{X^2\}/[E\{X\}]^2 \to 1$. $\qquad\square$

Larry Shepp showed that, using the Cauchy–Schwarz inequality, a stronger result than (4.7) holds. The Schwarz inequality says that for any two non-negative random variables X and Y, $(E\{XY\})^2 \leq E\{X^2\}E\{Y^2\}$. Thus

$$(E\{X\})^2 = (E\{I_{[X \neq 0]} X\})^2$$

$$\leq E\{I_{[X \neq 0]}^2\} E\{X^2\}$$

$$= P\{X \neq 0\} E\{X^2\}$$

$$= E\{X^2\} - P\{X = 0\} E\{X^2\}. \tag{4.9}$$

Hence, assuming that $E\{X^2\} \neq 0$:

$$P\{X = 0\} \leq \frac{\text{var}\{X\}}{E\{X^2\}}. \tag{4.10}$$

The difference between (4.7) and (4.10) lies in their denominators. In (4.7) the denominator $E\{X\} \neq 0$, whereas in (4.10) $E\{X^2\} \neq 0$.

4.3.3 Chernoff Bounds

Chernoff bounds provide bounds on the amount of probability that lies in the tail of the distribution of a random variable, where by "tail" we mean that part of the distribution that is far away from the mean. Herman Chernoff showed that these bounds are exponentially decreasing for both the lower tail and the upper tail, and are sharper than either Markov's inequality or Chebychev's inequality.

We need the following definition of a moment generating function:

Definition 4.3 The *moment generating function* (mgf) of the random variable X is defined as $M_X(t) = E\{e^{tX}\}$, assuming that the expectation exists for t in some neighborhood of 0.

Theorem 4.3 (Chernoff, 1952) *Let $X_1, X_2, \ldots, X_n$ be independent Bernoulli trials with $X_i = 1$ with probability p_i and $X = 0$ with probability $1 - p_i$, $i = 1, 2, \ldots, n$. Let $X = \sum_{i=1}^{n} X_i$ be their sum. Then, $X \sim \text{Bin}(n, p)$. Let $\mu = E\{X\} = \sum_{i=1}^{n} p_i$. The Chernoff bounds are*

1. (Upper tail) $P\{X \geq (1 + \beta)\mu\} \leq \exp\left\{\dfrac{-\beta^2 \mu}{2 + \beta}\right\}, \quad \beta > 0. \tag{4.11}$

2. (Lower tail) $P\{X \leq (1 - \beta)\mu\} \leq \exp\left\{\dfrac{-\beta^2 \mu}{2 + \beta}\right\}, \quad \beta > 0. \tag{4.12}$

Proof (a) Upper tail. Let $E\{e^{tX}\}$ be the *moment generating function* of X, where t is a parameter that will be chosen to provide the tightest possible bound. Then,

$$P\{X \geq (1 + \beta)\mu\} = P\left\{e^{tX} \geq e^{(1+\beta)t\mu}\right\} \leq \frac{E\{e^{tX}\}}{e^{(1+\beta)t\mu}}. \tag{4.13}$$

Next, we establish an upper bound on $E\{e^{tX}\}$:

$$E\left\{e^{tX}\right\} = E\left\{e^{t\sum_i X_i}\right\} = E\left\{\prod_i e^{tX_i}\right\} = \prod_i E\left\{e^{tX_i}\right\}$$

$$= \prod_i (p_i e^t + (1 - p_i) \cdot 1)$$

$$= \prod_i (1 + p_i(e^t - 1)). \tag{4.14}$$

Now, we use the approximation $1 + x \le e^x$, for all $x \in \mathbb{R}$, and that $\mu = \sum_i p_i$:

$$E\{e^{tX}\} \le \prod_i e^{p_i(e^t - 1)} = e^{\sum_i p_i(e^t - 1)} = e^{(e^t - 1)\mu}. \tag{4.15}$$

Substituting $E\{e^{tX}\} \le e^{(e^t - 1)\mu}$ into (4.15), we get, for all $t \ge 0$:

$$P\{X \ge (1 + \beta)\mu\} \le \frac{e^{(e^t - 1)\mu}}{e^{(1+\beta)t\mu}} = e^{(e^t - 1)\mu - (1+\beta)t\mu}. \tag{4.16}$$

Minimizing $((e^t - 1) - (1 + \beta)t)\mu$ wrt t, we get that $t = \log(1 + \beta)$. Substituting this value for t into (4.16), we get

$$P\{X \ge (1 + \beta)\mu\} \le \frac{e^{(e^{\log(1+\beta)} - 1)\mu}}{e^{(1+\beta)\log(1+\beta)\mu}}$$

$$= e^{((e^{\log(1+\beta)} - 1) - (1+\beta)\log(1+\beta))\mu}$$

$$= (e^{\beta - (1+\beta)\log(1+\beta)})^\mu \tag{4.17}$$

$$= \left(\frac{e^\beta}{(1 + \beta)^{1+\beta}}\right)^\mu. \tag{4.18}$$

This last result is not convenient for practical use. Instead, we use (4.17). We combine Taylor series expansions of $\log(1 + x)$ and $(1 - x)^{-1}$ to get

$$\log(1 + \beta) = \sum_{i \ge 1} (-1)^{i+1} \frac{\beta^i}{i} \ge \frac{\beta}{1 + \beta/2}, \tag{4.19}$$

whence,

$$\mu(\beta - (1 + \beta)\log(1 + \beta)) \le \mu\left(\beta - (1 + \beta)\frac{\beta}{1 + \beta/2}\right) = -\frac{\mu\beta^2}{2 + \beta}. \tag{4.20}$$

Putting (4.16) and (4.20) together, we have that

$$P\{X \ge (1 + \beta)\mu\} \le e^{-\beta^2\mu/(2+\beta)}. \tag{4.21}$$

(b) Lower tail. We use a similar argument as for the upper tail. Analogous to (4.13), we have that

$$P\{X \leq (1 - \beta)\mu\} = P\{e^{-tX} \geq e^{-(1-\beta)t\mu}\} \leq \frac{E\{e^{-tX}\}}{e^{-(1-\beta)t\mu}}. \tag{4.22}$$

As before, we establish an upper bound on $E\{e^{tX}\}$:

$$E\left\{e^{-tX}\right\} = \prod_i E\left\{e^{-tX_i}\right\}$$

$$= \prod_i (p_i e^{-t} + (1 - p_i))$$

$$= \prod_i (1 - p_i(1 - e^{-t})). \tag{4.23}$$

From the approximation $1 - x \leq e^{-x}$, where $x = p_i(1 - e^{-t})$, we have that

$$E\{e^{-tX}\} \leq \prod_i e^{p_i(e^{-t}-1)} = e^{\sum_i p_i(e^{-t}-1)} = e^{(e^{-t}-1)\mu}. \tag{4.24}$$

Substituting (4.23) into (4.21), we have that

$$P\{X \leq (1 - \beta)\mu\} \leq \frac{e^{(e^{-t}-1)\mu}}{e^{-(1-\beta)t\mu}} = e^{(e^{-t}-1)\mu+(1-\beta)t\mu}. \tag{4.25}$$

Minimizing $[(e^{-t} - 1) + (1 - \beta)t]\mu$ wrt t yields $t = -\log(1 - \beta)$. Substituting this value of t into (4.24) yields

$$P\{X \leq (1 - \beta)\mu\} \leq \frac{e^{(e^{\log(1-\beta)}-1)\mu}}{e^{(1-\beta)\log(1-\beta)\mu}}$$

$$= (e^{-[\beta+(1-\beta)\log(1-\beta)]})^\mu \tag{4.26}$$

$$= \left(\frac{e^{-\beta}}{(1 - \beta)^{1-\beta}}\right)^\mu. \tag{4.27}$$

We can show that

$$\beta + (1 - \beta)\log(1 - \beta) \geq \beta + (1 - \beta)\left(-\beta - \frac{\beta^2}{2}\right)$$

$$= \beta - (1 - \beta)\beta\left(1 + \frac{\beta}{2}\right)$$

$$= \beta - (1 - \beta)\frac{\beta}{(1 + \beta/2)^{-1}}$$

$$\geq \beta - (1 - \beta)\frac{\beta}{1 - \beta/2}$$

$$= \frac{\beta(1 - \beta/2) - (1 - \beta)\beta}{(2 - \beta)/2}$$

$$= \frac{\beta^2}{2 + \beta}, \tag{4.28}$$

where we use the inequalities $\log(1 - x) \geq -x - x^2/2$ and $(1 + x)^{-1} \geq 1 - x$. Thus

$$P\{X \leq (1 - \beta)\mu\} \leq e^{-\beta^2 \mu/(2+\beta)}. \tag{4.29}$$

$\square$

4.3.4 Cayley's Formula

We will need the following definitions:

Definition 4.4 A graph is referred to as a *tree* of order k if it has k nodes and $k - 1$ edges, is connected, and none of its subgraphs forms a cycle.

Definition 4.5 A *forest* of order k is a graph over k nodes all of whose connected components are trees, and where each tree is labeled by the sets of some disjoint partition of the k nodes.

Definition 4.6 A *rooted forest* is a forest whose trees each have a designated root node.

One of the classic results in graph theory is the following formula due to Arthur Cayley for the number of possible trees that can be constructed over k nodes.

Theorem 4.4 (Cayley, 1889) *For any positive integer k, the number of different trees that can be formed on k labeled nodes is $T_k = k^{k-2}$.*

Proof There are many different ways to prove this formula. The simplest and "most beautiful" (Aigner and Ziegler, 2010, pp. 145–146) is the "double-counting" method due to Jim Pitman (1999). This method entails two separate counting arguments for the same thing and then equating the two results.

First argument. Let T_k be the number of possible unrooted trees (i.e., trees having no designated root) that can be formed from the k given nodes. Pick one of those trees and select one of its k nodes to be the root. Then, to form a tree, $k - 1$ edges can be sequentially appended to that root node in $(k-1)!$ possible ways. So, the total number of such sequences is $T_k \cdot k \cdot (k - 1)! = T_k \cdot k!$.

Second argument. An alternative way of constructing such a sequence is to start with an empty graph (k isolated nodes and no edges) and add edges one at a time. Pick the first edge, e_1, to join up any two of the k nodes; this can be done in $k(k - 1)$ possible ways (k for the starting node and $k - 1$ for the end node). At the ith step, the edge we add must cross from one component to another: otherwise, it would create a cycle. Adding such an edge, therefore, decreases the number of components by one. Suppose that we have already added $k - i$ edges to an empty graph. Then, the graph will consist of i components (i.e., a rooted forest of i trees). As the next edge to add, e_i can start at any of the k nodes and end at any one of the $i - 1$ root nodes except the root of the tree containing the starting node; this can be done in $k(i - 1)$ ways. So, the total number of such sequences is $\prod_{i=2}^{k} k(i - 1) = k^{k-1} \cdot (k - 1)! = k^{k-2} \cdot k!$.

Equating these two results yields $T_k \cdot k! = k^{k-2} \cdot k!$, whence, $T_k = k^{k-2}$. $\square$

4.3.5 Branching Processes

Branching processes are Markov chains that have been applied to model various types of scientific processes. Some of these examples include the propagation of bovine spongiform encephalopathy (BSE) in Britain, infectious diseases and epidemics, cell proliferation kinetics, electron multipliers, neutron chain reactions, survival of family names, and survival of mutant genes. A good introduction to the theory of branching processes is Karlin (1966, Chapter 11).

We define the *Galton–Watson branching process* as follows. Let Z_k be the number of individuals present in the kth generation, where $k = 0, 1, 2, \ldots$. We start with a single individual at the zeroth generation (so that $Z_0 = 1$). All the offspring of that individual comprise the first generation, all offspring of the first generation comprise the second generation, and so on. Assume that each individual has a random number of children in the next generation, where the numbers of children are independent of each other and of the history of the process.

We visualize the branching process, $Z_0 = 1, Z_1, Z_2, \ldots$, as a random rooted tree. The offspring distribution can be described by a non-negative discrete random variable ξ with $P\{\xi = k\} = p_k$, $k = 0, 1, 2, \ldots$, $\sum_k p_k = 1$. We can view $p_0, p_1, p_2, \ldots$ as the respective probabilities of each individual in each generation producing $0, 1, 2, \ldots$ offspring. Then,

$$E\{\xi\} = \sum_k k p_k = \mu, \quad \text{var}\{\xi\} = \sum_k (k - \mu)^2 p_k = \sigma^2 < \infty. \tag{4.30}$$

Probability Generating Functions for Branching Processes
Let

$$\psi(s) = \psi_\xi(s) = \sum_{k=0}^{\infty} p_k s^k = E\{s^\xi\} \tag{4.31}$$

be the probability generating function for the number of offspring per individual. Note that $\psi(0) = 0$ and $\psi(1) = 1$. Further, $\psi'(s) = E\{\xi s^{\xi-1}\}$, whence,

$$\psi'(1) = \sum_k k p_k = \mu. \tag{4.32}$$

Also, $\psi(s)$ is a strictly increasing, strictly convex function; that is,

$$\psi''(s) = E\{\xi(\xi - 1)s^{\xi-2}\} \geq 0, \quad \text{for all } s \in [0, 1], \tag{4.33}$$

so that

$$\psi''(1) = E\{\xi(\xi - 1)\} = E\{\xi^2\} - E\{\xi\} = \sigma^2 - \mu + \mu^2. \tag{4.34}$$

It follows that $\text{var}\{\xi\} = \sigma^2 = \psi''(1) + \mu - \mu^2$.

Suppose $\xi_1, \xi_2, \ldots, \xi_{Z_j}$ are iid copies of the random variable ξ. Then,

$$Z_0 = 1, \quad Z_1 = \xi, \quad Z_2 = \sum_{i=1}^{Z_1} \xi_i, \quad \ldots, \quad Z_t = \sum_{i=1}^{Z_{t-1}} \xi_i. \tag{4.35}$$

The expected sizes of the first two generations are given by

$$E\{Z_1\} = \mu, \quad E\{Z_2\} = E\{E\{Z_2|Z_1\}\} = E\{\mu Z_1\} = \mu^2. \tag{4.36}$$

We use the same conditioning argument as in (4.36) to find the expected size of the tth generation:

$$E\{Z_t\} = E\{E\{Z_t|Z_{t-1}\}\} = E\{\mu Z_{t-1}\} = \mu E\{Z_{t-1}\}, \tag{4.37}$$

so that, repeating the conditioning $t - 1$ more times, we get

$$E\{Z_t\} = \mu^t. \tag{4.38}$$

Next, we calculate $\sigma_t^2 = \text{var}\{Z_t\}$. Let

$$
\begin{aligned}
\psi_t(s) &= \sum_{k=0}^{\infty} P\{Z_t = k\}s^k \\
&= \sum_{k=0}^{\infty}\sum_{j=0}^{\infty} P\{Z_t = k|Z_{t-1} = j\}P\{Z_{t-1} = j\}s^k \\
&= \sum_{k=0}^{\infty} s^k \sum_{j=0}^{\infty} P\{Z_{t-1} = j\} \cdot P\{\xi_1 + \cdots + \xi_j = k\} \\
&= \sum_{j=0}^{\infty} P\{Z_{t-1} = j\} \cdot \sum_{k=0}^{\infty} P\{Z_j = k\}s^k \\
&= \sum_{j=0}^{\infty} P\{Z_{t-1} = j\}[\psi(s)]^j \\
&= \psi_{t-1}(\psi(s)), \tag{4.39}
\end{aligned}
$$

where we used the fact that the moment generating function of a sum of iid random variables, $Z_j = \xi_1 + \cdots + \xi_j$, is the product of the individual moment generating functions; that is,

$$\psi_j(s) = \prod_{i=1}^{j} \psi(s) = [\psi(s)]^j. \tag{4.40}$$

In particular, $\psi_1(s) = \psi(s)$. Note that we can also write

$$\psi_t(s) = \psi(\psi_{t-1}(s)). \tag{4.41}$$

Differentiate (4.39) to get

$$\psi_t'(s) = \psi_{t-1}'(\psi(s))\psi'(s). \tag{4.42}$$

Differentiating again yields

$$\psi_t''(s) = \psi_{t-1}''(\psi(s))[\psi'(s)]^2 + \psi_{t-1}'(\psi(s))\psi''(s). \tag{4.43}$$

Setting $s = 1$ and noting that $\psi(1) = 1$ yields

$$\psi_t''(1) = \psi_{t-1}''(1)[\psi'(1)]^2 + \psi_{t-1}'(1)\psi''(1), \tag{4.44}$$

where $\psi'(1) = \mu$, $\psi''(1) = \sigma^2 - \mu + \mu^2$, and $\psi_t''(1) = \sigma_t^2 - \mu^t + \mu^{2t}$. Substituting these derivatives into this equation, we get

$$\sigma_t^2 - \mu^t + \mu^{2t} = (\sigma_{t-1}^2 - \mu^{t-1} + \mu^{2t-2})\mu^2 + \mu^{t-1}(\sigma^2 - \mu + \mu^2), \tag{4.45}$$

whence, multiplying out and canceling like terms, we have the following recursion relation:

$$\sigma_t^2 = \sigma_{t-1}^2 \mu^2 + \mu^{t-1}\sigma^2. \tag{4.46}$$

Repeatedly substituting for σ_{t-1}^2 in the above equation yields

$$
\begin{aligned}
\sigma_t^2 &= \sigma_0^2 \mu^{2t} + \sigma^2 \mu^{t-1}(1 + \mu + \mu^2 + \mu^3 + \cdots + \mu^{t-1}) \\
&= \begin{cases} \sigma^2 \mu^{t-1}\left(\frac{1-\mu^t}{1-\mu}\right), & \text{if } \mu \neq 1 \\ t\sigma^2, & \text{if } \mu = 1 \end{cases},
\end{aligned} \tag{4.47}
$$

where $\sigma_0^2 = \text{var}\{Z_0\} = 0$.

Extinction Probabilities

Now, we aggregate all individuals over all generations so that

$$Z = \sum_{t=1}^{\infty} Z_t. \tag{4.48}$$

If $Z < \infty$, then the population goes *extinct*, while if $Z = \infty$, then the population *survives forever*. Under what conditions will the population go extinct or survive? There are three cases to consider.

Subcritical case: $\mu < 1$. Because $\text{E}\{Z_t\} = \mu^t$:

$$\text{E}\{Z\} = \sum_{t=1}^{\infty} \text{E}\{Z_t\} = \frac{1}{1-\mu} < \infty. \tag{4.49}$$

So, by Markov's inequality:

$$\text{P}\{Z < \infty\} = 1 \tag{4.50}$$

and hence, *extinction is certain* (i.e., P{extinction} = 1).

Supercritical case: $\mu > 1$. If $p_0 = \text{P}\{\xi = 0\} = 0$, then the probability that an individual produces no offspring is zero, and so extinction becomes impossible (i.e., P{extinction} = 0). So, we will assume that $0 < p_0 < 1$. Setting $s = 0$ in (4.41) yields

$$q_t = \psi_t(0) = \psi(\psi_{t-1}(0)) = \psi(q_{t-1}). \tag{4.51}$$

Note that $\psi_t(0)$ is the probability $\text{P}\{Z_t = 0\}$ of extinction at or before the tth generation. The sequence $q_1, q_2, \ldots, q_t, \ldots$ is monotone increasing and, as probabilities, is bounded by 1. Therefore, the limit $q = \lim_{t \to \infty} q_t \leq 1$ exists. If we let $t \to \infty$ in (4.50), then q has to satisfy the equation

$$q = \psi(q). \tag{4.52}$$

We therefore look for the roots of the equation $x = \psi(x)$, where $x \in [0, 1]$. Actually, q is the smallest positive root of that equation. To see this, we use induction. Let s_0 be a positive root of (4.52). Then, because $\psi(s)$ is a strictly increasing function,

$$q_1 = \psi_1(0) = \psi(0) < \psi(s_0) = s_0. \tag{4.53}$$

Now, assume $q_{t-1} < s_0$. Then, by (4.51),

$$q_t = \psi(q_{t-1}) < \psi(s_0) = s_0. \tag{4.54}$$

So, $q_t < s_0$ for all t. Thus,

$$q = \lim_{t \to \infty} q_t \leq s_0, \tag{4.55}$$

so that q is the smallest positive root of (4.52). The graph of $y = \psi(x)$, which is convex, intersects the $45°$ line $y = x$ in at most two points. Because $\psi(1) = 1$, the curve and the line intersect at the point $(1, 1)$. Hence, the equation $x = \psi(x)$ can have at most one root in the interval $0 < x < 1$. Because $\mu = \psi'(1) > 1$, the slope of the tangent at $x = 1$ is greater than 1. It follows that there exists a unique root $0 < q < 1$, so that the probability of extinction in this case is less than 1, and *extinction is not certain*.

Critical case: $\mu = 1$. In this case, the slope of the tangent at $x = 1$ is equal to 1 and so the only solution of $q = \psi(q)$ is $q = 1$, in which case *extinction is certain* (i.e., $P\{\text{extinction}\} = 1$).

4.4 Graph Properties

In this section, we examine the conditions for the existence of various structural properties of random graphs. Specifically, we look at how large $p = p(N)$ should be for certain phase transitions, such as the emergence of trees or cycles, and also graph connectedness, to take place. These structural properties occur with a suddenness that often comes as a surprise to anyone not expecting them.

4.4.1 Emergence of Trees

In the beginning, there were N nodes and no edges, and $p = 0$. As we increase $p = p(N)$, edges start to appear, generating a random graph that evolves into an ensemble of components most of which form trees. This graph is often referred to as a *forest of trees*, which starts to appear when $p(N) = o(1/N)$.

Theorem 4.5 *Let X_k denote the number of components of $G \in \mathcal{G}_{N,p}$ that are trees having k nodes, where $k = aN^{2/3}$ and $a > 0$ is a constant. Let $p = \lambda/N$, where $\lambda = 1 + cN^{-1/3}$, $c \leq 1$. Then, as $N \to \infty$, the expected number of trees with k nodes is given by*

$$E\{X_k\} \to \frac{e^{[(c-a)^3 - c^3]/6}}{a^{5/2} N^{2/3} \sqrt{2\pi}}. \tag{4.56}$$

Thus, $E\{X_k\} \to 0$ as $N \to \infty$.

Proof We build a tree with k nodes as follows. Let X_k denote the number of trees each of which are formed from k nodes. Call them k-trees. First, we choose k out of the N nodes. We can do this in $\binom{N}{k}$ ways. The total number of possible edges that can be placed between the k nodes is $\binom{k}{2}$. Using the k selected nodes, Cayley's formula tells us that there are k^{k-2} possible ways of forming a tree from those nodes. Given such a tree, each of the $k - 1$ edges of the tree is present with probability p. Each of the other $\binom{k}{2} - (k - 1)$ edges among the k nodes is absent with probability $1 - p$. Also, each of the $k(N-k)$ edges from the k nodes of the tree to the remaining $N-k$ nodes is absent with probability $1-p$. Then, the expected number of trees that have k nodes is

$$E\{X_k\} = \binom{N}{k} k^{k-2} p^{k-1} (1-p)^{\binom{k}{2}-(k-1)+k(N-k)}. \tag{4.57}$$

Consider each individual term in (4.57). Stirling's formula yields

$$\binom{N}{k} = \frac{(N)_k}{k!} \sim \frac{N^k e^k}{k^k \sqrt{2\pi k}} \prod_{i=1}^{k-1}\left(1-\frac{i}{N}\right), \tag{4.58}$$

where $(N)_k = N(N-1)\cdots(N-k+1)$. For $i < k$,

$$\log\left(1-\frac{i}{N}\right) = -\frac{i}{N} - \frac{i^2}{2N^2} + \mathcal{O}\left(\frac{i^3}{N^3}\right). \tag{4.59}$$

Summing and taking exponents,

$$\prod_{i=1}^{k-1}\left(1-\frac{i}{N}\right) = \exp\left\{-\frac{k^2}{2N} - \frac{k^3}{6N^2} + o(1)\right\}. \tag{4.60}$$

Combining (4.58) and (4.60), and setting $k = aN^{2/3}$, we have that

$$\binom{N}{k} \sim \frac{N^k e^k}{k^k \sqrt{2\pi k}} \exp\left\{-\frac{k^2}{2N} - \frac{a^3}{6} + o(1)\right\}. \tag{4.61}$$

We set $p = \frac{\lambda}{N}$, where $\lambda = 1 + cN^{-1/3}$, so that

$$p = \frac{1}{N} + \frac{c}{N^{4/3}}, \tag{4.62}$$

where $c \le 1$. The form of p in (4.62) is often referred to as a *fine parametrization*, whilst $p = \lambda/N$ with λ fixed is viewed as a *coarse parametrization*. See, for example, Durrett (2007, Sections 2.6 and 2.7) and Grimmett (2010, p. 210). Then,

$$p^{k-1} = \left(\frac{1}{N} + \frac{c}{N^{4/3}}\right)^{k-1} = \frac{1}{N^{k-1}}\left(1 + \frac{c}{N^{1/3}}\right)^{k-1}. \tag{4.63}$$

Taking logarithms of (4.63) and setting $k = aN^{2/3}$, we have that

$$(k-1)\log\left(1+\frac{c}{N^{1/3}}\right) = (k-1)\left(\frac{c}{N^{1/3}} - \frac{c^2}{2N^{2/3}} + \mathcal{O}(N^{-1})\right)$$
$$= \frac{ck}{N^{1/3}} - \frac{c^2 a}{2} + o(1). \tag{4.64}$$

Taking logarithms of (4.57) yields the following term,

$$\log\left((1-p)^{\binom{k}{2}-(k-1)+k(N-k)}\right) = \left(\binom{k}{2} - (k-1) + k(N-k)\right)\log(1-p). \tag{4.65}$$

Now,

$$\log(1-p) = -p + \mathcal{O}(N^{-2}) = -\left(\frac{1}{N} + \frac{c}{N^{4/3}}\right) + \mathcal{O}(N^{-2}) \tag{4.66}$$

and setting $k = aN^{2/3}$, we have

$$\binom{k}{2} - (k-1) + k(N-k) = kN - \frac{k^2}{2} + \mathcal{O}(N^{2/3}). \tag{4.67}$$

Multiplying (4.66) and (4.67) together and then taking exponents yields

$$(1-p)^{\binom{k}{2}-(k-1)+k(N-k)}$$

$$= \exp\left\{-k - \frac{kc}{N^{1/3}} + \frac{k^2}{2N} + \frac{a^2 c}{2} + o(1)\right\}. \tag{4.68}$$

Putting it all together, we have that

$$E\{X_k\} \sim \frac{N^k}{k^k \sqrt{2\pi k}} \frac{k^{k-2}}{N^{k-1}} e^{-A_k(N)}$$

$$= \frac{e^{-A_k(N)}}{a^{5/2} N^{2/3} \sqrt{2\pi}}, \tag{4.69}$$

where

$$A_k(N) = k - \frac{k^2}{2N} - \frac{a^3}{6} + \frac{ck}{N^{1/3}} - \frac{c^2 a}{2} - k - \frac{kc}{N^{1/3}} + \frac{k^2}{2N} + \frac{a^2 c}{2}$$

$$= -\frac{a^3}{6} - \frac{c^2 a}{2} + \frac{a^2 c}{2}$$

$$= \frac{1}{6}[(c-a)^3 - c^3]. \tag{4.70}$$

The result follows. $\qquad\square$

It follows that $P\{X_k \geq 1\} \to 0$ as $N \to \infty$. In other words, if we define the threshold probability as $p_c(N) = N^{-1} + cN^{-4/3}$, then $P\{X_k = 0\} \to 1$ as $N \to \infty$, so that any tree of order k will, almost surely, not be a component of almost any graph.

See **Exercises 4.4, 4.5,** and **4.6** for the computation of $E\{X_k\}$ when the threshold probability is different from (4.63).

4.4.2 Emergence of Cycles

Theorem 4.6 *Let X denote the number of cycles in $G \in \mathcal{G}_{N,p}$. When N is large, with high probability, no graph has cycles.*

Proof To show that almost no graph has cycles, we need to show that $P\{X \neq 0\} \to 0$.

Let X_k denote the number of cycles with k nodes (i.e., k-cycles). Then, X_k can be expressed as a product of independent indicator functions summed over all possible subsets of k distinct nodes:

$$X_k = \sum_{v_1, \ldots, v_k \text{ distinct } \in \mathcal{V}} I_{[(v_1, v_2) \in \mathcal{E}]} \cdots I_{[(v_{k-1}, v_k) \in \mathcal{E}]} I_{[(v_k, v_1) \in \mathcal{E}]}, \tag{4.71}$$

where $I_{[(v_i, v_j) \in \mathcal{E}]}$ equals 1 if there is an edge between node v_i and node v_j, and 0 otherwise. The number of cycles in G is, therefore, $X = \sum_{k=3}^{N} X_k$. By linearity of expectation, the expected number of cycles is given by $E\{X\} = \sum_{k=3}^{N} E\{X_k\}$, which by independence of the indicator functions is

$$E\{X_k\} = \sum_{v_1,\ldots,v_k \text{ distinct } \in \mathcal{V}} E\{I_{[(v_1,v_2)\in\mathcal{E}]}\} \cdots E\{I_{[(v_{k-1},v_k)\in\mathcal{E}]}\} E\{I_{[(v_k,v_1)\in\mathcal{E}]}\}, \quad (4.72)$$

where $E\{I_{[(v_i,v_{i+1})\in\mathcal{E}]}\} = P\{(v_i,v_{i+1}) \in \mathcal{E}\} = p$ for all $i = 1,2,\ldots,k$, and $v_{k+1} = v_1$. So, for a specific set of distinct k nodes, the product of the expected values of the indicator functions is p^k.

Next, we need to determine how many k-cycles are in X_k. There are $\binom{N}{k}$ ways of selecting k nodes from N nodes to create a cycle of order k. The nodes of such a k-cycle can be ordered in various ways: choose any starting node, then choose a second node of the cycle in $k-1$ ways, a third node in $k-2$ ways, and so on. There are, therefore, $\frac{1}{2}\binom{N}{k}(k-1)!$ cycles of order k. We divide by 2 to compensate for double-counting because the reverse of a cycle is identical to the cycle itself. Hence, the expected number of k-cycles is

$$E\{X_k\} = \frac{1}{2}\binom{N}{k}(k-1)!\, p^k. \quad (4.73)$$

Because $\binom{N}{k}(k-1)! = \frac{1}{k}N(N-1)(N-2)\cdots(N-k+1) \le N^k/k$, the expected number of cycles in G is

$$E\{X\} = \sum_{k=3}^{N} \frac{1}{2}\binom{N}{k}(k-1)!\, p^k \le \sum_{k=3}^{N} \frac{N^k}{2k} p^k \le \sum_{k=3}^{N} (Np)^k. \quad (4.74)$$

If p is asymptotically smaller than $1/N$, then $Np \to 0$ as $N \to \infty$, and $\sum_{k=3}^{N}(Np)^k \to 0$. Hence, $E\{X\} \to 0$ as $N \to \infty$. From the first-moment method (see Section 4.3.1), we have that $P\{X \ne 0\} = P\{X \ge 1\} \le E\{X\} \to 0$. So, *when N is large, almost surely no graph has cycles.* □

More specific information can be obtained regarding the behavior of $E\{X\}$, the expected number of cycles in G. Let $(N)_k = N(N-1)\cdots(N-k+1)$ and let $p = c/N$, for some constant c. Then, from (4.74),

$$E\{X\} = \frac{1}{2}\sum_{k=3}^{N} \frac{(N)_k}{k!}(k-1)!\, p^k$$

$$= \frac{1}{2}\sum_{k=3}^{N} \frac{(N)_k}{N^k}\frac{c^k}{k}, \quad (4.75)$$

where we substituted $p = c/N$ in the last step. As we will see, the behavior of $E\{X\}$ depends heavily on the behavior of $(N)_k/N^k$. Note that

$$\frac{(N)_k}{N^k} = \left(1 - \frac{1}{N}\right)\left(1 - \frac{2}{N}\right)\cdots\left(1 - \frac{k-1}{N}\right). \quad (4.76)$$

Taking logs of (4.76) and expanding $\log(1-x)$ in a Taylor series, we get that

$$\log\frac{(N)_k}{N^k} = \sum_{i=1}^{k-1}\log\left(1 - \frac{i}{N}\right) = -\sum_{i=1}^{k-1}\sum_{m=1}^{\infty}\frac{(i/N)^m}{m}, \quad (4.77)$$

whence,

$$\frac{(N)_k}{N^k} = \exp\left\{-\sum_{m=1}^{\infty}\frac{1}{mN^m}\left(\sum_{i=1}^{k-1}i^m\right)\right\}$$

$$= \exp\left\{-\frac{1}{N}\sum_{i=1}^{k-1}i - \frac{1}{2N^2}\sum_{i=1}^{k-1}i^2 - \sum_{m=3}^{\infty}\frac{1}{mN^m}\sum_{i=1}^{k-1}i^m\right\}$$

$$= \exp\left\{-\frac{k^2}{2N} - \frac{k^3}{6N^2} + \mathcal{O}\left(\frac{k}{N} + \frac{k^4}{N^3}\right)\right\}, \tag{4.78}$$

where we used

$$\sum_{i=1}^{k-1}i = \frac{1}{2}k(k-1), \quad \sum_{i=1}^{k-1}i^2 = \frac{1}{6}k(k-1)(2k-1), \quad \sum_{i=1}^{k-1}i^m = \mathcal{O}(k^{m+1}),$$

and

$$\sum_{m=3}^{\infty}\frac{1}{m}\mathcal{O}\left(\frac{k^{m+1}}{N^m}\right) = \mathcal{O}\left(\frac{k^4}{N^3}\right). \tag{4.79}$$

If $k = \mathcal{O}(N^{3/4})$, then $\mathcal{O}\left(\frac{k}{N} + \frac{k^4}{N^3}\right) = \mathcal{O}\left(\frac{1}{N^{1/4}}\right) + \mathcal{O}(1) = \mathcal{O}(1)$ and, hence,

$$\frac{(N)_k}{N^k} \sim \exp\left\{-\frac{k^2}{2N} - \frac{k^3}{6N^2}\right\}. \tag{4.80}$$

We now divide the question of emergence of cycles into two cases: $0 < c < 1$ (i.e., $p < \frac{1}{N}$) and $c \geq 1$ (i.e., $p \geq \frac{1}{N}$).

> **Theorem 4.7** *If X denotes the number of cycles in $G \in \mathcal{G}_{N,p}$, and if $0 < c < 1$, then $E\{X\}$ will converge to a nonzero constant as $N \to \infty$.*

Proof We split the sum in (4.75) into two parts: $k \leq \sqrt{N}/w_N$ and $k > \sqrt{N}/w_N$, where $w_N \to \infty$ arbitrarily slowly.

Let $k = \mathcal{O}(N^{1/2}/w_N)$. Substituting this value of k into the big-O term in the exponent of (4.78) yields

$$\mathcal{O}\left(\frac{k}{N} + \frac{k^4}{N^3}\right) \sim \mathcal{O}\left(\frac{1}{w_N N^{1/2}} + \frac{1}{w_N^4 N}\right) = o(1).$$

Repeating the substitution into the first two terms in the exponent in (4.78) yields

$$\frac{k^2}{2N} + \frac{k^3}{6N^2} \sim \frac{1}{2w_N^2} + \frac{1}{6N^{1/2}w_N^3} = A_N,$$

say. Then, $A_N \to 0$ and so $e^{-A_N} \to 1$. Putting these last two results together gives us that $(N)_k/N^k = 1 + o(1)$. From (4.78), we have that

$$E\{X\} = \frac{1}{2} \sum_{k=3}^{\sqrt{N}/w_N} (1 + o(1)) \frac{c^k}{k} + \mathcal{O}(1) \sum_{k>\sqrt{N}/w_N}^{N} \frac{c^k}{k} \tag{4.81}$$

$$\sim \frac{1}{2} \sum_{k=3}^{N} \frac{c^k}{k}$$

$$= -\frac{1}{2} \log(1-c) - \frac{c}{2} - \frac{c^2}{4}, \tag{4.82}$$

where, because $c < 1$, the $\mathcal{O}(1)$ term is the remainder term of a convergent series. Thus, (4.82) shows that the expected number of cycles in graphs $G \in \mathcal{G}_{N,p}$ with $p < 1/N$ will converge to a constant as $N \to \infty$, and that constant will be independent of the size, N, of the graph. □

Theorem 4.8 *If X denotes the number of cycles in $G \in \mathcal{G}_{N,p}$, and if $c \geq 1$, then $E\{X\}$ will diverge.*

Proof First, let $c = 1$. Then, (4.81) is still valid. We split the second summation in (4.81) over $\sqrt{N}/w_N < k \leq N$ into two summations, one over $\sqrt{N}/w_N < k \leq \sqrt{N}w_N$ and one over $\sqrt{N}w_N < k \leq N$. We have the following approximations:

$$E\{X\} \sim \frac{1}{2} \sum_{k=1}^{\sqrt{N}/w_N} \frac{1}{k} + \sum_{k>\sqrt{N}/w_N}^{N} \frac{(N)_k}{N^k} \frac{1}{k}$$

$$\sim \frac{1}{2} \log\left(\frac{\sqrt{N}}{w_N}\right) + \mathcal{O}\left(\sum_{k>\sqrt{N}/w_N}^{\sqrt{N}w_N} \frac{1}{k} + \sum_{k>\sqrt{N}w_N}^{N} \frac{1}{k} e^{-k^2/2N} \right)$$

$$= \frac{1}{2} \log\left(\frac{\sqrt{N}}{w_N}\right) + \mathcal{O}\left(2\log(w_N) + e^{-w_N^2/2} \log(N) \right). \tag{4.83}$$

If $\log(w_N) = o(\log(N))$, the first term in (4.83) can be written as

$$\frac{1}{2} \log\left(\frac{\sqrt{N}}{w_N}\right) = \frac{1}{4} \log(N) - \frac{1}{2} \log(w_N) \sim \frac{1}{4} \log(N)). \tag{4.84}$$

As for the second term in (4.83), let $B_N = 2\log(w_N) + e^{-w_N^2/2} \log(N)$. We have that

$$B_N = 2\log(\log(N)) + \frac{\log(N)}{e^{(\log(N))^2/2}}, \tag{4.85}$$

and so the second term reduces to $o(\log(N))$. Therefore,

$$E\{X\} \sim \frac{1}{4} \log(N) \to \infty, \tag{4.86}$$

and an appeal to the second-moment method (see Section 4.3.1) shows that *almost every graph has a cycle* (see **Exercise 4.7**). □

If $c > 1$, then it can be shown that the number of cycles will tend to infinity as $N \to \infty$ (see **Exercise 4.8**).

4.4.3 Graph Connectivity

Suppose we wish to show, for example, that if the threshold function for connectivity is $c \log(N)/N$, then, when $c < 1$, P{connectivity} $\to 0$, while if $c > 1$, then P{connectivity} $\to 1$ as $N \to \infty$. Let p denote the probability that any two nodes are connected by an edge.

Define the indicator variable $I_i = 1$ if the ith node is isolated, and 0 otherwise. Note that the $\{I_i\}$ are dependent random variables. Then, let

$$s = \text{P\{a given node is isolated\}} = \text{P}\{I_i = 1\} = (1 - p)^{N-1}. \qquad (4.87)$$

Because $(1 - p)^{N-1}$ is roughly the same as $(1 - p)^N$ if $p = p(N) \to 0$ as $N \to \infty$, we have that

$$(1 - p)^N = \exp\{N \log(1 - p)\}$$
$$= \exp\left\{N\left(-p - \frac{p^2}{2} - \frac{p^3}{3} - \cdots\right)\right\}$$
$$= \exp\{-Np\}\exp\left\{-Np^2\left(\frac{1}{2} + \frac{p}{3} + \cdots\right)\right\}. \qquad (4.88)$$

As long as $Np^2 \to 0$, then

$$(1 - p)^N \approx e^{-Np}. \qquad (4.89)$$

If we let $p = c \log(N)/N$, then

$$\text{P\{a given node is isolated\}} = e^{-c \log(N)} = N^{-c}. \qquad (4.90)$$

Let $X = \sum_i I_i$ denote the total number of isolated nodes. Then,

$$\text{E}\{X\} = N \cdot N^{-c} = N^{1-c}. \qquad (4.91)$$

We now break this down to two cases: $0 < c < 1$ and $c \geq 1$.

Theorem 4.9 *If $0 < c < 1$, then the expected number of isolated nodes diverges to infinity.*

Proof It is sufficient to show that

$$\text{P\{there exists at least one isolated node\}} \to 1.$$

From (4.91), it follows that $\text{E}\{X\} \to \infty$ as $N \to \infty$. We now claim that this implies that $\text{P}\{X = 0\} \to 0$. In general, this is not true; furthermore, we cannot use a Poisson approximation because the $\{I_i\}$ are not independent. So, instead we show that $\text{var}\{X\} \leq 2\text{E}\{X\}$, which is an example of the second-moment method. Consider $\text{var}\{X\}$:

$$\text{var}\{X\} = \sum_i \text{var}\{I_i\} + \sum_i \sum_j \text{cov}\{I_i, I_j\}$$
$$= N\text{var}\{I_1\} + N(N - 1)\text{cov}\{I_1, I_2\}$$
$$= Ns(1 - s) + N(N - 1)[\text{E}\{I_1 I_2\} - \text{E}\{I_1\}\text{E}\{I_2\}]. \qquad (4.92)$$

Now, from (4.87), $E\{I_1\} = P\{I_1 = 1\} = (1 - p)^{N-1}$ and

$$
\begin{aligned}
E\{I_1 I_2\} &= P\{I_1 = 1, I_2 = 1\} \\
&= P\{\text{both node 1 and node 2 are isolated}\} \\
&= P\{\text{node 1 is isolated}\}P\{\text{node 2 is isolated}|\text{node 1 is isolated}\} \\
&= (1 - p)^{N-1}(1 - p)^{N-2} \\
&= (1 - p)^{2N-3}.
\end{aligned}
\tag{4.93}
$$

It follows that

$$
E\{X\} = N(1 - p)^{N-1}
\tag{4.94}
$$

$$
\begin{aligned}
E\{X^2\} &= \sum_i E\{I_i\} + \sum_{i \neq j} E\{I_1 I_j\} \\
&= E\{X\} + N(N - 1)(1 - p)^{2N-3}.
\end{aligned}
\tag{4.95}
$$

Thus,

$$
\begin{aligned}
\text{var}\{X\} &= E\{X^2\} - [E\{X\}]^2 \\
&= E\{X\} + N(N - 1)(1 - p)^{2N-3} - N^2(1 - p)^{2N-2} \\
&\leq E\{X\} + N^2 p(1 - p)^{2N-3} \\
&= E\{X\}[1 + Np(1 - p)^{N-2}] \\
&\leq 2 \cdot E\{X\},
\end{aligned}
\tag{4.96}
$$

because $Np(1 - p)^{N-2}$ is maximized wrt p at $p^* = 1/(N - 1)$, whence, for all N, $Np^*(1 - p^*)^{N-2} = \frac{N}{N-1}(\frac{N-2}{N-1})^{N-2}$. This function of N is monotone decreasing from 0.75 at $N = 3$ to a limiting value of $e^{-1} = 0.368$ as $N \to \infty$ (see **Exercise 4.5**). For convenience, we take the upper bound on this function as one. Now,

$$
\text{var}\{X\} \geq (0 - E\{X\})^2 P\{X = 0\},
\tag{4.97}
$$

whence, from (4.96) and (4.91),

$$
P\{X = 0\} \leq \frac{\text{var}\{X\}}{(E\{X\})^2} \leq \frac{2 \cdot E\{X\}}{(E\{X\})^2} = \frac{2}{E\{X\}} = \frac{2}{N^{1-c}} \to 0,
\tag{4.98}
$$

as claimed. In other words, P{there exists at least one isolated node} $\to$ 1 or P{G is disconnected} $\to$ 1 as $N \to \infty$. $\square$

Theorem 4.10 *If $c \geq 1$, then the expected number of isolated nodes converges to zero.*

Proof This case is more complicated. If $c > 1$, then $E\{X\} = N^{1-c} \to 0$. In other words, as $N \to \infty$, the expected number of isolated nodes in $G \in \mathcal{G}_{N,p}$, where $p = c \log(N)/N$, converges to zero. Does it follow that, as $N \to \infty$, P{G has isolated nodes} $= P\{X \neq 0\} \to 0$? We argue by contradiction. Suppose not. Suppose instead that P{G has isolated nodes} $\to \gamma$, where $\gamma > 0$ is a constant. If that were true, then there would be at least one isolated node and so $E(X) \geq 1 \cdot \gamma > 0$, which contradicts the fact that $E(X) \to 0$. $\square$

4.5 Giant Component

The most startling of the results of Erdős and Rényi (1959, 1960) on phase transitions is that for certain values of the critical probability $p_c(N)$, a "giant component" will emerge as the number of nodes N increases. For example, even though the 1999 telephone call-graph example of Section 3.5.1, like all real-life networks, could not be considered to be a random graph, it had one giant connected component, with 44,989,297 nodes (out of 53,767,087 nodes), which accounted for about 84% of the total number of nodes. Also, in the 2011 Facebook social network graph of Section 3.5.2, the vast majority (99.91%) of its members were found to belong to a single, giant, connected component, whilst the remaining components were tiny.

To derive the conditions for the emergence of a giant component in a random graph, we use the basic ideas behind branching processes (Section 4.3.4) and graph-searching algorithms (Section 3.7).

4.5.1 The Emergence of a Giant Component

Perhaps the most remarkable property of a random graph $\mathcal{G}$ is the emergence of a "giant" component when the probability hits a critical value $p = p_c(N)$. First, some definitions.

Definition 4.7 A *component* of $\mathcal{G}$ is a subset of nodes of $\mathcal{G}$, each node of which is reachable from all other nodes in the component by some path through the graph.

Definition 4.8 A *giant component* is a connected component of $\mathcal{G}$ that contains a very high proportion of all N nodes.

If a network contains a giant component, then every node within that component will be connected through a sequence of successive edges to every other node in that component. Furthermore, that component will almost always be the only giant component. Otherwise, we would experience a highly unstable situation. Indeed, if two giant components coexisted in some network, all it would take would be the addition of an edge joining one giant component to the other; this would lead to a merging of the two components into a single giant component. In general, any large component in the graph will be swallowed up by the giant component.

Consider the random graph $\mathcal{G}_{N,p}$, where N is the number of nodes and $p = p(N)$ is the probability of connecting any pair of nodes with an edge. Let $p(N) = c/N$, where $c > 0$ is a fixed constant. We break this problem into three phases depending upon the value of c.

1. *Subcritical phase.* If $c < 1$ (i.e., $p(N) < 1/N$), then, for large N and with high probability, all of the connected components of the graph will be very small, with the largest component, $\mathcal{C}_1$, having a tree-like structure containing only $|\mathcal{C}_1| = \mathcal{O}(\log(N))$ nodes (Erdős and Rényi, 1959).
2. *Critical phase.* If $c \sim 1$ (i.e., $p(N) \sim 1/N$), then, for large N and with high probability, the largest component, $\mathcal{C}_1$, will have $|\mathcal{C}_1| = \Theta(N^{2/3})$ nodes (Bollobás, 1984; Luczak, 1990).

3. *Supercritical phase.* If $c > 1$ (i.e., $p(N) > 1/N$), then there is a constant $\theta(c) \in (0, 1)$ given by

$$\theta(c) = 1 - \frac{1}{c} \sum_{k=1}^{\infty} \frac{k^{k-1}}{k!} (ce^{-c})^k,$$

with $\lim_{c \to \infty} \theta(c) = 1$, so that for large N and with high probability, the size of the unique giant component, C_1, will be linear in N, with $|C_1| = (\theta(c)+o(N))N$ nodes. The second largest component, C_2, will be very small, having tree-like structure with only $|C_2| = \mathcal{O}(\log(N))$ nodes (Erdős and Rényi, 1959).

The constant $\theta(c)$ can be viewed as $P_c\{X = 0\} = 1 - \sum_{k=1}^{\infty} P_c\{X = k\}$, where X is the total number of descendants of a Poisson branching process with rate $c > 0$. The probability distribution $P\{X = k\}$ was found by Emile Borel in 1942 using a geometrical argument (Borel, 1942). Tanner (1953) proved it is valid more generally. To summarize:

Theorem 4.11 *At the critical probability, $p_c(N) = 1/N$, the topological structure of a random graph suddenly switches from being a loose collection of small components, of which most of them are trees, to having one giant component and a lot of relatively small components.*

This change in the size of the largest component as c passes the value 1 was called a "double jump" by Erdős and Rényi.

The above statements on the emergence of the giant component have been proved in several different ways. Erdős and Rényi used a combinatorial argument together with the first- and second-moment method, while others prefer to use the theory of branching processes (Bollobás and Riordan, 2012), martingale methods (Nachmias and Peres, 2007), empirical distributions of independent random variables (Janson and Luczak, 2008), or graph search (Krivelevich and Sudakov, 2013). We follow the treatment of Janson, Luczak, and Rucinski (2000, Chapter 5).

Proof Let v be a particular node in $\mathcal{G}_{N,p}$. We are interested in identifying all nodes of the component C_v of which v is a member. Consider a *Galton–Watson branching process* starting at v that examines all of v's neighbors in the graph in the manner of a breadth-first search. In other words, we first visit all of v's neighbors, $v_1, v_2, \ldots, v_q$, say; then, we visit all of v_i's neighbors that differ from the previously visited nodes, $i = 1, 2, \ldots, q$; then, we visit all the new neighbors of each of the nodes found in the previous stage; and so on. The number of newly examined neighbors of v is distributed as a binomial random variable, $\text{Bin}(N - k, p)$, where $N - k$ is the number of *unexamined* nodes in the graph. The proof is divided into three phases.

(a) Subcritical phase. First, assume $c < 1$. Suppose the node v belongs to a component $C(k)$ having *at least* $k = k(N)$ nodes. This will occur only if the branching process, which starts at v, finds at least $k - 1$ new nodes after saturating k nodes. Let $X_1, \ldots, X_k \overset{iid}{\sim} \text{Bin}(N, p)$, where X_i is the degree of node v_i. Then, $X = \sum_{i=1}^{k} X_i \sim \text{Bin}(Nk, p)$ and

$$P\{v \in \mathcal{C}(k)\} \le P\{X \ge k - 1\}$$
$$= P\{X \ge ck + (1 - c)k - 1\}$$
$$= P\left\{X \ge \left(1 + \frac{(1 - c)k - 1}{ck}\right)ck\right\}$$
$$\le \exp\left\{-\frac{\left(\frac{(1-c)k-1}{ck}\right)^2 ck}{2 + \frac{(1-c)k-1}{ck}}\right\}$$
$$= \exp\left\{-\frac{[(1 - c)k - 1]^2}{(c + 1)k - 1}\right\}$$
$$\le \exp\left\{-\frac{(1 - c)^2 k}{2}\right\}, \tag{4.99}$$

where the fourth line is obtained from the third by applying Chernoff's bound, $P\{X \ge (1 + \beta)\mu\} \le \exp\{-\beta^2\mu/(2 + \beta)\}$, with $\mu = ck$ and $\beta = \frac{(1-c)k-1}{ck}$. Note that for $\beta > 0$, we need $k > \frac{1}{1-c}$. To get from the fifth line to the sixth, we have that because $(1 - c)k > 1$, then $(c + 1)k - 1 > 2ck$, and then we use $c < 1$.

We now choose k so that this upper bound converges to zero as $N \to \infty$. So, we set

$$k = \frac{3}{(1 - c)^2} \cdot \log N, \tag{4.100}$$

where k was chosen so that $(1 - c)^2$ cancels out and the "3" is needed to make the convergence rate work. So, the term in the exponent of (4.99) reduces to $-\frac{3}{2}\log N$ or $\log(N^{-3/2})$, and the log disappears with the exp. Then, the probability that the node v is contained in a component of size k or larger is $P\{v \in \mathcal{C}(k)\} \le N^{-3/2}$. Take a union bound over v to get

$$P\left\{\text{there exists } v \in \mathcal{C}\left(\frac{3\log N}{(1 - c)^2}\right)\right\} \le N \cdot N^{-3/2} = N^{-1/2}, \tag{4.101}$$

which converges to zero as $N \to \infty$. This implies that v cannot be a member of this component. □

(b) Supercritical phase. Next, assume $c > 1$. This is more complicated than the case of $c < 1$. Again, let k denote the number of nodes in the component. Let $k_- = c' \log(N)$, where c' is to be determined, and $k_+ = N^{2/3}$. We proceed in three stages.

The first stage:

Theorem 4.12A *Almost surely, as $N \to \infty$, for every $k \in [k_-, k_+]$, and all nodes v of $G \in \mathcal{G}_{N,p}$, either (a) the branching process that starts at v terminates before it takes k_- steps, or (b) at the kth step starting from v, there are at least $\frac{(c-1)k}{2}$ nodes in the component containing v that have been examined but are not yet saturated.*

Proof This theorem says that, in a very large random graph, with high probability, each component has either at most k_- nodes or at least k^+ nodes. To show this, we only need show that there are at most $k + \frac{(c-1)k}{2} = \frac{(c+1)k}{2}$ nodes in the component.

For every i, we can bound the degree, X_i, of node v_i from below by

$$X_i^- \overset{iid}{\sim} \text{Bin}\left(N - \frac{(c+1)k_+}{2}, p\right), \quad i = 1, 2, \ldots, k. \qquad (4.102)$$

Let $X = \sum_{i=1}^k X_i \sim \text{Bin}(Nk, p)$ and $X^- = \sum_{i=1}^k X_i^- \sim \text{Bin}(N', p)$, where $N' = \left(N - \frac{(c+1)k_+}{2}\right)k$. It will be convenient to say that the node v is "terminal" if, after k steps, the branching process at v dies or has found fewer than $\frac{(c+1)k}{2}$ nodes. Then,

$$P\{v \text{ is terminal after } k \text{ steps}\} \leq P\left\{X \leq \frac{(c+1)k}{2} - 1\right\}$$

$$\leq P\left\{X^- \leq ck - \frac{(c-1)k}{2}\right\}$$

$$= P\left\{X^- \leq \left(1 - \frac{c-1}{2c}\right)ck\right\}$$

$$\leq \exp\left\{-\frac{(c-1)^2 k}{8c}\right\}. \qquad (4.103)$$

We go from the first line to the second because X_i is stochastically bounded below by X_i^-. We go from the third line to the fourth line using the Chernoff bound $P\{X \leq (1-\beta)\mu\} \leq \exp\{-\beta^2\mu/2\}$, with $\beta = \frac{c-1}{2c}$ and $\mu = \mu(X^-) = ck$. Applying a union bound over k yields

$$P\{v \text{ is terminal}\} \leq \sum_{k=k_-}^{k_+} \exp\left\{-\frac{(c-1)^2 k}{8c}\right\} \qquad (4.104)$$

$$\leq k_+ \exp\left\{-\frac{(c-1)^2 k_-}{8c}\right\}. \qquad (4.105)$$

If we choose

$$k_- = \frac{16c}{(c-1)^2} \cdot \log N, \quad k_+ = N^{2/3}, \qquad (4.106)$$

then $P\{v \text{ is terminal}\} \leq N^{-4/3} \to 0$. Applying a union bound over v, we have that

$$P\{\text{there exists } v \text{ such that } v \text{ is terminal}\} \leq N \cdot N^{-4/3} = N^{-1/3}, \qquad (4.107)$$

which converges to zero as $N \to \infty$. □

We will find it convenient to use the following terminology. If the branching process that starts at v terminates within $k_- = \mathcal{O}(\log(N))$ steps, we will refer to the component containing v as a "small" component and its nodes as "small" nodes; if the process continues on for *at least* $k^+ = N^{2/3}$ steps, we will call the component containing v a "large" component and its nodes "large" nodes. Thus, the first stage shows that *each node belongs, almost surely, either to a small component (having fewer than k_- nodes) or to a large component (having more than k_+ nodes)*.

Next, the second stage:

Theorem 4.12B *Almost surely, as $N \to \infty$, there exists a unique component containing all of the large nodes.*

———— 98 ————

Proof Let v' and v'' be a fixed pair of large nodes that belong to components C' and C'', respectively, each component of which contains at least k_+ nodes. What is the probability that these components are different?

Suppose we start two separate branching processes, one from v' and the other from v'', and run them both for k_+ steps. We have just shown in part 1 that if we start from v' in C', there will be a set of nodes, V', say, such that at least $\frac{(c-1)k_+}{2}$ nodes from C' will be unexplored. A similar statement holds if we start a branching process from v'' in C''. There are now two possible outcomes: either these branching processes share some nodes over k_+ steps (in which case, we are done) or, with high probability, there exists an edge between the two sets of nodes V' and V''.

Consider the converse of the latter possibility, that there are no edges between the as-yet-unexplored nodes of V' and V''. The probability of the latter is

$$P\{\text{there is no edge between } V' \text{ and } V''\} \leq (1-p)^{|V'| \cdot |V''|}$$

$$= (1-p)^{((c-1)k_+/2)^2}$$

$$\leq \exp\left\{ -p\left(\frac{(c-1)k_+}{2} \right)^2 \right\}$$

$$= \exp\left\{ -\frac{(c-1)^2 c}{4} N^{1/3} \right\}. \qquad (4.108)$$

We go from the second line to the third by using the inequality $(1-p)^a \leq e^{-pa}$, and we go from the third line to the fourth by substituting $k_+ = N^{2/3}$ and $p = c/N$.

Taking a union bound (i.e., multiplying (4.108) by N^2) over pairs v' and v'' yields the statement that for any pair of large vertices, $v' \in V'$ and $v'' \in V''$, both of size at least k_+, the probability that there exists no edge between sets V' and V'' converges to zero as $N \to \infty$. This establishes that *the giant component is almost surely unique, having at least k_+ nodes*.

Furthermore, all other nodes are members of components with at most $\mathcal{O}(\log(N))$ nodes. □

Finally, the third stage:

Theorem 4.12C *Almost surely, as $N \to \infty$, there are $(1 + o(1))(1 - \theta)N$ small nodes, where $\theta \in (0, 1)$.*

Proof The last step is to assess the size of the large component. The way we do this is to determine the size of its complement, the number of small nodes. If we show that there are $(1 - \theta)N$ small nodes, where $\theta \in (0, 1)$, then the number of large nodes has to be θN. Accordingly, we derive an upper bound to the probability that a node belongs to a small component.

Abbreviate "branching process" to bp. Now, the bp can become extinct within k_- steps or it can become extinct after k_- steps. Define the following probabilities:

$$p_- = P\{\text{bp with } \text{Bin}(N, p) \text{ from } v \text{ goes extinct within } k_- \text{ steps}\}, \qquad (4.109)$$

$$p_+ = P\{\text{bp with } \text{Bin}(N - k_-, p) \text{ from } v \text{ goes extinct}\}, \qquad (4.110)$$

where $\text{Bin}(n, p)$ represents the offspring distribution of the branching process, and "steps" refers to the number of nodes that have been discovered. Then, for a node v

that belongs to a small component (which we abbreviate to "$v \in$ small"), we have the bounds

$$p_- \leq \mathrm{P}\{v \in \text{small}\} \leq p_+. \tag{4.111}$$

If a branching process goes extinct, it does so, almost surely, in a finite number of steps. Further, as $N \to \infty$, then $\mathrm{Bin}(N, p) \to \mathrm{Poisson}(c)$ and $k_- = c' \log(N) \to \infty$. Then,

$$p_- = \mathrm{P}\{\text{bp with } \mathrm{Bin}(N, p) \text{ from } v \text{ goes extinct}\} - o(1), \tag{4.112}$$

where the $o(1)$ term bounds the probability that the branching process goes extinct after k_- or more nodes have been discovered. Thus,

$$\mathrm{P}\{\text{bp with } \mathrm{Bin}(N, p) \text{ from } v \text{ goes extinct}\} \to 1 - \theta(c), \quad \text{as } N \to \infty, \tag{4.113}$$

where $\theta(c)$ is the unique solution in $(0, 1)$ to the equation

$$\theta + e^{-\theta c} = 1 \tag{4.114}$$

(Erdős and Rényi, 1959). This equation, however, does not have a simple, closed-form solution for θ.

A graphical solution can easily be obtained by plotting $y = 1 - e^{-\theta c}$ against θ and then superimposing upon that graph a $45°$ plot of $y = \theta, 0 < y, \theta < 1$. See **Figure 4.1**, where we have drawn curves for $c = 0.5, 1.0$, and 1.5. The graphical solution occurs when the two graphs intersect. It turns out that there can be either one or two possible solutions, depending upon the value of c. For the top curve ($c = 1.5$) in **Figure 4.1**, there are two solutions, one where $\theta = 0$ and the other where $\theta > 0$, which implies that a giant component exists for this latter value of c. For the lowest of the three curves ($c = 0.5$), the only solution is $\theta = 0$, which shows that for this value of c, there is no giant component. This shows that larger values of c produce a giant component, whereas small values of c do not.

Because $k_- \ll N$, the same holds for p_+. So, from (4.113), $\mathrm{P}\{v \in \text{small}\} \to 1 - \theta(c)$ as $N \to \infty$.

We now provide a probabilistic result regarding convergence. Let $Z_v = 1$ if v belongs to a small component and 0 otherwise. Then, $Z = \sum_{v=1}^{N} Z_v$ denotes the total number of small nodes. We need to find the first and second moments of Z and then use Chebychev's inequality.

From our previous result,

$$\mathrm{E}\{Z_v\} = \mathrm{P}\{Z_v = 1\} = \mathrm{P}\{v \in \text{small}\} \to 1 - \theta(c), \quad \text{as } N \to \infty. \tag{4.115}$$

Hence,

$$\mathrm{E}\{Z\} = \sum_v \mathrm{E}\{Z_v\} = (1 + o(1))(1 - \theta(c))N \to \infty, \quad \text{as } N \to \infty. \tag{4.116}$$

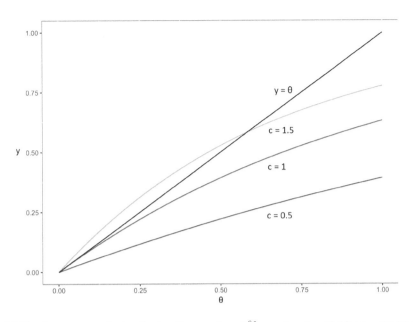

Figure 4.1 The three curves display the function $y = 1 - e^{-\theta c}$ as a function of θ for $c = 0.5, 1.0, 1.5$, and $0 < y, \theta < 1$. The blue curve is $c = 0.5$, the red curve is $c = 1$, and the green curve is $c = 1.5$. The black line is $y = \theta$. For a given value of c, the solution of the equation $\theta = 1 - e^{-\theta c}$ is given by the intersection of the corresponding c-curve and the line.

Next, consider the second moment of Z:

$$
\begin{aligned}
E\{Z^2\} &= E\left\{\left(\sum_v Z_v\right)^2\right\} \\
&= \sum_v E\{Z_v^2\} + \sum_{v \neq v'} E\{Z_v Z_{v'}\} \\
&= E\{Z\} + \sum_{v \neq v'} P\{v \in \text{small}, v' \in \text{small}\} \\
&= E\{Z\} + \sum_{v'} P\{v' \in \text{small}\} \sum_{v \neq v'} P\{v \in \text{small} \mid v' \in \text{small}\}.
\end{aligned}
$$

$$(4.117)$$

Consider the conditional probability, $\sum_{v \neq v'} P\{v \in \text{small} \mid v' \in \text{small}\}$. This probability can be divided into the sum of two probabilities, one term in which v and v' live in a single component ($v \neq v'$), and the other term in which v and v' live in two different components. The former term can be bounded from above by k_- because there are at most k_- nodes in the same component. Now, consider the latter term. Define

$$
\mathcal{G}' = \mathcal{G} \backslash \{u : u \text{ is in the same component as } v'\} \tag{4.118}
$$

to be the random graph that consists of all those nodes in $\mathcal{G}$ that live in a different component than that containing v'. If v and v' live in different components, then

$$
P\{v \in \text{small} \mid v' \in \text{small}\} = P\{v \in \text{small in } \mathcal{G}'\} \sim \rho(N - k_-, p), \tag{4.119}
$$

where $\rho(n, p)$ denotes the probability that a branching process with offspring distribution $\text{Bin}(n, p)$ goes extinct. Thus,

$$\sum_{v \neq v'} P\{v \in \text{small} \mid v' \in \text{small}\} \leq k_- + N\rho(N - k_-, p). \tag{4.120}$$

Because $k_- \ll N$, we have that

$$\rho(N - k_-, p) \sim \rho(N, p) \to 1 - \theta(c), \quad \text{as } N \to \infty. \tag{4.121}$$

Substituting (4.121) into (4.120) and then the result into (4.117) yields

$$E\{Z^2\} \leq E\{Z\} + N^2 [\rho(N, p)]^2 (1 + o(1)). \tag{4.122}$$

Dividing (4.122) through by $(E\{Z\})^2$ yields

$$\frac{E\{Z^2\}}{(E\{Z\})^2} \leq \frac{1}{E\{Z\}} + 1 + o(1) = 1 + o(1). \tag{4.123}$$

So, $E\{Z^2\} \sim (E\{Z\})^2$. Now, we use Chebychev's inequality,

$$P\{|X - E\{X\}| \geq a\} \leq \frac{1}{a^2} \text{var}\{X\}$$

to get

$$\begin{aligned}
P\{|Z - E\{Z\}| \geq \alpha E\{Z\}\} &\leq \frac{\text{var}\{Z\}}{\alpha^2 (E\{Z\})^2} \\
&= \frac{1}{\alpha^2} \left(\frac{E\{Z^2\}}{(E\{Z\})^2} - 1 \right) \\
&= \frac{1}{\alpha^2} \cdot o(1) = o(1), \tag{4.124}
\end{aligned}$$

as long as $\alpha = \alpha(N)$ does not grow too quickly. Thus, with high probability, we have that $Z = (1 + o(1))E\{Z\}$ (i.e., Z is concentrated about its mean), as claimed. $\quad\square$

(c) Critical phase. The case $c \sim 1$ is the most complicated and delicate, and, so far, no simple treatment is available in the literature. Erdős and Rényi's original work on phase transition for random graphs included the behavior of components when $p(N) = 1/N$. More detailed study (Bollobás, 1984; Luczak, 1990) of this case was accomplished when a second-order term was added to $p(N)$, so that the parametrization becomes

$$p(N) = \frac{1}{N} + \lambda N^{-4/3} = \frac{1 + \lambda N^{-1/3}}{N}, \quad -\infty < \lambda < \infty. \tag{4.125}$$

This case resolves into the following regions (Luczak, 1990).

Theorem 4.13 *Let $L(\mathcal{C}_k)$ denote the order of the kth largest component in $G \in \mathcal{G}_{N,p}$. Assume $p = p(N)$ is given by (4.126). Then,*

- *barely subcritical: if $\lambda \to -\infty$, then, almost surely, as $N \to \infty$, $L(\mathcal{C}_1) \ll N^{2/3}$;*
- *barely supercritical: if $\lambda \to \infty$, then, almost surely, as $N \to \infty$, $L(\mathcal{C}_1) \gg N^{2/3} \gg L(\mathcal{C}_2)$. Furthermore, $L(\mathcal{C}_1) = (2 + o(1))\lambda N^{2/3}$.*

When $\lambda \to$ constant, the probability that the largest component remains the largest is bounded away from 0 and 1; the larger is λ, the closer the probability is to 1.

Recent work on this problem has been carried out by Nachmias and Peres (2008). None of these arguments can be viewed as elementary, however.

4.6 Further Reading

Random graphs are covered in detail in the books by Palmer (1985), Janson, Luczak, and Rucinski (2000), Bollobás et al. (2001), Durrett (2007), Kolaczyk (2009, Chapter 6), Grimmett (2010, Chapter 11) and Newman (2010, Chapter 12). Other books include Wasserman and Faust (1994) and Doreian, Batagelj, and Ferligoj (2005).

4.7 Exercises

Exercise 4.1 Show that the random graph $\mathcal{G}_{N,M}$ is equivalent in distribution to $\mathcal{G}_{N,p}$ conditioned on the event $\{\mathcal{G}_{N,p}$ has M edges$\}$.

Exercise 4.2 Find a simple asymptotic formula for the quantity $\binom{\binom{N}{2}}{M}$ of graphs with N nodes and M edges, where $M = N - 1$.

Exercise 4.3 Let X be a non-negative integer-valued random variable. Show that whilst $E\{X\}$ can be arbitrarily large, $P\{X > 0\}$ can be arbitrarily small. Give your interpretation of this result, comparing it with the first-moment method.

Exercise 4.4 Let X_k denote the number of components of $G \in \mathcal{G}_{N,p}$ that are trees having k nodes, where k is fixed, and let $p_\lambda(N) = \lambda/N$, where $\lambda > 0$ is a constant. Show that, as $N \to \infty$, the expected number of trees with k nodes is given by

$$E\{X_k\} \to \frac{N\lambda^{k-1}e^{(1-\lambda)k}}{k^{5/2}\sqrt{2\pi}}. \tag{4.126}$$

Exercise 4.5 Let X_k denote the number of trees with k nodes in the random graph. Let $p_\lambda(N) = \lambda N^{-(k-2)/(k-1)}$ be the threshold function with $\lambda > 0$. Find $E\{X_k\}$. What if $\lambda = 1$?

Exercise 4.6 Let X_k denote the number of trees with k nodes in the random graph. Let $p_\lambda(N) = 2\lambda N^{-k/(k-1)}$ be the threshold function with $\lambda > 0$. Find $E\{X_k\}$. What if $\lambda = 1$?

Exercise 4.7 Complete the argument of Section 4.4.2 on the emergence of cycles for the case $c = 1$. *Hint:* Find an expression for $E\{X^2\}$ and repeat the argument that was used to find $E\{X\}$. Then compute $E\{X^2\}/[E\{X\}]^2$ and show that it converges to 1, so that $P\{X = 0\} \to 0$.

Exercise 4.8 Let X be a non-negative random variable. Show that $P\{X > 0\} \geq (E\{X\})^2/E\{X^2\}$. Give your interpretation of this result, comparing it with the second-moment method.

Exercise 4.9 Using Markov's inequality (i.e., $P\{Z \leq a\} \leq E\{Z\}/a$), show that $P\{Z < \infty\} = 1$.

Exercise 4.10 Use the second-moment method to show that the probability that a random graph will have a cycle will tend to 1.

Exercise 4.11 Let $c > 1$. Show that the number of cycles in a random graph will tend to infinity as $N \to \infty$.

Exercise 4.12 Show that

(a) $Np(1 - p)^{N-2}$ is maximized wrt p at $p^* = 1/(N - 1)$.

(b) For all N, $Np^*(1 - p^*)^{N-2} = \frac{N}{N-1}\left(\frac{N-2}{N-1}\right)^{N-2}$.

(c) As a function of N, $\frac{N}{N-1}\left(\frac{N-2}{N-1}\right)^{N-2}$ is monotone decreasing from 0.75 at $N = 3$, and has a limiting value of $e^{-1} = 0.368$ as $N \to \infty$.

Exercise 4.13 Let $\psi(s) = 1 - \alpha(1 - s)^\beta$, where $\alpha, \beta \in (0, 1)$. Show that $\psi(s)$ is a probability generating function.

Exercise 4.14 Show that the Taylor series expansion of $\log(1 - p)$ is given by $\log(1 - p) = -p - \frac{1}{2}p^2 - \frac{1}{3}p^3 - \cdots \approx -p$, for $p \to 0$.

Exercise 4.15 For the graphical solution to (4.114) (see **Figure 4.1**), at which value of θ does the curve $(y = 1 - e^{-\theta c})$ for $c = 1.5$ intersect the 45°-degree line $(y = \theta)$?

Exercise 4.16 In the proof of **Theorem 4.2**, show that the event $\{|X - E\{X\}| \geq E\{X\}\}$ contains the event $\{X = 0\}$.

Exercise 4.17 In the proof of **Theorem 4.3(a)**, show that $P\{e^{tX} \geq e^{(1+\beta)t\mu}\} \leq E\{e^{tX}\}/e^{(1+\beta)t\mu}$.

Exercise 4.18 In the proof of **Theorem 4.3(b)**, show that $P\{e^{-tX} \geq e^{-(1-\beta)\mu}\} \leq E\{e^{tX}\}/e^{-(1-\beta)t\mu}$.

Exercise 4.19 Let X be a non-negative, integer-valued, random variable (as in the first-moment method). Show that its expectation $E\{X\}$ can be arbitrarily large, whilst $P\{X > 0\}$ can be arbitrarily small.

Exercise 4.20 (Chebyshev) Let X be a real-valued random variable and let $f, g: \mathbb{R} \to \mathbb{R}$ be bounded, nondecreasing functions. Show that $E\{f(X)g(X)\} \geq E\{f(X)\}E\{g(X)\}$.

CHAPTER FIVE

Percolation on $\mathbb{Z}^d$

There is a close relationship between random graphs and percolation. In fact, percolation and random graphs have been viewed as "the same phenomenon expressed in different languages" (Albert and Barabási, 2002). Early ideas on percolation (although not under that name) in molecular chemistry can be found in the articles by Flory (1941) and Stockmayer (1943). This chapter (and the next) is about percolation theory and its applications. We show the major developments and flow of the subject and give proofs of the main results. We provide references to the articles and books in the percolation literature so that readers can search out the relevant theory, other fine details, and recent generalizations.

5.1 Introduction

The modern theory of percolation originated in the discussion following the presentation of a paper[1] by John Hammersley (and K.W. Morton) at the 1954 Symposium on Monte Carlo Methods at the Royal Statistical Society in London. As a discussant, Simon R. Broadbent, who worked at the British Coal Utilization Research Association, which sponsored the Symposium, drew attention to the following problem:

A square (in two dimensions) or cubic (in three) lattice consists of 'cells' at the interstices joined by 'paths' which are either open or closed, the probability that a randomly-chosen path is open being p. A 'liquid' which cannot flow upwards or a 'gas' which flows in all directions penetrates the open paths and fills a proportion $\lambda_r(p)$ of the cells at the rth level. The problem is to determine $\lambda_r(p)$ for a large lattice. Clearly, it is a non-decreasing function of p and takes the values 0 at p = 0 and 1 at p = 1. Its value in the two-dimensional case is not greater than in three dimensions.

This problem was motivated by his study of the flow of coal-dust particles through gas masks of coal miners. Broadbent wondered whether Monte Carlo methods might be a useful tool for solving the gas-mask problem.

Hammersley and Broadbent started working together and realized that the gas-mask problem was actually a new class of problems in which the random properties of the medium influence the percolation of the fluid through it, contrary to conventional

---

[1]The paper's title was "Poor Man's Monte Carlo," which described running a Monte Carlo simulation without using large computers (Domb).

diffusion theory, where the randomness is associated with the fluid. Hammersley introduced the term *percolation process* by analogy to the process of water flowing through a coffee percolator. This collaboration led to the introduction of the mathematical theory of percolation (Broadbent and Hammersley, 1957). As a practical example, they discussed the following question:

> If a lump of porous material (such as a large stone) were placed in a bucket of water, what would be the probability that the water would reach the center of the material (i.e., the origin)?

Clearly, if v is a node near the center of the stone, it gets wet iff there is an open passageway inside the stone for water to flow from a node on the surface of the stone to v.

5.2 Examples of Percolation

Since the publication of the Broadbent–Hammersley paper, percolation theory became a key building block for stochastic geometry and statistical mechanics, and has been applied, for example, to solid-state physics, biology, groundwater hydrology, the spread of oil in water, the modeling of forest fires, the spread of infection through an orchard, and the communication of messages in an unreliable network. Most recently, percolation has been used to predict the fluid permeability of first-year porous Antarctic sea ice during climate change (Golden, 2009; Pringle et al., 2009). New techniques are being devised to deal with the questions posed by those working in percolation theory.

In this section, we describe a few applications of percolation that touch upon the different types of percolation we describe.

5.2.1 Impurity Doping of Semiconductors

Semiconductors are amongst the most technologically important materials today. They can be found, for example, in computers, smart phones, CD players, TV remotes, satellite dishes, optical fiber networks, traffic signals, and automobiles. The aspect of semiconductors that makes them so important and useful is that we can control how electrical current passes through them, unlike metal, which conducts electricity under any conditions, or glass or plastics that do not conduct electricity at all.

A *semiconductor* is composed of individual atoms bonded together as sites in a regular, periodic, crystal lattice. Each site contains at most one atom and, in an infinite, two-dimensional lattice, each atom is bonded to four other atoms. In *elemental semiconductors*, the atoms are all of the same type (e.g., silicon (Si) or germanium (Ge)), while *compound semiconductors* are made up of two or more elements of the periodic table. See the right image in **Figure 5.1**.

Silicon, in its role as the basis for integrated circuit (IC) chips, has become the most popularly used semiconductor material. A silicon atom has 14 electrons orbiting in three shells (2–8–4) around the nucleus, where the further the shells are from the nucleus the greater the energy. See the left image in **Figure 5.1**. The outer four electrons (called *valence electrons*) form the "bonding" shell. A silicon atom bonds to each of its four neighboring silicon atoms in the lattice by sharing the four valence electrons with the four silicon atoms to create four double bonds. The bonding shell for each atom

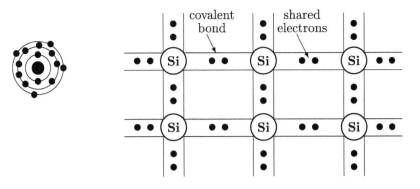

Figure 5.1 Left: Silicon atom with 14 electrons orbiting in three shells (2–8–4) around the nucleus. Right: Schematic representation of an elemental semiconductor, where each pair of silicon atoms (Si) is connected by a covalent bond formed from two electrons that are shared by the two silicon atoms.

now has eight electrons. Silicon is a remarkably stable material and can be heated to high temperatures while retaining its properties.

Now, consider a single atom. Electrons cannot move around freely but are subject to certain rules of movement. They occupy two types of *energy bands*: a *valence band* (located at a lower-energy level) and a *conduction band* (located at a higher-energy level), which are separated from each other by an *energy gap* E_g (also called a *band gap* or *forbidden region*), which is the energy difference between the highest occupied state in the valence band and the lowest occupied state in the conduction band. If the energy gap is relatively small ($E_g < 3$ eV), we have a semiconductor; otherwise, it is an insulator. In the case of metals, there is no energy gap because the valence band and the conduction band overlap; this permits free electrons to participate in the conduction process.

Suppose the conduction band is empty. The electrons are able to move around freely in the valence band until all states are occupied by electrons, at which point no further movement can take place. Because electrons in the valence band do not participate in the conduction process, no transportation of an electrical charge can occur, and the semiconductor is an insulator. To get a semiconductor to conduct an electrical current, an electron has to be "promoted" from the valence band to the conduction band; this can only happen if its energy is increased by at least that of the energy gap, which for semiconductors is a small amount of additional energy.

This situation can be changed by a process called "doping." *Doped semiconductors* are semiconductors that contain impurities. An extremely pure semiconductor is called *intrinsic*, a lightly or moderately doped semiconductor is called *extrinsic*, and a heavily doped semiconductor, which makes it act more like a conductor, is called *degenerate*. The impurities used depend upon the type of semiconductor. Doping of semiconductors takes place when a minute number of carefully chosen foreign impurity atoms (*dopands*) are added in a controlled fashion into the crystal structure of the semiconductor to alter its electrical properties while improving its conductivity. There are two types of doping processes:

- In *p-doping*, a silicon atom in the lattice is replaced by an "impurity" atom with fewer valence electrons, so that there is now a shortage of electrons in the crystal bonds. Assume there is a single "missing" electron that creates a "hole" (which

—— **107** ——

can be viewed as positively charged due to the absence of the negatively charged electron) that can transport current. A neighboring atom can provide an electron to fill the hole, which, in turn, creates a new hole, which can be filled by another of the neighboring electrons, and so on. Therefore, both electron and hole move around, but in opposite directions. In this case, conductivity takes place through the holes, not the electrons. As a result, this is called *acceptor impurity*.

- In *n-doping*, a silicon atom is replaced by an impurity atom with a greater number of valence electrons. This means that there is now an extra valence electron, which is loosely bound to the crystal lattice. The valence band of the atom is completely full, and so the extra electron has to be "donated" to the conduction band by boosting its energy level just a small amount so that current may flow. In this case, conductivity takes place through the negative electrons. Hence, this is called *donor impurity*.

A doped semiconductor is called either an *n-type semiconductor*, where "*n*" stands for negative, or a *p-type semiconductor*, where "*p*" stands for positive, if it is doped with donor or acceptor impurities, respectively.

When the conduction band is fully occupied by electrons, all movement stops. If all energy bands are completely filled or are completely empty, no electrical current can pass through the crystal, and so we have an *electrical insulator*. On the other hand, a crystal whose energy band is only partially filled (and whose electrons, therefore, can move fairly freely throughout the crystal lattice) will be an *electrical conductor*. For a given solid crystal, whether we have an insulator or a conductor depends upon the number of electrons per atom that are available to fill the energy bonds of that crystal.

How can we model the ability of a semiconductor to transport an electrical charge from one end of the crystal lattice to the other end? In the *site-percolation model*, we assume that each site of the lattice is occupied with probability p or is empty with probability $1 - p$, independent of its neighbors. In terms of the semiconductor, the occupied sites are the electrical conductors and the empty sites are the insulators. When p is small, the semiconductor is an insulator, while when p is large, the semiconductor becomes a conductor. Thus, a critical probability p_c has to exist such that for $p > p_c$, an infinite cluster of conduction atoms emerges and long-range electrical current can be transported across the lattice.

5.2.2 Infectious Diseases and Epidemics

Ebola, HIV/AIDS, severe acute respiratory syndrome (SARS), the West Nile virus, and the COVID-19 coronavirus are all infectious diseases or epidemics of various kinds that have emerged in recent years. Other epidemics include the bovine spongiform encephalopathy (BSE or "mad-cow" disease) outbreak, which began in 1986, and the 2001 foot-and-mouth outbreak, both of which occurred in the United Kingdom. We give a brief description of four of these examples.

The Ebolavirus

This disease was named after the Ebola River in Zaire (now called the Democratic Republic of the Congo) where it was first discovered. Initial outbreaks occurred during 1976 in the Sudan and Zaire. It is characterized as a human disease that is actually a collection of five known viruses with different infectious rates; the Zaire Ebolavirus is the worst with a 90% fatality rate in some epidemics and an average fatality rate of 83% over 27 years. The Reston Ebolavirus, discovered in Reston, Virginia, seems not to affect humans, but is terrifyingly lethal to monkeys. Primarily a disease restricted to Africa, outbreaks of the Ebolavirus have also occurred in the United Kingdom and Russia. No approved vaccine or treatment for the Ebolavirus is currently available. An excellent, but chilling, account of the Ebolavirus can be found in the 1995 Random House book *The Hot Zone: A Terrifying True Story* by Richard Preston (Preston, 1995). In March 2014, the largest Ebola outbreak in history was reported in Guinea, Sierra Leone, Liberia, and a few other contries, where over 28,500 people were infected and over 11,000 died. Even though the virus is now under control (the World Health Organization declared the end of Ebola in Guinea on December 29, 2015, in Liberia on January 14, 2016, and in Sierra Leone on March 17, 2016), the risk of Ebola outbreaks still remains, and sporadic flare-ups have since occurred.

HIV/AIDS

This disease is a retrovirus that primarily infects the human immune system. The HIV virus, which causes AIDS, has a high mortality rate and spreads quickly. By 2010, approximately 34 million people were estimated to have HIV worldwide. HIV infection is transmitted primarily by unprotected sexual intercourse, contaminated blood transfusions, hypodermic needles, and mother-to-child pregnancy, delivery, or breastfeeding. It is believed to have originated in nonhuman primates from west-central Africa and transmitted to humans in the early twentieth century. Currently, there is no cure for HIV, only a cocktail of various medications that help to relieve the symptoms. The 1987 book entitled *And the Band Played On: Politics, People, and the AIDs Epidemic* by Randy Shilts is worth consulting for further details (Shilts, 1987).

The SARS Epidemic

This started in Guangdong Province, southern China, during November 2002. Early reports called it an "influenza outbreak" because of its initial symptoms. It is a viral respiratory illness caused by a coronavirus. The 2003 outbreak in Hong Kong had an estimated 8000 cases and 750 deaths (mainly in older patients). Transmission of the disease was mostly due to a few individuals, and outbreaks occurred in a hospital and a crowded apartment building where individual contacts were very high. Otherwise, contact patterns proved to be very heterogeneous and uneven. Travel warnings were issued by the World Health Organization and other agencies, and some schools and universities were closed or attendance was denied to students from SARS-hit regions. Although SARS spread worldwide to nearly 40 countries, it did not activate a global pandemic, and is considered to be a relatively rare disease (about 10% fatality rate). Success in containing the disease has been attributed to fast action by public health officials. In fact, the Center for Disease Control has stated that no known cases since 2004 have been reported anywhere in the world. That is indeed fortunate because there is currently no cure or protective vaccine for SARS that is safe for use in humans.

The COVID-19 Coronavirus

The epicenter of the COVID-19 coronavirus (a virus related to the SARS virus and designated as SARS-CoV-2) was the city of Wuhan in the province of Hubei, China, with a population of 11 million in 2018. The outbreak has been traced to October or November 2019. It is unclear how the outbreak occurred, with different possibilities proposed, including the following:

- An initial suggestion was that the virus was transmitted from bats to live or dead wild animals, such as mink, raccoon dogs, ferret badgers, or pangolins, that were sold illegally in the Hua'nan Seafood "wet" Market in Wuhan.
- Another possible explanation is that it originated from six miners who were infected while working in a bat-infested cave in Mojiang, Yunnan Province, Southern China in April 2012 and took sick with COVID-19-like symptoms (three of them died); viral samples were then taken to a laboratory at the Wuhan Institute of Virology (WIV) where it was leaked by an accidentally infected worker, an infected laboratory animal, or a technical mishap.
- Another claim is that the virus was man-made in the WIV, possibly by merging the genetic material of two bat coronaviruses (ZC45 and/or ZXC21) and making the result easier for the virus to latch onto human cells, but then was accidentally leaked out of the laboratory.

The first possibility has been dismissed because of the lack of evidence and the second possibility of a laboratory leak is thought to be a reasonable explanation, but has been rated as "extremely unlikely" by a March 2021 World Health Organization report. The third possibility has been rejected by Chinese authorities, but is the most realistic explanation because it fits the evidence (not a single intermediary mutation has been found, unlike the case with SARS), and is currently being taken more seriously. An international investigation to study this issue has since been launched.

Local political officials had identified the coronavirus by the end of December 2019 and mapped its genetic sequence by the start of January 2020. However, during the first half of January, officials repeatedly tried to prevent news of coronavirus infections getting out to the public, whilst also denying that human-to-human infection could occur. Celebration of the Chinese New Year around mid-January saw huge parties and banquets taking place in the city of Wuhan, and over five million residents of Wuhan were allowed to leave the city without any screening. All these mistakes led to the unregulated spread of the coronavirus. In late January, President Xi ordered the quarantining of Wuhan and three other cities, which affected over 20 million people.

Clusters of the disease have since appeared in over 220 countries outside mainland China. The World Health Organization declared COVID-19 to be a pandemic in early March 2020. The current worldwide number of infected cases (as of October 2021) is around 240 million, with over 4.8 million deaths and over 217 million recovered.

Each of these infectious diseases can be modeled using a percolation process, where individuals are represented as sites on a lattice, and bonds, which represent contacts between individuals, indicate connections along which a disease might spread. Each site is viewed as either *susceptible* to the disease or not, and each bond is open with a probability p representing the infectiousness of the disease; otherwise, the bond is closed with probability $1 - p$. Sites that are grouped into a connected cluster are regarded as infected. In the case of such viruses, each individual would be viewed as susceptible and percolation would mean bond percolation. (In contrast, because not

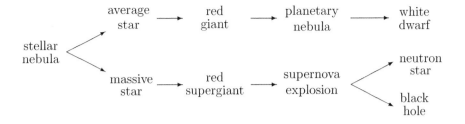

Figure 5.2 Flowchart of stellar evolution. The upper sequence deals with low- and medium-mass stars (including the Sun). The lower sequence deals with high-mass stars, which subdivides (after supernova) into a high-mass star (neutron star) and very high-mass star (black hole).

all computers are susceptible, computer viruses would more likely be modeled as site percolation.) An important question would be the determination of the distribution of infected cluster sizes (i.e., the number of sites reached by disease transmission along the bonds in the network).

5.2.3 Galactic Structure and Star Formation

The percolation process has been used to model star formation and the structure of spiral galaxies. The *phase transition* associated with percolation plays an important role in the stabilization and control of star formation. Several articles have appeared relating to percolation models of star formation, including computer simulations of those models. See, for example, Schulman and Seiden (1982, 1986) and Seiden and Schulman (1990). Movies of models of star formation based upon percolation theory were shown at the Smithsonian National Air and Space Museum (in Washington, D.C.) for many years.

There are *disk galaxies, elliptical galaxies, flocculent galaxies, spiral galaxies,* and *two-armed galaxies*. The most common type of galaxy is a *disk galaxy*, which is essentially a two-dimensional structure, the third dimension being so thin as to be almost non-existent. Disk galaxies contain stars of all ages, with young stars making up the spiral arms that initiate active star formation.

It takes an external force, such as a shock wave from a supernova explosion, to trigger star formation within a galactic disk, where the supernova is itself the product of an earlier (~10 million years ago) star formation event. There is no guarantee, however, that a supernova will lead to star formation in all neighboring regions; it depends upon many different factors. The star is completely destroyed by the supernova; most of its mass is returned to the interstellar medium, with only a neutron star or black hole remaining as evidence of the event. See **Figure 5.2**. Permeating the galaxy is a gas of atomic hydrogen. If a shock wave passes through the gas, a molecular cloud will probably form from the gas; within such clouds, stars are formed. The remainder of the cloud is dispersed into atomic hydrogen. The more massive stars go on to repeat the process, thereby providing a chain-reaction of star formation. So, a supernova explosion in one region of the galaxy has a good chance of giving rise to star formation in a neighboring region.

With this in mind, simulations of this star-formation process have been run with surprisingly good results. A disk galaxy may be idealized as follows.

1. Partition the galaxy into regions, where the size of a region depends upon how far a shock wave will have to travel to initiate the star-formation process.
2. Within each region, form a polar grid having C concentric rings, where the number depends upon the size of the galaxy being modeled.
3. Divide each ring by radially directed lines into a number of smaller, nearly square cells of roughly the same size. Think of a cell as a region of space whose size is on the order of a giant molecular cloud or galactic stellar cluster. A typical cell has six immediate neighboring cells, defined as those having contiguous boundaries, where two cells are in the same ring and two cells in each of the two adjacent rings, although five and seven neighbors are also possible.
4. Randomly seed a small number of cells with young star clusters.
5. With probability p, a region having an active star cluster at time t will create a new star cluster in its immediate neighboring cells at time $t + 1$.
6. After doing that for each young cluster, rotate each ring with velocity corresponding to that of the galaxy being modeled.
7. Repeat the process as many times as desired to simulate the evolution of a galaxy.

Simulations have used $C = 49$ rings and $p_c = 1/6$. The results are quite startling in that the simulations with p slightly bigger than p_c show excellent resemblance to actual galaxy structure.

5.3 Discrete Percolation

5.3.1 Lattices, Sites, and Bonds

We introduce notation and terminology by first discussing the d-dimensional infinite hypercubic lattice.

Definition 5.1 Let $\mathbb{Z} = \{\ldots, -1, 0, 1, \ldots\}$ be the set of all integers. In its most general form, percolation takes place on a d-dimensional infinite hypercubic lattice denoted by

$$\mathbb{L}^d = (\mathcal{V}, \mathcal{E}) = (\mathbb{Z}^d, \mathbb{E}^d), \tag{5.1}$$

where $\mathcal{V} = \mathbb{Z}^d$ is the set of *sites* (nodes or vertices) and $\mathcal{E} = \mathbb{E}^d$ (where $\mathbb{E} = \mathbb{Z} \times \mathbb{Z}$) is the set of all *bonds* (edges) between pairs of members of $\mathcal{V}$.

An important property of the lattice $\mathbb{L}^d$ is that it is translation invariant. We shall pay special attention to the two-dimensional infinite square lattice, $\mathbb{L}^2 = (\mathbb{Z}^2, \mathbb{E}^2)$.

Randomness is introduced into the system in the following way. Each bond is either *open* with probability $p \in [0, 1]$ or *blocked* (or *closed*) with probability $1 - p$, independently of all other bonds in the lattice. An open bond corresponds to the presence of an edge in a graph, whilst a blocked bond corresponds to the absence of an edge. The independence condition is a key feature of the binary-valued model. A *configuration* of $\mathbb{L}^d$ specifies which bonds are open and which are closed. In the lattice model, water (or any liquid) can flow along open bonds but is stopped from flowing through blocked bonds.

Bond percolation can be viewed as a collection of (iid) Bernoulli random variables, $X_e, e \in \mathbb{E}^d$, with parameter p, where

$$P_p\{X_e = 1\} = 1 - P_p\{X_e = 0\} = p, \tag{5.2}$$

for all $e \in \mathbb{E}^d$. The sample space is given by $\{0, 1\}^{\mathbb{E}^d}$. Because of the two possible states of the bonds ($X_e = 1$ indicates that bond e is open and $X_e = 0$ indicates that bond e is blocked or closed) and the independence condition, the model is often referred to as a *Bernoulli percolation model*. The probability P_p is the product probability measure on the configurations of open and closed bonds, and E_p denotes the corresponding expectation operator wrt P_p. The probability measure P_p is assumed to be invariant under both horizontal and vertical translation and under both horizontal and vertical axis reflection.

Definition 5.2 A *path* in the lattice is an alternating sequence of sites and bonds. An *open path* is a path all of whose bonds are open.

Definition 5.3 Two sites, $\mathbf{x}$ and $\mathbf{y}$, are *connected* (we write $\mathbf{x} \leftrightarrow \mathbf{y}$) if the site $\mathbf{y}$ can be reached from $\mathbf{x}$ by means of an open path from $\mathbf{x}$ to $\mathbf{y}$.

Definition 5.4 An *open cluster* of the site $\mathbf{x}$, denoted by $\mathcal{C}(\mathbf{x})$, is the collection of sites connected to $\mathbf{x}$ by open paths; it is maximally connected in the sense that it is not properly contained in any other subset of the set of sites.

If $\mathbf{0} = (0, 0, \ldots, 0)^\tau$ (with d zeroes) denotes the origin, then we write

$$\mathcal{C} = \mathcal{C}(\mathbf{0}) = \{\mathbf{x} \in \mathbb{Z}^d : \mathbf{0} \leftrightarrow \mathbf{x}\}. \tag{5.3}$$

An open cluster can be viewed as analogous to a component in random graph theory. Most of our attention will be focused on $\mathcal{C}$ or functions of $\mathcal{C}$.

Site percolation is similar to bond percolation but instead each site is designated as *open* (with probability p) or *closed* (with probability $1 - p$), independently of all other sites. Open sites are often called "occupied" and closed sites "unoccupied" or "empty." Site percolation is a generalization of bond percolation: every bond process can be translated to be a site process on a different lattice, but the converse is not necessarily true.

There are many similarities between random graph theory and percolation theory. The biggest difference between these two types of theories is that percolation occurs on a fixed *infinite* lattice where bonds can appear only between a site and its lattice neighbors. In the case of a finite rectangular lattice, each site can have two, three, or four nearest neighbors, depending upon whether that site is at the corner, boundary, or interior, respectively, of a finite lattice; see **Figure 5.3**. In a *finite* random graph, any node can be attached to any other node in the graph through an edge. Because of the differences in the underlying physical structures of the two scenarios, showing that similar properties hold does not mean that the methods used will be similar. An exception to this description is "continuum percolation," which is percolation on a lattice-free structure; see Section 6.4.

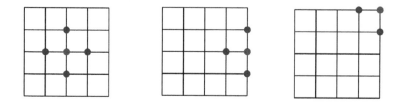

Figure 5.3 Nearest neighbors of a site at three locations on a finite lattice. Site in red, nearest neighbors in blue. Left: Interior site (four nearest neighbors). Center: Boundary point (three nearest neighbors). Right: Corner point (two nearest neighbors).

5.3.2 Percolation Probability

We are interested in finding the probability that a given site is contained in an infinite open cluster. Without loss of generality, we can take the site to be the origin because of the translation invariance of the lattice $\mathbb{L}^d$ and of the probability measure P_p. As the number of bonds tends to infinity, we seek the probability that the origin is part of an infinite, connected, open set in $\mathbb{L}^d$:

Definition 5.5 The *percolation probability* is defined as

$$\theta(p) = P_p\{\mathbf{0} \leftrightarrow \infty\} = P_p\{|\mathcal{C}| = \infty\}, \tag{5.4}$$

where $|\mathcal{C}|$ is the number of sites contained in $\mathcal{C}$ and

$$P_p\{|\mathcal{C}| = \infty\} = \lim_{k \to \infty} P_p\{|\mathcal{C}| \geq k\}. \tag{5.5}$$

Clearly, either $\theta(p) = 0$ or $\theta(p) > 0$.

In one dimension (i.e., $d = 1$), there obviously cannot be an infinite cluster unless $p = 1$. This follows because for any cluster $\mathcal{C}_R$ to the right of the origin, $P_p\{|\mathcal{C}_R| \geq k\} = p^k$, and similarly with any cluster $\mathcal{C}_L$ to the left of the origin. Thus, we focus on $d \geq 2$.

5.3.3 Phase Transition

When $p = 0$, every bond will be blocked in the lattice and so there cannot be any open path; hence, $\theta(0) = 0$. When $p = 1$, every bond will be open so that there exists an infinite open path; hence, $\theta(1) = 1$. So, what happens when $0 < p < 1$? When p is small and nonzero, there will be few open bonds, and there will exist some open paths, but they will be short. As p gets larger, more bonds become open and the possibility of long open paths emerges, especially ones starting at the origin. Thus, if $p < p'$, then $\mathcal{C}_p \subset \mathcal{C}_{p'}$ and, hence, $\theta(p) < \theta(p')$, which shows that $\theta(p)$ is a nondecreasing function of p. The general shape of the probability $\theta(p)$ as a function of p is conjectured to look like **Figure 5.4**.

The most important property of percolation theory is that there exists a nontrivial "critical" probability $0 < p_c < 1$ in two dimensions or higher (and for various types of lattices) that controls when percolation occurs and when it does not occur (Broadbent

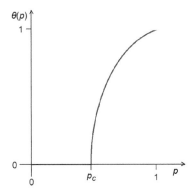

Figure 5.4 Graph sketch of the percolation probability $\theta(p)$ as a function of p. When $p < p_c$, percolation does not occur, whilst if $p > p_c$, percolation occurs.

and Hammersley, 1957; Hammersley, 1959).[2] Because $\theta(p)$ is nondecreasing, we have the following definition:

Definition 5.6 The *critical probability* is defined as

$$p_c = \sup\{p : \theta(p) = 0\} = \inf\{p : \theta(p) > 0\}, \tag{5.6}$$

which is the probability that $\theta(p)$ changes from being equal to zero to being positive.

See **Exercise 5.1**. As a result, the percolation model, like certain aspects of random graph theory, exhibits a *phase transition* showing abrupt changes in the behavior of the process as p changes. It has two phases:

- a *subcritical phase* ($p < p_c$), where $\theta(p) = 0$, so that each open cluster $\mathcal{C}$ is, almost surely, finite;
- a *supercritical phase* ($p > p_c$), where $\theta(p) > 0$, so that there is a nonzero probability that $\mathcal{C}$ is an infinite open cluster;

and

- a *critical point* at $p = p_c$.

When $\theta(p) > 0$, it is customary to say that "percolation occurs," whereas when $\theta(p) = 0$, we say that "percolation does not occur." The subcritical and supercritical phases have been heavily studied for two-dimensional lattices and are now well understood. We state the main results regarding these phases without proof; see Grimmett (1989) for details.

We note that for d-dimensional lattices, $d > 2$, it is still an open problem to show that either $\theta(p_c) = 0$ or $\theta(p_c) > 0$. What we do know is that $\theta(p_c) = 0$ for two-dimensional lattices (Harris, 1960; Kesten, 1980) and for d-dimensional lattices when $d \geq 19$ (Hara and Slade, 1990). The case of $3 \leq d \leq 18$ has not yet been resolved.

---

[2]The *critical probability* p_c is often written as p_H to honor Hammersley.

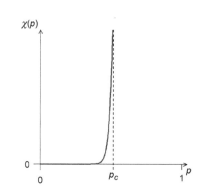

Figure 5.5 Graph sketch of the mean cluster size $\chi(p)$ as a function of p.

5.3.4 Mean Cluster Size

It is possible that the expected size of an open cluster will not be finite, where "size" refers to the number of open sites contained in the cluster.

Definition 5.7 The expected size of an open cluster $\mathcal{C}$ at site v is referred to as the *mean cluster size* and is denoted by

$$\chi_v(p) = \mathrm{E}_p\{|\mathcal{C}(v)|\} = \sum_{k=1}^{\infty} k \mathrm{P}_p\{|\mathcal{C}(v)| = k\}. \tag{5.7}$$

If the site of interest is the origin $\mathbf{0}$, then we write $\chi(p) = \mathrm{E}_p\{|\mathcal{C}|\}$. By translation invariance of the percolation process, $\chi(p) = \chi_v(p)$ for all sites v, so that $\chi(p)$ does not depend upon any site v. As p increases, so does $\chi(p)$. Furthermore

$$\chi(p) = \sum_{k=1}^{\infty} k \mathrm{P}_p\{|\mathcal{C}| = k\} + \infty \cdot \mathrm{P}_p\{|\mathcal{C}| = \infty\}. \tag{5.8}$$

If $\theta(p) = \mathrm{P}_p\{|\mathcal{C}| = \infty\} > 0$, then $\chi(p) = \infty$, which happens in the supercritical phase when $p > p_c$. On the other hand, in the subcritical phase when $p < p_c$, we have that $\theta(p) = 0$ and $\chi(p) < \infty$ (Aizenman and Newman, 1986). See **Figure 5.5**.

5.4 Subcritical Phase

Consider, first, the subcritical phase, $p < p_c$, in which there exist only small open clusters of sites. There are two questions of particular interest in the subcritical phase:

1. *What is the probability that there exists an open path joining two sites that are far apart in the lattice?*
2. *What can we say about the cluster size distribution?*

This section deals with these two questions.

5.4.1 Long Open Paths

The first result of this kind was given by Hammersley (1957). A site in $\mathbb{Z}^d$ can be represented by $\mathbf{x} = (x_1, \ldots, x_d)^\tau$. Let $\|\mathbf{x}\| = \max\{|x_i| : 1 \leq i \leq d\}$ be the distance between the site $\mathbf{x}$ and the origin.

Definition 5.8 Let B_k be the box,

$$B_k = [-k, k]^d = \{\mathbf{x} \in \mathbb{Z}^d : \|\mathbf{x}\| \leq k\}, \tag{5.9}$$

centered at the origin $\mathbf{0}$.

The primary use of the box is to approximate the infinite lattice $\mathbb{L}^d$, where the open bonds of the box are copied from the open bonds that are specified in $\mathbb{E}^d$.

Definition 5.9 The *external boundary* (or *surface*) ∂B_k of the box B_k is the set of sites that are not in B_k but are adjacent to some site inside B_k.

In other words, $\partial B_k = \{v \in \mathcal{V} : v \notin B_k, v \sim v' \text{ for some } v' \in B_k\}$.

Let $\mathbf{0} \leftrightarrow \partial B_k$ denote an open path from the origin $\mathbf{0}$ to a site in ∂B_k. Hammersley showed the following.

Theorem 5.1 (Hammersley, 1957) *If $\chi(p) < \infty$, then there exists $\eta(p) > 0$ such that, for all $k \geq 1$,*

$$\mathbf{P}_p\{\mathbf{0} \leftrightarrow \partial B_k\} \leq e^{-k\eta(p)}. \tag{5.10}$$

In other words, the probability of an open path joining the origin to a site on ∂B_k decays exponentially as $k \to \infty$. This result can be proved using the BK inequality (van den Berg and Kesten, 1985), although Hammersley had earlier employed a different argument involving branching processes. The condition $\chi(p) < \infty$ was later replaced by the less stringent condition $p < p_c$ to get a stronger form of Hammersley's result (Menshikov, Molchanov, and Sidorenko, 1986; Aizenman and Barsky, 1987; Menshikov, 1987):

Theorem 5.2 *If $p < p_c$, then there exists $\psi(p) > 0$ such that, for all $k \geq 1$,*

$$\mathbf{P}_p\{\mathbf{0} \leftrightarrow \partial B_k\} \leq e^{-k\psi(p)}. \tag{5.11}$$

The function $\psi(p)$ in (5.11) has the following properties (Grimmett, 1999, Theorem 6.14):

1. ψ is continuous and non-increasing in $(0, 1]$;
2. ψ is strictly decreasing and positive on $(0, p_c)$;
3. $\psi(p) \to \infty$ as $p \downarrow 0$;
4. $\psi(p_c) = 0$.

See **Figure 5.6** for a graph sketch of $\psi(p)$.

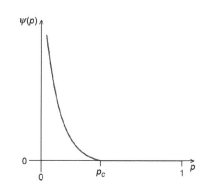

Figure 5.6 Graph sketch of $\psi(p)$ as a function of p.

There are also stronger forms of the above result. Because much of the behavior of the probability $P_p\{0 \leftrightarrow \partial B_k\}$ is unknown, current knowledge indicates that the probability behaves like an exponential function times k to some power; that is, approximately $Ck^a e^{-k\psi(p)}$, where $a = a(d)$ and C is some strictly positive constant.

5.4.2 Correlation Length

When p approaches p_c from below, $\psi(p)$ will be small (i.e., $\psi(p) \to 0$ as $p \uparrow p_c$). Hence, for p close to p_c, there is a scaling problem. One can show the following (Grimmett, 1999, p. 127):

Theorem 5.3 *There exists $\xi(p)$ such that, for all k and $0 < p < \frac{1}{2}$:*

$$P_p\{0 \leftrightarrow \partial B_k\} \approx e^{-k/\xi(p)}, \quad \text{as } k \to \infty. \tag{5.12}$$

It turns out that $\xi(p)$ is the reciprocal of $\psi(p)$: $\xi(p) = 1/\psi(p)$. Because $\psi(p_c) = 0$, it follows that $\xi(p_c) = \infty$. Other properties of $\xi(p)$ include:

1. $\xi(p)$ is continuous and strictly increasing on $(0, p_c)$;
2. $\xi(p) \to 0$ as $p \downarrow 0$;
3. $\xi(p) \to \infty$ as $p \uparrow p_c$;
4. $\xi(p) \leq \chi(p)$ if $p < p_c$.

See **Figure 5.7** for a sketch of the graph of $\xi(p)$. The quantity $\xi(p)$, $p < p_c = \frac{1}{2}$, is referred to as the *correlation length* of the percolation process (Chayes, Chayes, and Fröhlich, 1985). Note that $\xi(p)$ is only defined for $p \leq p_c$. Because sites are regarded as "correlated" if they are connected (i.e., if they belong to the same finite open cluster), $\xi(p)$ can be viewed as the average distance over which two sites are connected.

Let I_A be the indicator of the set A (i.e., $I_A = 1$ if A is true, and 0 if A is false).

Definition 5.10 The *mean size of a finite open cluster at the origin* is defined by

$$\chi^f(p) = E_p\{|C| I_{[|C| < \infty]}\}. \tag{5.13}$$

If $p < p_c$, then $\chi^f(p) = \chi(p)$. However, if $p > p_c$, this may not be true.

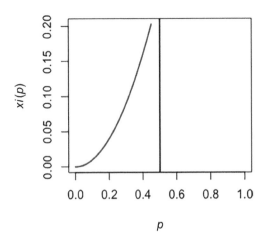

Figure 5.7 Graph sketch of the correlation length $\xi(p)$ of the percolation process as a function of p.

Let $\mathbf{y} = (y_1, \ldots, y_d)^\tau \in C$ and let $|\mathbf{y}| = \max_i |y_i|$.

Definition 5.11 The *correlation length* is defined as

$$\xi(p) = \left\{ \frac{1}{\chi^f(p)} \sum_{\mathbf{y}} |\mathbf{y}|^2 \cdot P_p\{\mathbf{y} \in C, |C| < \infty\} \right\}^{1/2}. \tag{5.14}$$

$\xi(p)$ can also be characterized as the "typical" radius of a finite cluster. Other definitions of correlation length have been proposed.

The exponential term in (5.12) becomes large only when k is of the order of $\xi(p)$, and when k exceeds that value, the probability is small, so that there exists an open path between points a distance k apart.

5.4.3 Cluster Size Distribution

In the subcritical phase, the cluster size distribution decays exponentially.

Theorem 5.4 (Kesten, 1981) *If* $p < p_c$, *there exists* $\phi(p) > 0$ *such that, for all* $k \geq 1$,

$$P_p\{|C| \geq k\} \leq e^{-k\phi(p)}. \tag{5.15}$$

Thus,

$$\chi(p) = \sum_{k=1}^{\infty} k P_p\{|C| = k\} = \sum_{k=1}^{\infty} P_p\{|C| \geq k\} < \infty. \tag{5.16}$$

This shows that the distribution of $|C|$ has exponential decay. A further result is the following.

Theorem 5.5 (Kesten, 1981; Aizenman and Newman, 1984) *If* $p < p_c$, *then* $\chi(p) < \infty$, *and if* $k > [\chi(p)]^2$, *then*

119

$$P_p\{|\mathcal{C}| \geq k\} \leq 2\exp\left\{-\frac{1}{2}k[\chi(p)]^{-2}\right\}. \tag{5.17}$$

See also Grimmett (1999, Section 6.3).

5.5 Supercritical Phase

In the supercritical phase, when $p > p_c$, the probability function $\theta(p)$ is continuous and infinitely differentiable (in any dimension d), which follows from the uniqueness of the infinite open cluster (van den Berg and Keane, 1984). All other clusters are, almost surely, small and finite, with large, finite clusters highly unlikely to appear.

With that in mind, suppose we ignore the single, infinite, open cluster and consider only the finite open clusters. Specifically, we are interested in the cluster size distribution, $P_p\{|\mathcal{C}| = k\}$, and the tail probability of the cluster size distribution, $P_p\{k \leq |\mathcal{C}| < \infty\}$.

Theorem 5.6 *For the d-dimensional lattice $\mathbb{L}^d$, if $p_c < p < 1$, there exist $\gamma = \gamma(p) < \infty$ and $\eta = \eta(p) > 0$ such that, for all k,*

$$e^{-\gamma k^{(d-1)/d}} \leq P_p\{|\mathcal{C}| = k\} \leq e^{-\eta k^{(d-1)/d}}. \tag{5.18}$$

Because the rate of decay to zero is slower than exponential decay, we say that the cluster size distribution decays *subexponentially*. Furthermore, there exists $\beta = \beta(p) > 0$ such that, for all k,

$$P_p\{k \leq |\mathcal{C}| < \infty\} \leq e^{-\beta k^{(d-1)/d}}. \tag{5.19}$$

5.6 Critical Point

Most of the focus in percolation theory has been on infinite d-dimensional lattices, $d \geq 2$, where it is of interest to know how large p has to be before a particular site is contained in an infinite open cluster.

Definition 5.12 The *critical point* p_c (also known as the *percolation threshold* or *singularity*) is the probability at which there is a sudden change as p moves past p_c: below p_c all open clusters are, almost surely, finite (and $\theta(p) = 0$), whilst above p_c, there is an infinite open cluster (and $\theta(p) > 0$), which is almost surely unique.

The unique critical point p_c at which an infinite open cluster emerges should be viewed as analogous to the critical probability, $p_c(N) = 1/N$, in random graph theory at which a unique giant component suddenly appears.

Early attention centered around studying p_c and deriving its value for different types of lattices. Although the problems were easy to state, solving them proved to be very difficult. For example, the exact determination of the critical point p_c for the bond-percolation model on the two-dimensional square lattice was not obtained until 1980. Hammersley (1957) discovered that $0 < p_c < 1$, and, more importantly, he gave lower and upper bounds for p_c in the form

$$\frac{1}{\lambda} \leq p_c \leq 1 - \frac{1}{\lambda}, \tag{5.20}$$

Table 5.1 Bond-percolation thresholds for various
two-dimensional lattices

Lattice type	Threshold p_c
Square	0.5000
Triangular	$2\sin(\pi/18) = 0.3473$
Honeycomb	$1 - 2\sin(\pi/18) = 0.6527$

where λ is referred to as the *connective constant* of the lattice $\mathbb{L}^2$. It was then shown that $\lambda \approx 2.639$ (Hiley and Sykes, 1961), yielding a lower bound of $\frac{1}{\lambda} \approx 0.379$. Harris (1960) improved on the lower bound by showing that $p_c \geq \frac{1}{2}$. Then, Kesten (1980) showed, in a remarkable paper, that the bond-percolation threshold was $p_c = \frac{1}{2}$. The key quantity in Kesten's argument involved another critical probability:[3]

$$p_T = \sup\{p : \chi(p) < \infty\} = \inf\{p : \chi(p) = \infty\}, \tag{5.21}$$

at which $\chi(p)$ suddenly passes from having a finite value to having an infinite value. It is clear that $p_T \leq p_c$ because, from (5.8), $\chi(p) \geq \theta(p) \cdot \infty$, which is infinite for $p > p_c$. A previous result (Russo, 1978; Seymore and Welsh, 1978) showed that $p_T + p_c = 1$ and $p_T \leq \frac{1}{2}$ for the square lattice $\mathbb{L}^2$. Kesten then showed that $p_T \geq \frac{1}{2}$, so that $p_c \leq \frac{1}{2}$. Putting these results together, including Harris's result, yielded, for $d = 2$, that

$$p_c = p_T = \frac{1}{2}. \tag{5.22}$$

Exact values of p_c for higher-dimensional square lattices, such as the cubic lattice on $\mathbb{Z}^3$, are currently still unknown. However, we do know that $p_c = p_T$ for all d (Menshikov, Molchanov, and Sidorenko, 1986; Aizenman and Barsky, 1987).

Some examples of bond-perculation thresholds are listed in **Table 5.1**. The results on *triangular* and *honeycomb* (also known as *hexagonal*) lattices were derived by Weirman (1981). Not much is known about p_c in dimensions higher than 2. We do know that $p_c < \frac{1}{2}$ for $d = 3$ (Kesten, 1982), but not much more. There is not much known regarding the question of whether an infinite open cluster exists, but we do know that if it does exist, it will be, almost surely, unique.

5.7 Power-Law Conjectures

Many physical systems are modeled by a set of *power laws*. We have seen, for example, that if $p < p_c$, $P_p\{|\mathcal{C}| \geq k\}$ decays exponentially in k, for any dimension, and this probability has been conjectured to decay as a power law with a given power. In this section, we describe the different types of critical exponents that provide the powers of the conjectured power laws.

5.7.1 Critical Exponents

In percolation models, a huge amount of work has been carried out in the theoretical physics literature to explain the behavior of percolation near its critical point

---

[3]The subscript "T" on p_T is used to honor H.N.V. Temperley.

$p = p_c$. The quantities $\theta(p)$, $\chi(p)$, $\chi^f(p)$, $\xi(p)$, and $\kappa'''(p)$ have been conjectured to behave like powers of $|p - p_c|$ when p is close to p_c. Specifically, it is believed that there exist *critical exponents* β, γ, γ', and ν, all believed to be strictly positive, and $-1 < \alpha < 0$, such that[4]

$$\theta(p) \approx (p - p_c)^\beta, \quad \text{as } p \downarrow p_c, \tag{5.23}$$

$$\chi(p) \approx (p - p_c)^{-\gamma}, \quad \text{as } p \uparrow p_c, \tag{5.24}$$

$$\chi^f(p) \approx (p - p_c)^{-\gamma'}, \quad \text{as } p \downarrow p_c, \tag{5.25}$$

$$\xi(p) \approx |p - p_c|^{-\nu}, \quad \text{as } p \uparrow p_c, \tag{5.26}$$

$$\kappa'''(p) \approx |p - p_c|^{-1-\alpha}, \quad \text{as } p \to p_c. \tag{5.27}$$

The values of $\beta, \gamma, \gamma', \nu$, and α are thought to be *universal*; that is, they depend only on the dimension d and not on the geometric properties of the type of graph $\mathcal{G}$. This is in contrast to p_c, which definitely does depend upon the lattice structure of $\mathcal{G}$. It has also been conjectured that $\gamma' = \gamma$. Because $\chi^f(p) = \chi(p)$ for $p < p_c$, (5.24) and (5.25) boil down to a conjecture that $\chi^f(p) \approx |p - p_c|^\gamma$ for $p \to p_c$.

Universality is an important feature in the theory of phase transitions. It is believed that all systems in nature are members of one of a small number of *universality classes*, where all problems in a given class have identical critical behavior (e.g., critical exponents), whilst different classes have different critical behavior. Unfortunately, universality has been rigorously proved in only a few cases.

Other conjectures for power laws include the following:

$$P_{p_c}\{|\mathcal{C}| = k\} \approx k^{-1-\frac{1}{\delta}}, \quad \text{as } k \to \infty, \tag{5.28}$$

$$P_{p_c}\{k \le |\mathcal{C}| < \infty\} \approx k^{-\frac{1}{\delta}}, \quad \text{as } k \to \infty, \tag{5.29}$$

$$P_{p_c}\{0 \leftrightarrow \partial B_k\} \approx k^{-\frac{1}{\delta r}}, \quad \text{as } k \to \infty, \tag{5.30}$$

$$P_{p_c}\{\text{rad}(\mathcal{C}) = k\} \approx k^{-1-\frac{1}{\rho}}, \quad \text{as } k \to \infty, \tag{5.31}$$

$$\frac{E_p\{|\mathcal{C}|^{k+1}I_{[|\mathcal{C}|<\infty]}\}}{E_p\{|\mathcal{C}|^k I_{[|\mathcal{C}|<\infty]}\}} \approx |p - p_c|^{-\Delta}, \quad \text{as } p \to p_c, \ k \ge 1, \tag{5.32}$$

$$P_{p_c}\{\mathbf{x} \in \mathcal{C}\} \approx \|\mathbf{x}\|^{2-d-\eta}, \quad \text{as } \|\mathbf{x}\| \to \infty, \tag{5.33}$$

where $\delta, \delta_r, \Delta, \rho$, and η are each believed to be strictly positive and are also regarded as universal critical exponents. In (5.33), $\|\mathbf{x}\|$ is the distance from the origin $\mathbf{0}$ to the node $\mathbf{x}$ in the lattice, and in (5.31), $\text{rad}(\mathcal{C}) = \max_{\mathbf{x} \in \mathcal{C}} \|\mathbf{x}\|$ is the *radius* of $\mathcal{C}$. The exponent Δ is often referred to as the *gap exponent*.

5.7.2 Scaling Relations

Using experimental data and nonrigorous arguments, physicists (e.g., Michael E. Fisher, John W. Essam) conjectured the existence of the above critical exponents.

---

[4]The logarithmic relationship $a(p) \approx [b(p)]^\alpha$ means that $\log a(p) / \log b(p) \to \alpha$ as $p \downarrow p_c$ or as $p \uparrow p_c$, depending upon context.

They further predicted that the critical exponents were not independent of each other, that the exponents possess the following *scaling relations*:

$$2 - \alpha = \gamma + 2\beta = \beta(\delta + 1) = \Delta + \beta, \ \gamma = \nu(2 - \eta). \tag{5.34}$$

From these relations, we can see that $\Delta = \beta\delta = \beta + \gamma$ holds. See, for example, Fisher (1998) for historical remarks and references. Furthermore, two *hyperscaling relations*, which relate the critical exponents to the dimension d, are conjectured to exist,

$$d \cdot \rho = \delta + 1, \ d \cdot \nu = 2 - \alpha, \tag{5.35}$$

and they are thought to hold for $d \leq 6$. From these relationships, it is not difficult to show that

$$\alpha = 2 - d \cdot \nu, \ \beta = \frac{d \cdot \nu}{\delta + 1}, \ \gamma = d \cdot \nu \left(\frac{\delta - 1}{\delta + 1}\right), \tag{5.36}$$

$$\eta = 2 - d \left(\frac{\delta - 1}{\delta + 1}\right), \ \Delta = d \cdot \nu \left(\frac{\delta}{\delta + 1}\right), \tag{5.37}$$

and these exponents depend only on d. For example, when $d = 2$, $\eta = \frac{4}{\delta+1}$. Some of these identities are due to Rushbrook, Widom, and Josephson. Furthermore, (5.29) and (5.30) are equivalent through the relation $2\delta_r = \delta + 1$.

Theoretical physicists used a combination of Coulomb gas methods, conformal field theory, and quantum gravity arguments to predict the exact values of these critical exponents for two-dimensional lattices or when d is very large. For $d = 2$, such predictions are given by

$$\alpha = -\frac{2}{3}, \ \beta = \frac{5}{36}, \ \gamma = \frac{43}{18}, \ \delta = \frac{91}{5}, \tag{5.38}$$

$$\eta = \frac{5}{24}, \ \nu = \frac{4}{3}, \ \rho = \frac{48}{5}, \ \Delta = \frac{91}{36} \tag{5.39}$$

(Nienhuis, Riedel, and Schick, 1980). The existence of some of these critical exponents and the conjectures as to their values have been proved to be correct for the two-dimensional triangular lattice, but have not been rigorously proved for other lattices (Smirnov and Werner, 2001).

The conjectures assume that the above quantities have asymptotic expansions for $p - p_c$ near zero and for large k, which are then used to derive the scaling and hyperscaling relations and form the core of what is known as *scaling theory*.

5.7.3 The Renormalization (Semi)Group

The *renormalization group* (commonly abbreviated as RG) is a powerful theoretical tool that is used to establish scale invariance under a group (actually, a semigroup) of transformations. RG reduces the scale of the percolation problem by techniques known as "coarse graining" and "rescaling." The RG method seeks to "divide and conquer" by taking a large problem, separating it into a collection of smaller, more tractable problems, and solving each in turn. The most important achievement of RG theory was to derive good approximations to the critical exponents (Fisher, 1998).

Some History

Starting in the late 1940s, renormalization theory was developed to solve problems in quantum field theory by means other than analytical techniques based upon mean field theory (MFT). In fact, MFT solutions were found to be completely wrong near the critical points in two and three dimensions, and disagreed with existing experimental results. Indeed, Lars Onsager's famous exact solution in 1944 to the two-dimensional Ising model flatly contradicted MFT.[5] By this time, even though renormalization had become accepted as a useful tool, physicists still had trouble making sense of it.

The 1950s saw attempts at demystifying renormalization theory and some success was achieved in this direction. In 1953, Stückelberg and Petermann gave a detailed formulation of the "renormalization group" that they had introduced in 1951; this was also independently discovered by Murray Gell-Mann[6] and Francis E. Low in 1954, who presented a more general formulation of the renormalization group. Further work in this area appeared by Nikolai N. Bogoliubov and Dmitry V. Shirkov in a series of articles in the late 1950s.

By the early 1960s, physicists, such as Leo P. Kadanoff, Benjamin Widom, Michael E. Fisher, and Cyril Domb, were studying, amongst other things, power laws at or near critical points, how to calculate values for the critical exponents, why and how the MFT-determined predictions were erroneous, and providing a derivation of the scaling relations.

Then, in 1966, Kadanoff introduced a radical new formulation of renormalization theory together with an heuristic derivation of the scaling relations, especially for the Ising model (Kadanoff, 1966).

Kadanoff's construction. Consider a square lattice in which the sites are spaced a distance of unity apart (the "lattice constant") and where the dimensionality is $d \geq 2$. Kadanoff's method consisted of the following three steps:

1. *Block formation.* Fix $b > 0$. Partition the lattice into disjoint blocks[7] each having b^d sites.
2. *Coarse graining.*[8] Associate with each block a *super-site* that, in effect, replaces the block of original sites (i.e., coarse graining) while creating a specific block characterization based upon relevant information extracted from that block. If the site values are binary (or even discrete), the principle of *majority rule* is followed whereby the most frequently occurring value in each block is taken as the value for the super-site. If the site values are continuous, then averaging the values of the sites in the block would be appropriate. This step creates a new lattice whose spacing is b times larger than the original lattice.
3. *Rescaling.* Rescale the new lattice of super-sites by dividing each dimension of the original lattice by b so that, under scale invariance, the new lattice would appear to be identical in its properties to the original lattice.

---

[5]For other work, Onsager was awarded the 1968 Nobel Prize for Chemistry.
[6]For other work, Gell-Mann was awarded the 1969 Nobel Prize for Physics.
[7]Usually referred to as *Kadanoff blocks*.
[8]This operation is very similar to the statistical idea of smoothing a function.

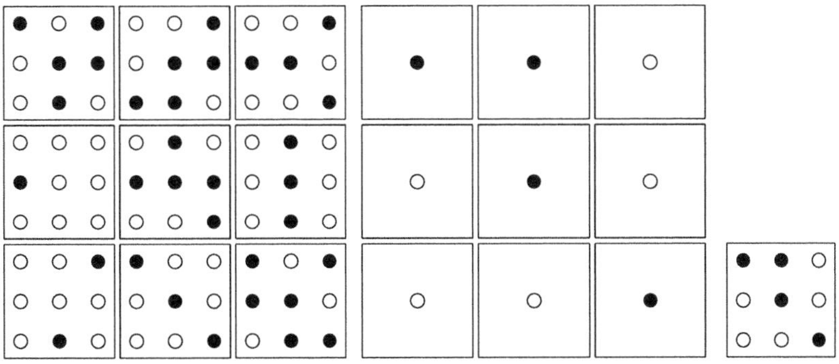

Figure 5.8 Kadanoff's three-step renormalization group transformation for a square 9×9 lattice. Left: Block formation step, where $b = 3$ and $d = 2$, and nodes are open (black circles) or closed (white circles). Center: Course graining step, where the single node in each block is a super-site, which is open or closed using the principle of majority rule. Right: Rescaling step, where the lattice from the second step is rescaled by the factor b.

These three steps comprise the *renormalization group transformation*. See **Figure 5.8.** Essentially, this transformation smooths out the variations in site occupancy/non-occupancy that are smaller than the block size. Kadanoff recommended that b should be large but smaller than the correlation length ξ. Kadanoff's incomplete and intuitive proposal was very appealing at first, but interest waned due to severe technical difficulties.

Wilson's breakthrough. For five years, no-one put the "finishing touches" to Kadanoff's work. Then, in 1971, the big breakthrough came when Kenneth G. Wilson, who had been struggling with the problem for several years, combined Kadanoff's heuristic, vague idea with the work of Gell-Mann and Low to produce a precise, workable program. Wilson formally developed the (real-space) renormalization group and applied it to problems in statistical physics involving critical phenomena and, later, percolation. His work was an attempt to justify the scaling relations and to calculate the critical exponents that entered into those relations, and he succeeded brilliantly. Wilson's modifications included iterating Kadanoff's block renormalization process until the system leads to a number of "fixed points," each of which could be interpreted as a different physical theory.[9]

Worked Example

Consider the following example of site percolation on the triangular lattice. A triangular lattice has its sites organized into triangles each having three sites, where each site is a member of exactly one triangle. Suppose the sites of this triangular lattice are each either occupied (with probability p_0) or unoccupied (with probability $1 - p_0$). Renormalize each triangle of sites (i.e., $b^2 = 3$) into a single *super-site*, which we position in the center of the triangle. These super-sites are also laid out in the form of a triangular lattice. See **Figure 5.9.**

---

[9]For his extensive and important theory of phase transitions and his computational breakthroughs in calculating the critical exponents, Wilson was awarded the 1982 Nobel Prize in Physics (Wilson, 1983). Fisher, Kadanoff, and Wilson were awarded the 1980 Wolf Prize in Physics for their work on phase transitions.

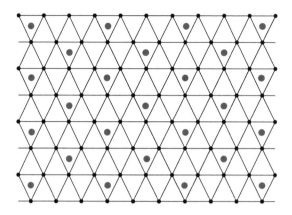

Figure 5.9 Worked Example. Site percolation on the triangular lattice. Each node (black dot) is surrounded by six triangles. The red dots are the super-sites, each representing three nodes of a triangle. The super-sites are occupied or unoccupied depending upon majority rule.

Using the principle of "majority rule," a super-site is designated as *occupied* iff at least two of the three sites in the super-site are occupied. Super-sites so formed are not independent of each other; this is because there is a dependence of the super-site on the sites that made up the block in the original lattice so that information is lost. However, if we assume that the blocks are similar to each other, then independence should be a reasonable approximation to make for this process.

In our example, there are eight possible configurations of three nodes, each of which can be occupied (white dot) or unoccupied (black dot). Of those, four configurations have two or three unoccupied sites in a super-site of three. These possible super-sites are as follows:

In each of these four configurations, the majority rule says that the super-site is unoccupied. Those configurations with zero or one unoccupied node have a super-site that is regarded as occupied.

Hence, the binomial probability that a super-site is occupied is

$$p_1 = p_0^3 + 3p_0^2(1 - p_0). \tag{5.40}$$

Now, iterate this procedure n times by grouping each set of three super-sites at level $n - 1$ into a new super-site at level n. Using the same rules as before, the probability that a level-n cluster is occupied is

$$p' = p^3 + 3p^2(1 - p) = 3p^2 - 2p^3 = R_3(p), \tag{5.41}$$

where $p' = p_n$ and $p = p_{n-1}$. **Figure 5.10** displays the S-shaped curve of the function $R_3(p) = 3p^2 - 2p^3$ and the line $R_3(p) = p$. The intersection of the line and the curve yields the fixed points.

The function $R_b(p)$ is called the *renormalization group transformation* for a super-site arrangement of b sites, and has the property that $R_b(R_b(p)) = R_{b^2}(p)$. It actually forms a semigroup of operators because there is no inverse function R_b^{-1}. What this

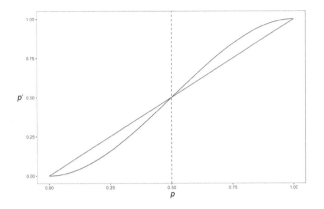

Figure 5.10 The blue curve displays the S-shaped function $R_3(p) = 3p^2 - 2p^3$ as a function of $p \in [0, 1]$ and the straight line $p' = R_3(p) = p$. The fixed point $p^* = \frac{1}{2}$ is where the line and the curve intersect.

means is that although one can take a block of b^d sites and create a super-site, the occupancy configuration of those original sites cannot be recovered because the information from those sites is now missing. In other words, one cannot return to the original lattice only knowing the lattice of super-sites. So, the use of the term "group" is a misnomer.

Definition 5.13 A *fixed point* p^* for $R_b(p)$ is a value of p such that

$$R_b(p^*) = p^*. \tag{5.42}$$

Worked Example (continued)
In the above example, (5.42) becomes $2p^3 - 3p^2 + p = 0$, which can be factored into $p(2p^2 - 3p + 1) = 0$. Thus, either $p = 0$ or $2p^2 - 3p + 1 = (2p - 1)(p - 1) = 0$, so that there are three fixed points, 0, $\frac{1}{2}$, and 1. The *trivial* fixed points $p = 0$ and 1 are *stable* because repeated transformations of points close to either 0 or 1 will move p' closer and closer to 0 or 1, respectively. For example, if we take $p = 0.9$, then $p' = 0.972$, which is closer to 1; if we take $p = 0.1$, then $p' = 0.028$, which is closer to 0. The *nontrivial* fixed point $p = \frac{1}{2}$ is *unstable* because repeated transformations of points close to $\frac{1}{2}$ will move p' further away from $\frac{1}{2}$. Taking $p = 0.45$ yields $p' = 0.42525$, while $p = 0.55$ yields $p' = 0.57475$, both of which are further away from $\frac{1}{2}$. Repeated transformations of a point near an unstable fixed point will move it towards a stable fixed point. In 1972, Wilson and Fisher showed that fixed points that are nontrivial in up to $4 - \epsilon$ dimensions (ϵ small) become trivial in four dimensions.

In percolation models, the renormalization transformation is used primarily to determine numerical values for the critical exponents. Suppose p_c is the unique nontrivial fixed point of $R_b(p)$. Consider the effect of renormalization on the exponent ν of the correlation length $\xi(p) \approx |p - p_c|^{-\nu}$. If $\xi'(p')$ denotes the correlation length for the new (renormalized) lattice, then the renormalized p' is defined by

$$p' = R_b(p). \tag{5.43}$$

Because all lengths in the renormalized lattice are reduced by a constant b in comparison to all lengths in the original lattice, we expect that the correlation length will be reduced by the same amount,

$$\xi'(p') = \frac{\xi(p)}{b} = |p' - p_c|^{-\nu}. \tag{5.44}$$

Note that if we begin at the critical point (e.g., if $p_0 = p_c = 1$), then $\xi(p_c) = \infty$, whence also $\xi'(p_c) = \infty$.

Expanding $R_b(p)$ about the fixed point $p = p^*$, we have that

$$R_b(p) = R_b(p^*) + \lambda(p - p^*) + \mathcal{O}((p - p^*)^2), \tag{5.45}$$

where

$$\lambda = R_b'(p^*) = \left.\frac{dR_b(p)}{dp}\right|_{p=p^*} = \left.\frac{dp'}{dp}\right|_{p=p^*} \tag{5.46}$$

is the first derivative of the renormalization function evaluated at $p = p^*$. In the above example, $R_b'(p) = 6p(1 - p)$, and at the fixed point $p = p^* = \frac{1}{2}$, we have $\lambda = \frac{3}{2}$. If we substitute (5.43) into (5.45), assume that p' is close to the fixed point p^*, and ignore quadratic terms, we obtain

$$p' - p^* \approx \lambda(p - p^*). \tag{5.47}$$

Thus, the difference between p' and the fixed point p^* in the new lattice is proportional to the difference between p and the fixed point p^* in the original lattice.

Substituting (5.26) into (5.44) yields

$$|p' - p_c|^{-\nu} \approx \frac{1}{b}|p - p_c|^{-\nu}. \tag{5.48}$$

Taking logarithms of both sides of (5.48) and solving for ν gives

$$\nu = \frac{\log b}{\log \lambda}, \tag{5.49}$$

where $\lambda = |p' - p_c|/|p - p_c|$.

Worked Example (continued)

For the above example, $b = \sqrt{3}$ and $\lambda = \frac{3}{2}$, so that

$$\nu = \frac{\log(\sqrt{3})}{\log(3/2)} = 1.35476, \tag{5.50}$$

which is very close to the "exact" result in (5.39) of $4/3 = 1.333$. Although (5.49) is an approximation to the true value of ν, the approximation should improve as $b \to \infty$.

5.8 The Number of Infinite Clusters

There are several reasons why we should be interested in how many infinite clusters occur in percolation on an infinite lattice. As we shall see, one of the most important results in percolation theory deals with the supercritical phase in which a unique infinite cluster occurs. The uniqueness of an infinite cluster is closely related to the development of renormalization techniques, and particularly for determining hyperscaling relations and the values of critical exponents. Uniqueness also implies the continuity of the percolation probability. Non-uniqueness can only mean that there are, almost surely, an infinite number of infinite clusters, and such a situation occurs only for percolation on Cayley trees or Bethe lattices.

5.8.1 Configuration Space

Recall the following notation for *configuration space*. Let $\mathbb{L}^d = (\mathbb{Z}^d, \mathbb{E}^d)$ be a d-dimensional hypercubic lattice, $d \geq 2$, where $\mathbb{Z}^d$ is the set of nodes and $\mathbb{E}^d$ is the set of edges.

Definition 5.14 The sample space for bond percolation is given by

$$\Omega = \{0, 1\}^{\mathbb{E}^d} = \prod_{b \in \mathbb{E}^d} \{0, 1\}. \tag{5.51}$$

The sample space Ω is a product space of all possible configurations of the lattice $\mathbb{L}^d$, and is endowed with a σ-field[10] $\mathcal{F}$ of subsets of Ω (generated by the finite-dimensional cylinder sets) and a product probability $\mathbf{P}_p$.

5.8.2 Shift Invariance and Tail Events

Let the vector $w \in \Omega$ denote a configuration of $\mathbb{L}^d$ and let A be an event in the sample space Ω. Define the translation of w by v as $T_v(w) = w(T_v^{-1})$. We have the following definitions of a shift (or translation) of w and A:

Definition 5.15 The *shift* (or *translation*) of the configuration $w \in \Omega$ by $v \in \mathbb{Z}^d$ is the configuration given by

$$(w + v)(e) = (T_v w)(e) = w(T_v^{-1}(e)) = w(T_{-v}(e)) = w(e - v), \tag{5.52}$$

for all $e \in \mathbb{E}^d$ (or $e \in \mathbb{Z}^d$).

Definition 5.16 The *shift* (or *translation*) of the event A by $v \in \mathbb{Z}^d$ is the event

$$A + v = T_v(A) = \{T_v(w) : w \in A\} = \{w : T_v^{-1}(w) \in A\}$$
$$= \{w : T_{-v}(w) \in A\} = \{w : w - v \in A\}. \tag{5.53}$$

Definition 5.17 The event A is *shift invariant* if

$$T_v(A) = A, \quad \text{for all } v \in \mathbb{Z}^d. \tag{5.54}$$

If A is the event that "there exists an infinite cluster," and all nodes are shifted simultaneously, then the infinite cluster will also shift, but it will still remain an infinite cluster. So, A is a shift-invariant event. Similarly, the event that "there exists exactly k infinite clusters" is also a shift-invariant event. Events that are not shift invariant are those that contain, for example, a reference to a specific node (such as the origin) being a member of an infinite open cluster.

Definition 5.18 Let $A_1, A_2, \ldots, A_n, A_{n+1}, \ldots$ be an infinite sequence of independent events. The σ-field of events generated by the events $A_{n+1}, A_{n+2}, \ldots$ is denoted by $\mathcal{T}_n = \sigma(A_{n+1}, A_{n+2}, \ldots)$. The *tail σ-field* is $\mathcal{T} = \cap_{n=1}^{\infty} \mathcal{T}_n$. Events that are determined by the infinite sequence of events $A_1, A_2, \ldots$ and not by a

---

[10] A σ-field $\mathcal{F}$ is a non-empty collection of subsets that is closed under all countable set operations of complements, unions, and intersections.

finite number of events such as $A_1, A_2, \ldots, A_n$ are in the tail σ-field $\mathcal{T}$ and are called *tail events*.

In terms of bond percolation, A is a tail event if it does not depend upon the state of any finite number of edges. An infinite open cluster, if it exists, is a tail event because it is unaffected by any change to a finite number of edges (i.e., open to closed, and vice versa) of that cluster. For example, if there is no infinite cluster, changing a finite number of edges will not bring one into being. Similarly, if there exist infinitely many infinite clusters, changing any finite number of edges is not going to alter more than those edges, and, hence, will not change the status of those infinite clusters.

We have the following result.[11]

Theorem 5.7 (Kolmogorov zero–one law) *If A is a tail event of an infinite sequence of independent events, then $\mathrm{P}\{A\} \in \{0, 1\}$.*

Proof Let $\mathcal{B}_\infty = \sigma(A_1, A_2, \ldots)$ be the σ-field generated by the entire sequence. The σ-field $\mathcal{B}_n = \sigma(A_1, \ldots, A_n)$ is independent of the σ-field $\mathcal{T}_n = \sigma(A_{n+1}, A_{n+2}, \ldots)$. Let $\mathcal{T}$ be the tail σ-field. Let $A \in \mathcal{T} \subset \mathcal{B}_\infty$ be a tail event. Then A is independent of $\mathcal{B}_n$, for any n. Now, a set of events is independent if every finite subset of that set is independent. So, $\mathcal{B}_\infty$ and $\sigma(A)$ are independent. But $A \in \mathcal{B}_\infty$ and $A \in \sigma(A)$. Hence, A is independent of itself. This implies that $\mathrm{P}\{A\} = \mathrm{P}\{A \cap A\} = [\mathrm{P}\{A\}]^2$, whence $\mathrm{P}\{A\} = 0$ or 1. $\qquad\square$

Using a similar argument as for the Kolmogorov zero–one law, we have the following zero–one law for shift-invariant events.

Theorem 5.8 *If A is a shift-invariant event, then $\mathrm{P}_p\{A\} \in \{0, 1\}$.*

If every shift-invariant event has probability 0 or 1, then the probability is called *(shift) ergodic*.

Definition 5.19 The probability measure P_p on Ω is *shift invariant* (or *stationary*) if, for all $A \subset \mathbb{N} = \{0, 1, 2, \ldots\}$ and $\nu \in \mathbb{E}^d$ (or $\nu \in \mathbb{Z}^d$),

$$(T_\nu \mathrm{P}_p)\{A\} = \mathrm{P}_p\{T_\nu^{-1} A\} = \mathrm{P}_p\{T_{-\nu} A\} = \mathrm{P}_p\{A\}. \tag{5.55}$$

In particular, the product measure P_p on the d-dimensional infinite square lattice $\mathbb{Z}^d$ is shift invariant.

5.8.3 The Newman–Schulman Result

Let $p \in [0, 1]$. One of the most important results in percolation theory is that of Newman and Schulman (1981a,b), who showed the following.

Theorem 5.9 (Newman and Schulman, 1981a,b) *There are, almost surely, either 0, 1, or an infinite number of distinct infinite open clusters in iid site or bond*

---

[11] See, e.g., Loève (1963, p. 229), Billingsley (1979, pp. 49–52).

percolation models on the d-dimensional infinite square lattice $\mathbb{Z}^d$. This result also holds more generally.

As we have seen, the first possibility, the zero number of infinite open clusters, occurs when $p < p_c$. We have also seen that when $p > p_c$, there exists an infinite open cluster and that the infinite cluster is, almost surely, unique (Harris, 1960, for $d = 2$; Aizenman and Barsky, 1987, for $d \geq 3$), which is the second possibility. The third possibility, an infinite number of infinite-size clusters, is impossible on $\mathbb{Z}^d$ (Aizenman and Barsky, 1987). We shall see that there are some classes of graphs, such as regular trees, in which an infinite number of infinite-size clusters can exist.

Proof Let $\mathcal{H}$ denote the set of all clusters and

$$\mathcal{H}_0 = \{C \in \mathcal{H} : |C| = \infty\} \tag{5.56}$$

the set of all infinite clusters. Let the random variable $N_0 = |\mathcal{H}_0|$ be the number of distinct infinite clusters. For each $k \in \mathbb{N} \cup \{\infty\}$, the event $A_k = \{N_0 = k\}$ is shift invariant. The events $\{A_k\}$ are disjoint, and their union is the entire probability space. By the zero–one law, $\mathbf{P}_p\{A_k\} \in \{0, 1\}$. Thus, there exists $k_0 \in \mathbb{N} \cup \{\infty\}$ such that $\mathbf{P}_p\{A_{k_0}\} = \mathbf{P}_p\{N_0 = k_0\} = 1$.

We proceed now by contradiction; that is, we assert that for some $k_0 \in [2, \infty)$, $\mathbf{P}_p\{A_{k_0}\} > 0$, so that, almost surely, $k_0 \notin \{0, 1, \infty\}$. We start by defining a d-dimensional cube.

Definition 5.20 Let n be a positive integer. A d-dimensional cube B_n centered at the origin is defined by

$$B_n = [-n, n]^d = \{v = (v_1, \ldots, v_d) \in \mathbb{Z}^d : v_\ell \in [-n, n], \ell = 1, 2, \ldots, d\}. \tag{5.57}$$

Let W_n be the set of lattice configurations for which there are exactly k_0 infinite clusters (i.e., $N_0 = k_0$) and each of them enters (i.e., has nonzero intersection with) the cube B_n. Then $\mathbf{P}_p\{W_n\} > 0$. This follows because $W_n \to A_{k_0}$ as $n \to \infty$, and if W_n had zero probability for all n, then so would A_{k_0}. Thus, $\mathbf{P}_p\{W_n\} \to \mathbf{P}_p\{A_{k_0}\} = 1$ as $n \to \infty$. Therefore, for some m, $\mathbf{P}_p\{W_m\} > 0$. Let T be the transformation that makes all edges in B_m open and leaves all edges outside B_m unchanged. Then $\mathbf{P}_p\{T(W_m)\} > 0$. When all the edges in W_n become open, there will be one large open cube plus a number of infinitely long tails emerging from the cube; all these tails will be connected to each other through the cube. So, there will no longer be k_0 infinite clusters, and $T(W_m) \subset \{N_0 = 1\}$. So, $\mathbf{P}_p\{N_0 = 1\} > 0$, which contradicts the assertion that $k_0 \notin \{0, 1, \infty\}$. $\square$

5.8.4 Uniqueness of the Infinite Cluster

So, we know that k_0 has to be either 0, 1, or ∞. All we need to show now is that an infinite number of infinite clusters (i.e., $k_0 = \infty$) is not possible on $\mathbb{Z}^d$. The uniqueness result says that

Theorem 5.10 *If N_0 denotes the number of infinite open clusters and if $\theta(p) > 0$ for some p, then $\mathbf{P}_p\{N_0 = 1\} = 1$.*

The condition $\theta(p) > 0$ expresses the idea that every node should have a strictly positive probability of being a member of an infinite open cluster. Demonstration of this result was simplified and shortened (Gandolfi, Keane, and Russo, 1988). A beautiful proof of this result subsequently appeared in Burton and Keane (1989), an argument that is now accepted as a landmark in percolation theory.

Proof We need to exclude the possibility that $k_0 = \infty$. So, we assume that $\mathbf{P}_p\{N_0 = \infty\} = 1$, and then show that it leads to a contradiction.
    We proceed in two steps.

Step I. Let $w \in \Omega$ be a configuration. Following Grimmett (1999, p. 199; 2006), we call a node $v \in \mathbb{Z}^d$ a *trifurcation* of w if it satisfies the following three conditions:

1. The node v belongs to an infinite open cluster C.
2. There are exactly three open edges that contain v.
3. Removing v and its three open edges splits C into exactly three disjoint infinite clusters (and no finite clusters).

If all three conditions hold, we write A_v for the event that "v is a trifurcation."[12] By shift-invariance of $\mathbf{P}_p$, the probability that v is a trifurcation is independent of v; that is

$$\mathbf{P}_p\{A_v\} = \mathbf{P}_p\{A_O\}, \quad \text{for all } v \in \mathbb{Z}^d, \tag{5.58}$$

where A_O is the event that "there is a trifurcation at the origin." We focus on A_O and show that $\mathbf{P}_p\{A_O\} > 0$.
    Let $M_{B_n,0}$ denote the number of infinite clusters that intersect the cube B_n defined in (5.57), when every edge in B_n is closed. Let M_{B_n} denote the number of infinite clusters that intersect the cube B_n. Suppose one cluster goes through B_n completely. If all of B_n were closed, then there would be two infinite clusters intersecting B_n. Thus, $M_{B_n,0} \geq M_{B_n}$. It follows that, as $n \to \infty$, $B_n \to \mathbb{Z}^d$ and

$$\mathbf{P}_p\{M_{B_n,0} \geq 3\} \geq \mathbf{P}_p\{M_{B_n} \geq 3\} \to \mathbf{P}_p\{N_0 \geq 3\} = 1. \tag{5.59}$$

The last equality follows because we assumed that $\mathbf{P}_p\{N_0 = \infty\} = 1$. So, there exists $m \in \mathbb{N}$ such that

$$\mathbf{P}_p\{M_{B_m,0} \geq 3\} \geq \frac{1}{2}, \tag{5.60}$$

where $\frac{1}{2}$ is taken as an arbitrary constant in $(0, 1)$. Thus, at least three infinite clusters intersect B_m (when all edges in B_m are closed) with probability at least $\frac{1}{2}$.
    Fix this value of m. We now show how to make the origin a trifurcation. If $M_{B_m,0} \geq 3$, we can find three nodes, x, y, and z, say, on ∂B_m, the boundary of B_m, each node of which lies in a distinct open infinite cluster of $\mathbb{E}^d \setminus B_m$. That is, $x \in \partial B_m \cap C_x$, $y \in \partial B_m \cap C_y$, and $z \in \partial B_m \cap C_z$. Within the cube B_m, draw three separate open paths from the origin to the points x, y, and z. See **Figure 5.11**. We can do this because the

---

[12] Burton and Keane (1989) had previously used the term *encounter point*.

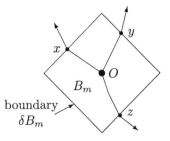

Figure 5.11 Trifurcation at the origin 0 in a two-dimensional cube, B_m, with three nodes, x, y, and z, on the boundary of B_m, each of which lies in a distinct infinite open cluster.

origin sits inside B_m. These paths can be chosen such that the origin is the unique node common to any two of them, and each path touches exactly one node on ∂B_m.

Consider the configuration of edges $w \in \{M_{B_m,0} \geq 3\}$. Recall that A_O is the event that there is a trifurcation at the origin. Let $J_{x,y,z}$ denote the event that "the only edges in B_m that are open are the paths connecting the origin to x, y, and z, while all other edges in B_m are closed." Now,

$$P_p\{A_O\} \geq P_p\{J_{x,y,z} \cap \{M_{B_m,0} \geq 3\}\}$$
$$= P_p\{J_{x,y,z}|M_{B_m,0} \geq 3\}P_p\{M_{B_m,0} \geq 3\}. \qquad (5.61)$$

Note that $J_{x,y,z}$ depends upon the edges of B_m, while $\{M_{B_m,0} \geq 3\}$ depends upon the edges outside B_m. Now,

$$P_p\{J_{x,y,z}|M_{B_m,0} \geq 3\} \geq [\min(p, 1-p)]^{R_m} = \gamma, \qquad (5.62)$$

say, where R_m is the total number of edges in B_m. Substituting (5.60) and (5.62) into (5.61), we have, for $p \in (0,1)$,

$$P_p\{A_O\} \geq \frac{\gamma}{2} > 0. \qquad (5.63)$$

Thus, assuming $P_p\{N_0 = \infty\} = 1$, it follows that $P_p\{A_O\} > 0$. In other words, there is a positive probability that the origin is a trifurcation.

Step II. We switch back from m to n to get the limiting results. Let $T_n = \sum_{v \in B_n} I_{A_v}$ denote the number of trifurcations in B_n, where $I_A = 1$ if A is true, and 0 otherwise. Then, using (5.58),

$$E_p\{T_n\} = \sum_{v \in B_n} P_p\{A_v\} = |B_n| \cdot P_p\{A_0\}. \qquad (5.64)$$

This shows that T_n grows like $|B_n|$, as $n \to \infty$. Choose a trifurcation $t_1 \in B_n$. Suppose the node $x_1 \in \partial B_n$ is connected to t_1 by an open path (i.e., $t_1 \leftrightarrow x_1$) in B_n. Now, choose a second trifurcation $t_2 \in B_n$. By definition of a trifurcation, there exists another node $x_2 \in \partial B_n$, $x_2 \neq x_1$, such that $t_2 \leftrightarrow x_2$ in B_n. Repeat this process by choosing at each stage a new trifurcation $t_i \in B_n$ and a new node $x_i \in \partial B_n$, where $t_i \leftrightarrow x_i$ in B_n. Because there are T_n trifurcations in B_n, there will be at least T_n distinct nodes $x_i \in \partial B_n$. It follows that $|\partial B_n| \geq T_n$.

Taking expectations of the last result and using (5.63), we have that

$$\frac{|\partial B_n|}{|B_n|} \geq \frac{\gamma}{2} > 0, \qquad (5.65)$$

which says that the surface/volume ratio of the cube is strictly positive. For large n, this is impossible: the surface $|\partial B_n| = |B_n \setminus B_{n-1}| = |B_n| - |B_{n-1}|$ grows like n^{d-1}, while the volume $|B_n|$ grows like n^d, so that their ratio converges to zero as $n \to \infty$ (i.e., as the cube gets bigger and bigger), which contradicts (5.65). Consequently, $P_p\{N_0 = \infty\} = 1$ cannot hold. So, either $P_p\{N_0 = 0\} = 1$ or $P_p\{N_0 = 1\} = 1$. □

5.9 Further Reading

There are many books on percolation; we recommend those by Kesten (1982), Grimmett (1989, 1999), and Bollobás and Riordan (2006). An introductory approach to percolation from a physics perspective is given in the book by Stauffer and Aharony (1992). A book on trees and networks from a probability perspective is Lyons and Peres (2017). The book by Grimmett (2010) has two chapters on percolation (Chapters 3 and 5). Edited volumes that include chapters on percolation include Grimmett and Newman (1990) and Bramson and Durrett (1999).

A different interpretation of percolation on a network is discussed by Newman (2010, Chapter 16), who defines (site) percolation as the result of removing from a network a fraction of nodes and the edges attached to those nodes, where different methods of removing nodes (e.g., uniformly at random, or ordered by degree) are discussed. In that sense, the study of site percolation is about the study of robustness of a network to performance failure. In the same approach, bond percolation is defined as the result of removing a fraction of the edges from a network. It is interesting that Durrett (2007), and also Newman (2010), both of whom discuss the topic of percolation, do not mention the contributions of most of the early pioneers of percolation theory.

Sections 5.4, 5.5, and 5.6. There are many excellent books on the physics of phase transitions, critical behavior, renormalization, and percolation; see, for example, Gitterman and Halpern (2004), McComb (2004), and Plischke and Bergersen (2006).

Section 5.7.3. An edited proceedings of the Renormalization Group Workshop RG2000 appeared in *Physics Reports* in 2001 (O'Connor and Stephens, 2001).

5.10 Exercises

Exercise 5.1 On a two-dimensional square lattice, form a 2×2 super-site with $b^2 = 4$.

(a) Show that for the majority rule to hold, there are five possible super-site configurations of three or four sites occupied out of four.

(b) Find the recursion equation, which is a fourth-degree polynomial in p.

(c) Find the fixed points.

(d) Take $\text{est}(p_c) = 0.7676$: unstable. Compare with the best known estimate of 0.5927 for the square lattice.

(e) Show that repeated transformations of p close to 0 or 1 moves p' closer to 0 or 1.

(f) Find the value of ν.

Exercise 5.2 On a two-dimensional square lattice, form a 3×3 super-site with $b^2 = 9$.

(a) List all the possible super-site configurations. A super-site is occupied if a cluster spans the super-site vertically and horizontally.
(b) Find the probability of each configuration.
(c) Find $R(p)$.
(d) Write a computer program to solve for the fixed point p^*.
(e) Find the exponent ν of the correlation length.
(f) Plot the difference $R(p) - p$ against p.
(g) Find the value of p at which $R(p) - p$ crosses the x-axis.
(h) Compare with the "exact" value and with the result for $b = 2$.

Exercise 5.3 On a two-dimensional triangular lattice, form a seven-site block with $b^2 = 7$.

(a) List all possible super-site configurations. A super-site is occupied if the majority rule holds.
(b) Find the recursion equation, which is a seventh-degree polynomial in p.
(c) Find the trivial fixed points and show that the nontrivial fixed point is 1/2.
(d) Find the exponent ν for the correlation length and compare your result with the value $\nu = 4/3$.

Exercise 5.4 Show that $\theta(p) \neq 1$ for any value of $p < 1$.

Exercise 5.5 Show that (a) for $p < 1/3$, $\theta(p) = 0$; (b) for $p > 2/3$, $\theta(p) > 0$.

Exercise 5.6 Consider a two-dimensional square lattice of size 2×2 and let $P_p\{X_e = 1\} = p$. Show that the percolation probability is $\theta(p) = p^2(2 - p^2)$. Graph $\theta(p)$ against p.

Exercise 5.7 Give an example of a shift-invariant event that is not a tail event, so that Kolmogorov's zero–one law does not apply.

Exercise 5.8 Give an example of a tail event that is not shift invariant.

Exercise 5.9 Let A denote the event that there exists an infinite component. Show that A is shift invariant.

Exercise 5.10 On a finite lattice, why is the critical probability p_c not well-defined?

Exercise 5.11 Let $B_k = [-k,k]^d$ be a box of side length $2k + 1$, centered at the origin (see (5.9)). Let $\partial B_k = B_k \backslash B_{k-1}$ denote the external boundary of the box B_k. Show that $\lim_{k \to \infty} P_p\{0 \leftrightarrow \partial B_k\} = \theta(p)$.

CHAPTER SIX

Percolation Beyond $\mathbb{Z}^d$

Percolation can be defined more generally than as a process on $\mathbb{Z}^d$, $d \geq 2$. In this chapter, we motivate the main ideas and theory of percolation on more general graphs by application to polymer gelation and amorphous computing. We describe various types of percolation processes on trees, transitive graphs, and continuum percolation.

6.1 Introduction

We now entertain the idea that percolation can occur on more exotic graphs than the hypercubic lattice $\mathbb{Z}^d$, $d \geq 2$. Specifically, we will describe percolation on trees, such as Cayley trees and Bethe lattices, and percolation on transitive and quasi-transitive graphs. Then, we study continuum percolation, which has attracted a lot of research attention in recent years.

6.2 Example: Polymer Gelation

Gelation is the process of forming a gel (i.e., a "soft" solid) from a viscous liquid. For example, boiling an egg, hardening of cement, or the vulcanization of rubber. There are two types of gels, chemical (or *strong*) gels and physical (or *weak*) gels. A chemical gel is a permanent gel, formed when molecules are *crosslinked* (i.e., joined) by covalent bonds, which are irreversible, to create a three-dimensional network of molecules. On the other hand, a physical gel forms when molecules are joined by hydrogen or ionic bonds that are reversible: for example, changing the temperature may reverse the gelation process back to a liquid state. In fact, there are many types of actions by which reversability of physical gels may occur. In the following, we only consider chemical gels.

The gelation process starts with a huge number of molecules (known as *monomers*), which join irreversibly to form clusters (known as *macromolecules* or *polymers*). Two molecules in the same cluster are connected (directly or indirectly through other molecules in the same cluster) by such bonds, whereas two molecules in two different clusters are not connected by such bonds.

The molecules can be modeled as sites of a lattice. In this model, molecules do not loop back on themselves, there is no interaction between different molecules, and the network branches out indefinitely. Thus, bonds are formed at random on an infinite

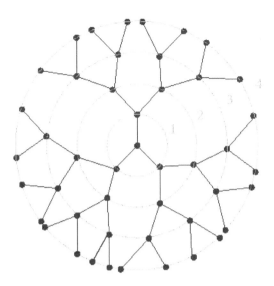

Figure 6.1 A Bethe lattice is an infinite cycle-free graph. It is actually a pseudo-lattice because it cannot be embedded in any real lattice. The root node lies at the center of the graph and the other nodes are arranged in shells around the root node. The Cayley tree (pictured here), which is finite, is a special case of the Bethe lattice. Each node in a Cayley tree is connected to r neighbors, where r is the coordination number, and continues for *n* generations. Here, r = 3 and n = 4.

Cayley tree (also known as a *Bethe lattice*); see **Figure 6.1** for a graph of a finite Cayley tree. Any two molecules on a branched polymer are linked by a single linear path, which makes the problem one-dimensional and easily solvable. A bond between two molecules exists with a certain probability, and the formation of a bond is independent of all other bonds. The effects of all other variables that contribute to bond formation (e.g., temperature, bond concentration) are absorbed into that probability.

There are two types of *aggregation* (or merging) processes: *monomer–cluster aggregation* (where a monomer is engulfed by a cluster) and *cluster–cluster aggregation* that produces a larger cluster whose mass equals the sum of the masses of the two component clusters. As these clusters multiply, both in size and number, a *sol–gel phase transition* occurs, which controls whether or not gelation occurs. The *gel point* is a probability that separates two phases:

- *Sol phase.* When the probability is smaller than the gel point, the system is said to be in its *sol phase* (referring to any polymer that can be dissolved in a good solvent and has good fluidity like a liquid) and consists of a collection of finite clusters, called the *sol*.
- *Gel phase.* When the probability is higher than the gel point, the system is in the *gel phase*, and the aggregation process continues until an infinite giant cluster (called the *gel*) is formed; the giant cluster expands by absorbing finite clusters and then grows to dominate the whole system.

Unlike the sol, the term "gel" refers only to those types of polymers that are insoluble in all solvents. In real systems, a sol and a gel can coexist, but if that happens, the finite clusters will be entirely confined within the gel. Gelation occurs when the system transitions from no-gel to gel.

The percolation model is considered to be the basic model of gelation (Flory,[1] 1941; Stockmayer, 1943) and offers more information than does kinetic theory. Experiments and simulations show that this theory has good agreement with reality. A criticism of the percolation theory for gelation is that irreversible chemical bonds are not really formed in a random fashion, but are formed by a kinetic process containing both deterministic and random components (Herrmann, Stauffer, and Landau, 1983).

Pierre-Gilles de Gennes[2] was the first to discover the equivalence between the gel point and the phase transition for percolation (de Gennes, 1976a,b).

6.3 Percolation on Trees

The Bethe lattice occupies an important place in percolation theory. Unlike the lattice $\mathbb{L}^d$, whose properties get more complicated as the size of the lattice increases, many of the most interesting percolation quantities can be computed explicitly for the Bethe lattice.

6.3.1 Cayley Trees, Bethe Lattices

Let $\mathbb{T}$ be a labeled binary tree, in which each node has two daughter nodes (and a single parent node). So, each node has three edges (or branches), except the root node, which only has two. The *degree* of a node is the number of other nodes to which it is connected. If every node of a tree has the same degree, then the tree is called *regular*.

Definition 6.1 If a regular tree has *degree r* and continues for *n* generations, then the tree is often called an *r-regular Cayley tree*[3] (or just an *r-Caley tree*) and is denoted by $\mathbb{T}_r$. See **Figure 6.1**.

The value of r in $\mathbb{T}_r$ is referred to as either *the number of neighbors* of each node, or as the *coordination number* of a node. There are no loops or cycles in $\mathbb{T}_r$, and $\mathbb{T}_r$ is said to be *locally finite* if the degree of every node is finite. Note that in any finite r-Cayley tree, every boundary node is connected to only one other node. For a tree to be an r-Cayley tree, another branch at the root node can be added; because this does not affect the critical exponents of percolation on $\mathbb{T}_r$, we choose not to do this.

Physicists use the term *Bethe lattice*[4] to describe an infinite Cayley tree, although an infinite Cayley tree enjoys different critical properties than the Bethe lattice.[5] Strictly

---

[1]For his theoretical and experimental work in the physical chemistry of the macromolecule, Paul John Flory received the 1974 Nobel Prize in Chemistry.

[2]For discovering that methods developed for studying order phenomena in simple systems can be generalized to more complex forms of matter, in particular to liquid crystals and polymers, de Gennes received the 1991 Nobel Prize in Physics.

[3]Such trees are named after Arthur Cayley, who coined the word "tree" in 1857 to describe the logical branching that occurs when producing a sequence of derivatives applied in a very specific order, starting at a "root" node.

[4]Bethe lattices are based upon work on phase transition of atoms in a lattice by Hans A. Bethe in 1935. For other work, Bethe was awarded the 1967 Nobel Prize for Physics.

[5]Some references (e.g., Stauffer and Aharony, 1991, Section 2.4) describe Cayley trees as being infinite and the same as Bette lattices.

speaking, a Bethe lattice, which contains no cycles, is actually a "pseudo-lattice" because it cannot be embedded in a real lattice (e.g., square lattice, triangular lattice, honeycomb lattice) that has cycles. The study of percolation on a Bethe lattice was initiated by Flory (1941). Bethe lattices have since been used to represent a chemical percolation model of coal structure (Grant et al., 1989; Hayashi and Miura, 2004), as well as many applications in solid-state and condensed-matter physics.

In the case of binary trees, we label the nodes of $\mathbb{T}_3$ by strings of 1s and 2s. The root node is labeled as $\emptyset$ and the two offspring of the root node are labeled 1 and 2. The offspring of 1 are labeled 11 and 12, whilst the offspring of 2 are labeled 21 and 22. The second generation has $2 \times 2 = 2^2 = 4$ nodes (or sites) and the third generation has $4 \times 2 = 2^3 = 8$ nodes (sites). And so on. In the nth generation, there are 2^n offspring, and the offspring of $\ell_1 \ell_2 \ldots \ell_n$, where $\ell_i = 1$ or 2, are labeled either $\ell_1 \ell_2 \ldots \ell_n 1$ or $\ell_1 \ell_2 \ldots \ell_n 2$. Thus, a sphere of two generations contains $1 + 2 + 4 = 7$ nodes and a sphere of three generations contains $1 + 2 + 4 + 8 = 15$ nodes.

If we start out with three offspring of the root node, and each of these nodes has two offspring, then we see from **Figure 6.1** that the first generation has three nodes, the second generation has $3 \times 2 = 6$ nodes, the third generation has $6 \times 2 = 3 \times 2^2 = 12$ nodes, and so on. In the nth generation, there are $3 \times 2^{n-1}$ nodes. Thus, a sphere of two generations contains $1 + 3 + 6 = 1 + 3(1 + 2) = 10$ nodes and a sphere of three generations contains $1 + 3 + 6 + 12 = 1 + 3(1 + 2 + 4) = 22$ nodes. For n generations, the sphere contains $1 + 3(1 + 2 + 2^2 + 2^3 + \cdots + 2^{n-1}) = 3 \times 2^n - 2$ nodes. Of those, the $3 \times 2^{n-1}$ nodes along the surface of the sphere are called *surface nodes*. The fraction of total nodes that are surface nodes is, therefore,

$$\frac{3 \times 2^{n-1}}{3 \times 2^n - 2} = \frac{1}{2} \times \frac{3 \times 2^{n-1}}{3 \times 2^{n-1} - 1}.$$

For $n = 2$, this fraction is 0.6; for $n = 3$, it is 0.55; and for $n = 4$, it is 0.52. For large n, this fraction tends to $1/2$.

For r-Cayley trees $\mathbb{T}_r$, the offspring of the root node $\emptyset$ are labeled $1, 2, \ldots, r$. The offspring of 1 are labeled $11, 12, \ldots, 1r$, and so on, and the offspring of z are labeled $z1, z2, \ldots, zr$. This labelling continues, with the nth generation, which has r^n offspring, having labels $\ell_1 \ell_2 \ldots \ell_n$, where, for $i \geq 1$, $\ell_i \in \{1, 2, \ldots, r\}$.

As in bond percolation on a square lattice, if $0 < p < 1$, each edge of the Bethe lattice $\mathbb{T}_r$ is *open* with probability p and *closed* with probability $1 - p$, independently of all other edges. One can view p as the fraction of edges that survive following a random failure of the remaining edges.

6.3.2 Critical Probability

Let us take a walk on $\mathbb{T}_r$, where each node has common degree $r, r \geq 3$. If we start from the center node (i.e., the origin) and work our way outwards, we first encounter $r - 1$ edges, a proportion of which will be open, the rest closed. At each step in this branching process, assuming that we do not retrace our steps, we see $r - 1$ new edges. On average, we encounter $p(r - 1)$ open edges along which we can continue our journey. If $p(r - 1) < 1$, we will see fewer and fewer open edges at each generation, the process will, almost surely, die out (i.e., go extinct), and, almost surely, only finite clusters will be formed. On the other hand, if $p(r - 1) > 1$, the process will continue indefinitely and, almost surely, will produce at least one infinitely large cluster.

Thus, the value of the critical probability $p_c(\mathbb{T}_r)$ for percolation satisfies $p_c(\mathbb{T}_r) \cdot (r - 1) = 1$. The same argument works for site percolation – just change "edge" to "node." The exact solution for bond or site percolation is therefore as follows.

Theorem 6.1 *The critical probability for percolation on an r-regular tree $\mathbb{T}_r$, $r \geq 3$, is given by*

$$p_c(\mathbb{T}_r) = \frac{1}{r - 1}. \tag{6.1}$$

Thus, for a Bethe lattice with $r = 3$, which is a binary tree, $p_c(\mathbb{T}_3) = \frac{1}{2}$. Note that $r = 2$ corresponds to a one-dimensional graph, where an infinite cluster at the origin will only appear if $p_c(\mathbb{T}_2) = 1$. Clearly, as the common degree r increases, $p_c(\mathbb{T}_r)$ decreases, which violates the universality principle of being invariant to the local structure of the graph.

6.3.3 An Infinite Number of Infinite Clusters

Recall that in the case of percolation on $\mathbb{Z}^d$, if an infinite open cluster exists, it is a.s. unique. When it comes to supercritical percolation on a regular tree $\mathbb{T}_r$, however, the number of infinite open clusters is quite a different story. As before, $\mathbf{P}_p$ is the probability measure on the set of configurations of open and closed bonds. Grimmett and Newman (1990) show that, using a branching-process argument, regular trees (just like lattices) possess two distinct phases.

Theorem 6.2 *Let $\mathbb{T}_r$ be an r-regular tree and suppose that $p_c < p < 1$ (i.e., the supercritical phase), where $p_c = (\mathbb{T}_r) = 1/(r - 1)$.*

1. *If there exists an infinite open cluster, then there are, almost surely, infinitely many infinite open clusters. In other words, if N_0 is the number of infinite open clusters, then $\mathbf{P}_p\{N_0 = \infty\} = 1$.*
2. *In the case that $p \leq p_c(\mathbb{T}_r)$, there is no infinite open cluster, all open clusters are, almost surely, finite.*

6.3.4 Critical Exponents

For percolation on a Bethe lattice, the critical exponents can be determined exactly by expressing the problem in terms of branching processes. We follow Durrett's, (1985) approach in this section.

Let $\mathbb{T}_3$ be an infinite binary tree. Consider the following Galton–Watson branching process on $\mathbb{T}_3$. Let $\theta_k(p)$ denote the probability that there exists an open path from the root node $\emptyset$ to a node k levels down the tree. Note that $\theta_0(p) = 1$. Suppose that the path passes through v_1, one of the two daughter nodes of the root node. The subtree with root node v_1 is identical to its parent's tree starting at $\emptyset$ (except with one fewer level).

The probability that there exists an open path from $\emptyset$ to a node k levels down the tree and passing through v_1 is the probability of an open edge from $\emptyset$ to v_1, multiplied by the probability of an open path from v_1 to a node $k - 1$ levels down the tree, namely,

$p\theta_{k-1}(p)$. Hence, the probability that there is no open path starting at $\emptyset$ to a node k levels down the tree and passing through v_1 is $1 - p\theta_{k-1}(p)$. Any path from $\emptyset$ down the tree has to pass through one of the two daughter nodes of the root node. Thus, the probability that there is no open path from $\emptyset$ to a node k levels down the tree is $(1 - p\theta_{k-1}(p))^2$. So,

$$\theta_k(p) = 1 - (1 - p\theta_{k-1}(p))^2. \tag{6.2}$$

Now, consider the function

$$f_p(\theta) = 1 - (1 - p\theta)^2, \tag{6.3}$$

so that $\theta_k(p) = f_p(\theta_{k-1}(p))$. Note that $f_p(0) = 0$ and $f_p(1) \leq 1$. The derivative of $f_p(\theta)$ wrt θ is $f_p'(\theta) = 2p(1 - p\theta)$ and so $f_p'(0) = 2p$. Also, $f_p''(\theta) < 0$. There exists a fixed point θ_* that satisfies $\theta_* = f_p(\theta_*)$. That is, if $1 - \theta_* = (1 - p\theta_*)^2 = 1 - 2p\theta_* + p^2\theta_*^2$, or if $(1 - 2p)\theta_* + p^2\theta_*^2 = 0$. Thus, we have

$$\theta_*[(1 - 2p) + p^2\theta_*] = 0. \tag{6.4}$$

The two possible solutions of this equation are $\theta_* = 0$, or if $\theta_* \neq 0$, then $(1 - 2p) + p^2\theta_* = 0$. If $2p < 1$ (i.e., $p < \frac{1}{2}$), the graph of $f_p(\theta)$ is given on the left side of **Figure 6.2**, and the only fixed point is at $\theta_* = 0$. If $2p > 1$ (i.e., $p > \frac{1}{2}$), the graph of $f_p(\theta)$ is given on the right side of **Figure 6.2**, and the fixed point is at $\theta_* = \frac{2p-1}{p^2} > 0$. For example:

p	0.6	0.7	0.8	0.9
θ_*	0.556	0.816	0.938	0.988

To summarize:

$$\theta(p) = \begin{cases} 0, & \text{if } p \leq \frac{1}{2} \\ \frac{2p-1}{p^2}, & \text{if } p \geq \frac{1}{2} \end{cases}. \tag{6.5}$$

Thus, $p_c = \frac{1}{2}$. Differentiating $\theta(p) = 2p^{-1} - p^{-2}$ wrt p for $p \geq \frac{1}{2}$ yields $\theta'(p) = -2(p^{-2} - p^{-3})$, which at $p = \frac{1}{2}$ equals 8; thus, $\theta(p) \sim 8(p - p_c)$ as $p \downarrow p_c$.

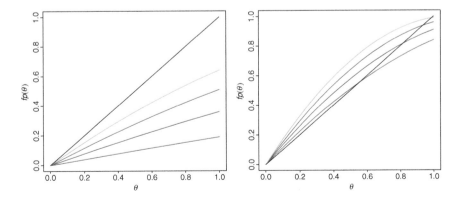

Figure 6.2 Graphs of $f_p(\theta) = 1 - (1 - p\theta)^2$, $0 \leq p, \theta \leq 1$. Left: $p = 0.1$ (*red*), 0.2 (*blue*), 0.3 (*purple*), 0.4 (*green*). Right: $p = 0.6$ (*red*), 0.7 (*blue*), 0.8 (*purple*), 0.9 (*green*). The x-axis is θ and the y-axis is $f_p(\theta)$. The black line at 45° intersects with the curves to determine the fixed points θ_* for each value of p.

For percolation on a binary tree, the critical exponent β for $\theta(p) \approx (p - p_c)^\beta$ as $p \downarrow p_c$ is given by $\beta = 1$.

Next, let $p < p_c$, and consider the critical exponent γ, as in

$$\chi(p) = E_p\{|C|\} \approx (p_c - p)^{-\gamma}, \tag{6.6}$$

where C is the cluster containing the origin, which on a tree is the root node $\emptyset$. Any node $x \in \mathbb{T}$ is a member of C iff the unique path from $\emptyset$ to x is open (i.e., every edge along that path is open). In the first generation from $\emptyset$, there are two nodes; in the second generation, there are $2^2 = 4$ nodes; in the nth generation, there are 2^n nodes. So, the mean cluster size when $p \uparrow p_c$ is

$$\chi(p) = E_p\{|C|\} = \sum_{n=0}^{\infty} 2^n p^n = \sum_{n=0}^{\infty}(2p)^n = (1 - 2p)^{-1} = \tfrac{1}{2}(\tfrac{1}{2} - p)^{-1}. \tag{6.7}$$

Comparing (6.7) with (6.6), we see that $p_c = \tfrac{1}{2}$ and $\gamma = 1$.

When $p > p_c$, the critical exponent γ' is defined as

$$\chi^f(p) = E_p\{|C|; |C| < \infty\} \approx (p - p_c)^{-\gamma'}, \tag{6.8}$$

which is the mean cluster size for the finite clusters (i.e., excluding the unique infinite cluster). In terms of the branching process on $\mathbb{T}$, the finite clusters are those processes that become extinct. Let Z_n denote the number of open nodes at the nth generation of the branching process, and let $Z = \{Z_n, n \geq 0\}$. Then, Z is a Markov chain with transition probabilities given by

$$p_{ij} = P_p\{Z_{n+1} = j | Z_n = i\} = \binom{2i}{j} p^j (1 - p)^{2i-j}. \tag{6.9}$$

To see this, note that the i open nodes in the nth generation yield $2i$ nodes in the $(n+1)$st generation; of those $2i$ nodes, j of them will be open, each with probability p, and $2i - j$ nodes will be closed, each with probability $1 - p$.

Let $\bar{Z}_n$ denote Z_n conditioned upon the event $A = \{|C| < \infty\}$ that the process becomes extinct, and let $\bar{Z} = \{\bar{Z}_n, n \geq 0\}$. Then, for $i, j \geq 0$, $\bar{Z}$ is a Markov chain with transition probabilities

$$
\begin{aligned}
\bar{p}_{ij} = P_p\{\bar{Z}_{n+1} = j | \bar{Z}_n = i\} &= \frac{P_p\{\bar{Z}_{n+1} = j, \bar{Z}_n = i\}}{P_p\{\bar{Z}_n = i\}} \\
&= \frac{P_p\{Z_{n+1} = j, Z_n = i, A\}/P_p\{A\}}{P_p\{Z_n = i, A\}/P_p\{A\}} \\
&= \frac{P_p\{Z_{n+1} = j, A | Z_n = i\}P_p\{Z_n = i\}}{P_p\{A | Z_n = i\}P_p\{Z_n = i\}} \\
&= \frac{P_p\{Z_{n+1} = j | Z_n = i\}[P_p\{A\}]^j}{[P_p\{A\}]^i}.
\end{aligned} \tag{6.10}
$$

Substituting (6.9) into (6.10) and setting

$$P_p\{A\} = 1 - \theta(p) = 1 - \frac{2}{p} + \frac{1}{p^2} = \left(\frac{1 - p}{p}\right)^2, \tag{6.11}$$

we have that

$$\bar{p}_{ij} = P_p\{\bar{Z}_{n+1} = j | \bar{Z}_n = i\} = \binom{2i}{j} p^{2i-j}(1-p)^j. \tag{6.12}$$

If we compare (6.12) for the chain conditioned on extinction with (6.9) for the unconditioned chain, we see that the two expressions are similar but that $p > \frac{1}{2}$ is switched with $1 - p$. It follows from (6.7) and (6.11) that

$$\begin{aligned}
\chi^f(p) &= E_p\{|C|; |C| < \infty\} \\
&= E_p\{|C| \mid |C| < \infty\}P_p\{|C| < \infty\} \\
&= \chi(1-p)(1-\theta(p)) \\
&= \tfrac{1}{2}(p - \tfrac{1}{2})^{-1}\left(\frac{1-p}{p}\right)^2.
\end{aligned} \tag{6.13}$$

Note that as $p \downarrow p_c$, $((1-p)/p)^2 \to 1$. Thus, $\chi^f(p) \approx (p - p_c)^{-1}$ when $p > p_c$, so that $\gamma' = 1$.

We now deal with the moments of $|C|$ when C is a finite cluster. Specifically, we are interested in $E_p\{|C|^k; |C| < \infty\}$, or, more usefully, the following ratio and its critical exponent Δ:

$$\frac{E_p\{|C|^{k+1}; |C| < \infty\}}{E_p\{|C|^k; |C| < \infty\}} \approx |p - p_c|^{-\Delta}, \text{ as } p \to p_c, \ k \geq 1. \tag{6.14}$$

Note that, from above, for $p > p_c$:

$$E_p\{|C|^k; |C| < \infty\} = E_{1-p}\{|C|^k\}P_p\{|C|^k < \infty\}, \tag{6.15}$$

and so we need only look at $p < p_c = \frac{1}{2}$. Then

$$E_p\{|C|^k; |C| < \infty\} = E_p\{|C|^k\} = \sum_{n=1}^{\infty} n^k P_p\{|C| = n\}. \tag{6.16}$$

It can be shown (Durrett, 1985) that

$$P_p\{|C| = n\} = \frac{1}{n}\binom{2n}{n-1}p^{n-1}(1-p)^{n+1}. \tag{6.17}$$

We approximate $\binom{2n}{n-1}$ by Stirling's formula[6] as follows:

$$\binom{2n}{n-1} = \frac{n}{n+1}\binom{2n}{n} \sim \frac{n}{n+1}\frac{2^{2n}}{\sqrt{n\pi}} \sim \frac{4^n}{\sqrt{n\pi}}. \tag{6.18}$$

Let $\epsilon = \frac{1}{2} - p$, so that $4p(1-p) = 1 - 4\epsilon^2$. Then, from (6.16), (6.17), and (6.18),

$$\begin{aligned}
E_p\{|C|^k\} &= \sum_{n=1}^{\infty} n^k \frac{1}{n}\binom{2n}{n-1}p^{n-1}(1-p)^{n+1} \\
&= \sum_{n=1}^{\infty} n^{k-1}\frac{4^n}{\sqrt{n\pi}}(\tfrac{1}{2} - \epsilon)^{n-1}(\tfrac{1}{2} + \epsilon)^{n+1}
\end{aligned}$$

[6]Stirling's formula says that $n! \sim (n/e)^n\sqrt{2\pi n}$.

$$= A \sum_{n=1}^{\infty} n^{k-1} \frac{(1-4\epsilon^2)^n}{n^{1/2}}$$

$$\sim A \int_1^{\infty} n^{k-1} \frac{(1-4\epsilon^2)^n}{n^{1/2}} dn, \tag{6.19}$$

where we have approximated the sum by an integral and A is a constant not depending upon n. Now, let $x = -n \log(1 - 4\epsilon^2)$. Then $e^{-x} = (1 - 4\epsilon^2)^n$ and $dx/dn = -\log(1-4\epsilon^2)$. Substituting these expressions into (6.19) and using the approximation $\log(1-x) \sim x$ yields

$$E_p\{|C|^k\} \sim A \int_0^{\infty} n^{k-1} \frac{e^{-x}}{n^{1/2}} \frac{1}{\log(1-4\epsilon^2)} dx$$

$$= A \left(\int_0^{\infty} x^{k-\frac{3}{2}} e^{-x} dx \right) [\log(1-4\epsilon^2)]^{-k+\frac{1}{2}}$$

$$\sim A \cdot \Gamma(k - \tfrac{1}{2})(4\epsilon^2)^{-k+\frac{1}{2}}, \tag{6.20}$$

where $\Gamma(a) = \int_0^{\infty} x^{a-1} e^{-x} dx$ is the gamma function. Now, if we substitute (6.20) into (6.14) and use $\epsilon = \frac{1}{2} - p$, we get

$$\frac{E_p\{|C|^{k+1}\}}{E_p\{|C|^k\}} \sim \frac{\Gamma(k+\frac{1}{2})(4\epsilon^2)^{-k-\frac{1}{2}}}{\Gamma(k-\frac{1}{2})(4\epsilon^2)^{-k+\frac{1}{2}}} = B_k 4^{-1}(\tfrac{1}{2} - p)^{-2}, \tag{6.21}$$

where B_k is a constant depending upon k. Comparison of (6.21) with (6.14) shows that $\Delta = 2$.

To summarize:

Theorem 6.3 *The critical exponents for percolation on a binary tree have the following values:*

$$\alpha = -1, \quad \beta = 1, \quad \gamma = \gamma' = 1, \quad \delta = 2, \quad \eta = 0, \quad \nu = \tfrac{1}{2}, \quad \rho = \tfrac{1}{2}, \quad \Delta = 2. \tag{6.22}$$

6.4 Percolation on Transitive Graphs

As we have seen, lattices and regular trees share the fact that they each have two separate phases divided by a critical point p_c: in the subcritical phase (i.e., $p < p_c$) for both, there are no open infinite clusters; for the supercritical phase (i.e., $p > p_c$) of the lattice, $\mathbb{Z}^d$, there is a unique infinite open cluster, while for an r-regular tree, $\mathbb{T}_r$, there are an infinite number of infinite open clusters. These situations are consistent with the Newman–Schulman (1981a) result that certain types of graphs have, almost surely, either 0, 1, or ∞ number of infinite open clusters. These percolation phases have been studied under a variety of conditions. However, new ideas appeared 30 years ago that changed the way that those involved in the study of percolation viewed the subject.

6.4.1 Uniqueness or Non-uniqueness, or Both?

The study of uniqueness of infinite clusters was given a severe jolt when Grimmett and Newman (1990) discovered that there are classes of graphs $\mathcal{G}$ in which an infinite number of infinite clusters can exist for certain values of $p \in (p_c(\mathcal{G}), 1)$ but not for other $p \in (p_c(\mathcal{G}), 1)$. The example studied in that pioneering article involved the direct product

$$\mathbb{L} = \mathbb{T}_r \times \mathbb{Z} \tag{6.23}$$

of an r-regular tree $\mathbb{T}_r$ and the line $\mathbb{Z} = \{z : z = \cdots, -1, 0, 1, \ldots\}$.

Definition 6.2 The *direct product* $\mathcal{G} = \mathcal{G}_1 \times \mathcal{G}_2$ of two graphs $\mathcal{G}_1 = (\mathcal{V}_1, \mathcal{E}_1)$ and $\mathcal{G}_2 = (\mathcal{V}_2, \mathcal{E}_2)$ has the node set $\mathcal{V} = \mathcal{V}_1 \times \mathcal{V}_2$, and an edge set $\mathcal{E}$ consisting of pairs (u_1, u_2) and (v_1, v_2) such that either $u_1 = v_1$ and $\{u_2, v_2\} \in \mathcal{E}_2$ or $u_2 = v_2$ and $\{u_1, v_1\} \in \mathcal{E}_1$. The resulting graph $\mathcal{G}$ is often called a *product graph*.

Thus, for every integer in $\mathbb{Z}$, there exists a tree $\mathbb{T}_r$, and every tree is connected to the trees above and below it on each plane of $\mathbb{Z}$. To be more specific:

Definition 6.3 Let $\mathcal{V}(\mathbb{T}_r)$ denote the node set of the r-regular tree. Then, the product graph $\mathbb{L} = \mathbb{T}_r \times \mathbb{Z}$ has node set

$$\mathcal{V}(\mathbb{L}) = \{(t, z) \in \mathbb{L} : t \in \mathcal{V}(\mathbb{T}_r), z \in \mathbb{Z}\}, \tag{6.24}$$

with origin $(\emptyset, 0)$ and edge set $\mathcal{E}(\mathbb{L})$ defined by

$$\mathcal{E}(\mathbb{L}) = \{((t_1, z_1), (t_2, z_2)) : (t_1, z_1) \sim (t_2, z_2)\}, \tag{6.25}$$

where $(t_1, z_1) \sim (t_2, z_2)$ means that the nodes (t_1, z_1) and (t_2, z_2) are adjacent to each other.

Definition 6.4 The distance, $\delta_{\mathbb{T}_r}(t_1, t_2)$, between two nodes, $t_1, t_2 \in \mathcal{V}(\mathbb{T}_r)$, is defined as the number of edges in the unique path from t_1 to t_2; also, define $\delta_{\mathbb{Z}}(z_1, z_2) = |z_1 - z_2|$ to be the distance between $z_1, z_2 \in \mathbb{Z}$. Then, $\mathcal{E}(\mathbb{L})$ is defined by the relation

$$(t_1, z_1) \sim (t_2, z_2) \text{ iff } \delta_{\mathbb{T}_r}(t_1, t_2) + \delta_{\mathbb{Z}}(z_1, z_2) = 1. \tag{6.26}$$

Definition 6.5 Two nodes of $\mathbb{L}$ are said to be *adjacent* iff either their $\mathbb{T}_r$-components are equal and their $\mathbb{Z}$-components are adjacent in $\mathbb{Z}$, or vice versa.

Grimmett and Newman showed that, for sufficiently large r, $\mathbb{L}$ has *three* nontrivial phases:

1. When $p < p_c(\mathbb{L})$, there are, almost surely, *no* infinite open clusters.
2. When p is larger than but close to $p_c(\mathbb{L})$, there are, almost surely, *infinitely many* infinite open clusters.
3. When p is close to 1, the infinite open cluster is, almost surely, *unique*.

Similar findings were also shown to hold for the graph $\mathbb{L}_d = \mathbb{T}_r \times \mathbb{Z}^d$, $d \geq 2$. Schonmann (1999) showed that r need not be large to produce this phenomenon, that it actually holds for all $r \geq 3$.

The Grimmett–Newman examples clearly demonstrated that hitherto undiscovered structure could be found if one looked at percolation on graphs that were different from the classical square lattice $\mathbb{Z}^d$ (or even trees). Attention of researchers quickly focused on the issue of uniqueness vs. non-uniqueness of infinite clusters for very general graphs. See, for example, the survey article by Häggström and Jonasson (2006).

6.4.2 Transitive and Quasi-Transitive Graphs

In a highly influential article, Benjamini and Schramm (1996) followed up the Grimmett–Newman examples with a fascinating collection of research questions, conjectures, and results. In particular, they proposed that percolation should be studied through graphs and lattices more exotic than $\mathbb{Z}^d$. To go beyond $\mathbb{Z}^d$, new types of graphs have to be defined using additional structure. With this in mind, Benjamini and Schramm focused on studying percolation on *transitive graphs* and, more generally, on *quasi-transitive graphs*.

First, we need the following definitions:

Definition 6.6 Let $\mathcal{G}_1 = (\mathcal{V}_1, \mathcal{E}_1)$ and $\mathcal{G}_2 = (\mathcal{V}_2, \mathcal{E}_2)$ be two graphs.

- An *isomorphism* between $\mathcal{G}_1$ and $\mathcal{G}_2$ is a bijection[7] $f : \mathcal{V}_1 \to \mathcal{V}_2$ such that $(v, v') \in \mathcal{E}_1$ iff $(f(v), f(v')) \in \mathcal{E}_2$, for all $v, v' \in \mathcal{V}_1$. In other words, iff the graph structure is preserved.
- An *automorphism* of a graph $\mathcal{G}$ is an isomorphism between $\mathcal{G}$ and itself. The set of all automorphisms of $\mathcal{G}$ forms a group, $\text{Aut}(\mathcal{G})$, under the operation of composition.

Now we can define transitive and quasi-transitive graphs.[8]

Definition 6.7 A graph $\mathcal{G} = (\mathcal{V}, \mathcal{E})$ is called *transitive* if nodes cannot be distinguished from each other; that is, for every pair of nodes $v, v' \in \mathcal{V}$, there exists $\gamma \in \text{Aut}(\mathcal{G})$ such that $\gamma(v') = v$.

Examples of transitive graphs include $\mathbb{Z}^d$, the Grimmett–Newman example $\mathbb{T}_r \times \mathbb{Z}$, and Cayley graphs.

Definition 6.8 Let H be a countable group and suppose that $S \subseteq H$ is a finite generating set of H (i.e., every element of H can be expressed as a sum or product of elements of S). Assume S is symmetric (i.e., $S = S^{-1}$). A graph $\mathcal{G} = (\mathcal{V}, \mathcal{E})$ is a (right) *Cayley graph* with node set $\mathcal{V} = H$ such that $(v, v') \in \mathcal{E}$ iff there exists an element $s \in S$ such that $v' = vs$.

Although all Cayley graphs are transitive, not all transitive graphs are Cayley graphs (see, e.g., Häggström and Jonasson, 2006). Note that S is finite iff $\mathcal{G}$ is locally finite. Furthermore, every Cayley graph is connected.

---

[7]A *bijection* is a mapping that is one-to-one and onto.
[8]We have defined node-transitive graphs. Similarly, edge-transitive graphs can also be analogously defined.

Definition 6.9 A graph $\mathcal{G} = (\mathcal{V}, \mathcal{E})$ is called *quasi-transitive* if there are finitely many *orbits* (i.e., equivalence classes) of nodes defined by graph automorphisms. In other words, $\mathcal{G}$ is quasi-transitive if $\mathcal{V}$ can be partitioned into a finite number of node subsets, $V_1, V_2, \ldots, V_k$, such that for any pair of nodes $v, v' \in V_i$, there exists $\gamma \in \mathrm{Aut}(\mathcal{G})$ such that $\gamma(v') = v$.

If the quasi-transitive graph $\mathcal{G}$ has only one orbit, then $\mathcal{G}$ is a transitive graph.

Let $\mathcal{G}$ be a quasi-transitive graph and consider bond percolation on $\mathcal{G}$. The usual percolation critical value is given by

$$p_c = p_c(\mathcal{G}) = \inf\{p : \mathrm{P}_p\{\text{there exists an infinite cluster}\} = 1\}. \qquad (6.27)$$

In light of the Grimmett–Newman results, Benjamini and Schramm (1996) defined a second critical value:

$$p_u = p_u(\mathcal{G}) = \inf\{p : \mathrm{P}_p\{\text{there exists a unique infinite cluster}\} = 1\}. \qquad (6.28)$$

We have already seen that when $\mathcal{G} = \mathbb{Z}^d$, an infinite cluster, if it exists, is, almost surely, unique (Aizenman and Barsky, 1987; Burton and Keane, 1989), so that $p_u(\mathbb{Z}^d) = p_c(\mathbb{Z}^d)$. In the case of percolation on the tree $\mathbb{T}_r$, $r \geq 3$, we have that $p_u(\mathbb{T}_r) = 1$. For more general graphs, uniqueness of the infinite cluster may not hold.

Currently, a great deal of attention is being paid to the problem of how to characterize those graphs $\mathcal{G}$ that have

$$0 < p_c(\mathcal{G}) \leq p_u(\mathcal{G}) \leq 1. \qquad (6.29)$$

These relations hold for the r-regular tree $\mathbb{T}_r$ because

$$p_c(\mathbb{T}_r) = \frac{1}{r-1} < 1 = p_u(\mathbb{T}_r). \qquad (6.30)$$

For the transitive graph $\mathbb{T}_r \times \mathbb{Z}$, Grimmett and Newman (1990) showed that $p_c < p_u < 1$. Generalizing that example, Benjamini and Schramm (1996) showed that if $\mathcal{G}$ is a quasi-transitive graph, then the product graph $\mathbb{T}_r \times \mathcal{G}$ has $p_c < p_u$, assuming r is large enough. To construct further generalizations, we introduce the following definition:

Definition 6.10 The *edge-isoperimetric constant* of $\mathcal{G}$ is defined as

$$h_\mathcal{E}(\mathcal{G}) = \inf_S \frac{|\partial_\mathcal{E} S|}{|S|}, \qquad (6.31)$$

where the infimum ranges over all finite non-empty subsets S of $\mathcal{V}$ and

$$\partial_\mathcal{E} S = \{(u, v) \in \mathcal{E} : u \in S, v \in \mathcal{V} \backslash S\} \qquad (6.32)$$

is the *edge boundary* of S. The quantity $h_\mathcal{E}(\mathcal{G})$ is also referred to as the *Cheeger constant*.

The edge-isoperimetric constant $h_{\mathcal{E}}(\mathcal{G})$ reminds us of the surface-to-volume ratio in (5.65) that played a role in the Burton–Keane proof of uniqueness of the infinite cluster on $\mathbb{Z}^d$. A similar notion of *node-isoperimetric constant*, $h_{\mathcal{V}}(\mathcal{G}) = \inf_S |\partial_{\mathcal{V}} S|/|S|$, may also be defined, but where

$$\partial_{\mathcal{V}} S = \{u \in S : \text{there exists } v \in \mathcal{V} \backslash S \text{ such that } (u, v) \in \mathcal{E}\}. \tag{6.33}$$

From either definition, we have the following types of graphs.

Definition 6.11 A graph $\mathcal{G}$ is said to be *amenable* if $h(\mathcal{G}) = 0$. If $h(\mathcal{G}) > 0$, the graph $\mathcal{G}$ is said to be *non-amenable*.

Note that if $\mathcal{G}$ is transitive, then $h_{\mathcal{V}}(\mathcal{G}) = 0$ iff $h_{\mathcal{E}}(\mathcal{G}) = 0$; as a result, it makes no difference which definition of isoperimetric constant, $h_{\mathcal{E}}(\mathcal{G})$ or $h_{\mathcal{V}}(\mathcal{G})$, is used for amenability, which is why we have omitted the subscript in Definition 6.11.

Some examples:

- The lattice $\mathbb{Z}^d$ is an amenable graph: just take $S_n = \{-n, \ldots, n\}^d$, and observe that $|\partial_{\mathcal{E}} S_n|/|S_n| \to 0$ as $n \to \infty$.
- A Bethe lattice, which does not have a boundary and, hence, does not possess boundary points connected by occupied bonds to the center of the lattice, is an example of an amenable graph.
- An r-regular Cayley tree $\mathbb{T}_r$ with $r \geq 3$ has a boundary even as the lattice increases in size to its limit and is, therefore, an example of a non-amenable graph.

Percolation on a Bethe lattice has an exact solution corresponding to classical mean-field theory (Stauffer and Aharony, 1991, Section 2.4). For non-amenable graphs, such as the Cayley tree, percolation uses a different approach (Benjamini et al., 1999).

Benjamini and Schramm conjectured that if $\mathcal{G}$ were quasi-transitive and non-amenable, then, $p_c < p_u$; however, this conjecture is still open. They noted that if their conjecture were true, it would provide a useful characterization of amenability. They also stated that the arguments of Newman–Schulman and Burton–Keane regarding the number of infinite clusters (0, 1, or ∞) could be extended to quasi-transitive graphs. Let N_0 denote the number of infinite clusters in the graph $\mathcal{G}$. Häggström and Peres (1999) proved that if $\mathcal{G}$ were quasi-transitive and satisfied a certain condition, then, almost surely:

$$N_0 = \begin{cases} 0, & \text{if } p \in [0, p_c) \\ \infty, & \text{if } p \in (p_c, p_u). \\ 1, & \text{if } p \in (p_u, 1] \end{cases} \tag{6.34}$$

Schonmann (1999) then proved the result in full generality by removing the additional condition. See also Häggström and Peres (1999) and Kesten (2002).

The result (6.34) does not specify the value of N_0 at either of the critical points p_c or p_u. However, it has been shown (Benjamini et al., 1999) that for percolation on non-amenable Cayley graphs, there is, almost surely, no infinite cluster (i.e., $N_0 = 0$) at the critical point p_c.

6.5 Continuum Percolation

There are situations in which a regular lattice-based structure would be inappropriate to impose upon a model (e.g., when points are randomly distributed in space) and a lattice-free model would be more appropriate. In such instances, space is viewed as being continuous, not discrete. Fortunately, much of the development of the topic has been able to use discretization methods based upon the corresponding results from discrete percolation.

Many of the early applications of continuum percolation were in physics, and were concerned with modeling electrical networks and amorphous semiconductors; more recently, continuum percolation[9] has been applied to wireless communication networks (Penrose, 1997; Gupta and Kumar, 1998, 2000; Booth et al., 2003), which include *smart homes* that connect all devices and appliances in a home so that they can communicate with each other and the homeowner. Furthermore, continuum percolation has been found useful for single-linkage cluster analysis and minimum spanning trees (Ramey, 1982; Kesten, 1987).

6.5.1 Example: Amorphous Computer Network

First, we motivate the idea of continuum percolation through an example on amorphous computing, whose importance and relevance are due to new types of technology that could possibly produce impressive amounts of information having great power and at very little cost.

An amorphous computer network consists of a vast number ($\sim 10^6$–10^{12}) of *computational particles*: tiny (millimeter-scale), extremely cheap, asynchronous computer microprocessors (*nodes*) that can communicate via radio waves and memory, over a short range and with limited reliability. All nodes are identical and are controlled by the same program (although they each have an independent random number generator that provides a unique identifier of that particle recognizable to its neighbors), they encode and decode messages in identical fashion, they compute at similar speeds, and they have a modest amount of computing power and limited memory (so they need to be energy efficient). Although they are capable of sending and receiving data to/from a remote operator, we assume that a sender does not know if a message has been received. Particles also have no knowledge of their geographical positions or orientations.

The particles are mixed with bulk materials, such as gels or concrete, thereby creating a "smart paint," and are randomly distributed (or painted) onto a surface. Using their natural ability for massive parallelism, the particles enter into a discovery phase in which they can sense data, recognize their neighboring nodes, and self-assemble to form a giant parallel computer. Possible implementations of amorphous computing systems include microfabrication and cellular engineering (Abelson, Beal, and Sussman, 2007).

---

[9]Continuum percolation is also known as the *Boolean model* in stochastic geometry (see, e.g., Meester and Roy, 1996, Section 1.4).

- *Microfabrication* involves painting walls, buildings, or bridges with smart paint so that traffic can be monitored, buildings can be protected from intruders, and small cracks in foundations or walls can be repaired. There is now interest in making the particles mobile.
- *Cellular engineering* was inspired by computational advances in systems biology that support the emerging field of *synthetic biology*. The idea is to embed digital-logic circuits of significant complexity into individual living cells through their DNA, which could function, for example, as drug delivery vehicles. There is a concern that cellular computing tends to be very slow and would not be competitive for this type of computing. Possibilities exist, however, that cellular engineering could be successful in controlling fabrication processes at the molecular level.

Current work centers on the construction and implementation of various languages (e.g., Daniel N. Coore's *Growing-Point Language, GPL*; see Coore, 1999) that are proposed for programming amorphous computers.

One goal of amorphous computing is to propagate a message throughout the system (via *diffusion wave propagation*). Suppose that a specific particle A broadcasts a new message while setting its *hop count* to 0. When a neighboring particle B receives the message, the hop count will be incremented by one, and then the message will be rebroadcast to B's neighbors. If B receives another message with a higher hop count, it will ignore it in favor of A's message. That way, the hop count operates as an estimate of a particle's distance from A. Under what conditions (and with what probability) could A's message be broadcast so that a large portion of an entire network receives the message?

This is a question for which the *continuum percolation* model can provide an answer. In this "no-lattice" model, circular disks with a fixed radius (the range of a particle) are randomly sprinkled around the infinite plane. The location of the center point (i.e., the position of the particle) of each disk is modeled by a two-dimensional homogeneous Poisson point process. The disks are allowed to overlap with each other. We are interested in the smallest density of disks that would guarantee an infinite cluster and, hence, enable a message to be propagated through the network via a path of overlapping disks. Percolation theory shows that there exists a *critical density* of the disks; for a density larger than the critical density, there will emerge, almost surely, a unique infinite cluster for which percolation occurs.

6.5.2 Gilbert's Disk Percolation Model

The basic idea behind what we know today as *continuum percolation* was provided by Edgar N. Gilbert. His article (Gilbert, 1961) is generally credited with starting the study of continuum percolation and describing its critical properties. It appeared just after the Erdős and Rényi (1959, 1960) articles on random graphs and his own article on the same subject (Gilbert, 1959). He realized that the traditional random-graph models of Erdős and Rényi could not represent networks of short-range, wireless-transmitting stations accurately enough, and that a new type of theory was necessary. Gilbert's article proved to be innovative and well ahead of its time; yet, it took 25 years for his ideas and contributions to be recognized by mathematicians and probabilists.

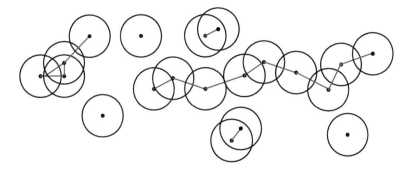

Figure 6.3 Part of the graph in two dimensions of Gilbert's disk model. Pairs of Poisson-distributed points in the plane (i.e., centers of circles, black dots) are said to be adjacent to each other if the corresponding disks overlap, in which case the centers of the disks are joined up with red lines.

Gilbert's motivation was to provide an alternative model of a communications network, in which an infinite number of wireless broadcasting stations are randomly distributed over an infinite plane. The wireless transmitting stations are represented by an infinite number of points $\mathbf{X}_i \in \mathbb{R}^2, i = 1, 2, \ldots$, which are distributed according to a two-dimensional homogeneous Poisson point process $\mathcal{P}_\lambda$ with *intensity* (or *density*) λ. Thus, the expected number of points in any region is λ times the volume of that region. A disk $D = D(r, \lambda)$ with a fixed radius r is centered at each point, where r is the same for all points. The radius represents the signal strength of each transmitter.[10] Gilbert's question was whether the system of stations could provide long-distance, multi-hop communication.

Let $\mathcal{D} = \mathcal{D}(r, \lambda)$ represent the union of all the disks. Pairs of disks belong to the same cluster if they overlap each other. See **Figure 6.3**. The model is said to *percolate* if there exists an infinite cluster in $\mathcal{D}$. Gilbert demonstrated that there exists a critical value, λ_c, for the density of the Poisson point process (usually called the *critical density*). He showed that if $\lambda < \lambda_c$, then the model, almost surely, does not percolate, while if $\lambda > \lambda_c$, there emerges, almost surely, an infinite cluster in the infinite plane, and the model, almost surely, percolates. In that latter case, long-distance, multi-hop communication would, almost surely, be possible.

Note that, as opposed to percolation on a lattice, where there is a critical probability, in continuum percolation, there is a critical density and the state of percolation depends upon λ and r, and not on a probability p. However, because the expected number of points in a circle of radius r is $\pi r^2 \lambda$, the important parameter is $a = \pi r^2 \lambda$.

6.5.3 Origins

Gilbert never used the term "continuum percolation" in his 1961 article, where he called his model a *random plane network*, or in any of his publications, and he never

---

[10] In stochastic geometry, this model has been called the *Poisson "blob" model*. A generalization, called the *Poisson Boolean model* (see, e.g., Meester and Roy, 1996, Section 1.4), assumes that the disk radius is a non-negative random variable, independent of the location of its center, so that, for example, different stations can have different signal strengths. Another possible generalization is to leave the radius fixed but vary the Poisson intensity parameter. More general random shapes than disks in two dimensions or spheres in three dimensions are studied by Hall (1985) and others.

returned to the subject. So, if Gilbert never used the term continuum percolation, who was the first who did?

The first such use of the term "continuum percolation" (as in "percolation on a continuum") appears to have been Zallen and Scher (1971) in an article in *Physical Review B*. They were concerned about the motion of a classical particle in a random potential V; in particular, they studied the transition from localized to delocalized states in amorphous semiconductors. They recognized that this was not a standard lattice-type problem. They introduced a density function $\phi(E)$ that specified the fraction of space that is accessible (i.e., satisfying $V < E$) to particles of energy E. They then appealed to percolation theory to deduce that when $\phi(E) < \phi_c = \phi(E_c)$, where E_c is some critical energy level, the system behaves like an electric insulator (and, therefore, is of no further interest in studying its electrical properties), whilst when $\phi(E) > \phi_c$, an infinite cluster of electrons is formed that can conduct electric current. The density ϕ_c was called the *critical density*, and it was estimated in one, two, and three dimensions. This article had very little in common with Gilbert's article, other than the use of percolation in a continuum setting.

A more recognizable version of continuum percolation was provided by Pike and Seager (1974) in *Physical Review B*, followed by Haan and Zwanzig (1977) in the *Journal of Physics A*. In both articles, points are randomly distributed over some homogeneous material, and a geometrical region, such as a circle or sphere, is centered at each point; if two regions overlap, their centers are joined to form a cluster, so that a collection of clusters of varying size correspond to a given configuration of points. No specification or model was given as to how the points were spatially distributed over the material. The Pike–Seager article cited Gilbert's article, but only to criticize his Monte Carlo results; the Haan–Zwanzig article did not cite Gilbert.

These two articles started an industry. A large number of articles on continuum percolation (i.e., randomly distributed points serving as centers of possibly overlapping, fixed- or random-radius circles or spheres) began to appear in the physics literature, including those by Gawlinski and Stanley (1981), Vicsek and Kertész (1981), both in *Journal of Physics A*, and Kertész and Vicsek (1982) in *Zeitschrift für Physik B*. In particular, either Monte Carlo experiments or series expansions were used to estimate the critical exponents of the continuum percolation system.

Thus, by the early 1980s, the concepts and the term "continuum percolation" had become fairly well-established in the physics literature, mostly without any recognition or knowledge of Gilbert's 1961 article. The Pike–Seager article appears to be an anomaly, being the only article in the physics literature that referenced Gilbert (1961) in any way, but without discussing his novel ideas.

The first statistical references to Gilbert (1961) were Roberts (1967) and Roberts and Storey (1968) in two related articles in *Biometrika*, but neither used the term continuum percolation. The big breakthrough came in 1985 with the article entitled "On continuum percolation" by Peter Hall (1985). Hall referenced Gilbert's article (as well as the Roberts, Pike–Seager, Haan–Zwanzig, and Gawlinski–Stanley articles, the last two of which explicitly used the term continuum percolation). Hall developed the theory of continuum percolation, credited Gilbert's role in that development, and caused the term to enter the mathematics and probability lexicon. In related articles, Zuev and Sidorenko (1985a,b) developed an approximation theory of continuum percolation, but without using the term "continuum" and without referencing Gilbert.

6.5.4 The d-Dimensional Poisson Point Process

As we have seen, Gilbert's disk model uses the concept of a two-dimensional Poisson point process as a way of generating points in the plane. The following representation is a generalization of the Poisson process to d dimensions.

Let $\lambda > 0$ and let $\mathcal{P}_\lambda \in \mathbb{R}^d$ be a countably infinite set of random points in d-dimensional space. Let A be a bounded region (i.e., Borel set) of that space with size $|A|$ (usually taken to be the Lebesgue measure of A in $\mathbb{R}^d$). Let the random variable $N_\lambda(A)$ be the number of points of $\mathcal{P}_\lambda$ that fall into A. Then, we have the following definition:

Definition 6.12 $\mathcal{P}_\lambda$ is a *d-dimensional homogeneous Poisson point process with intensity (density)* λ if:

1. For any finite collection, $A_1, A_2, \ldots, A_n$, of pairwise disjoint regions of d-dimensional space, $N_\lambda(A_1), N_\lambda(A_2), \ldots, N_\lambda(A_n)$ are mutually independent random variables.
2. $N_\lambda(A) \sim \text{Poisson}(\lambda|A|)$; that is

$$P\{N_\lambda(A) = k\} = e^{-\lambda|A|} \frac{(\lambda|A|)^k}{k!}, \quad k = 1, 2, \ldots. \tag{6.35}$$

In the case of two dimensions (i.e., $d = 2$), such a process can be created by dividing the plane into unit squares and then, independently for each square, uniformly sprinkling a Poisson-distributed number of points into each square.

6.5.5 Disks and Clusters

Definition 6.13 In two dimensions, $D = D(\lambda, r)$ represents a disk with radius $r > 0$ and whose center $\mathbf{X} \in \mathbb{R}^2$ is located in the plane with Poisson intensity $\lambda > 0$.

Definition 6.14 Two disks, D_i and D_j, are said to be *adjacent* if they overlap each other (i.e., if $D_i \cap D_j \neq \emptyset$), in which case, they are also said to be members of the same cluster.

If two disks, D_i and D_j, overlap, we join up their center points, $\mathbf{X}_i$ and $\mathbf{X}_j$, and this provides an *edge* in the graph; see **Figure 6.3**.

Note. There are differences regarding the meaning of the word "overlap." Some articles and texts say that the centers of two disks have to be a distance of at most r (i.e., one radius) apart for them to be joined (see, e.g., Gilbert, 1961; Gupta and Kumar, 1998; Booth et al., 2003; Bollobás and Riordan, 2006, p. 242), while other texts are less severe, saying that the distance has to be at most $2r$ (i.e., one diameter) apart (Grimmett, 1989, Section 10.5; implicit in Hall, 1985; see also Gawlinski and Stanley, 1981; Kesten, 1987).

Definition 6.15 Two disks, D_i and D_j are said to *communicate* with each other (we write $D_i \leftrightarrow D_j$) if there exists a sequence of disks, $D_{i_1}, D_{i_2}, \ldots, D_{i_k}$, such that $D_{i_1} = D_i$, $D_{i_k} = D_j$, and D_{i_m} is adjacent to $D_{i_{m+1}}$, $m = 1, 2, \ldots, k - 1$.

Definition 6.16 A *cluster* (or *connected component*) of disks is a set of disks, $\{D_i : i \in I\}$, which is maximal with $D_i \leftrightarrow D_j$ for all $i, j \in I$.

If we have an infinite number of such disks, which we denote by $D_1, D_2, \ldots,$ then $\mathcal{D} = \cup_{i=1}^{\infty} D_i$ will denote the union of all the disks.

6.5.6 Critical Intensity

Let $\mathcal{G}_{r,\lambda}$ denote the graph with a collection of sites (nodes) and a bond (edge) between a pair of sites is present if the sites are at most a certain distance (either r or $2r$, depending upon the model) apart from each other. We note that although λ and r form the parameters of Gilbert's disk system, their importance lies in their being combined into the single parameter, $a = \pi r^2 \lambda$. The parameter a is called the *connection area* and the graph can be expressed as $\mathcal{G}(a)$, where a is a constant. Thus, the graph $\mathcal{G}(a)$ will change as we change the values of r and λ, but a has to remain fixed. For example, if the density λ of Poisson points is scaled up by the constant $c > 0$, then because $a = \pi(r/\sqrt{c})^2(c\lambda)$, the common radius r of each disk has to get proportionately smaller. For convenience and without loss of generality, we take the disks to have unit radius (i.e., $r = 1$). Thus, λ is the sole parameter of the continuum percolation problem.

Let P_λ represent the probability measure corresponding to a Poisson process with intensity λ. Define $W(\mathbf{u})$ as those points generated by the Poisson process that are centers of disks or spheres in the cluster containing $\mathbf{u} \in \mathbb{R}^d$.

Definition 6.17 The *continuum percolation probability* $\gamma(\lambda)$ is defined by

$$\gamma(\lambda) = P_\lambda\{|W| = \infty\}, \tag{6.36}$$

where $W = W(\mathbf{0})$ is the cluster at the origin.

It follows that

$$P_\lambda\{\text{there exists } \mathbf{u} \text{ such that } |W(\mathbf{u})| = \infty\} = \begin{cases} 0, & \text{if } \gamma(\lambda) = 0 \\ 1, & \text{if } \gamma(\lambda) > 0 \end{cases}. \tag{6.37}$$

There is, almost surely, a unique infinite cluster of overlapping disks or spheres whenever $\gamma(\lambda) > 0$, in which case, the system is said to *percolate*. What we need is to show that there exists a *critical intensity* λ_c of λ, defined by

$$\lambda_c = \sup_\lambda \{\gamma(\lambda) = 0\}, \tag{6.38}$$

which is the largest value of λ for which the probability that the cluster size $|W|$ is infinite changes from being zero to being positive.

6.5.7 Approximation by Discrete Percolation

The basic idea is to represent the continuum percolation problem as a limit of a sequence of discrete percolation problems. We superimpose over $\mathbb{R}^d$ a hypercubical lattice, drawing a box around each node in the lattice, and then checking to see which boxes contain at least one point generated by the Poisson process. Nodes with boxes that contain a point from that process are regarded as "open" (the remainder are "closed"), and bonds are drawn between pairs of open lattice points iff they are "close" to each other. This reduces the problem to one of site percolation on a regular lattice, so that we can apply previous discrete percolation results.

The following development is due to Zuev and Sidorenko (1985a,b) (and reproduced in Menshikov, Molchanov, and Sidorenko, 1986 and Grimmett, 1989, Section 10.5). A similar approximation in which an "imaginary covering mesh" is placed over the Poisson-distributed points was previously discussed by Gawlinski and Stanley (1981).

First, we define a *rescaled lattice*. Let $\mathbf{k} = (k_1, k_2, \ldots, k_d)^\tau \in \mathbb{Z}^d$ be a site in a d-dimensional regular lattice graph, and let n be a positive integer. Rescale the site $\mathbf{k}$ by setting

$$\mathbf{x} = (x_1, x_2, \ldots, x_d)^\tau = \left(\frac{k_1}{n}, \frac{k_2}{n}, \ldots, \frac{k_d}{n} \right)^\tau = n^{-1} \mathbf{k}, \qquad (6.39)$$

and let $\mathcal{G}_n$ denote the rescaled graph with sites

$$\mathbb{Z}_n^d = n^{-1} \mathbb{Z}^d = \{ n^{-1} \mathbf{k}; \; \mathbf{k} \in \mathbb{Z}^d \}. \qquad (6.40)$$

To turn the set of sites $\mathbb{Z}_n^d$ into a lattice $\mathcal{L}_n$, we need to define what we mean by an edge. We draw the d-dimensional box

$$B_n(\mathbf{x}) = \left[x_1 - \frac{1}{2n}, x_1 + \frac{1}{2n} \right) \times \cdots \times \left[x_d - \frac{1}{2n}, x_d + \frac{1}{2n} \right), \quad \mathbf{x} \in \mathbb{Z}_n^d, \qquad (6.41)$$

around each site $\mathbf{x}$ in the rescaled graph so that $\mathbf{x}$ is at the center of the box. Note that, if $\mathbf{x}$ and $\mathbf{y}$ are two sites:

$$B_n(\mathbf{x}) \cap B_n(\mathbf{y}) = \emptyset. \qquad (6.42)$$

Two sites, $\mathbf{x}$ and $\mathbf{y}$, in the graph are said to be *adjacent* (written $\mathbf{x} \sim \mathbf{y}$) iff there exist points $\mathbf{u} \in B_n(\mathbf{x})$ and $\mathbf{v} \in B_n(\mathbf{y})$ such that the Euclidean distance between them is small; that is

$$\| \mathbf{u} - \mathbf{v} \|_2 = \left\{ \sum_{i=1}^{d} (u_i - v_i)^2 \right\}^{1/2} \leq 2, \qquad (6.43)$$

in which case they are joined by a bond (edge). If $\mathbf{u}$ and $\mathbf{v}$ are at most a distance 2 apart, then their disks (or spheres), each having unit radius, overlap each other.[11] Thus, we have constructed the rescaled lattice $\mathcal{L}_n$ with sites and bonds well-defined. See **Figure 6.4**.

Next, we consider site percolation on $\mathcal{L}_n$. A site $\mathbf{x}$ is said to be *open* if $B_n(\mathbf{x})$ contains at least one point of the Poisson process; otherwise, the site $\mathbf{x}$ is said to be *closed*. Assume that the origin $\mathbf{0}$ is a point generated by the Poisson process. From (6.41), where $|B_n(\mathbf{0})| = n^{-d}$, the Poisson probability that $B_n(\mathbf{0})$ is empty is $e^{-\lambda/n^d}$. Thus, a site of $\mathbb{Z}_n^d$ is open with probability

$$p = p_n(\lambda) = 1 - e^{-\lambda/n^d}, \qquad (6.44)$$

---

[11] In (6.43), we adopt the rule that disks overlap if their centers are at most $2r$ apart. Zuev and Sidorenko use the more restrictive definition of "overlap," so that their upper limit of distance in (6.43) is 1.

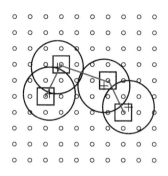

Figure 6.4 The small circles represent nodes on the lattice. A box is drawn around each node. If a box contains a point + from a Poisson process, then the site is "open." Larger circles each with a center + are drawn. Bonds (red lines) are drawn between pairs of open nodes whose circles overlap each other.

and closed with probability $1 - p$, independently of all other sites (because of (6.41)). Thus, for each point $\mathbf{u}$ of the Poisson process, there exists a unique open node $\mathbf{x} \in \mathbb{Z}_n^d$ such that $\mathbf{u} \in B_n(\mathbf{x})$. This implies that there is a many-to-one mapping, $\mathbf{x} = f_n(\mathbf{X})$, of the points $\mathbf{X}$ into the set $\mathbb{Z}_n^d$ of sites $\mathbf{x}$ of $\mathcal{L}_n$. So, an infinite cluster $W^{(c)}$ of disks (or spheres), whose centers are in the Poisson process, is associated with an infinite open cluster $W_n^{(d)}$ in $\mathcal{L}_n$, where the superscripts "c" and "d" refer to the continuum and discrete percolation problems, respectively. In other words, percolation of the points in the Poisson process implies percolation on $\mathcal{L}_n$. This, in turn, means that for any n,

$$\lambda > \lambda_c \implies p_n(\lambda) \geq p_c(\mathcal{L}_n), \tag{6.45}$$

where $p_c(\mathcal{L}_n)$ denotes the critical probability for site percolation on the rescaled lattice $\mathcal{L}_n$. Because $p_n(\lambda)$ is an increasing function, p_n^{-1} exists and so, from (6.44),

$$\lambda_c \geq p_n^{-1}(p_c(\mathcal{L}_n)) = -n^d \log(1 - p_c(\mathcal{L}_n)). \tag{6.46}$$

Although this proves only that $\lambda_c > 0$, the result (6.46) will be helpful.

Let

$$\lambda_T = \sup_\lambda \{ \mathrm{E}_\lambda \{ |W| \} < \infty \} \tag{6.47}$$

denote the largest value of λ for which the expected cluster size changes from being finite to being infinite. Now, if $\mathrm{E}_\lambda\{|W|\} < \infty$, then $\mathrm{P}_\lambda\{|W| = \infty\} = 0$. So,

$$\lambda_T \leq \lambda_c. \tag{6.48}$$

Actually, as we shall see, equality holds in (6.48).

Let $W_n^{(d)}$ denote the set of open sites of $\mathcal{L}_n$ that form a cluster containing the origin or a site in the set $\{ \mathbf{x} \in \mathbb{Z}_n^d : \mathbf{x} \sim \mathbf{0} \}$ of neighbors of the origin. Suppose $|W_n^{(d)}| = m$. Recall that every $\mathbf{x} \in W_n^{(d)}$ is the center of a box that also contains at least one point generated by the Poisson process. Consider one such box. The mean number of such points in the box is $\mathrm{E}_\lambda\{N|N \geq 1\}$, where $N \sim \mathrm{Poisson}(\lambda/n^d)$. We have that

$$\mathrm{E}_\lambda\{N|N \geq 1\} = \frac{\mathrm{E}_\lambda\{N\}}{\mathrm{P}_\lambda\{N \geq 1\}} = \frac{\lambda/n^d}{1 - e^{-\lambda/n^d}}. \tag{6.49}$$

Because $W^{(c)} \subseteq f_n^{-1}(W_n^{(d)})$, it follows that

$$
\begin{aligned}
\mathrm{E}_\lambda\{|W^{(c)}|\} &\le \mathrm{E}_\lambda\{|f_n^{-1}(W_n^{(d)})|\} \\
&= \sum_{m=0}^{\infty} m \mathrm{E}_\lambda\{N|N \ge 1\} \mathrm{P}_\lambda\{|W_n^{(d)}| = m\} \\
&= \frac{\lambda/n^d}{1 - e^{-\lambda/n^d}} \sum_{m=0}^{\infty} m \mathrm{P}_\lambda\{|W_n^{(d)}| = m\} \\
&= \frac{\lambda/n^d}{1 - e^{-\lambda/n^d}} \, \mathrm{E}_\lambda\{|W_n^{(d)}|\}.
\end{aligned} \tag{6.50}
$$

The next step is to simplify $\mathrm{E}_\lambda\{|W_n^{(d)}|\}$.

Consider the one-dimensional case, in which a disk is an interval and its diameter is the width of that interval. How many lattice sites surrounding the origin in $\mathcal{L}_n$ are contained within an interval of width 2 whose center is at the origin? The sites surrounding the origin within a distance of 1 are situated at distances $\pm\frac{1}{2n}, \pm\frac{2}{2n}, \ldots,$ $\pm\frac{2n}{2n}$ from the origin. So, an interval of width 2 centered at the origin will contain up to $2n$ sites on one side of the origin and up to $2n$ sites on the other side. Hence, it will contain at most $4n$ *neighboring sites*. In d dimensions, therefore, the origin in $\mathcal{L}_n$ will possess at most $(4n)^d$ neighbors.

Now, the mean number of open sites in $\mathcal{L}_n$ that form a cluster containing the origin is bounded above by the mean size of an open cluster in $\mathcal{L}_n$ at the origin times the number of neighbors of the origin (plus one for the origin itself). In other words,

$$
\mathrm{E}_\lambda\{|W_n^{(d)}|\} \le (1 + (4n)^d) \mathrm{E}_\lambda\{|C_n^{(d)}|\}, \tag{6.51}
$$

where $C_n^{(d)}$ is the open cluster of $\mathcal{L}_n$ at the origin. Thus,

$$
\mathrm{E}_\lambda\{|W^{(c)}|\} \le \frac{(\lambda/n^d)(1 + (4n)^d)}{1 - e^{-\lambda/n^d}} \mathrm{E}_\lambda\{|C_n^{(d)}|\}. \tag{6.52}
$$

From this, we see that if $\mathrm{E}_\lambda\{|W^{(c)}|\} = \infty$, then $\mathrm{E}_\lambda\{|C_n^{(d)}|\} = \infty$. Furthermore, $\mathrm{E}_\lambda\{|C_n^{(d)}|\} = \infty$ only if $p \ge p_c(\mathcal{L}_n)$. Hence,

$$
p_n(\lambda_T) \ge p_c(\mathcal{L}_n), \tag{6.53}
$$

or, using (6.44),

$$
\lambda_T \ge p_n^{-1}(p_c(\mathcal{L}_n)) = -n^d \log(1 - p_c(\mathcal{L}_n)). \tag{6.54}
$$

Next, we replace our disk (or sphere) of radius 1 by a slightly larger disk (or sphere) with radius[12] $\ell_n \le 1 + \frac{\sqrt{d}}{n}$. Then, we place the center of the enlarged disk (or sphere) at a Poisson point. We now revise our definition of adjacency as follows.

Definition 6.18 Two open sites, $\mathbf{x}, \mathbf{y} \in \mathcal{L}_n$, are said to be *adjacent*, written $\mathbf{x} \sim \mathbf{y}$, if there exist Poisson points $\mathbf{u}$ and $\mathbf{v}$ with $\mathbf{u} \in B_n(\mathbf{x})$, $\mathbf{v} \in B_n(\mathbf{y})$, and

$$
\| \mathbf{u} - \mathbf{v} \|_2 \le 2\ell_n. \tag{6.55}
$$

---

[12] Zuev and Sidorenko (1985a) use a sphere with *diameter* $\ell_n \le 1 + 2\sqrt{d}/n$ because their disks have diameter 1. Here, we use disks with *radius* 1. The reason for the specific radius of $1 + \sqrt{d}/n$ is that the diagonal of a box is $\sqrt{d}/n$, and so the radius enlargement is by a box diagonal.

Using this revised definition of adjacency, the enlarged disks (or spheres) with centers at **u** and **v** always overlap, whilst the disks (or spheres) with radius 1 and centers at **u** and **v** may not overlap. So, if there is an infinite open cluster $W_n^{(d)}$ at the origin of $\mathcal{L}_n$, then the corresponding set of disks (or spheres) whose centers are Poisson points also forms an infinite cluster $\widetilde{W}^{(c)}$. In other words, percolation of sites in $\mathcal{L}_n$ implies percolation of Poisson points. Now, the set of disks (or spheres) each having radius ℓ_n and Poisson intensity λ is equivalent to a set of disks (or spheres) each having radius 1 and increased Poisson intensity $\ell_n^d \lambda$. It thus follows that

$$p_c(\mathcal{L}_n) \le p_n(\ell_n^d \lambda) \implies \ell_n^d \lambda \ge \lambda_c \implies \lambda \ge \ell_n^{-d} \lambda_c, \tag{6.56}$$

whence, from (6.43),

$$\lambda_c \ell_n^{-d} \le -n^d \log(1 - p_c(\mathcal{L}_n)). \tag{6.57}$$

Thus,

$$\lambda_c \le -\ell_n^d n^d \log(1 - p_c(\mathcal{L}_n)). \tag{6.58}$$

Because $\lim_{n \to \infty} \ell_n = 1$, it follows from (6.48), (6.54), and (6.58) that

$$\lambda_T = \lambda_c = \lim_{n \to \infty} \{-n^d \log(1 - p_c(\mathcal{L}_n))\}. \tag{6.59}$$

6.6 Further Reading

Section 6.2. A book that deals with the use of percolation in the modeling of polymers is Vanderzande (1998), especially Chapter 6. See also Stauffer, Coniglio, and Adam (1982) and Ben-Naim and Krapivsky (2005).

Section 6.3. An excellent book on probability (including percolation) on trees and networks is by Lyons and Peres (2017).

Section 6.5. Continuum percolation is described in the book by Meester and Roy (1996), and in Grimmett (1989, Section 10.5), Grimmett (1999, Section 12.10), and Bollobás and Riordan (2006, Chapter 8). A review of developments in continuum percolation was given by Balberg (1987).

Section 6.5.1. Much of the research into amorphous computing is carried out at the Amorphous Computing Project in MIT's AI Laboratory (Abelson et al., 2000; Abelson, Beal, and Sussman, 2007). For a recent reference that discusses the relationship between amorphous computing and percolation, see Petru and Wiedermann (2007).

6.7 Exercises

Exercise 6.1 Let $p_n \in [0, 1]$, where $n \in \mathbb{N}$, be such that $np_n \to \lambda \in (0, \infty)$, as $n \to \infty$. Under these conditions, show that the Poisson distribution, $\mathcal{P}_\lambda$, can be viewed as a limit of binomial distributions, $\text{Bin}(p_n)$.

Exercise 6.2 Let $d = 2$. Construct an algorithm to draw a two-dimensional Poisson process on the plane by dividing the plane into unit squares and then, independently for each square, uniformly sprinkle a Poisson-distributed number of points into each square.

Exercise 6.3 Let $\mathbb{T}_r$ be a Bethe lattice with coordination number r (i.e., each node has r neighbors). Consider site percolation. Assume that an arbitrary node is occupied with probability p (often called the "concentration" of the infinite network). Let P be the probability that an arbitrary node (including the origin) is connected to the infinite cluster (i.e., P shows the "strength" of the infinite cluster).

(a) Explain why, in an infinite lattice, $P = 0$ when $p < p_c$. With this in mind, you need only consider $p > 1/(r-1)$.
(b) Let Q be the probability that an arbitrary node is not connected to the infinite cluster through one of the edges emanating from that node. Show that if Q satisfies the relation $Q = (1 - p) + pQ^{r-1}$, then $P = p(1 - Q^r)$.

Exercise 6.4 Let $\mathbb{T}_r$ be a Bethe lattice with coordination number r. The distance, $\ell(v)$, of the node v to the root node is the *level* (or *generation number*) of that node. Consider all those nodes v for which $\ell(v) \leq L$. The set of nodes at level L is the *boundary* of $\mathbb{T}_r$. The number of nodes, $\mathcal{S}_r(L)$, on the boundary is the *surface area* of $\mathbb{T}_r(L)$. The *volume*, $\mathcal{V}_r(L)$, is the number of nodes in $\mathbb{T}_r(L)$.

(a) Show that the number of nodes at the Lth level of $\mathbb{T}_r$ is exactly $\mathcal{S}_r(L) = (r-1)^L$.
(b) Show that
$$\mathcal{V}_r(L) = \frac{(r-1)^{L+1} - 1}{r - 2}.$$
(c) From (a) and (b), show that, as $L \to \infty$, the fraction of surface nodes has the property
$$\frac{\text{number of surface nodes}}{\text{total number of nodes}} = \frac{\mathcal{S}_r(L)}{\mathcal{V}_r(L)} \to \frac{r-2}{r-1}.$$
(d) What conclusions can be drawn from this result?

Exercise 6.5 Use Stirling's formula to show that $\binom{2n}{n-1}$ can be approximated by $4^n/\sqrt{n\pi}$ (see (6.18)).

Exercise 6.6 Show that a graph is connected and acyclic iff it is a tree.

Exercise 6.7 Let $\mathcal{G}$ be a graph. Suppose $\mathcal{G}$ is symmetric and connected. Show that it is also transitive.

Exercise 6.8 Show that a Cayley graph is transitive.

Exercise 6.9 Use a computer package to program the following experiment, which simulates continuum percolation. The basic idea is that you have a square of side length L and you randomly throw k disks each of radius r onto the square. Specify L and r so that it makes sense for this simulation (i.e., make r much smaller than L). For each disk, use a random number generator to set up a coordinate point (x, y) inside the square. With probability p, place the center of a disk at this coordinate. Consider different values of p. Repeat this process k times for each experiment, where k can be 10, 20, or 30. Now, try walking across the square from one side to its opposite side by walking only on adjacent (overlapping) disks. If you can do this, then there exists a cluster (or connected component) of disks. Set the computer program to print out your simulations. Looking at the results, does there exist a cluster of disks? Run this experiment T times, where T can be 100, 500, 1000. What proportion of the T runs resulted in clusters of disks?

CHAPTER SEVEN

The Topology of Networks

This chapter describes the small-world phenomenon and the Watts–Strogatz model, degree distributions, power-law distributions, and scale-free networks.

7.1 Introduction

Let $\mathcal{G} = (\mathcal{V}, \mathcal{E})$ be a graph, where $\mathcal{V}$ is the set of vertices or nodes and $\mathcal{E}$ is a set of edges, each edge of which connects a pair of nodes in $\mathcal{V}$. Let $N = |\mathcal{V}|$ and $M = |\mathcal{E}|$ denote the number of nodes and edges, respectively. In statistical network models, the graph $\mathcal{G}$ is considered to be a *random graph* in that the nodes in $\mathcal{V}$ are fixed objects which are then randomly connected pairwise by edges in $\mathcal{E}$. Network edges are viewed as random variables, where each edge represents the connection strength between two nodes; an edge is either present or absent (with a certain probability) depending upon whether there is a statistical connection between the nodes.

7.2 It's a Small World

One of the most fascinating ideas that was discovered by those who study large social networks is that the world is actually very small: one can construct a chain of about six friends to get a letter from one person to another person, neither of whom know each other. Moreover, such short chains are in great supply. Although the *small-world phenomenon*, as it came to be known, was mentioned several times before the 1960s, it was not until Stanley Milgram carried out his audacious project that the academic world (and the general public) paid attention to how truly small is the world.

7.2.1 Milgram's Experiments

In 1967, Stanley Milgram published in the popular magazine *Psychology Today* an article entitled "The small-world problem," in which he reported on a series of experiments to study the minimum number of steps it would take for a letter to get from one person to another person (who was unknown to the first person) through a succession of intermediaries. He arranged for each of 296 "starters" (100 blue-chip stock owners and 96 volunteers from Omaha, Nebraska, and 100 volunteers from Boston, Massachusetts) to forward a letter to someone known on a first-name basis, and then, using the same instructions, for that person to forward the same letter to another person, and so on, until the letter arrived at the "target" individual, a stockbroker living in Sharon, a Boston suburb, as quickly as possible. After all was said and done, only 64 letters (21.6%) reached the target, with the median chain being of length six.

This appeared to be quite a striking accomplishment in those days, even though a majority of the letters never reached their destination. Thus was born the "small-world phenomenon," the idea that the world can appear small when viewed in terms of friendship chains.

The results of Milgram's experiments have been heavily criticised on several counts, including selection bias (none of the starters could be viewed as members of a random sample), nonresponse bias, and the potential for overestimating or underestimating the chain length. Results from further similar experiments by Milgram and others were never published because of the low chain-completion rates (Kleinfeld, 2002).

7.2.2 Six Degrees of Separation

The idea (but not the term itself) of "six degrees of separation" is due to Hungarian author Frigyes Karinthy, who wrote the 1929 short story *Chain-Links*, about a conversation between two people, in which one of the characters discusses the following game:

> One of us suggested performing the following experiment to prove that the population of the Earth is closer together now than they have ever been before. We should select any person from the 1.5 billion inhabitants of the Earth – anyone, anywhere at all. He bet us that, using no more than five individuals, one of whom is a personal acquaintance, he could contact the selected individual using nothing except the network of personal acquaintances.

So, Karinthy had hit upon the idea of six degrees of separation in 1929.

In 1990, building upon Karinthy's story and the now-famous published accounts of Milgram's experiments, the playright John Guare débuted the New York play *Six Degrees of Separation*, where he had the character Ouisa Kittredge speak the play's most famous lines:

> I read somewhere that everybody on this planet is separated by only six other people. Six degrees of separation. Between us and everyone else on this planet. The president of the United States. A goldolier in Venice. Fill in the names. I find that (A) tremendously comforting that we're so close and (B) like Chinese water torture that we're so close. Because you have to find the right six people to make the connection.

This play was later (1993) turned into a feature-length movie. Although the public perceived Guare's choice of the number "six" to be derived from Milgram's results, Guare attributed his use of "six" to the 1909 Nobel Laureate Guglielmo Marconi, who supposedly conjectured that, even without his wireless telegraphy invention, he could be connected to anyone in the world through an average of 5.83 other people.

Table 7.1 Breakdown of Bacon numbers amongst all actors/actresses (as of August 6, 2016)

Bacon number	# of people
0	1
1	3,303
2	381,495
3	1,383,150
4	356,429
5	30,815
6	3,640
7	584
8	116
9	26
10	1
Total	2,159,560

Unfortunately, there is no evidence that Marconi ever made such a statement (even though some claim, erroneously, that it was made in his Nobel Lecture).

The Bacon Number

This idea found its way to the parlor game "Six Degrees of Kevin Bacon," in which every movie actor and actress has his/her own "Bacon number."[1] The *movie collaboration graph* has actors/actresses as nodes, and two nodes have a common edge if the corresponding actors/actresses appeared in a movie together. The Bacon number is the shortest path that links that actor or actress through a sequence of movies to Kevin Bacon.[2] For example, Patrick Billingsley (the author of *Probability and Measure*) appeared in *The Untouchables* (1987) with Robert de Niro, and de Niro appeared in *Sleepers* (1996) with Kevin Bacon; so Patrick Billingsley has a Bacon number of 2.

An *average personality number* (APN) for a person is a weighted average of the degree of separation of all those who link to that particular person. The breakdown of Bacon numbers amongst all actors/actresses (as of August 6, 2016) is given in **Table 7.1**. So, Kevin Bacon's APN is given by

$$\begin{aligned} APN = & [1(0) + 3303(1) + 381495(2) + 1383150(3) + 356429(4) \\ & + 30815(5) + 3640(6) + 584(7) + 116(8) \\ & + 26(9) + 1(10)]/2159560 \\ = & 3.02. \end{aligned}$$

[1]The following conversation occurred in the novel *Deadly Heat* by Richard Castle (Castle, 2013, p. 74: "We're all connected one way or another," said Rook. "You can trace anyone to anyone in six hops. It's like playing Six Degrees of Marsha Mason."
Detective Rymer said, "You mean Six Degrees of Kevin Bacon."
Rook said, "Please. I grew up with a mom who's a Broadway diva. In our house, it was always Marsha Mason."

[2]Brett Tjaden and Glenn Wasson of the University of Virginia started the website `oracleofbacon.org`, which has been rebuilt and maintained by Patrick Reynolds.

An interesting fact is that Kevin Bacon is not the "Center of the Hollywood Universe," a title given to that actor/actress whose APN is a minimum.[3] Previous centers include Rod Steiger, Donald Sutherland, Eric Roberts, Dennis Hopper, Christopher Lee, and Harvey Keitel.

The Erdős Number

In a similar vein, a mathematician's "Erdős number" is derived from the distance to Paul Erdős (who had about 1500 publications and 509 coauthors). The *mathematics collaboration graph*[4] has about 401,000 mathematicians as nodes and about 676,000 edges (as of July 2004), where two nodes are connected by an edge if the mathematicians have coauthored an article together. The average number of collaborators per person over the entire graph is 3.36. There is one large component in the graph consisting of about 268,000 nodes. An Erdős number is the shortest path through a chain of coauthored publications to Paul Erdős. The largest Erdős number is 13, and the average is 4.65 (which is the smallest average distance to the other nodes).

Facebook Survey

In May 2011, the social network Facebook carried out a huge study to discover the average number of degrees of separation of all its active members (i.e., those who logged on at least once over the past month). At the time of the study, there were on the order of 721 million members (or nodes in the graph) and around 69 billion friendships (or edges). The study found an average of 3.74 degrees of separation[5] between any two members of Facebook, and that this number appears to have stabilized with the growth of the network.

Oracle of Baseball

There is a similar game played out in baseball in the form of the Oracle of Baseball[6] in which any two major league players are linked by the shortest list of teammates. The most linkable player is Minnie Minoso with the smallest average connectedness (AC) of 3.02, and the least linkable player is Ed Duffy with an AC of 6.92. More recognizable players include Lou Gehrig at 3.591, Cal Ripken Jr 3.619, Babe Ruth 3.712, Greg Maddux 3.730, Barry Bond 3.755, Ty Cobb 3.775, Bobby Cox 3.801, and Derek Jeter 3.963.

7.3 The Watts–Strogatz Model

7.3.1 Imitating Real-World Networks

How well do random graphs mimic real-world networks? In a study of real-world networks (Watts and Strogatz, 1998), it was found that random graphs qualified as "small-world" networks, in that nodes tend to be highly clustered and have small diameters similar to the topological structure of random graphs. The random graphs generated by the Erdős–Rényi model have a very small element of clustering of nodes (because they are defined by a probability that is constant and independent

---

[3]See oracleofbacon.org.

[4]See www.oakland.edu/enp.

[5]In reporting the results of this study, the November 21, 2011 issue of the *New York Times* misquoted the number as 4.74.

[6]See www.baseball-reference.com/oracle.

for all edges joining pairs of nodes) and do not take account of the "small-world phenomenon" as evidenced by real-world networks.

A one-parameter "small-world" model that was introduced by Duncan J. Watts and Steven H. Strogatz is characterized by two components:

- Short average paths linking pairs of individuals (a characteristic of the Erdős–Rényi model).
- A "clustering coefficient" that measures how closely the nodes and all their nearest neighbors are all linked directly to each other.

The Watts–Strogatz model was specifically designed to overcome the limitation on clustering of the Erdős–Rényi model. The motivation for the Watts–Strogatz model was taken from social networks where people have friends who are very similar to themselves (referred to as *homophily*), but usually also have other friends and acquaintances who live far away from them, perhaps even in other countries.

7.3.2 The Watts–Strogatz Algorithm

Definition 7.1 A *(regular) ring lattice* is defined as having a fixed number N of nodes, $v_1, \ldots, v_N$, spaced uniformly around a circle.

Suppose each node on a ring lattice is connected to its k nearest-neighbor nodes ($k/2$ on each side) in the ring (assuming k is even). Thus, an edge, (v_i, v_j), between nodes v_i and v_j is present iff $1 \leq |i - j| \leq k/2$. The value of k is chosen by the user, but to ensure sparsity (i.e., relatively few edges) and a connected network at all times, assume that $1 \ll \log N \ll k \ll N$.

The Watts–Strogatz algorithm randomly "rewires" each edge by connecting each node in the graph to another node that is randomly chosen with probability p ($0 < p < 1$); with probability $q = 1 - p$, the edge is left in place. At one extreme, $p = 0$ corresponds to "regularity," whilst at the other extreme, $p = 1$ corresponds to "disorder." This rewiring process, which changes one end of a connected pair of nodes whilst keeping the other node fixed, therefore introduces the possibility of long-range connections into the graph. The Watts–Strogatz rewiring process is displayed in **Figure 7.1**, where we used $N = 15$, $k = 4$, and $p = 0, 0.25$, and 0.5.

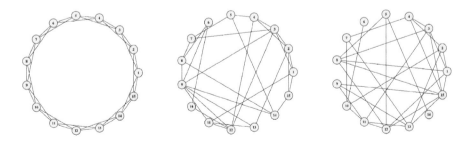

Figure 7.1 Watts–Strogatz graph rewiring of $N = 15$ nodes on a ring lattice, with $k = 4$ nearest neighbors ($k/2$ on each side). Left: $p = 0$. Middle: $p = 0.25$. Right: $p = 0.5$.

The Watts–Strogatz rewiring method appealed to social scientists because it incorporated weak links, and edges that were randomly generated by homophily, which, in turn, led to an approximation of a small-world network. The algorithm is given as **Algorithm 7.1**.

Algorithm 7.1 *Watts–Strogatz algorithm*

---

1. Fix k and N so that $1 \ll \log N \ll k \ll N$.
2. Initialize a graph as a ring lattice with N nodes, $v_1, \ldots, v_N$, in which each node is symmetrically connected to its k nearest neighbors (i.e., $k/2$ on one side and $k/2$ on the other side, where k is even).
3. Consider the nearest neighbor of the ith node v_i. With probability p, replace the edge (v_i, v_j), $i < j$, by a new edge (v_i, v_ℓ), where v_ℓ is a randomly chosen node in the graph, not including v_i (which would introduce a loop) and not duplicating an existing edge. Otherwise, with probability $q = 1 - p$, leave the edge (v_i, v_j) in place.
4. Repeat the process by moving clockwise around the ring, visiting each node in turn until one lap is completed.
5. Carry out a second lap of the ring by rewiring edges to their second-nearest neighbors in the clockwise direction.
6. Repeat the clockwise process by considering more distant nearest neighbors until each edge in the original graph is considered once.
7. Stop rewiring after $k/2$ laps.

As p varies from 0 to 1, the original graph ($p = 0$) becomes closer to the Erdős–Rényi random graph $\mathcal{G}_{N,p}$, where $p = Nk/2\binom{N}{2}$ and all edges are rewired randomly. A small-world network appears for relatively small values of p. In step 1, the further away is the rewired node v_ℓ, the smaller the average path length becomes. This algorithm can be generalized to a different number of nearest neighbors of each node.

7.3.3 Average Path Length

Of special interest in network theory is the determination of *path length*. This issue is important in applications such as transportation, communications on the Internet, and the spread of diseases. Path length is also one of the most useful tools in analyzing brain networks, where it is used as a measure of the capacity of the network to pass information between its nodes and, hence, acts as an indicator of network integration. The shorter the path length, the easier it is for neurological information to be transmitted between all pairs of nodes.

Definition 7.2 The *path length* between any two nodes is the number of edges in a sequence of "hops" from beginning node to end node.

There may be more than one such path through the network.

Definition 7.3 The *geodesic path* between any two nodes is that path between the two nodes that has the shortest length (known as the *geodesic distance*).

Let $L_k(p)$ be the average distance between any two randomly chosen nodes in the ring. Let k be the number of nearest neighbor nodes.

Definition 7.4 When $p = 0$ (the original ring lattice), the *average path length* is

$$L_k(0) = \frac{N}{2k} \gg 1; \tag{7.1}$$

when $0 < p < 1$, the average path length decreases quickly as p increases; and when $p \to 1$ (a random graph), the average path length is

$$L_k(1) = \frac{\log N}{\log k}. \tag{7.2}$$

If $L_k(p)$ is large, then deleting some edges to near neighbors and replacing them by random edges will drop the value of $L_k(p)$ significantly. It is recommended to plot $L_k(p)/L_k(0)$ against $\log p$.

Comment. The average path length is very sensitive to "outlying" nodes, such as those that are not connected to any other nodes.

7.3.4 Clustering Coefficient

"Clustering" the nodes of a graph or network is an important statistical technique for analyzing the underlying structure of a network. In a brain network, for example, clustering plays a vital role because brain regions need to communicate a great deal of shared information, thereby contributing to a healthy brain system. Nodes that are members of different brain clusters tend not to share all that information and, therefore, become functionally separated from other nodes. This and similar types of scenarios led to the formulation of a *clustering coefficient*, which is a number that lies between 0 (no clustering) and 1 (maximal clustering), and is designed to measure the extent of clustering in the network. Several definitions of this measure have been proposed.

Let E_{ik} be the number of edges that actually exist amongst the k neighboring nodes of node v_i. There are $\binom{k}{2}$ possible edges between these k neighbors of node v_i. Define C_{ik} as the fraction

$$C_{ik} = \frac{E_{ik}}{\binom{k}{2}} = \frac{2E_{ik}}{k(k-1)}. \tag{7.3}$$

Thus, C_{ik} is a measure of how much node v_i and its immediate k neighbors are directly connected to each other. Averaging C_{ik} over all N nodes yields

Definition 7.5 The *clustering coefficient* for the entire network is

$$C_k(p) = \frac{1}{N}\sum_{i=1}^{N} C_{ik} = \frac{2}{k(k-1)}\frac{1}{N}\sum_{i=1}^{N} E_{ik}. \tag{7.4}$$

Ring lattice. Consider a ring lattice (i.e., $p = 0$), where each node has k neighbors, $\frac{k}{2}$ to the left and $\frac{k}{2}$ to the right, assuming k is even. Let v_i be an arbitrary node on the ring lattice. Then, we have the following computations:

- Starting with the leftmost neighboring node that is $\frac{k}{2}$ nodes away from the node v_i, the number of pairwise connections between this node and the other nearest neighbors of v_i (excluding any connections with v_i) is $\frac{k}{2} - 1$.
- Consider the neighboring node that is $\frac{k}{2} - 1$ nodes away from v_i. The number of pairwise connections between this node and the other nearest neighbors of v_i (excluding any connections with v_i, but including connections with the neighbor to the immediate right of v_i) is $1 + \frac{k}{2} - 1$.
- Consider the neighboring node that is $\frac{k}{2} - 2$ nodes away from v_i. The number of pairwise connections between this node and the other nearest neighbors of v_i (excluding any connections with v_i, but including connections with the two neighbors to the immediate right of v_i) is $2 + \frac{k}{2} - 1$.
- And so on.
- Finally, consider the neighboring node that is 1 node away from v_i. The number of pairwise connections between this node and the other neighboring nodes of v_i is $2(\frac{k}{2} - 1)$.

Adding these numbers together gives the total number of actual pairwise connections amongst the k neighboring nodes of node v_i in the ring lattice:

$$E_{ik} = \sum_{j=0}^{\frac{k}{2}-1} \left(j + \frac{k}{2} - 1 \right) = \frac{3}{8}k(k-2). \tag{7.5}$$

We could have also started from the rightmost neighboring node of v_i and worked our way leftwards; this would have repeated the numbers given above, thereby multiplying E_{ik} by 2. However, these edges are undirected, and so we would also have had to divide E_{ik} by 2. Hence, these 2s cancel out and E_{ik} is unaffected. Because E_{ik} is independent of i, averaging over all nodes as in (7.4) does not change the number. Dividing this quantity by $\binom{k}{2}$ yields the clustering coefficient for the ring lattice,

$$C_k(0) = \frac{3(k-2)}{4(k-1)}, \tag{7.6}$$

which is independent of the number of nodes N. The graph has a nonzero clustering coefficient iff $k \geq 3$. As $k \to \infty$, we see that $C_k(0) \to \frac{3}{4}$, which shows the graph to be highly clustered. For example, $C_4(0) = 0.5$, $C_{10}(0) = 0.667$, $C_{30}(0) = 0.724$, and $C_{50}(0) = 0.735$.

It was shown (Barrat and Weigt, 2000) that, as $N \to \infty$ and for a regular ring lattice,

$$C_k(p) = C_k(0)q^3, \tag{7.7}$$

where $q = 1 - p$. So, for this model, dependence on N is very small. It is recommended to plot $C_k(p)/C_k(0)$ against $\log p$, where we take $-4 \leq \log p \leq 0$. For a range of values of p, $C_k(p)/C_k(0) \approx 1$, and then it declines around $p \approx 0.1$ (i.e., around $\log p = -1$).

The above definition (7.6) of a clustering coefficient assumes that every node has the same number k of nearest neighbors. Although this definition can easily be adapted to the case of a different number of neighbors for each node (see **Exercise 7.4**), the new formula will lead to unsatisfactory results if nodes are severely unbalanced (i.e., by the number of neighbors to which they can connect).

Consider, for example, the following scenario (Newman, Strogatz, and Watts, 2001). Let v_i and v_j be any two nodes in the graph and suppose v_i has two connected neighbors whilst v_j has 100 neighbors but none are connected. Then $C_{i2} = 1$ and $C_{j,100} = 0$, so that $C(p) = 0.5$. This value makes little sense. Node v_j has $\binom{100}{2} = 4950$ possible edges between the 100 neighbors. So, for nodes v_i and v_j, we have a total of 4951 possible edges, only one of which is connected. The clustering coefficient (i.e., the average probability of a pair being connected) should, therefore, be $\frac{1}{4951} = 0.0002$, not 0.5. The difference in these two numbers is attributed to the denominator $k(k-1)$, which biases the clustering coefficient in favor of nodes with a small number of connections.

Triangles. With this scenario in mind, an alternative clustering coefficient was proposed (Newman, Strogatz, and Watts, 2001) in which the total number of pairs of nodes that have a common neighbor is divided by the total number of such pairs that are connected. This ratio has been expressed as

$$C'_k(p) = \frac{3N_\triangle}{N_3}, \tag{7.8}$$

where $N_\triangle$ is the number of triangles in the graph and N_3 is the number of connected triples. A "triangle" refers to three nodes, each of which is connected to the other two nodes, and a "connected triple" refers to three nodes at least one of which is connected to both the other two nodes. For example, if node v_i is connected to node v_j, and v_j is connected to node v_k, then we can state that, with high probability, v_i is connected to v_k, which illustrates the social network notion of *transitivity*. The "3" in the numerator of (7.8) accounts for the fact that each triangle contributes three separate connected triples of nodes, one for each of its three nodes. Note that $0 \leq C'_k(p) \leq 1$. Furthermore, the difference between $C_k(p)$ and $C'_k(p)$ turns out to be very small and converges to zero as $N \to \infty$.

7.4 Degree Distributions

Much of the earliest research in social networks was descriptive. One of the most interesting descriptive features was the shape of the *degree distribution* of the network. The *node degree* (i.e., the number of edges that a node has) can be viewed as the "popularity" of that node. We call the set of other nodes that directly link to that node by edges the *in-links* to that node. Clearly, the greater the number of in-links, the more popular the node. The variability of the node degree is specified by a probability distribution $P(k)$, which gives the probability that a randomly chosen node will have exactly k edges.

7.4.1 Erdős–Rényi Model

For the Erdős–Rényi random graph model $\mathcal{G}_{N,p}$, the degree distribution is quite straightforward (Erdős and Rényi, 1959; Bollobás, 1981).

Let K_i be the degree of node v_i. Because there are at most $N-1$ nodes in the network that are available to link with node v_i, then K_i has the binomial distribution, $K_i \sim \mathrm{Bin}(N-1, p)$, with $N-1$ trials and probability of success p. Thus

$$P_p\{K_i = k\} = \binom{N-1}{k} p^k q^{N-1-k}, \quad k = 0, 1, 2, \ldots, N-1. \tag{7.9}$$

If v_i and v_j are different nodes, then K_i and K_j are approximately independent random variables.

Let X_k be the number of nodes having degree k. We wish to find $P_p\{X_k = a\}$ as a function of k. The expected number of nodes having degree k is

$$E\{X_k\} = N \cdot P_p\{K_i = k\} = \lambda_k, \tag{7.10}$$

where

$$\lambda_k = N \binom{N-1}{k} p^k q^{N-1-k}. \tag{7.11}$$

The asymptotic distribution of X_k is given by

Theorem 7.1 *As $N \to \infty$ and $p \to 0$, so that λ_k remains bounded, the degree distribution of an Erdős–Rényi random graph converges to a Poisson distribution,*

$$P_p\{X_k = a\} = \frac{e^{-\lambda_k} \lambda_k^a}{a!}, \tag{7.12}$$

with mean (or average degree) λ_k and variance λ_k.

See, for example, Feller (1964, Section VI.5).

7.4.2 Watts–Strogatz Model

When $p = 0$, every node has the same degree k and, hence, the degree distribution for a ring lattice is a delta function centered at k. When $p \in (0, 1)$, this introduces disorder into the network, which, in turn, affects the degree distribution.

Suppose $p \in (0, 1)$. The Watts–Strogatz algorithm has the following effect on the network. Randomly rewiring each edge with probability p introduces $pNk/2$ rewired edges that link to nodes from different subcultures. If v_i is a node in the ring lattice, its degree k_i will consist of two kinds of edges: (1) edges that remain attached to v_i, and (2) edges that were rewired to point towards v_i. Edges leaving clockwise from v_i will remain attached whether they are rewired or not, while edges leaving counterclockwise could be rewired to point towards any node other than v_i.

Let S_i denote the contribution to the degree from edges that, with probability $1-p$, were not rewired, and let R_i denote the contribution to the degree from edges that, with probability $1/N$, were rewired to point towards v_i. Then, the degree of v_i can be expressed as

$$k_i = S_i + R_i + \frac{k}{2}. \tag{7.13}$$

So, we need to find the probability that $S_i = s$ and the probability that $R_i = r$, and then combine the two probabilities.

The probability of the number of edges that were connected to v_i in the original ring lattice, but were not rewired, is given by the binomial probability, $\mathrm{Bin}(k/2, 1-p)$,

$$P_p\{S_i = s\} = \binom{k/2}{s} (1-p)^s p^{\frac{k}{2}-s}, \quad s \in \left[0, \frac{k}{2}\right], \tag{7.14}$$

and zero otherwise. The probability of the number of edges that were rewired towards v_i is given by the binomial probability, $\text{Bin}(pNk/2, 1/N)$,

$$P_p\{R_i = r\} = \binom{pNk/2}{r}\left(\frac{1}{N}\right)^r\left(1 - \frac{1}{N}\right)^{\frac{pNk}{2}-r}, \quad r \in \left[0, \frac{pNk}{2}\right],$$

$$\approx \frac{(pk/2)^r\, e^{-\frac{pk}{2}}}{r!}, \tag{7.15}$$

for large N, and zero otherwise, which is approximated by the Poisson probability, $\text{Poi}(pk/2)$. Combining these two independent probabilities yields the degree distribution (Barrat and Weigt, 2000),

$$P_p\{X_k = a\} = \sum_{j=0}^{\min(a-k/2,k/2)} P\{S_i = j\} \cdot P\{R_i = a - k/2 - j\}$$

$$= \sum_{j=0}^{\min(a-k/2,k/2)} \binom{k/2}{j}(1-p)^j p^{\frac{k}{2}-j}\frac{(pk/2)^{a-\frac{k}{2}-j}e^{-\frac{pk}{2}}}{(a-k/2-j)!}, \tag{7.16}$$

where $a \geq \frac{k}{2}$, and zero otherwise. The degree distribution has a shape that is similar to that of a random graph; it has a pronounced peak at $a = k$ and decays exponentially for large $|a - k|$. All nodes have roughly the same degree.

As $p \to 1$, the degree distribution becomes

$$P_p\{X_k = a\} = \frac{(k/2)^{a-\frac{k}{2}}e^{-\frac{k}{2}}}{(a-k/2)!}, \tag{7.17}$$

which is a Poisson distribution for the variable $a - \frac{k}{2}$ with mean $\frac{k}{2}$.

7.5 The Power Law and Scale-Free Networks

Real-world networks have degree distributions that are extremely unbalanced: a few nodes are extremely popular and act as hubs for the network, whilst the vast majority of nodes appear to be somewhat isolated. Early empirical results (Moreno and Jennings, 1938) indicated that such skewed degree distributions having very large values occur universally in networks, and that these distributions bear no resemblance to those formed from the random connections of the Erdős–Rényi graphs.

7.5.1 Power-Law Distributions

Studies of the degree distributions of large real-world networks (de Solla Price, 1976; Barabási and Albert, 1999) confirmed the observations of Moreno and Jennings. Furthermore, it was shown that if the quantity being measured appears to be a type of popularity, then the fraction of nodes that have k in-degrees (for varying values of k) follows a curve that decays as the *power law*. The power-law distribution is not symmetric and decays much more slowly to zero than the Gaussian distribution, which implies that, under the power law, there is a greater chance of extreme outcomes. A well-known example of a power law is the *Pareto distribution*.

Table 7.2 Examples of the power-law exponent β estimated for different types of networks. No details of the estimation methods were described. Sources: [a]Barabási and Albert (1999), [b]Albert and Barabási (2002), [c]Wang and Chen (2003), [d]Zhou (2004), [e]Chung et al. (2003), [f]Barabási et al. (2003). N is the number of nodes in the network

Network	N	β
Nonbiological networks		
Internet (domain level)[c]	32,711	2.1
Internet (router level)[c]	118,298	2.1
Internet (autonomous systems)[d]	6,374	2.22
WWW[a,b]	325,729	2.1 (in), 2.45 (out)
U.S. power grid[a]	4,941	4
Movie actors[a,c]	225,226	2.3
Coauthorship (math)[b,c,f]	70,975	2.5
Coauthorship (neuroscience)[b,f]	209,293	2.1
Citation[a,b,f]	783,339	3
Words, synonyms[b,f]	22,311	2.8
Long-distance telephone calls[b,f]	53×10^6	2.1–2.3
Biological networks		
Metabolic system (*E. coli*)[b,c,f]	778	1.7–2.2
Ythan estuary[f]	134	1.05
Silwood Park food web[b,c,f]	154	1.13
Protein–protein interaction		1.5–2.5
Yeast gene expression[e]		1.4–1.7

Definition 7.6 If X_k is the number of nodes with degree k (where $\sum_k X_k = N$), then the power law has the general form

$$X_k \sim c_N k^{-\beta}, \quad \beta > 1, \tag{7.18}$$

where β is the *power-law* (or *scaling*) *exponent* and c_N is a normalizing constant.

We have listed in **Table 7.2** some of the examples of networks and their estimated power-law exponents as published in several articles. The estimated value, $\widehat{\beta}$, of β for each of these networks could have been determined by simply regressing the logarithm, $\log X_k$, of the degree distribution against $\log k$ and taking $\widehat{\beta}$ to be the negative slope of the least-squares line; see the discussion following (7.19) below. However, *none of the articles listed in the table explained their estimation methods for determining the values of β, leaving the impression that those values were probably determined in a purely ad hoc manner rather than by a rigorous statistical method.*

If we take this table at face value, we see that there is a major difference between the value of β for nonbiological and biological networks. For nonbiological networks, such as the Internet, collaboration, citation, movie actors, U.S. power grids, and social networks, the power-law exponent typically falls into the range $2 < \beta < 4$, while biological networks often have exponents with $1 < \beta < 2$. This difference was noted by Chung et al. (2003). Examples of biological networks include the yeast protein–protein interaction network, which has its exponent ranging from 1.5 to 2.5 (depending upon the data used); the *Escherichia coli* metabolic network, which has an exponent ranging from 1.7 to 2.2; the yeast gene expression network, which has an exponent ranging from 1.4 to 1.7; and gene functional interactions, with exponent 1.6. If we accept the validity of these results, this suggests that biological and nonbiological networks may evolve differently.

7.5.2 Scale-Free Networks

Networks that have degree distributions characterized by the power law are called *scale-free* (or *scale-invariant*) networks (Barabási and Albert, 1999). Unfortunately, the term "scale-free" has not been well-defined. Most of the literature uses the following definition of "scale-free" (Broido and Clauset, 2018):

Definition 7.7 A network is said to be *scale-free* if the fraction of nodes $P(k)$ with degree k follows a power-law distribution with exponent $\beta > 1$.

In other words, the power law states that some nodes should have a lot more connections than other nodes. As a result, there is no uniformity of scale that would characterize the network. Hence, the characteristics of a scale-free network do not depend upon the size, N, of the network. There are stronger (e.g., that $\beta \in [2,3]$ or that the node-degree distribution is formed by preferential attachment) and weaker versions of this definition.

Taking logarithms of (7.18) yields

$$\log X_k \sim \alpha - \beta \log k, \tag{7.19}$$

where $\alpha = \log c_N$. Note that α can be viewed as the logarithm of the number of nodes with degree 1. If we plot $\log X_k$ against $\log k$, the points should fall along a straight line with a slope of $-\beta$. An estimate of the exponent β can, therefore, be obtained from the slope of the line using simple linear regression (or some robust variation of it). Note also that because $\log X_k \geq 0$, then $\alpha \geq \beta \log k$, or $\frac{\alpha}{\beta} \geq \log k$. Thus, $1 \leq k \leq e^{\alpha/\beta}$.

Comment. The work of Barabási and Albert proved to be very exciting to some researchers, while many others remained skeptical. Their work came at a time when statistical physicists viewed power laws as the key to understanding universal laws in many different physical systems. Apparently, statistical physicists, who had been searching for organizing principles in their research, began to see power laws everywhere. Following this development, Barabási and Albert then transferred that idea to explaining why power laws were turning up in complex networks. It started an industry, which later became known as "network science." Widespread attention on scale-free networks was encouraged by articles published on the topic in popular scientific magazines and in the national media.

7.6 Properties of a Scale-Free Network

7.6.1 Diameter

A measure of distance between two nodes, $v_i, v_j \in \mathcal{V}$, is usually defined as the number of edges defining the shortest path connecting the two nodes. This distance is called the *shortest path length* and is denoted by l_{ij}. The following definition of the diameter of a graph is the traditional one:

Definition 7.8 The *diameter*, $d_\mathcal{G}$, of a graph $\mathcal{G}$ is defined as the longest distance between two nodes:

$$d_\mathcal{G} = \max_{i,j} l_{ij}. \tag{7.20}$$

It has been shown through simulations and heuristic arguments (Albert, Jeong, and Barabási, 1999; Barabási, Albert, and Jeong, 2000; Newman, Strogatz, and Watts, 2001) that, if the degree sequence has a desired distribution, then the diameter of the graph $\mathcal{G}$ should have the form $A + B \log N$, where N is the number of nodes in $\mathcal{G}$, and A and B are constants.

7.6.2 Average Path Length

Another effective definition of the diameter of a network is the *average shortest path length*:

$$l = \frac{1}{N(N-1)} \sum_{i \neq j} l_{ij}. \tag{7.21}$$

There is no theoretical formula that yields a good approximation for the average path length in a scale-free network. Simulations indicate that the average path length in a scale-free network is smaller than that for a random graph, for any network size (see, e.g., Albert and Barabási, 2000). See **Table 7.3** for a list of average path lengths l for certain types of networks.

7.6.3 Clustering Coefficient

The clustering coefficient C measures how different a real network is from a completely random graph. See **Table 7.3** for a list of clustering coefficients C for certain types of networks. In terms of coauthorship networks, C tells us "how much a node's collaborators are willing to collaborate with each other, and it represents the probability that two of a node's collaborators wrote a paper together" (Barabási et al., 2002).

No statistical prediction formula exists so far for the clustering coefficient for a scale-free model. Some simulation results are available indicating that, for a scale-free network, (1) the clustering coefficient follows a power-law model, decreasing with network size, and (2) the clustering coefficient is much larger than that for a random graph, and the difference slowly increases with network size (see, e.g., Albert and Barabási, 2000).

7.6.4 Resilience of Scale-Free Networks

One issue that has become increasingly important relates to the reliability of complex networks (Albert, Jeong, and Barabási, 2000; Callaway et al., 2000; Schneider et al., 2011).

The Internet and the World Wide Web (WWW) were among the first scale-free networks to be examined as "case studies" for robustness to certain types of node (and/or edge) failures (Cohen et al., 2000). Node failure, which implies the removal

Table 7.3 Published values of the average path length l and clustering coefficient C for different types of networks. N is the number of nodes in the network. Sources: [a]Watts and Strogatz (1998), [b]Albert and Barabási (2002), [c]Wang and Chen (2003), [d]Newman (2005)

Network	N	l	C
Nonbiological networks			
Internet (domain level)[c]	32,711	3.56	0.24
Internet (router level)[c]	32,711	9.51	0.03
Internet (autonomous systems)[d]	6,374		0.24
WWW (undirected)[b]	153,127	3.1	0.11
WWW[c,d]	153,127	3.1	0.11 (in), 0.11 (out)
U.S. power grid[a,b,d]	4,941	18.7	0.08
Movie actors[a,b,c]	225,226	3.65	0.79
Coauthorship (biology)[d]	1,520,251		0.081
Coauthorship (math)[d]	253,339		0.15
Coauthorship (neuroscience)[b]	209,293	6.0	0.76
Company directors[d]	7,673		0.59
Words, synonyms[b]	22,311	4.5	0.7
Biological networks			
Metabolic system[c]	778	3.2	
Silwood Park food web[b,c]	154	3.4	0.15
C. elegans[a]	282	2.65	0.28

of all edges counted in the degree of that node, is, therefore, much more harmful to the network than edge deletion. It was discovered that although Internet routers randomly fail from time to time, the network keeps on running and rarely suffers any major disruption of service. On the other hand, during the COVID-19 pandemic, supply-chain networks were severely disrupted (due mainly to broken links) so that many companies could not carry out the production and delivery of their goods and services, which, in turn, affected the duration and severity of the crisis.

Scale-free networks have been shown to be amazingly resilient (i.e., robust) against random failures, such as random deletions of nodes and random errors. This type of robustness of scale-free networks is due to the inhomogeneity of the network, the fact that the vast majority of nodes will have low degrees (and so will be more likely to be chosen) while a few nodes will have very high degrees (and, thus, be less likely to be chosen). Removing nodes with low degree will not change the overall structure of the network. A network can fall apart, however, only if a significant percentage of the nodes are removed.

On the other hand, scale-free networks are extremely fragile when big hubs (i.e., nodes with high degrees) are maliciously attacked and deleted. This type of vulnerability has been portrayed as the "Achilles heel" of the Internet and the WWW. Assuming that attackers have full knowledge about the topology of the network, a malicious attack would not delete nodes randomly, but would, first, deliberately remove the largest hub, then the second largest hub, then the third, and so on. This strategy will quickly lead to a total network collapse because it is this small fraction of hubs that holds the network together. Such an attack would break the network into many very small, disconnected groups of nodes.

Removing a major hub from the network may also result in a "cascading" effect on other nodes. The Internet provides a good example of this phenomenon (Albert and Barabási, 2002). Suppose, for example, a highly connected router is removed. This could then cause Internet traffic, which would normally pass through that router, to be rerouted through a different router. The substitute router, however, may not have the capacity to handle the increased traffic, which would, in turn, lead to a possible "denial-of-service" response. This search for a suitable conduit would then continue, disrupting the operation of some fraction of other nodes.

Consider, for example, the case of transportation networks (see Section 3.7.1). If a network remains connected even when disruptions to nodes or edges occur, then it is considered to be resilient. A hub-and-spoke transportation network works very efficiently for air traffic, but should such a large hub fail, the network could turn out to be not very resilient. An interesting problem for transportation systems is whether changes to one network would affect another network. For example, air and sea transportation networks, where certain nodes (e.g., large cities) are common to both networks, may be especially vulnerable if those common nodes are targeted through labor disputes, natural breakdowns, damaged infrastructure, seismic event, or political disagreements. Models of transportation resilience using percolation theory are described by Dall'Asta et al. (2006) and Ganin et al. (2017).

7.6.5 Not All Networks are Scale-Free

Although many articles in the scientific literature have claimed that a wide variety of networks are scale-free (see, e.g., Barabási and Bonabeau, 2003), researchers in different fields are now re-examining that general claim. We know, for example, that Erdös–Rényi random networks do not obey the power law and, hence, are not scale-free. Furthermore, recent empirical evidence has appeared in the literature showing that the scale-free model does not characterize many real-world networks. As a result, whether or not any particular network is deemed to be strictly scale-free has become a controversial topic.

Many Networks are Not Scale-Free
Amaral et al. (2000) conducted an empirical study of several different types of "real-world" networks. They studied the following classes of networks.

Technological and transportation networks. They showed that the electric-power grid of Southern California (in which the nodes are generators, transformers, and sub-stations, and the edges are high-voltage transmission lines) and the network of the world's largest airports by traffic, cargo, and number of passengers (in which the nodes are the airports, and the edges are the non-stop connections) are not strictly scale-free, and that these networks exhibit instead exponentially decaying tails (that decay faster than a power-law tail).

Social networks. Social networks that are shown not to be completely scale-free include the movie-actors network (in which the edges show which pair of actors were cast at least once in the same movie), the acquaintance network of Utah Mormons, and the friendship network of students in a certain high-school. The actors' network is sharply truncated by an exponential tail. The Mormon and friendship networks, on the other hand, do not exhibit a power-law distribution; instead, they show results consistent with a Gaussian distribution.

Biological networks. Biological networks have many of the features of small-world networks. They typically contain "hubs" (i.e., nodes with a high degree of edges when compared to other nodes), which are also common to social networks and transportation networks. An unstructured model would, therefore, fail to fit a power-law degree distribution to highly structured observed data in which hubs play a prominent role.

In the case of the neuronal network of the worm *Caenhorhabditis elegans*, the nodes are the neurons and the edges are the connections between neurons. For the conformation space of a lattice polymer chain, the nodes are the possible conformations of the polymer chain and the edges are the possibility of connecting a pair of conformations through local movements of the chain. Both these networks show tails that decay approximately as exponential, rather than have power-law tails.

In a study of the protein–protein interaction network of yeast (*Saccharomyces cerevisiae*), which has 1870 proteins (nodes) and 2240 direct physical interactions (edges) – information that was obtained through the Y2HS (yeast two-hybrid screen) system (see Section 2.6.3) – Jeong et al. (2001) showed that yeast PPI has a degree distribution that is not strictly scale-free, but rather follows a power-law distribution with an exponential cutoff at degree 20. The exponential cutoff shows that for degrees of 20 and more, the number of protein interactions is fewer than would be expected from a pure scale-free network.

In a study of the fruit fly, *Drosophila melanogaster*, protein–protein interaction network, Bader (2006) showed that, when reliable, biologically relevant interactions are considered, the entire network (involving over 7000 proteins (the nodes) and over 20,000 pairwise interactions (the edges)) may have neither power law nor scale-free structure; specifically, the study showed that the degree distribution decayed faster than a power-law distribution, converging instead to an exponential distribution as the network increases in size.

How Common are Scale-Free Networks?

In a recent study of a "large and diverse corpus" of 927 network datasets from a number of different domains, Broido and Clauset (2018)[7] approach the problem from a different direction: they claim that very little rigorous statistical testing has appeared in the literature of real-world networks that compares a power-law distribution to alternative non-scale-free distributions.

So, they set out to provide a formal statistical test of the power-law model for degree distribution as compared with alternative non-scale-free distributions. The latter category was taken to consist of the exponential distribution, the log-normal distribution, the Weibull distribution, and the power-law distribution with exponential cutoff. See **Table 7.4** for these various distributions. The exponential distribution has a thin right-hand tail and relatively low variance, the log-normal has a very heavy tail and is not scale-free, and the Weibull distribution, which generalizes

---

[7]Their provocative article was discussed by Klarreich (2018) in *Quanta Magazine*, which was then reproduced in *The Atlantic* magazine.

Table 7.4 Distributions used by Broido and Clauset (2018) as alternatives to the power-law distribution. Note that $f(x)$ is the probability density without the normalizing constant

Distribution	$f(x)$
Exponential	$e^{-\lambda x}$
Log-normal	$\frac{1}{x}e^{-(\log x - \mu)^2/2\sigma^2}$
Weibull	$e^{-(x/b)^a}$
Power law with exponential cutoff	$x^{-\alpha}e^{-\lambda x}$

the exponential distribution, can have a thin or heavy tail. Discrete versions of each of these continuous distributions were obtained for the study because node degrees have integer values.

The 927 real network datasets (which we will call "source" networks) were divided into five types: biological (500), informational (15), social (145), technological (200), and transportation (67) networks. By sequentially stripping away any complex property possessed by each such network, they transformed each of them into a set of multiple "simple" (i.e., static, unipartite, unweighted, and undirected) networks. For example, a directed network is replaced by three degree sequences: an in-degree, an out-degree, and a total-degree network sequence. These transformations yielded a total of 23,999 simple networks. (Note that there is no claim that the simple networks were generated by randomly sampling from the source networks.) Extremely sparse and extremely dense graphs, which were likely not to be scale-free, were excluded from their study. This left 4477 simple networks, each of which was considered as a vote for whether the particular source network is scale-free or not.

The authors used a likelihood-ratio test, the difference between the log-likelihoods of the power law and of particular alternative models, as the test statistic R. The competing hypotheses were as follows:

- The null hypothesis, $\mathcal{H}_0$, states that there is no difference between the two models (i.e., the results are non-informative about choice of model).
- The alternative hypothesis, $\mathcal{H}_1$, states that there is a real difference between the two models (i.e., one model is preferred over the other).

To determine how weak or strong the evidence is of scale-free structure for any given network, p-values were computed against a null model of $R = 0$, that the two models cannot be distinguished from each other. A positive value of R indicates that the power law fits the data better, while a negative value indicates that the alternative distribution is a better fit to the data. $\mathcal{H}_0$ is rejected if the p-value is smaller than 10%, and the sign of R is interpreted as evidence of the more plausible model of the degree distribution, while $\mathcal{H}_0$ is not rejected if the p-value is larger than 10%.

They then divided the source networks into five scale-free categories: super-weak, weakest, weak, strong, and strongest. Classification of a source network into one of those categories was based upon the percentage of the corresponding simple networks whose p-value exceeded 10%. If 90% or more of the simple networks passed the scale-free test, then the corresponding source network was placed in the strongest category, while if only 50% passed the test, then the source network was placed into the super-weak category.

Over all datasets, the power law was found to be a poor fit for two-thirds of the degree distributions, while a power law could not be ruled out for the remaining one-third. The exponential distribution was equally as good as the power law, 36–37%, for modeling the degree distribution. In 88% of the networks, the log-normal distribution was as good or a better fit than the power law, reflecting the difficulty of distinguishing between the two distributions using a finite dataset. In 42% of all networks, the Weibull proved to be a better model of degree distributions than the power law, which was better in 33% of cases. Furthermore, only 4% were shown to be strongly scale-free (meaning that the power law should have an exponent between 2 and 3, and be superior to each of the alternative distributions). From these results, and more, the authors conclude that "in general, a scale-free distribution is rarely the best model of a network's degrees" and that

> It is remarkably rare for a network data set to exhibit the strongest form of direct evidence of scale-free structure ... we find essentially no empirical evidence to support the special status that the power law has held in network science as a starting point for modeling and analyzing the structure of real networks. Instead, it is an empirical fact that real-world networks exhibit a rich variety of degree structures, relatively few of which are convincingly scale-free.

Unfortunately, there are too many problems with this "statistical" analysis of network datasets to be persuaded by their results. The results of their study can hardly be regarded as "rigorous," as Broido and Clauset intended, especially given the manner in which they massaged the datasets.

Barabási's response. Indeed, the work by Broido and Clauset on the scale-free model of the degree distributions of networks led to a spirited response by Barabási (2018). He felt that their five categories of scale-free networks (super-weak, weakest, weak, strong, strongest) were arbitrary and did not reflect the nature of scale-free networks. He did not like the methodology used, especially the idea of creating and testing multiple simple networks instead of studying each original source network. He also noted that Broido and Clauset tried to force a pure power-law fit to every network in the study, when theory says it was not appropriate to do so.

Barabási pointed to a number of counterintuitive results from Broido and Clauset. In particular, he noted that none of the networks known to be scale-free (e.g., preferential attachment networks[8]) were placed in the strongest category. At the other extreme, only about half of the Erdős–Rényi networks, which are known not to be scale-free, were classified as weak or weakest. Barabási explained that

> By 2001, it was pretty clear that there is no one-size-fits-all formula for the degree distribution for networks driven by the scale-free mechanism. A pure power law only emerges in simple idealized models, driven only by growth and preferential attachment, and free of any additional effects,

such as the disappearance of nodes, addition of new edges linking already existing nodes, edge deletion, and so on. He argues that one should create a generative model that predicts the shape of the degree distribution of a given real-world network.

---

[8]See Section 8.4,

Furthermore, that model could be made to be more complicated than just a power law by incorporating corrections to the degree exponent β and taking into account a possible expansion of the network.

7.7 Further Reading

Section 7.2. The small-world phenomenon is described in the books by Durrett (2007, Chapter 5), Jackson (2008, Chapter 4), Kolaczyk (2009, Section 6.3), Easley and Kleinberg (2010, Chapter 20), Newman (2010, Section 3.6 and 15.1), and van der Hofstad (2017b, Chapters 3, 5, and 7).

Section 7.3. The Watts–Strogatz model is described in Barrat, Barthélemy, and Vespignani (2013, Section 3.1.3).

Section 7.4. Degree distributions are described in Kolaczyk (2009, Section 4.2.1), Barrat, Barthélemy, and Vespignani (2013, Section 1.3.1).

Section 7.5. Power law and scale-free networks with a focus on genome biology are described in the edited volume by Koonin, Wold, and Karev (2006). See also Durrett (2007, Chapter 4) and Newman (2010, Section 8.4).

Section 7.6.4. Resilience in networks is described in Durrett (2007, Section 4.7), Newman (2010, Chapter 16), and Barrat, Barthélemy, and Vespignani (2013, Chapter 6).

7.8 Exercises

Exercise 7.1 Choose a favorite actor or actress and compute his/her Bacon number using the website `oracleofbacon.org`.

Exercise 7.2 Choose a well-known statistician/probabilist/mathematician and find his/her Erdős number by using the website `www.oakland.edu/enp`.

Exercise 7.3 Choose a favorite baseball player (any era) and find his Oracle of Baseball number using the website `www.baseball-reference.com/oracle`.

Exercise 7.4 Write a computer routine to carry out the Watts–Strogatz rewiring algorithm (**Algorithm 7.1**) and then apply it to a ring-shaped network with $N = 50$ nodes, $k = 3$, and $p = 0.6$. (a) Draw the corresponding graph. (b) Draw the histogram of the degree distribution (use logs on each axes). (c) Compute the clustering coefficient.

Exercise 7.5 Choose a network dataset of your interest (not one of the datasets in **Table 7.2**). (a) Draw the network graph. (b) Draw the histogram representing the node-degree distribution of the data. (c) Draw the scatterplot of the logarithm of node-degree frequency (y-axis) against the logarithm of node degree (x-axis). (d) Does the scatterplot look like a scale-free network? (e) Compute the linear regression and estimate the power-law exponent β. (f) Compare your result with those listed in **Table 7.2**.

Exercise 7.6 Suppose X and Y are independent Poisson variables. Show that (a) $X+Y$ has the Poisson distribution, (b) X given $X + Y$ has the binomial distribution.

Exercise 7.7 Draw the degree distributions for (a) Sampson's monk data, (b) Zachary's karate club data, and (c) the terrorist network data. Use log scales on both axes. In each case, estimate the power-law exponent β.

Exercise 7.8 Download the 30,288 majority opinions of the Supreme Court citation network from `jhfowler.ucsd.edu/judicial.htm` (see Section 2.3.2) and convert them to an undirected network.

(a) Draw the degree distribution (i) using a linear scale on each axis, and (ii) using a log scale on each axis.
(b) Compute the shortest path lengths between all pairs of nodes and the average shortest path length (l). Compare your results with those listed in **Table 7.3**.

CHAPTER EIGHT

Models of Network Evolution and Growth

There have been several attempts at incorporating real-world components into network generation and growth. Most attention has centered on trying to create a desired structure for the node-degree distribution, such as clustering and the power-law property. This chapter discusses the advantages and disadvantages of the "configuration" and the "expected-degree" models, and describes how the growth of a network can be formulated through the "preferential-attachment" and "random-copying" (or "duplication") models.

8.1 Introduction

It is well known that real-world networks are not consistent with the Erdős–Rényi random graph model in which edges between any pair of existing nodes in the network occur independently with uniform probability. One of the main differences lies in the way in which node degrees are distributed amongst the nodes. In the Erdős–Rényi model, node degrees are Poisson distributed with a peak around the mean and rapidly decaying tails, whilst in many large, real-world networks, we see evidence of power-law degree structure with heavy-tailed degree distributions.

This difference raises a number of questions. What kind of role should node-degree distributions play in forming accurate models for networks? What if we made sure that the model for a network mimicked that of a real network by prespecifying the degree of every node in the network so that it agrees with the degree distribution of the target network? Would this prespecification be sufficient to simulate networks that had the same properties as the real network?

Another difference between real-world networks and the random graph model is that real-world networks exhibit a tendency towards strong clustering, whereas random graphs do not. Clustering occurs when the probability that any two nodes are neighbors of each other is higher when those nodes have a common neighbor. We saw in Chapter 7 that clustering can be measured by the clustering coefficient, which is the probability that two neighbors of a given node are also neighbors of each other, averaged over the entire network. Comparisons of the clustering coefficient for various real networks with that computed from a random graph show that the latter is a really poor approximation of the former.

In this chapter, we first describe two attempts to create a model in which the node-degree distribution is prespecified (or, at least, hinted at). We introduce the "configuration" model and the "expected-degree" model, and we discuss the advantages and disadvantages of both types of models.

Following those descriptions, we then describe two models for growing a network by adding nodes and edges using different strategies: the "preferential-attachment" and the "random-copying" (or "duplication") models. The preferential-attachment model adds nodes and edges in such a way that new edges will favor a few existing nodes that already have large degrees. The random-copying model copies edges from a randomly chosen node and grafts those edges onto a newly added node to the network. Although both models enhance our knowledge of how networks grow, neither is considered to be the final word on this topic.

8.2 The Configuration Model

The *configuration model* (Bender and Canfield, 1978; Wormald, 1978) is a widely used model for generating undirected random graphs with arbitrary non-Poisson degree distributions. The name "configuration model" was coined by Bollobás (1980). See also Bollobás (1985) and Molloy and Reed (1995).

In this model,

$$\mathcal{D} = \{k_1, k_2, \ldots, k_N\} \tag{8.1}$$

is taken to be a given node-degree sequence, where $k_i \geq 1$ is the degree of node v_i, $i = 1, 2, \ldots, N$. To construct such a degree sequence, we create a collection of elements" by listing the node v_i k_i times,

$$\underbrace{v_1, \ldots, v_1}_{k_1 \text{ times}} \underbrace{v_2, \ldots, v_2}_{k_2 \text{ times}} \cdots \underbrace{v_N, \ldots, v_N}_{k_N \text{ times}},$$

and then choose any two elements at random from the list and place an edge between them. We then delete those two elements from the list and repeat the process until no elements remain; this assumes there are an even number of elements. If, for any ℓ, $k_\ell = 0$, this means that the ℓth node is isolated and can, therefore, be ignored. In other words, we prespecify the degree of each of the N nodes, possibly to explore how closely the degree distribution captures the complexities of a real-world network, such as a power law.

The model can also be viewed as a fixed set of N nodes, with no edges, where the ith node v_i has k_i "spokes" (or "stubs" or "half-edges") emanating from it. Two spokes are selected uniformly at random from the network and joined to form an edge of the network. This process is repeated on the remaining spokes until no unattached spokes remain. The total number of spokes (i.e., $\sum_{i=1}^{N} k_i = 2M$, where $M = |\mathcal{E}|$ is the total number of edges in the network) has to be even so that no spoke is left unattached. When all spokes are accounted for, the resulting network is a member of the set of all networks that possess the specified degree sequence. The resulting graph is called the *configuration model* with degree sequence $\mathcal{D} = \{k_1, k_2, \ldots, k_N\}$, which we represent by $\mathrm{CM}_N(\mathcal{D})$.

We have the following result.

Theorem 8.1 *For the configuration model* $\text{CM}_N(\mathcal{D})$*, the probability of an edge between nodes* v_i *and* v_j *is*

$$p_{ij} = \frac{k_i k_j}{2M - 1} \approx \frac{k_i k_j}{2M}, \tag{8.2}$$

where the approximation holds if the total number of edges $M = |\mathcal{E}|$ *is very large.*

Proof Clearly, for the configuration model, the larger the degrees k_i and k_j, the greater the probability that the corresponding nodes v_i and v_j will be connected. The probability p_{ij} can also be viewed as the expected number of edges between nodes v_i and v_j. To see this, note that the number of possible edges in the network is the number of ways of choosing two spokes from the total of $2M$ spokes, which is $\binom{2M}{2}$ or $M(2M - 1)$. Thus, the probability of an edge between v_i and v_j is $k_i k_j / M(2M - 1)$. Multiplying this by M for the number of edges yields the expected number of edges between v_i and v_j, namely, (8.2). □

The resulting network may contain multiple edges (which would occur if pairs of spokes are selected more than once from the same two nodes) and self-loops (which would occur if two spokes are selected from the same node).

8.2.1 Multiple Edges

Suppose there is an edge between nodes v_i and v_j. What is the probability that, under the configuration model, a *second* edge would appear between those same two nodes? In the following, X denotes the degree of a randomly chosen node and p_k denotes the degree distibution of that node; that is

$$p_k = \text{P}\{X = k\}, \quad k = 0, 1, 2, \ldots . \tag{8.3}$$

Denote by M_N the number of multiple edges.

We have the following result.

Theorem 8.2 *Consider the configuration model* $\text{CM}_N(\mathcal{D})$ *with degree sequence* $\mathcal{D} = \{k_1, k_2, \ldots, k_N\}$*. Let* X *denote the degree of a randomly chosen node and let* M_N *denote the number of multiple edges. The expected number of multiple edges is*

$$\text{E}\{M_N\} = \frac{1}{2} \left(\frac{\text{E}\{X^2\} - \text{E}\{X\}}{\text{E}\{X\}} \right)^2 .$$

Proof Suppose we used one spoke from v_i and one spoke from v_j to create the first edge between those two nodes. So, the probability that a second edge appears joining the same two nodes is $(k_i - 1)(k_j - 1)/2M$. The probability of two such edges is, therefore, $k_i k_j (k_i - 1)(k_j - 1)/(2M)^2$. But this probability is for two specific nodes, v_i and v_j. Summing this last expression over all distinct pairs of nodes in the network, we have that the expected number of multiple edges is

$$E\{M_N\} = \sum_{i \neq j} \frac{k_i k_j (k_i - 1)(k_j - 1)}{(2M)^2}$$

$$= \frac{1}{2} \frac{1}{(2M)^2} \sum_{i=1}^{N} k_i (k_i - 1) \sum_{j=1}^{N} k_j (k_j - 1)$$

$$= \frac{1}{2(E\{X\})^2 N^2} \sum_i (k_i^2 - k_i) \sum_j (k_j^2 - k_j)$$

$$= \frac{1}{2(E\{X\})^2} \left(\frac{1}{N} \sum_i k_i^2 - \frac{1}{N} \sum_i k_i \right) \left(\frac{1}{N} \sum_j k_j^2 - \frac{1}{N} \sum_j k_j \right)$$

$$= \frac{(E\{X^2\} - E\{X\})^2}{2(E\{X\})^2}$$

$$= \frac{1}{2} \left(\frac{E\{X^2\} - E\{X\}}{E\{X\}} \right)^2, \tag{8.4}$$

where we set

$$E\{X^m\} = \sum_k k^m p_k = \frac{1}{N} \sum_i k_i^m, \quad m = 1, 2, \tag{8.5}$$

p_k is the degree distribution (i.e., proportion of nodes having degree k), and

$$2M = \sum_{i=1}^{N} k_i = E\{X\}N. \tag{8.6}$$

From this result, we conclude that if $E\{X\}$ and $E\{X^2\}$ are finite, then the expected number of multiple edges is a constant, which for large N, quickly becomes negligible. □

8.2.2 Self-Loops

A similar argument can be used to find the expected number of self-loops.

Theorem 8.3 *Consider the configuration model* $CM_N(\mathcal{D})$ *with degree sequence* $\mathcal{D} = \{k_1, k_2, \ldots, k_N\}$. *Let* X *denote the degree of a randomly chosen node and let* S_N *denote the number of self-loops in the network. The expected number of self-loops is*

$$E\{S_N\} = \frac{E\{X^2\} - E\{X\}}{2E\{X\}}.$$

Proof Let k_i be the number of spokes at node v_i. The number of possible self-loops for node v_i is $\binom{k_i}{2}$. So, the probability p_i of a self-loop for v_i is

$$p_i = \frac{k_i(k_i - 1)/2}{2M} = \frac{k_i(k_i - 1)}{4M}. \tag{8.7}$$

Now, we sum p_i over i to get the expected number of self-loops in the network $CM_N(\mathcal{D})$:

$$E\{S_N\} = \sum_i \frac{k_i(k_i - 1)}{4M}$$

$$= \frac{1}{2} \cdot \frac{1}{2M} \left(\sum_i k_i^2 - \sum_i k_i \right)$$

$$= \frac{1}{2E\{X\}} \left(\frac{1}{N} \sum_i k_i^2 - \frac{1}{N} \sum_i k_i \right)$$

$$= \frac{E\{X^2\} - E\{X\}}{2E\{X\}}. \tag{8.8}$$

So, for large N and $E\{X^2\} < \infty$, self-loops will be a negligible proportion of the total number of edges in the network. □

If a graph does not contain either self-loops or multiple edges, it is referred to as a *simple graph*. If the node degrees of a simple graph are ordered as $k_1 \geq k_2 \geq \cdots \geq k_N$, then the degree sequence $\mathcal{D}$ is called *graphic*. A degree sequence $\mathcal{D}$ will be graphic iff $\sum_{i=1}^{N} k_i$ is even and

$$\sum_{i=1}^{d} k_i \leq d(d-1) + \sum_{i=d+1}^{N} \min\{d, k_i\}, \tag{8.9}$$

for each integer $d \leq N - 1$ (Erdős and Gallai, 1960).

If, however, self-loops and multiple edges are allowed to be present, the result is known as a *multigraph* having the given degree sequence. Some authors recommend either removing self-loops and merging multiple edges into single edges, or running the process over and over again until a simple graph is generated (with no self-loops or multiple edges). This last option is so computationally inefficient that it would be difficult for the algorithm to terminate within a reasonable time. In a large network, features such as self-loops and multiple edges are not that important anyway. In fact, as $N \to \infty$, the probability that the multigraph $CM_N(\mathcal{D})$ will become a simple graph (with a vanishingly small proportion of self-loops or multiple edges) is asymptotically proportional to $\exp\{-\frac{\alpha}{2} - \frac{\alpha^2}{4}\}$, where

$$\alpha = \frac{E\{X^2\} - E\{X\}}{E\{X\}} \tag{8.10}$$

(van der Hofstad, 2017a, Section 7.4). In other words, as $N \to \infty$, the number of self-loops and the number of multiple edges are asymptotically independent Poisson random variables with means $\alpha/2$ and $(\alpha/2)^2$, respectively.

8.2.3 Sharing a Common Neighbor

We can obtain another property of the configuration model.

Theorem 8.4 *For the configuration model* $\mathrm{CM}_N(\mathcal{D})$, *let* N_{ij} *denote the number of common nodes that the nodes* v_i *and* v_j *share. Let* p_{ij} *be the probability that node* v_i *connects to node* v_j. *Then, the expected number of neighboring nodes that nodes* v_i *and* v_j *share is*

$$E\{N_{ij}\} = p_{ij}\frac{E\{X^2\} - E\{X\}}{E\{X\}}.$$

Proof Note that if v_i is connected to v_l, then there are $k_l - 1$ spokes of v_l still open. So, the probability that node v_j connects to node v_l is $k_j(k_l - 1)/2M$ (not $p_{jl} = k_jk_l/2M$). Multiplying the probabilities for the two connections and summing, we get

$$E\{N_{ij}\} = \sum_l \frac{k_ik_l}{2M} \cdot \frac{k_j(k_l - 1)}{2M}$$

$$= \frac{k_ik_j}{2M}\sum_l \frac{k_l(k_l - 1)}{2M}$$

$$= p_{ij}\frac{1}{E\{X\}}\left(\frac{1}{N}\sum_l k_l^2 - \frac{1}{N}\sum_l k_l\right)$$

$$= p_{ij}\frac{E\{X^2\} - E\{X\}}{E\{X\}}, \tag{8.11}$$

where we used $2M = NE\{X\}$. So, the expected number of neighboring nodes[1] that v_i and v_j share is proportional to p_{ij}, where the constant of proportionality depends only on the first two moments of the degree distribution and not on v_i or v_j. □

8.2.4 Excess-Degree Distribution

The *excess-degree distribution* is the degree distribution of a randomly chosen node that is connected to a randomly chosen neighbor, excluding the edge between them. In other words, if we take a randomly chosen node and travel along one of its (randomly chosen) edges to another node, what would be the probability that the neighboring node has degree k?

Theorem 8.5 *For the configuration model* $\mathrm{CM}_N(\mathcal{D})$, *let* X *be the degree of a randomly chosen node and let* p_k *denote the proportion of nodes with degree* k. *Then, the probability that a neighbor has degree* k *is*

$$q_k = \frac{(k + 1)p_{k+1}}{E\{X\}}, \quad k = 0, 1, 2, \ldots.$$

In order that the node in question is attached by an edge to a node of degree k, the edge must be attached to one of the Np_k nodes of degree k. In the configuration model, two randomly chosen spokes are connected (given each node's degree) to create an edge between the corresponding nodes. Thus, for large N, the endpoint of every edge has equal probability $k/2M$ of connecting to one of the spokes attached to the given node.

---

[1] Recall that a pair of nodes that are connected to each other are referred to as "neighbors."

We have the following result.

Theorem 8.6 *Under the same conditions as Theorem 8.5, the degree distribution of a randomly chosen neighbor is*

$$P\{\text{neighbor has degree } k\} = \frac{k}{2M} N p_k = \frac{k p_k}{E\{X\}}. \tag{8.12}$$

The expected degree of such a neighbor is, therefore, given by

$$E\{\text{neighbor has degree } k\} = \sum_k k P\{\text{neighbor has degree } k\}$$

$$= \frac{E\{X^2\}}{E\{X\}} > E\{X\} \tag{8.13}$$

because $E\{X^2\} - (E\{X\})^2 > 0$. This implies that the neighbors of a given node each have a greater (expected) degree than does the node itself. In terms of social networks: "Your friends have more friends than you do."

Suppose we are located at a randomly chosen node v, which has degree at least one. What is the probability that a randomly chosen neighbor of node v has degree k? This question is answered by the *excess-degree distribution*, which yields the probability of the number of edges (excluding the edge from the given node) attached to the neighboring node.

From v, suppose we travel along one of its edges to a neighboring node, say, v'. Once at v', we look at its edges to get to a second neighbor. But one of those edges leads back to v, so that edge does not lead to a second neighbor of v. Thus, the number of second neighbors of v is one fewer than the degree of the node. The correct probability distribution is, therefore, $q_{k-1} = k p_k / E\{X\}$, which is the above result (8.12) for the probability that a neighbor has degree k. If we replace k by $k+1$ in this expression, this is equivalent to

$$q_k = \frac{(k+1) p_{k+1}}{E\{X\}}, \tag{8.14}$$

which is the excess-degree distribution. Note that $\sum_{k=0}^{\infty} q_k = 1$.

8.2.5 Clustering Coefficient

Now that we have the excess-degree distribution, we can determine an expression for its clustering coefficient. The clustering coefficient C is the average probability that two nodes, v_i and v_j, say, that are each connected to a given third node, v_l, say, which has degree at least 2, are themselves connected to each other, so that the three nodes, v_i, v_j, and v_l, form a triangle.

Theorem 8.7 *For the configuration model* $CM_N(\mathcal{D})$, *let X be the degree of a randomly chosen node. Then, the clustering coefficient C is*

$$C = \frac{(E\{X^2\} - E\{X\})^2}{N(E\{X\})^3}. \tag{8.15}$$

Proof Let v_i be a node of degree at least two, and let v_j and v_l be two neighbors of v_i. We know that nodes v_j and v_l are connected with probability $k_j k_l / 2M$. The clustering coefficient C is given by this probability times the probability, q_{k_j}, that v_j has excess degree k_j and the probability, q_{k_l}, that v_l has excess degree k_l, summed over all choices of k_j and k_l:

$$
\begin{aligned}
C &= \sum_{k_j=0}^{\infty} \sum_{k_l=0}^{\infty} q_{k_j} q_{k_l} \frac{k_j k_l}{2M} \\
&= \frac{1}{2M} \left(\sum_{k=0}^{\infty} q_k k \right)^2 \\
&= \frac{1}{2M(\mathrm{E}\{X\})^2} \left(\sum_{k=0}^{\infty} k(k+1) p_{k+1} \right)^2 \\
&= \frac{1}{2M(\mathrm{E}\{X\})^2} \left(\sum_{k=0}^{\infty} k(k-1) p_k \right)^2 \\
&= \frac{(\mathrm{E}\{X^2\} - \mathrm{E}\{X\})^2}{N(\mathrm{E}\{X\})^3},
\end{aligned}
\tag{8.16}
$$

where we used (8.13) to get the third equality. $\square$

As with the expected number of multiple edges, the clustering coefficient becomes negligible as $N \to \infty$. Unfortunately, this result does not reflect what we know of real-world networks, because such networks typically enjoy a high amount of clustering.

As a result, some authors (Ball, Sirl, and Trapman, 2009, 2010; Miller, 2009; Newman, 2009) introduced various types of community structure into the configuration model (e.g., to understand the spread of epidemics) to make its clustering feature more realistic.

8.2.6 Probability Generating Functions

As before, let X denote the degree of a randomly chosen node and let p_k be the probability that the randomly chosen node has degree k. Recent studies on the configuration model in the physics literature (e.g., Newman, Strogatz, and Watts, 2001) have emphasized the use of probability generating functions in determining features of the model.

Definition 8.1 The *probability generating function* of X is defined by

$$
G_0(s) = \mathrm{E}\{s^X\} = \sum_{k=0}^{\infty} p_k s^k,
\tag{8.17}
$$

where $G_0(1) = 1$ so that p_k, the probability that a node has degree k, is correctly normalized. The generating function $G_0(s)$ converges absolutely for all $s \leq 1$.

We have the following properties.

- The probability p_k is found by differentiating $G_0(s)$ k times:

$$p_k = \frac{1}{k!} \left. \frac{d^k G_0(s)}{ds^k} \right|_{s=0}. \tag{8.18}$$

- The moments of X are given by

$$\mu = E\{X\} = \sum_k k p_k = G_0'(1), \tag{8.19}$$

$$E\{X^m\} = \sum_k k^m p_k = s^m \left. \frac{d^m G_0(s)}{ds^m} \right|_{s=1}. \tag{8.20}$$

Theorem 8.8 *Suppose we randomly select r nodes from a network with N nodes. Let X_i denote the degree of node v_i. Suppose also that we assume those nodes have degrees $k_1, k_2, \ldots, k_r$. Then, the probability distribution of the total number of degrees of those nodes, $\sum_{i=1}^r X_i$, has generating function $[G_0(x)]^r$.*

Proof Let $G(z)$ be the generating function corresponding to p_k, the probability that a node has degree k. The probability that the degrees take the values $k_1, k_2, \ldots, k_r$ is $\prod_{i=1}^r p_{k_i}$. The probability, $p(m)$, that the degrees add up to m is the sum of these probabilities over all sets of the k_i that sum to m:

$$p(m) = P\left\{ \sum_{i=1}^r X_i = m \right\} = \sum_{k_1=0}^\infty \cdots \sum_{k_r=0}^\infty \delta\left(m, \sum_{i=1}^r k_i\right) \prod_{i=1}^r p_{k_i}, \tag{8.21}$$

where $\delta(u, v) = 1$ if $u = v$, and 0 otherwise. Thus, only those sets of degrees that sum to m contribute to $p(m)$. The generating function $H(s)$ corresponding to $p(m)$ is given by

$$H(s) = \sum_{m=0}^\infty p(m)s^m$$

$$= \sum_{m=0}^\infty s^m \sum_{k_1=0}^\infty \cdots \sum_{k_r=0}^\infty \delta\left(m, \sum_i k_i\right) \prod_{i=1}^r p_{k_i}$$

$$= \sum_{k_1=0}^\infty \cdots \sum_{k_r=0}^\infty s^{\sum_i k_i} \prod_{i=1}^r p_{k_i}$$

$$= \sum_{k_1=0}^\infty \cdots \sum_{k_r=0}^\infty \prod_{i=1}^r p_{k_i} s^{k_i}$$

$$= \left(\sum_{k=0}^\infty p_k s^k \right)^r$$

$$= [G(s)]^r. \tag{8.22}$$

$\square$

- The probability generating function for the excess-degree distribution (8.14) is

$$
\begin{aligned}
G_1(s) &= \sum_{k=0}^{\infty} q_k s^k = \sum_{k=0}^{\infty} \left(\frac{(k+1)p_{k+1}}{\mu} \right) s^k \\
&= \frac{1}{\mu} \sum_{k=0}^{\infty} k p_k s^{k-1} \\
&= \frac{1}{\mu} G_0'(s) = \frac{G_0'(s)}{G_0'(1)},
\end{aligned}
\tag{8.23}
$$

where the expected node degree μ is given by (8.18). So, from $G_0(s)$, we can find $G_0'(s)$ (and, hence, $G_0'(1)$), which yields $G_1(s)$. In other words, we do not need to compute the generating function $G_1(s) = \sum_{k=0}^{\infty} q_k s^k$ directly from the definition of the excess-degree distribution.

8.2.7 Distributional Examples

We have the following specific examples.

Binomial-Distributed Networks

Let

$$
p_k = \binom{n}{k} p^k (1-p)^{n-k}, \quad k = 0, 1, 2, \ldots, n.
\tag{8.24}
$$

Then,

$$
G_0(s) = \sum_{k=0}^{n} p_k s^k = \sum_{k=0}^{\infty} \binom{n}{k} (ps)^k (1-p)^{n-k} = ((1-p) + ps)^n.
\tag{8.25}
$$

Note that $G_0'(s) = np(1 - p + ps)^{n-1}$, whence $\mu = G_0'(1) = np$.

Poisson-Distributed Networks

Let

$$
p_k = \frac{e^{-\lambda} \lambda^k}{k!}, \quad k = 0, 1, 2, \ldots.
\tag{8.26}
$$

Here, $\mu = \lambda$. Then,

$$
q_{k-1} = \frac{k p_k}{\mu} = \frac{k \cdot e^{-\lambda} \lambda^k}{k! \lambda} = \frac{e^{-\lambda} \lambda^{k-1}}{(k-1)!},
\tag{8.27}
$$

which is also Poisson. Let $G_0(s) = \sum_k p_k s^k$ and $G_1(s) = \sum_k q_k s^k$ be the two generating functions corresponding to p_k and q_k, respectively. Then,

$$
G_0(s) = \sum_{k=0}^{n} p_k s^k = e^{-\lambda} \sum_{k=0}^{\infty} \frac{(\lambda s)^k}{k!} = e^{\lambda(s-1)}.
\tag{8.28}
$$

Furthermore, $G_0(s)$ and $G_1(s)$ are related:

$$
\frac{G_0'(s)}{\mu} = \frac{1}{\mu} \sum_{k=1}^{\infty} k p_k s^{k-1} = \sum_{k=1}^{\infty} \left(\frac{k p_k}{\mu} \right) s^{k-1} = \sum_{k=1}^{\infty} q_{k-1} s^{k-1} = G_1(s).
\tag{8.29}
$$

Table 8.1 Probability generating functions for some distributions

Distribution	$G_0(s)$	$\mu = G_0'(1)$
Binomial	$((1-p)+ps)^n$	np
Poisson	$e^{\lambda(s-1)}$	λ
Exponential	$\dfrac{1-e^{-\lambda}}{1-se^{-\lambda}}$	$\dfrac{1}{e^{\lambda}-1}$

Because $G_0'(s) = \lambda e^{\lambda(s-1)}$, it follows that $\mu = G_0'(1) = \lambda$, and so $G_0(s) = G_1(s)$. Thus, the generating functions corresponding to the degree distribution and the excess-degree distribution are identical for the Poisson case.

Exponential-Distributed Networks

Let

$$p_k = (1 - e^{-\lambda})e^{-\lambda k}, \quad k = 0, 1, 2, \ldots, \tag{8.30}$$

where λ is a constant. The generating function is

$$
\begin{aligned}
G_0(s) &= \sum_k p_k s^k = (1 - e^{-\lambda}) \sum_k e^{-\lambda k} s^k \\
&= (1 - e^{-\lambda}) \sum_k (e^{-\lambda} s)^k \\
&= \frac{1 - e^{-\lambda}}{1 - se^{-\lambda}}.
\end{aligned}
\tag{8.31}
$$

We also have

$$G_0'(s) = \frac{(1 - e^{-\lambda})e^{-\lambda}}{(1 - se^{-\lambda})^2}, \tag{8.32}$$

so that

$$\mu = G_0'(1) = \frac{e^{-\lambda}}{1 - e^{-\lambda}} = \frac{1}{e^{\lambda} - 1}. \tag{8.33}$$

Thus,

$$G_1(s) = \frac{G_0'(s)}{G_0'(1)} = \left(\frac{1 - e^{-\lambda}}{1 - se^{-\lambda}} \right)^2. \tag{8.34}$$

A summary of these probability generating functions is given in **Table 8.1**.

8.2.8 Neighbors of Neighbors

Many properties of the configuration model $\mathrm{CM}_N(\mathcal{D})$ have been studied, such as the probability that a given node has k second neighbors (i.e., neighbors of neighbors of a given node), or third neighbors, or, more generally, dth neighbors. These probabilities can be found from the generating function for the degree distribution (Newman, Strogatz, and Watts, 2001).

For example, suppose we wish to derive the probability, $p_k^{(2)}$, that there are k second neighbors of a given node v. This probability can be written as

$$p_k^{(2)} = \sum_{t=0}^{\infty} p_t \, p^{(2)}(k|t), \tag{8.35}$$

where $p^{(2)}(k|t)$ is the conditional probability that the node v has k second neighbors, given that it has t first neighbors. We are implicitly assuming here that the network has a tree structure, so that second neighbors are assumed not to be first neighbors. Then, the probability $p_k^{(2)}$ is the conditional probability $p^{(2)}(k|t)$ averaged over all possible values of t. Whilst this probability can be obtained in a straightforward way, we show how it can be obtained using generating functions.

The generating function for the probability distribution, $p_k^{(2)}$, of the number of second neighbors of a given node v is

$$\begin{aligned}
G^{(2)}(s) &= \sum_{k=0}^{\infty} p_k^{(2)} s^k \\
&= \sum_{k=0}^{\infty} \sum_{t=0}^{\infty} p_t \, p^{(2)}(k|t) s^k \\
&= \sum_{t=0}^{\infty} p_t \sum_{k=0}^{\infty} p^{(2)}(k|t) s^k \\
&= \sum_{t=0}^{\infty} p_t [G_1(s)]^t \\
&= G_0(G_1(s)), \tag{8.36}
\end{aligned}$$

where, from (8.21), the generating function of $p^{(2)}(k|t)$ is the tth power of the generating function for a first neighbor of the node v:

$$\sum_{k=0}^{\infty} p^{(2)}(k|t) s^k = [G_1(s)]^t. \tag{8.37}$$

For example, in a Poisson-distributed network, we showed that $G_0(s) = G_1(s) = e^{\lambda(s-1)}$, whence,

$$G^{(2)}(s) = e^{\lambda(e^{\lambda(s-1)}-1)}. \tag{8.38}$$

Similarly, the distribution of third neighbors is generated by

$$G^{(3)}(s) = G_0(G_1(G_1(x))). \tag{8.39}$$

These results for second and third neighbors of a given node can be generalized to dth neighbors. Using (8.35),

$$\begin{aligned}
G^{(d)}(s) &= \sum_{k=0}^{\infty} \sum_{t=0}^{\infty} p_t^{(d-1)} p^{(d)}(k|t) s^k \\
&= \sum_{t=0}^{\infty} p_t^{(d-1)} \sum_{k=0}^{\infty} p^{(d)}(k|t) s^k
\end{aligned}$$

$$= \sum_{t=0}^{\infty} p_t^{(d-1)}[G_1(s)]^t$$

$$= G^{(d-1)}(G_1(s)), \tag{8.40}$$

so that

$$G^{(d)}(s) = G_0(G_1(\cdots G_1(s) \cdots)). \tag{8.41}$$

Thus, the generating function $G^{(d)}(s)$ is actually $d - 1$ copies of G_1 hierarchically placed inside a single G_0.

8.2.9 Phase Transition

Just as we showed that the Erdős–Rényi Poisson random-graph model exhibits a phase transition, it can also be shown that a similar phenomenon occurs for the configuration model. Recall that, for the random-graph model, below the phase-transition point, there can be no giant component, just many small components that are not connected to each other. On the other hand, above the phase-transition point, there is a giant component, which consists of a sizeable proportion of the nodes of the entire network, plus a number of small components that fill up the regions of the network not consumed by the giant component.

A demonstration that a phase transition exists for the configuration model was given by Molloy and Reed (1995). See also Newman, Strogatz, and Watts (2001) and Durrett (2007, Section 3.2).

Theorem 8.9 (Molloy and Reed, 1995) *Let X denote the degree of a randomly chosen node. Then, there exists a giant component for the configuration model iff*

$$Q = E\{X^2\} - 2E\{X\} = \sum_{k=0}^{\infty} k(k-2)p_k > 0. \tag{8.42}$$

Note that Q can also be expressed as $\frac{1}{N} \sum_{i=1}^{N} k_i(k_i - 2)$. The phase transition for the network occurs when $Q = 0$. This condition is trivially true if $E\{X^2\} = \infty$.

We first make a few remarks about condition (8.42). The terms in the sum with $k = 0$ or $k = 2$ will vanish and, hence, will provide no contributions to the sum. This condition implies that nodes of degree 0 or 2 can be added or removed from the network without having any effect on the condition for the existence of the giant component. How can we explain this statement? Clearly, nodes with degree 0 (i.e., nodes that are isolated, having no edges) have no effect on whether or not there is a giant component. A node with degree 2 has two edges and can be viewed as residing between two other nodes. So, if we add a node with degree 2 to a network, this is equivalent to adding an extra node to the middle of an edge that already exists. However, such nodes have no effect on whether the network contains a giant component, and so they are ignored in condition (8.42).

The basic idea behind condition (8.42) is that, starting from a randomly chosen node, we move through the network by exploring its path structure, which, as we mentioned above, is assumed to possess a tree structure (i.e., second neighbors cannot be first neighbors). If the node happens to be in a giant component, such a path should lead to an increasing number of neighboring nodes that have not yet been visited. On the other hand, if the node is in a small component, then at a certain point, the path will terminate and additional neighboring nodes will no longer exist.

To derive condition (8.42), we follow the argument based upon generating functions that is given in Newman (2010, Chapter 13).

Proof Let X denote the degree of a randomly selected node. In general, the expected value of X can be found by taking the first derivative of its generating function, $G(s)$, and then evaluating the result at $s = 1$; that is, $E\{X\} = G'(1)$ (see (8.19)). In this case, we need to find the first derivative of the generating function, $G^{(2)}(s)$, of the number of second neighbors of a given node. Differentiating (8.36), we have that

$$G^{(2)\prime}(s) = G_0'(G_1(s))G_1'(s). \tag{8.43}$$

Set $s = 1$. Then,

$$G^{(2)\prime}(1) = G_0'(G_1(1))G_1'(1) = G_0'(1)G_1'(1) = E\{X\}G_1'(1), \tag{8.44}$$

where we used $G_1(1) = 1$ from (8.23), and $G_0'(1) = E\{X\}$ from (8.19). Also, if $G_1(s) = \sum_k q_k s^k$, then $G_1'(s) = \sum_k k q_k s^{k-1}$. Using (8.14),

$$
\begin{aligned}
G_1'(1) &= \sum_{k=0}^{\infty} k q_k \\
&= \sum_{k=0}^{\infty} k \frac{(k+1)p_{k+1}}{E\{X\}} \\
&= \frac{1}{E\{X\}} \sum_{k=0}^{\infty} k(k+1)p_{k+1} \\
&= \frac{1}{E\{X\}} \sum_{k=0}^{\infty} (k-1)k p_k \\
&= \frac{E\{X^2\} - E\{X\}}{E\{X\}}.
\end{aligned} \tag{8.45}
$$

Putting (8.44) and (8.45) together, we have that the expected number of second neighbors is

$$G^{(2)\prime}(1) = E\{X^2\} - E\{X\}. \tag{8.46}$$

For the special case of a Poisson random network, the expected number of second neighbors is $G^{(2)\prime}(1) = (E\{X\})^2$, the square of the expected number of first neighbors.

Now, we generalize this result to dth neighbors. Differentiating (8.40), we have that

$$G^{(d)\prime}(s) = G^{(d-1)\prime}(G_1(s))G_1'(s). \tag{8.47}$$

As before, set $s = 1$. Then, the expected number of dth neighbors is given by

$$
\begin{aligned}
G^{(d)\prime}(1) &= G^{(d-1)\prime}(G_1(1))G_1'(1) = G^{(d-1)\prime}(1)G_1'(1) \\
&= G^{(d-1)\prime}(1)\frac{E\{X^2\} - E\{X\}}{E\{X\}},
\end{aligned} \tag{8.48}
$$

where we used (8.45). This has the form of $u_d = u_{d-1}\alpha$, where α is a constant and $u_d = G^{(d)\prime}(s)$. The rhs of this expression can be iteratively expanded $d - 1$ times to become $u_d = u_1\alpha^{d-1}$. Hence, using $G'(1) = E\{X\}$, the expected number of dth neighbors is

$$G^{(d)\prime}(1) = E\{X\}\left(\frac{E\{X^2\} - E\{X\}}{E\{X\}}\right)^{d-1}. \tag{8.49}$$

Note that this result is a function of just $E\{X\}$ and $E\{X^2\}$, the first and second moments of the degree distribution.

Depending upon the value of the ratio $(E\{X^2\} - E\{X\})/E\{X\}$, the expression (8.49) could grow exponentially or decrease exponentially as d increases. If (8.49) grows exponentially, then a giant component must exist in the network; otherwise, the set of neighboring nodes available to be explored will quickly diminish in size until it becomes empty, so that no giant component can exist. So, a giant component will exist iff $E\{X^2\} - E\{X\} > E\{X\}$, which leads to the condition given by (8.42). □

Example: Power-Law Distribution

Suppose that the configuration network has a "pure" power-law degree distribution of the form

$$p_k = ck^{-\beta}, \ k \geq 1, \tag{8.50}$$

and 0 for $k = 0$, where $c^{-1} = \zeta(\beta) = \sum_{k=1}^{\infty} k^{-\beta}$ is the Riemann zeta function. The condition for a giant component to exist is that $E\{X^2\} - 2E\{X\} > 0$. Now,

$$E\{X\} = \sum_{k=0}^{\infty} kp_k = \frac{1}{\zeta(\beta)}\sum_{k=1}^{\infty} k^{-(\beta-1)} = \frac{\zeta(\beta - 1)}{\zeta(\beta)} \tag{8.51}$$

and

$$E\{X^2\} = \sum_{k=0}^{\infty} k^2 p_k = \frac{1}{\zeta(\beta)}\sum_{k=1}^{\infty} k^{-(\beta-2)} = \frac{\zeta(\beta - 2)}{\zeta(\beta)}. \tag{8.52}$$

We substitute (8.51) and (8.52) into the condition for a giant component and solve for the critical value β_c. The critical value, β_c, is defined as the solution to the equation $E\{X^2\} - 2E\{X\} = 0$, which translates to

$$\zeta(\beta_c - 2) = 2\zeta(\beta_c - 1). \tag{8.53}$$

Numerical methods show that $\beta_c = 3.47875\ldots$. See **Figure 8.1**. A unique giant component exists a.s. iff $\beta < \beta_c$, whilst if $\beta \geq \beta_c$, the configuration model a.s. has no giant component, only a large number of small components. These results were first presented in Aiello, Chung, and Lu (2000), who gave more specific results regarding the values of $\beta \in [0, \beta_c)$ for which there exists a giant component:

1. When $0 < \beta < 1$, the network is a.s. connected, and there is a giant component, which contains a fraction of the nodes that tends to 1 as $N \to \infty$.
2. When $1 < \beta < 2$, the network is a.s. not connected, there is a giant component, and all smaller components have size $\mathcal{O}(1)$.
3. When $\beta = 2$, there is a giant component, and the second largest component a.s. has size $\mathcal{O}\left(\frac{\log N}{\log\log N}\right)$.

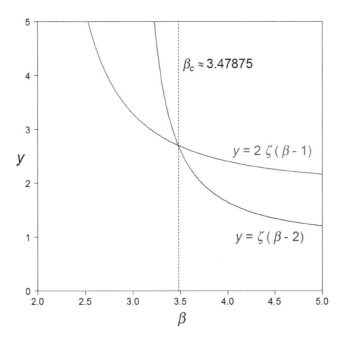

Figure 8.1 Solution of the equation $\zeta(\beta_c - 2) = 2\zeta(\beta_c - 1)$ for the power-law distribution of the configuration network. The blue curve is the lhs and the red curve is the rhs of the equation. They cross at the critical value $\beta_c \approx 3.47875$.

4. When $2 < \beta < \beta_c$, there is a giant component, which does not contain the entire network, and all smaller components have size $\mathcal{O}(\log N)$.
5. The cases $\beta = 1$ and $\beta = \beta_c$ are a lot more complicated. When $\beta = 1$, the network can be either connected or disconnected, and a nontrivial probability is attached to each such situation.

A modified power-law degree distribution having exponential cutoff,

$$p_k = ck^{-\beta}e^{-\alpha k}, \quad k \geq 1, \tag{8.54}$$

where α, β, and c are constants, was studied by Newman, Strogatz, and Watts (2001). The exponential cutoff has been observed amongst various types of networks.

In the case of general degree distributions with power-law tails, a giant component exists iff $2 < \beta \leq 3$. This condition derives from the fact that $E\{X\}$ will be finite if $\beta > 2$, whilst $E\{X^2\}$ will not be finite if $\beta \leq 3$, in which case $E\{X^2\} - 2E\{X\} > 0$. As we have seen, real-world networks with power-law degree distributions tend to have values of β in the range $2 < \beta \leq 3$, which implies that all such networks will most likely have a giant component. These results (and more) for the configuration model were derived using probability generating functions.

8.2.10 Diameter of the Giant Component

One of the interesting questions asked of a network is how fast information (or even perhaps a computer virus or an infectious disease) can be passed along the paths in the network. This obviously depends upon the structure of the network and its size. If we are dealing with a giant component, for example, then we are interested in the "diameter" of that component.

Recall that the *diameter* of a network (or of a component of a network) is defined as the average length of the shortest path between any pair of nodes, assuming that such a connected path exists. For example, the diameter of the World Wide Web is the smallest number of URL links that must be followed to navigate from one document to a second document (Albert, Jeong, and Barabási, 1999).

Theorem 8.10 *Let X denote the degree of a randomly selected node. The diameter of the giant component is*

$$D = \frac{\log[(N-1)(E\{X^2\} - 2E\{X\}) + (E\{X\})^2] - \log((E\{X\})^2)}{\log(E\{X^2\} - E\{X\}) - \log E\{X\}}. \tag{8.55}$$

Proof To derive an expression for the diameter, we start with the result (8.49) for the expected number of dth neighbors. We sum that expression out to a Dth neighborhood and then apply the formula for the sum of D terms of an arithmetic progression,[2]

$$\sum_{d=1}^{D} E\{X\} \left(\frac{E\{X^2\} - E\{X\}}{E\{X\}}\right)^{d-1} = E\{X\} \left(\frac{1 - \left(\frac{E\{X^2\} - E\{X\}}{E\{X\}}\right)^D}{1 - \frac{E\{X^2\} - E\{X\}}{E\{X\}}}\right). \tag{8.56}$$

If this sum equals $N-1$, it tells us that, starting from a randomly chosen node, we have reached every other node in the network, so that D gives us a clue as to the diameter of the giant component. The diameter will be approximately that value of D that solves the equation

$$E\{X\} \left(\frac{1 - \left(\frac{E\{X^2\} - E\{X\}}{E\{X\}}\right)^D}{1 - \frac{E\{X^2\} - E\{X\}}{E\{X\}}}\right) = N - 1. \tag{8.57}$$

After some algebra, we have

$$\left(\frac{E\{X^2\} - E\{X\}}{E\{X\}}\right)^D = \frac{(E\{X\})^2 + (N-1)E\{X^2\} - 2E\{X\}}{(E\{X\})^2}. \tag{8.58}$$

Taking logarithms of both sides yields (8.55). □

Thus, the diameter is a function only of the first and second moments of the underlying distribution of X. See, for example, Newman, Strogatz, and Watts (2001).

Example: Poisson-Distributed Network

We can simplify this expression quite a bit in the case of a Poisson network. If X has a Poisson distribution, then $\text{Var}\{X\} = E\{X^2\} - (E\{X\})^2 = E\{X\}$, so that $E\{X^2\} - E\{X\} = (E\{X\})^2$. Then,

---

[2] $\sum_{d=1}^{D} ar^{d-1} = \frac{a(1-r^D)}{1-r}$, where a is a constant.

$$D = \frac{\log[(N-1)((E\{X\})^2 - E\{X\}) + (E\{X\})^2] - \log((E\{X\})^2)}{\log((E\{X\})^2) - \log E\{X\}}$$

$$= \frac{\log\left(\frac{(N-1)((E\{X\})^2 - E\{X\}) + (E\{X\})^2}{(E\{X\})^2}\right)}{\log\left(\frac{(E\{X\})^2}{E\{X\}}\right)}$$

$$= \frac{\log\left((N-1)\left(\frac{E\{X\}-1}{E\{X\}}\right) + 1\right)}{\log E\{X\}}. \tag{8.59}$$

If $E\{X\} \gg 1$, then $(E\{X\} - 1)/E\{X\} \approx 1$, whence our estimate of the diameter is

$$D \approx \frac{\log(N)}{\log E\{X\}}. \tag{8.60}$$

Thus, for a Poisson random network, the diameter D is proportional to $\log N$. Note that if this result were used for a more complicated network, it could yield a very poor estimate of its diameter.

8.3 Expected-Degree Random Graph

The configuration model has proved to be very popular for generating networks with specific degree sequences or distributions. However, critics have pointed to a number of disadvantages of the model, especially complications due to the fact that the edges are not independently chosen, and that the configuration model generates a multigraph, not a simple graph. As a result, *the Erdős–Rényi model is not a special case of the configuration model.*

Furthermore, the configuration model does not fit the power-law structure very well. On the one hand, the maximum degree for a power-law model can be quite large. However, Molloy and Reed (2000, Theorem 1) showed that, because of the possibility of multiple edges, one of the conditions for the existence of a giant component in a configuration model is that the degree of any node is bounded from above; specifically, they showed that the maximum node degree can be at most $N^{\frac{1}{4}-\epsilon}$, where $\epsilon > 0$.

These concerns led to the construction of the "expected-degree" random graph model.

8.3.1 Chung–Lu Model

The *expected-degree model*, $\mathcal{G}_N(\mathbf{w})$, is a relaxation of the configuration model (Chung and Lu, 2002a,b). It is defined as follows.

Definition 8.2 Let $G \in \mathcal{G}_N(\mathbf{w})$. Let $\mathbf{w} = (w_1, \ldots, w_N)^\tau$ be an N-vector of weights, where w_i is the weight attached to node v_i. Note that w_i need not be an integer. An edge between the pair of nodes v_i and v_j occurs with probability

$$p_{ij} = \rho w_i w_j, \quad \rho = \frac{1}{\text{Vol}(G)}, \tag{8.61}$$

where

$$\text{Vol}(G) = \sum_{l=1}^{N} w_l \tag{8.62}$$

is the *volume* of G. In this model, the edges are selected independently of each other and the total number of edges is taken to be a Poisson random variable with mean p_{ij} for an edge between nodes v_i and $v_j, i = 1, 2, \ldots, N$. If we assume that $\max_i w_i^2 < \sum_l w_l$, then $p_{ij} \leq 1$, for all i, j. The *expected* degree of node v_i is, therefore,

$$\text{E\{degree of node } v_i\} = \sum_{j=1}^{N} p_{ij} = \sum_{j=1}^{N} \rho w_i w_j = w_i, \quad i = 1, 2, \ldots, N, \tag{8.63}$$

which is the reason the model is referred to as the *expected-degree model*, and w_i is referred to as the *expected degree* of node v_i.

If the weight of each node is taken to be the same, that is, $w_i = w, i = 1, 2, \ldots, N$, then $\rho = 1/(Nw)$, and so, $p_{ij} = \rho w^2 = w/N$. Thus, the Erdős–Rényi model, where all edges occur with the same probability, is a special case of the expected-degree model.

The probability of an edge between nodes v_i and v_j can be written as

$$p_{ij} = \frac{w_i w_j}{2m}, \tag{8.64}$$

where $m = \frac{1}{2} \sum_l w_l$ is now the expected number of edges.

Self-loops (i.e., an edge from v_i to v_i) are permitted in this model. In fact, the probability of a self-loop occurring at node v_i is given by

$$p_{ii} = \frac{w_i^2}{4m}, \tag{8.65}$$

where the denominator contains a multiplier of 2 to adjust for the diagonal element of the adjacency matrix being defined as twice the number of self-loops at node v_i. Self-loops are actually necessary for this process to work: if self-loops were not permitted, then the expected degree of node v_i would be

$$\text{E\{degree of node } v_i\} = \sum_{j \neq i} p_{ij} = \sum_{j \neq i} \rho w_i w_j = w_i(1 - \rho w_i), \tag{8.66}$$

which is not equal to w_i as required. This problem tends to be of little practical importance, however, because, as $N \to \infty$, the difference between the "self-loop" model and the "no-loop" model becomes negligible.

This model, as defined by the expected-degree sequence **w** and the expected number of edges m, can be used to generate networks with different degree sequences, which are selected from a distribution having a specified expected-degree sequence. Because the degree sequence $k_1, k_2, \ldots, k_N$ is not the same as the expected-degree distribution **w**, we cannot select the degree distribution (i.e., the proportion, p_k, of nodes having degree k) of the network. This is an undesirable feature of the model because the degree distribution tends to play a major role in modeling real-life networks.

8.3.2 Giant Component

Chung and Lu (2002a) studied the structure of a network having a given expected-degree sequence. In particular, they derived conditions for the existence and size of a giant component and of the smaller components in a network with a given expected-degree sequence. Their main result is the following (Chung, 2008). Let

$$\bar{w} = \frac{1}{N} \sum_l w_l > 1 \tag{8.67}$$

be the *average expected degree* and let e be the base of the natural logarithm. Then,

Theorem 8.11 (Chung, 2008) *We have the following bounds on the volume of the largest component and the second-largest component:*

1. *Almost surely, G is not completely connected but has a unique giant component C with volume*

$$\mathrm{Vol}(C) \geq \begin{cases} \left(1 - \frac{2}{\sqrt{\bar{w}e}} + o(1)\right) \mathrm{Vol}(G) & \text{if } \bar{w} \geq \frac{4}{e} = 1.4715\ldots \\ \left(1 - \frac{1+\log \bar{w}}{\bar{w}} + o(1)\right) \mathrm{Vol}(G) & \text{if } \bar{w} < 2. \end{cases} \tag{8.68}$$

2. *Almost surely, the second-largest component C_2 has volume*

$$\mathrm{Vol}(C_2) \leq (1 + o(1))\mu(\bar{w}) \log(N), \tag{8.69}$$

where

$$\mu(\bar{w}) = \begin{cases} \frac{1}{1+\log(\bar{w})-\log(4)} & \text{if } \bar{w} > \frac{4}{e} = 1.4715\ldots \\ \frac{1}{\bar{w}-1-\log(\bar{w})} & \text{if } 1 < \bar{w} < 2. \end{cases} \tag{8.70}$$

8.4 Preferential Attachment

What are the consequences to the degree distribution of an increase in the number of nodes in a network? It has been shown empirically that adding nodes to a network does not lead to random edges and a degree distribution that is Gaussian. Instead, the added nodes exhibit a tendency known as "preferential attachment"; that is, new connections will overwhelmingly favor a few existing nodes that have large degrees. This has also been called the "rich-get-richer effect" (Barabási and Albert, 1999), in which a few nodes end up with too many connections and most nodes have too few. Preferential attachment has also been used as a method of generating a scale-free network (Broido and Clauset, 2018).

8.4.1 A Brief History

The concept of preferential attachment goes back to the urn models of György Pólya, who in 1923 tried to explain the nature of certain distributions, with the result that preferential attachment is also known as a *Pólya process*. In 1925, George Udney Yule used the idea (but not the name) "preferential attachment" to explain the power-law distribution of the number of species per genus of flowering plants, where new species in a genus are generated by the *Yule process*. Herbert Simon,[3] in a study on the distribution of wealth, gave a modern treatment (but also not the name) of preferential attachment (Simon, 1955), building on a 1949 study by George Kingsley Zipf, who rank-ordered the frequency with which words occur in English-language texts. Simon showed that when their frequencies were plotted against their rank, the resulting probability of a word occurring exactly r times is proportional to $1/r$, which is a power law. His results became known as *Zipf's law*, which he showed holds for many other situations, such as city sizes and assets of business firms. Inspired by Simon's work, Derek de Solla Price (1976) adopted the term *cumulative advantage*, in the sense of preferential attachment, to explain the citation network of scientific publications.

The term *preferential attachment* was introduced by Barabási and Albert (1999), motivated at first to explain the power law in the World Wide Web and then to explain the ubiquitous nature of power laws in real-world networks (Barabási, 2002, Chapter 7). The term has since become popular with those working in network science, and the preferential attachment property has been observed on many different types of networks.

8.4.2 Examples of Preferential Attachment

Citation networks. An article tends to be preferentially cited by other articles if it has already been well cited. A much-cited article means that it has been noticed as important and will, therefore, have even more citations in the future.

Movie actor networks. Actors and actresses who appear in many movies tend to be chosen for future movies because their demonstrated acting skills are well known, which helps casting directors consider them for new roles.

Online edit networks. These include *Wikipedia* and *Wiktionary*, where users edit and re-edit pages in the network. Pages that have been edited many times are likely to be re-edited many more times in the future.

Metabolic networks. The metabolic network of *Escherichia coli* was shown to contain certain enzymes that have a higher average connectivity than other enzymes, and new edges are added to the highly connected enzymes at a faster rate than those enzymes with lower connectivities (Light, Kraulis, and Elofsson, 2005).

8.4.3 Preferential Attachment Model

Consider the following *preferential attachment model*. A network can evolve in two ways, by adding a new node and its associated connections to existing nodes or by adding connections between pairs of existing nodes (Barabási and Albert, 1999). We follow the development by Chung and Lu (2006, Chapter 3), which is slightly more general than other published descriptions.

The preferential attachment model is defined through a parameter $p \in [0, 1]$ and an initial graph $\mathcal{G}_0$ at time 0. This initial graph $\mathcal{G}_0$ is usually taken to have a single node with one self-loop and degree 1. In this model, multiple edges and self-loops are

---

[3]Simon was awarded the 1978 Nobel Prize in Economic Science.

permitted, but both should be extremely rare as the size of the network increases. An edge is added to this graph in either of two ways:

- A *node-step* adds a new node v to the graph together with a new edge from v to an existing node v' (a *growth* factor), where v' is chosen randomly and independently in proportion to its degree in the current graph (a *preferential attachment* factor).
- An *edge-step* adds a new edge by independently and *preferentially* selecting a pair of existing nodes and linking them; that is, both nodes are independently chosen with probability proportional to their degrees. We allow the two nodes for an edge-step to be identical (i.e., we allow a self-loop to be added).

The random graph $\mathcal{G}(p)$ with preferential attachment is now constructed using **Algorithm 8.1**.

Algorithm 8.1 *Preferential attachment algorithm*

---

1. Start with an initial graph $\mathcal{G}_0$ having a single node with one self-loop and degree 1.
2. At time $t > 0$, the graph $\mathcal{G}_t$ is formed by modifying $\mathcal{G}_{t-1}$ as follows:

   - With probability p, take a node-step.
   - Otherwise, take an edge-step.

Krapivsky and Redner (2001) discuss a model similar to $\mathcal{G}(p)$ for a directed network. We note, however, that most published accounts of the Barabási–Albert preferential attachment model use only the node-step without any mention of the edge-step.

Let N_t and M_t denote the number of nodes and edges, respectively, of $\mathcal{G}(p)$ at time t. Then, $M_t = t + 1$ and $N_t = 1 + \sum_{i=1}^{t} s_i$, where $s_i = 1$ wpr p and $s_i = 0$ wpr $1 - p$. Because $E\{s_i\} = p$, it follows that $E\{N_t\} = 1 + \sum_{i=1}^{t} E\{s_i\} = 1 + tp$. Let N_{kt} denote the number of nodes with degree k at time t. Note that $N_{10} = 1$ and $N_{0t} = 0$.

The following argument uses the so-called *master equation* method for developing asymptotics for $E\{N_{kt}\}$ by considering what happens at a single time step t (Dorogovstev, Mendes, and Samukhin, 2000).

We note that a node with degree k at time t could have occurred from either of two possibilities:

1. The node had degree k at time $t - 1$ and no edge was added to it at time t.
2. The node had degree $k - 1$ at time $t - 1$ and then a new edge was added to it at time t.

At time t, the total number of edges is t and the total degree is $2t$. So, the probability that an edge links to an existing node is proportional to the degree of that node; that is, $k/2t$.

Let $\mathcal{B}_t$ denote the σ-field[4] associated with the probability space at time t. Let $t > 0$ and $k > 1$. Then $E\{N_{kt}|\mathcal{B}_{t-1}\}$ is the sum of two terms. The first term is the product of $N_{k,t-1}$, the number of nodes with degree k at time $t - 1$, and the probability, $1 - \left(\frac{kp}{2t} + \frac{2k(1-p)}{2t}\right)$, of 0 nodes added at time t. The second term is the product of

---

[4]See the footnote on p. 146 for a definition of a σ-field.

$N_{k-1,t-1}$, the number of nodes with degree $k-1$ at time $t-1$, and the probability, $\frac{(k-1)p}{2t} + \frac{2(k-1)(p-1)}{2t}$, of adding one node (which could appear by either a node-step or an edge-step) at time t. Putting this together, we have that

$$E\{N_{kt}|\mathcal{B}_{t-1}\} = N_{k,t-1}\left(1 - \left(\frac{kp}{2t} + \frac{2k(1-p)}{2t}\right)\right)$$
$$+ N_{k-1,t-1}\left(\frac{(k-1)p}{2t} + \frac{2(k-1)(p-1)}{2t}\right). \tag{8.71}$$

Let $k = 1$. Then,

$$E\{N_{1t}|\mathcal{B}_{t-1}\} = N_{1,t-1}\left(1 - \frac{2-p}{2t}\right) + p. \tag{8.72}$$

Taking expectations yields

$$E\{N_{1t}\} = E\{N_{1,t-1}\}\left(1 - \frac{2-p}{2t}\right) + p. \tag{8.73}$$

Next, we claim (and show by induction) that $E\{N_{kt}\}$ follows a power law as $t \to \infty$. First, we state the following general result (Chung and Lu, 2006, Section 3.3; Durrett, 2007, Section 4.1).

Theorem 8.12 *Let $\{a_t\}$ be a sequence that satisfies the recursive relation, $a_{t+1} = a_t\left(1 - \frac{b_t}{t}\right) + c_t$, where $\lim_{t\to\infty} b_t = b$ and $\lim_{t\to\infty} c_t = c$. Then, $\lim_{t\to\infty} \frac{a_t}{t}$ exists and has the value $\frac{c}{1+b}$.*

We now use this result to show that, for each k, $\frac{E\{N_{kt}\}}{t} \to n_k$ as $t \to \infty$. We first look at the case $k = 1$. Let $b_t = b = \frac{2-p}{2}$ and $c_t = c = p$. Then, from the above result, $\lim_{t\to\infty} \frac{E\{N_{1t}\}}{t}$ exists and has limiting value

$$\frac{E\{N_{1t}\}}{t} \to n_1 = \frac{2p}{4-p}, \quad as\ t \to \infty. \tag{8.74}$$

Next, we assume that $\lim_{t\to\infty} \frac{E\{N_{k-1,t}\}}{t}$ exists.

Let $b_t = b = \frac{k(2-p)}{2}$ and $c_t = \frac{1}{2t}E\{N_{k-1,t-1}\}(2-p)(k-1)$, so that $c = \frac{1}{2}n_{k-1}(2-p)(k-1)$. Applying the above result again, $\lim_{t\to\infty} \frac{E\{n_{kt}\}}{t}$ exists and, as $t \to \infty$, has value

$$\frac{E\{N_{kt}\}}{t} \to n_k$$
$$= n_{k-1} \cdot \frac{(2-p)(k-1)}{2+k(2-p)}$$
$$= n_{k-1} \cdot \frac{k-1}{k + \frac{2}{2-p}}. \tag{8.75}$$

Iterating this equation repeatedly yields

$$n_k = \frac{2p}{4-p} \cdot \prod_{j=2}^{k} \frac{j-1}{j + \frac{2}{2-p}}$$

$$= \frac{2p}{4-p} \cdot \frac{\Gamma(k)\Gamma(2 + \frac{2}{2-p})}{\Gamma(k+1+\frac{2}{2-p})}, \tag{8.76}$$

where $\Gamma(a) = \int_0^\infty x^{a-1} e^{-x} dx = (a-1)!$ is the gamma function.

If the preferential attachment model yields graphs obeying the power law, then $n_k \propto k^{-\beta}$, for some power β. It follows that

$$\frac{n_k}{n_{k-1}} = \frac{k^{-\beta}}{(k-1)^{-\beta}} = \left(1 - \frac{1}{k}\right)^\beta = 1 - \frac{\beta}{k} + \mathcal{O}\left(\frac{1}{k^2}\right). \tag{8.77}$$

From (8.75), we also have

$$\frac{n_k}{n_{k-1}} = \frac{k-1}{k + \frac{2}{2-p}} = 1 - \frac{1 - \frac{2}{2-p}}{k + \frac{2}{2-p}} = 1 - \frac{1 + \frac{2}{2-p}}{k} + \mathcal{O}\left(\frac{1}{k^2}\right). \tag{8.78}$$

So, equating (8.77) and (8.78), we have that the preferential attachment model almost surely generates graphs that have the power law with exponent

$$\beta = 1 + \frac{2}{2-p} = 2 + \frac{p}{2-p} \in [2,3]. \tag{8.79}$$

In other words,

$$n_k \propto k^{-(2 + \frac{p}{2-p})}, \quad k \geq 2. \tag{8.80}$$

Chung and Lu then show that the degree distribution (i.e., the number of nodes with degree k) at time t is, almost surely,

$$N_{kt} = n_k t + \mathcal{O}\left(2\sqrt{k^3 t \log(t)}\right), \quad k = 1, 2, \ldots. \tag{8.81}$$

Thus, the number of nodes with degree k, almost surely, grows linearly with time t.

We can express n_k in terms of the beta function. From (8.79), we have that $p = 2(2-\beta)/(1-\beta)$, so that $2p/(4-p) = p(\beta-1)/\beta$. It follows that

$$n_k = \frac{p(\beta-1)}{\beta} \cdot \frac{\Gamma(k)\Gamma(1+\beta)}{\Gamma(k+\beta)}$$

$$= p(\beta-1) \cdot \frac{\Gamma(k)\Gamma(\beta)}{\Gamma(k+\beta)}$$

$$= p(\beta-1) \cdot \mathcal{B}(k, \beta), \tag{8.82}$$

where $\mathcal{B}(a,b) = \int_0^1 x^{a-1}(1-x)^{b-1} dx = \frac{\Gamma(a)\Gamma(b)}{\Gamma(a+b)}$ is Euler's beta function.

As we previously noted, most accounts of the preferential attachment model consist only of the node-step without an edge-step. If we set $p = 1$ in n_k, the degree distribution (8.81) exhibits linear growth over time with

$$n_k = \frac{4}{k(k+1)(k+2)} \propto k^{-3}, \ as \ k \to \infty, \tag{8.83}$$

where we used $\Gamma(x) = (x-1)!$.

A generalization of the preferential attachment model adds more than one edge to the graph at each step. In introducing their model for adding c edges from a single newly added node at each step, Barabási and Albert (1999) set up the initial graph $\mathcal{G}_0$ to have c_0 connected nodes.[5] In this case, c edges ($c < c_0$) can be added at each step in either of two ways:

- A *node-c-step* is defined by adding a new node to the graph together with c new edges that link the new node to c existing nodes, each of which is chosen preferentially, with probability proportional to the degree of that node in the current graph.
- An *edge-c-step* is defined by adding c new edges to the graph, each edge linking a pair of existing nodes that are randomly and preferentially chosen by probability proportional to the degree of each node.

The new random graph, $\mathcal{G}(c, p)$, is defined in a similar way as was $\mathcal{G}(p)$ (see **Algorithm 8.1(c)**).

Algorithm 8.1(c) *Generalized preferential attachment algorithm*

---

1. Start with $\mathcal{G}_0$.
2. At time $t > 0$, the graph $\mathcal{G}_t$ is formed by modifying $\mathcal{G}_{t-1}$ as follows:

   - With probability p, take a node-c-step.
   - Otherwise, take an edge-c-step.

Carrying through a similar argument for $\mathcal{G}(c, p)$ as was done for $\mathcal{G}(p)$, it turns out that all appearances of c cancel out, and the results reduce to those obtained for $\mathcal{G}(p)$. This shows that the graphs generated by $\mathcal{G}(c, p)$ are invariant wrt c and have the same power-law distribution as graphs that are generated by $\mathcal{G}(p)$. Thus, the exponent β is independent of the scale c.

---

[5]It is nowhere stated in the Barabási–Albert article whether the c_0 nodes of $\mathcal{G}_0$ are isolated (so that each node has degree 0) or connected. This "rather imprecise" description of $\mathcal{G}_0$ led to criticism of the Barabási–Albert article by Bollobás et al. (2001) and Bollobás and Riordan (2002), who complained that the entire graph-creation process outlined by Barabási and Albert "does not make sense." One point they make is that "as the degrees are initially zero, it is not clear how the process is supposed to get started." A possible solution is for each node in $\mathcal{G}_0$ to be isolated with its own self-loop and, hence, degree 1 (generalizing the $c = 1$ case above). In Barabási (2002, Figure 7.1), however, it is stated explicitly that the nodes of $\mathcal{G}_0$ are connected. See also Barabási and Bonabeau (2003, p. 55).

8.4.4 Nonlinear Preferential Attachment

An empirical study of preferential attachment for 47 online networks (Kunegis, Blattner, and Moser, 2013), which included social networks, rating networks, communication networks, folksonomies, Wiki edit networks, and explicit and implicit interaction networks, found that they did not follow the standard preferential attachment model; instead, preferential attachment for these online networks was judged to be *nonlinear*, with the form of the nonlinear model specific to the type of network.

In the standard (linear) preferential attachment model, the attachment probability, $p(k_i)$, is proportional to k_i, the degree of node v_i. In a nonlinear model, the probability is proportional to k_i^γ; that is,

$$p_\gamma(k_i) = \frac{k_i^\gamma}{\sum_j k_j^\gamma}, \quad \gamma \geq 0, \tag{8.84}$$

where γ is a constant to be determined and $\sum_i p_\gamma(k_i) = 1$ (Krapivsky, Redner, and Leyvraz, 2000; Krapivsky and Redner, 2001). The degree distribution turns out to be highly dependent upon the value of γ. When $\gamma > 0$, this reflects the tendency for new nodes to gravitate towards those nodes with high degrees. It was found that a variety of behaviors occur when $\gamma < 1$, $\gamma = 1$, and $\gamma > 1$. When $\gamma \in (0, 1)$, the model is called *sublinear preferential attachment*, while, when $\gamma > 1$, it is referred to as *superlinear preferential attachment*. The case of $\gamma = 1$ is the linear preferential attachment model. Some of our treatment of this topic is adapted from that of Durrett (2007, Chapter 4).

Let $n_k(t) = E\{N_{kt}\}$ denote the expected number of nodes of degree k at time t. Assume there are no isolated nodes, so that $n_0(t) \equiv 0$, for any t. Also, $n_1(0) = 1$. Let

$$M_\gamma(t) = \sum_{j=1}^{k} j^\gamma n_j(t), \quad \gamma \geq 0 \tag{8.85}$$

denote the low-order moments of the degree distribution. Then, $M_0(0) = 1$, $M_1(0) = 0$,

$$M_0(t) = M_0(0) + t = t + 1, \tag{8.86}$$
$$M_1(t) = M_1(0) + 2t = 2t, \tag{8.87}$$

so that the first two moments increase linearly with t.

We divide the model into three cases according to the value of γ.

A. Sublinear Case $(0 < \gamma < 1)$

Theorem 8.13 *For large k, the degree distribution of the sublinear preferential attachment model follows a mixture of a power law and a stretched exponential distribution in k; that is,*

$$p_\gamma(k) \propto \frac{\mu}{k^\gamma} \exp\left\{-\left(\frac{\mu}{1-\gamma}\right) k^{1-\gamma}\right\}, \quad 0 < \gamma < 1. \tag{8.88}$$

A stretched exponential distribution occurs when the argument of the exponential function has a fractional power, which, if it lies in $(0,1)$, stretches the standard exponential distribution.

Proof Let $\gamma < 1$. Then $1 \leq j^\gamma \leq j$ for each $j = 1, 2, \ldots, k$, and, hence,

$$M_0(t) \leq M_\gamma(t) \leq M_1(t). \tag{8.89}$$

Dividing (8.89) through by t, we have that

$$1 + \frac{1}{t} \leq \frac{M_\gamma(t)}{t} \leq 2, \quad \text{for all } t. \tag{8.90}$$

So, $M_\gamma(t)/t$ is a bounded sequence that we assume has a limit:

$$\frac{M_\gamma(t)}{t} \to \mu \in [1, 2], \quad \text{as } t \to \infty. \tag{8.91}$$

We may write, for large t, that $M_\gamma(t) \sim \mu t$.

Now, a node will have k edges at time $t + 1$ if it has $k - 1$ edges at time t and one more edge is added to that node with probability $(k - 1)^\gamma / M_\gamma(t)$, so that n_k may increase by 1. However, if a node at time t already has k edges and one more edge is added with probability $k^\gamma / M_\gamma(t)$, then that node at time $t + 1$ no longer has k edges, and so n_k may decrease by 1. We can express this as the following *rate equation*:

$$n_k(t + 1) - n_k(t) = \frac{1}{M_\gamma(t)} \{(k - 1)^\gamma n_{k-1}(t) - k^\gamma n_k(t)\} + \delta_{k1}, \tag{8.92}$$

where $\delta_{k1} = 1$ if $k = 1$, and 0 otherwise, represents the introduction to the graph of a new node with degree 1 and with no incoming edges, thereby increasing n_1 by 1.

Setting $k = 1$ and, hence, $\delta_{k1} = 1$, and rearranging (8.92) yields

$$n_1(t + 1) = 1 + \left(1 - \frac{1}{M_\gamma(t)}\right) n_1(t). \tag{8.93}$$

Suppose, for any k, that

$$\frac{n_k(t)}{t} \to n_k, \quad \text{as } t \to \infty. \tag{8.94}$$

Then, from Durrett (2007, Lemma 4.1.1), a solution to (8.93) is given by

$$n_1 = \frac{\mu}{1 + \mu}, \tag{8.95}$$

where μ is defined by (8.91).

If $k > 1$, then $\delta_{k1} = 0$, and (8.91) becomes

$$n_k(t + 1) = \left(1 - \frac{k^\gamma}{M_\gamma(t)}\right) n_k(t) + \frac{(k - 1)^\gamma}{M_\gamma(t)} n_{k-1}(t). \tag{8.96}$$

From Durrett (2007, Lemma 4.1.2), a solution to (8.96) is given by

$$n_k = \frac{(k - 1)^\gamma}{k^\gamma + \mu} n_{k-1}. \tag{8.97}$$

Now, we iterate on k:

$$n_k = \prod_{j=2}^{k} \frac{(j-1)^\gamma}{j^\gamma + \mu} n_1$$

$$= \frac{\mu}{k^\gamma} \prod_{j=1}^{k} \frac{j^\gamma}{j^\gamma + \mu}$$

$$= \frac{\mu}{k^\gamma} \prod_{j=1}^{k} \left(1 + \frac{\mu}{j^\gamma}\right)^{-1}. \tag{8.98}$$

Recall that $\gamma < 1$. Taking logarithms of (8.98) and using the approximation $\log(1+x) \sim x$, the product becomes

$$\log \prod_{j=1}^{k} \left(1 + \frac{\mu}{j^\gamma}\right)^{-1} = -\sum_{j=1}^{k} \log\left(1 + \frac{\mu}{j^\gamma}\right)$$

$$\sim -\sum_{j=1}^{k} \frac{\mu}{j^\gamma}$$

$$\sim -\left(\frac{\mu}{1-\gamma}\right) k^{1-\gamma}. \tag{8.99}$$

Note that, from (8.86) and (8.95),

$$\mu = \lim_{t \to \infty} \frac{M_\gamma(t)}{t} = \lim_{t \to \infty} \sum_{j=1}^{k} j^\gamma \frac{N_j(t)}{t} = \sum_{j=1}^{k} j^\gamma n_j. \tag{8.100}$$

From (8.99), we have that

$$\frac{k^\gamma n_k}{\mu} = \prod_{j=1}^{k} \left(1 + \frac{\mu}{j^\gamma}\right)^{-1}. \tag{8.101}$$

Substituting (8.100) for μ into the lhs of (8.101) and summing over all k yields the implicit relation

$$1 = \sum_{k=1}^{\infty} \prod_{j=1}^{k} \left(1 + \frac{\mu}{j^\gamma}\right)^{-1}. \tag{8.102}$$

The last piece of the puzzle is to determine μ. We can find μ from condition (8.101), which is then used to compute the value of n_1. Equation (8.101) can be evaluated numerically for μ. The function $\mu = \mu(\gamma)$ is a monotone-increasing function that depends only weakly on γ. Specifically, $\mu(\gamma)$ increases smoothly between 1 and 2 as γ increases from 0 to 1. When $\gamma = 0$, then $\mu = 1$, and $p_0(k) \propto e^{-k}$, corresponding to an exponential model. $\qquad \square$

B. Linear Case ($\gamma = 1$)

To see that a power-law degree distribution occurs for $\gamma = 1$ (which is the linear preferential attachment model), we note that when $\gamma = 1$, then $\mu = 2$, and $k^{1-\gamma}/(1-\gamma) \to \log k$ as $\gamma \to 1$. Thus, we have that

Theorem 8.14 *The degree distribution of the linear preferential attachment model is given by*

$$p_1(k) \propto k^{-1} \exp\{-2 \log k\} = k^{-3}. \tag{8.103}$$

C. Superlinear Case ($\gamma > 1$)

The main characteristic of the model in the superlinear case is that it displays a "winner-takes-all" attitude. In other words,

Theorem 8.15 *When $\gamma > 1$, a single node emerges that has very high degree with edges that link to almost every other node in the network.*

Proof We follow Krapivsky, Redner, and Leyvraz (2000) and Krapivsky and Redner (2001). Suppose $1 < \gamma \le 2$. From (8.96), setting $t = k - 1$ and noting that $n_k(t) = 0$ whenever $k \ge t + 1$, we iterate to get

$$n_k(k) = \frac{(k-1)^\gamma n_{k-1}(k-1)}{M_\gamma(k-1)} = n_2(2) \prod_{j=2}^{k-1} \frac{j^\gamma}{M_\gamma(j)}. \tag{8.104}$$

Changing indexes from j to k, and summation limits from k to t, we have that

$$M_\gamma(t) = \sum_{k=1}^{t} k^\gamma n_k(t) \le t^{\gamma-1} \sum_{k=1}^{t} k n_k(t) = t^{\gamma-1} M_1(t) = t^{\gamma-1} 2t = 2t^\gamma. \tag{8.105}$$

So, $M_\gamma(t) \propto t^\gamma$, where $\gamma > 1$. We assume that

$$\frac{M_\gamma(t)}{t^\gamma} \to 1, \quad \text{as } t \to \infty. \tag{8.106}$$

Now, setting $k = 1$ in (8.92) (so that $\delta_{k1} = 1$) and letting $s = t + 1$,

$$n_1(s) - n_1(s-1) = 1 - \frac{n_1(s-1)}{M_\gamma(s-1)}. \tag{8.107}$$

The term n_1/M_γ on the rhs of (8.107) is

$$\frac{n_1}{M_\gamma} = \frac{n_1}{\sum_{j=1}^{k} j^\gamma n_j}, \tag{8.108}$$

which, for large t and hence large k, will be very small and can be ignored. Hence,

$$n_1(s) - n_1(s-1) = 1. \tag{8.109}$$

Evaluating this expression for $s = 2, 3, \ldots, t$, then summing and canceling like terms on the lhs, we have that $n_1(t) - n_1(1) = t - 1$. Because $n_1(1) = 1$, it follows that

$$n_1(t) = t. \tag{8.110}$$

Now, set $k = 2$ in (8.92), so that $\delta_{k1} = 0$. Because $n_1(t) \le t$ for $t \ge 2$, we have that

$$n_2(t+1) - n_2(t) = \frac{1}{M_\gamma(t)} \{n_1(t) - 2^\gamma n_2(t)\}$$

$$\le \frac{t}{M_\gamma(t)}$$

$$\sim \frac{1}{t^{\gamma-1}}. \tag{8.111}$$

Evaluating this expression for $s = 1, 2, 3, \ldots, t - 1$, then summing both sides, and canceling like terms on the lhs, yields

$$n_2(t) - n_2(1) \leq \sum_{s=1}^{t} \frac{1}{s^{\gamma-1}}, \tag{8.112}$$

where we added the positive term $\frac{1}{t^{\gamma-1}}$ to the rhs. Note that $n_2(1) = 0$. Dividing both sides by $t^{2-\gamma}$ and taking lim sups,

$$\limsup_{t \to \infty} \frac{n_2(t)}{t^{2-\gamma}} \leq \limsup_{t \to \infty} \frac{1}{t^{2-\gamma}} \sum_{s=1}^{t} \frac{1}{s^{\gamma-1}}$$

$$= \limsup_{t \to \infty} \frac{1}{t^{2-\gamma}} \int_{1}^{t} \frac{1}{s^{\gamma-1}} ds$$

$$= \frac{1}{2 - \gamma}. \tag{8.113}$$

Because dropping $2^{\gamma} n_2(t)$ in the calculation of (8.113) does not change the limit, it follows that

$$\frac{n_2(t)}{t^{2-\gamma}} \to \frac{1}{2 - \gamma}, \quad \text{as } t \to \infty. \tag{8.114}$$

Thus, $n_2(t) \sim \frac{t^{2-\gamma}}{2-\gamma} \propto t^{2-\gamma}$. Repeating the argument for $k = 3, 4, \ldots$, we have that $n_3 \propto t^{3-2\gamma}$, $n_4 \propto t^{4-3\gamma}$, and so on. The general result is

$$n_k(t) \sim c_k t^{k-(k-1)\gamma}, \quad k \geq 1, \tag{8.115}$$

where

$$c_k = \prod_{j=1}^{k-1} \frac{j^{\gamma}}{1 + j(1 - \gamma)}. \tag{8.116}$$

This result for $n_k(t)$ holds as long as the exponent, $k - (k - 1)\gamma$, is positive; that is, as long as $k - k\gamma + \gamma > 0$, or $\gamma > k(\gamma - 1)$, or $k < \frac{\gamma}{\gamma-1}$.

Because $n_1(t) = t$, the total number of edges is $2t = M_1(t) = t + \sum_{j=2}^{k} j n_j(t)$, whence $\sum_{j=2}^{k} j n_j(t) = t$. In other words, half of the total number of edges are accounted for by $n_1(t)$ (i.e., nodes with degree 1), while the remaining t edges are distributed amongst the other nodes. In fact, if $k > \frac{\gamma}{\gamma-1}$, the exponent of t in (8.115) is negative, and so, for each $k \geq 1$, $n_k(t) \to 0$ as $t \to \infty$. Hence, the remaining t edges are all connected to a single node of degree of order $\sim t$.

When $\gamma > 2$, almost all (i.e., all but a finite number of) nodes are connected to a single node that plays the role of a "gel" site. That is,

- $\gamma > 2$: $n_1 \propto t$, and for $k > 1$, N_k is finite.

For values of γ between 1 and 2, the following transitions occur.

- $\frac{3}{2} < \gamma < 2$: $n_1 \propto t$, $n_2 \propto t^{2-\gamma}$, and for $k > 2$, n_k is finite.
- $\frac{4}{3} < \gamma < \frac{3}{2}$: $n_1 \propto t$, $n_2 \propto t^{2-\gamma}$, $n_3 \propto t^{3-2\gamma}$, and for $k > 3$, n_k is finite.
- $\frac{5}{4} < \gamma < \frac{4}{3}$: $n_1 \propto t$, $n_2 \propto t^{2-\gamma}$, $n_3 \propto t^{3-2\gamma}$, $n_4 \propto t^{4-3\gamma}$, and for $k > 4$, n_k is finite.

And so on. These intervals for γ become smaller and smaller, as they inch towards the value 1. In general, for $m = 1, 2, 3, \ldots$,

- $\frac{m+1}{m} < \gamma < \frac{m}{m-1}$: $n_1 \propto t, \ldots, n_m \propto t^{m-(m-1)\gamma}$, and for $k > m$, n_k is finite. $\qquad \square$

8.5 Random Copying

A different explanation for the appearance of power-law degree distributions for networks is given by *stochastic copying* (or *duplication*) *models*, which were proposed for the World Wide Web (Kleinberg et al., 1999; Kumar et al., 2000), long-distance telephone calls (Aiello, Chung, and Lu, 2000), biological networks (Chung et al., 2003), and citation networks (Krapivsky and Redner, 2005).

In order to model random networks so that they exhibit the characteristics found in real networks, several types of copying mechanisms have been proposed, each motivated by a particular application.

8.5.1 World Wide Web

The first type of copying model deals with the evolution of the World Wide Web and combines two different edge-creation strategies (Kleinberg et al., 1999; Kumar et al., 2000). At each time step, a new node (i.e., webpage) is added to the Web graph and edges are assigned from that new node to a subset of all existing nodes in the current graph. To generate such edges, a stochastic choice is made: one possibility is to choose the edges independently and at random; alternatively, the new node duplicates the structure of the graph by "copying" edges from a randomly chosen node. Such a copying process is thought to be a fundamental feature of the growth of the Web.

A more formal description of this model is given by **Algorithm 8.2**.

For this copying model, let N_{kt} denote the number of nodes with in-degree k at time t. It was shown (Kumar et al., 2000) that $E\{N_{kt}\} = tk^{-\beta}$, where $\beta = (2-\alpha)/(1-\alpha)$. Thus, β varies between 2 (when α is close to 0) and ∞ (when α is close to 1). It follows that N_{kt} has a power-law in-degree distribution given by $N_{kt} \sim Ck^{-\beta}$. Extensions to this model have also been proposed.

8.5.2 Biological Networks

A second type of copying model was given for biological systems, such as genetic regulatory networks and protein–protein interaction networks, where there is strong evidence of widespread duplication of the information contained in the genome (Ohno, 1970; Wagner, 1994; Sidow, 1996; Wolfe and Shields, 1997; Bhan, Galas, and Dewey, 2002; Friedman and Hughes, 2003). Modeling a biological system by a network graph is carried out, for example, by setting the nodes to be proteins and

---

Algorithm 8.2 *Random copying model for WWW*

---

1. Let $\mathcal{G}_{t-1}$ be the current graph and let $m \geq 1$ be some constant.
2. At time $t > 0$, a single node v' is added to $\mathcal{G}_{t-1}$ to yield $\mathcal{G}_t$.
3. The node v' is linked to m existing nodes in $\mathcal{G}_{t-1}$. The destination nodes are chosen as follows:

   - Randomly choose an existing "prototype" node from $\mathcal{G}_{t-1}$.
   - With probability $\alpha \in (0, 1)$, the destination of the ith link from v' is chosen uniformly at random from the existing nodes of $\mathcal{G}_{t-1}$.
   - With probability $1 - \alpha$, the link from v' is copied from the ith link from v, $i = 1, 2, \ldots, m$, as follows:

     – If the degree of v is greater than m, we copy a random subset of m edges.
     – If the degree of v is smaller than m, we copy those edges from v and then pick another random node and copy those edges. Continue until m edges have been copied.

---

each (undirected) edge represents an interaction between a pair of proteins. There is evidence that the proteome network of the yeast *Saccharomyces cerevisiae* has a degree distribution in the form of a power law with exponential cutoff (see Jeong et al., 2001).

Three copying mechanisms are described by Chung et al., (2003).

1. Full duplication model. Given an initial graph $\mathcal{G}_0$, a new node v' is added and an existing node v is randomly selected from $\mathcal{G}_0$. Then, for every neighbor w of v, an edge from v' to w is added, so that w also becomes a neighbor of v'. In other words, v' becomes an exact duplicate of v because v' is linked to all the neighbors of v. Note that v does not become a neighbor of v'. This copying process is then repeated, so that at time step t, a new node and corresponding edges are added to the current graph $\mathcal{G}_{t-1}$, to yield $\mathcal{G}_t$, $t = 1, 2, \ldots$. This is the *full duplication* model.

2. Partial duplication model. A variation on this theme adds a stochastic element to the edge-creation process. Instead of taking every neighbor of v to be a neighbor of v', each neighbor of v is randomly chosen with probability $\alpha \in [0, 1]$ to be a neighbor of v', or not with probability $1 - \alpha$. This yields the *partial duplication* model. The full duplication model, which is obtained by setting $\alpha = 1$, does not result in a power-law distribution. An equivalent way of viewing the partial duplication process is by carrying out the full duplication process and then deleting each newly created edge with probability $1 - \alpha$ or retaining that edge with probability α.

Chung et al. showed that if the partial duplication model with probability α generates a power-law graph of the form $ck^{-\beta}$, where c is a constant, then the exponent β (which is a function of α) must satisfy the equation

$$\alpha(\beta - 1) = 1 - \alpha^{\beta-1}, \tag{8.117}$$

213

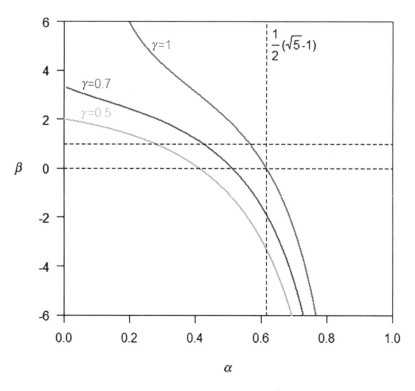

Figure 8.2 Solution of the equation $\beta(1-\gamma)+\alpha\gamma(\beta-1)=1-\gamma\alpha^{\beta-1}$ for the hybrid model, where $\alpha \in [0,1]$ and $\gamma = 0.5, 0.7$, and 1.0. Note that $\beta = 1$ is a solution for any value of α, and when $\alpha = \frac{1}{2}$ and $\gamma = 1$, then $\beta = 2$.

which can also be written as

$$\alpha\beta - \alpha + \alpha^{\beta-1} = 1. \tag{8.118}$$

There are two curves that characterize the solutions of this equation. Clearly, $\beta = 1$ is always a solution (for any α), and forms a line. The other curve is a monotonically decreasing function of α over most of the interval $(0, 1)$. The two curves intersect at the point $(\alpha, \beta) = (0.56714329\ldots, 1)$. When $\alpha > \alpha^* = 0.56714329\ldots$, $\beta = 1$ is the only solution. If $\alpha \in (\frac{1}{2}, 1)$, then $\beta < 2$. When $\alpha = \frac{1}{2}$, the solution is $\beta = 2$.

3. A hybrid model. A hybrid model that combines full duplication (selected with probability $1 - \gamma$) with partial duplication (selected with probability γ) of a node is also proposed, where $\gamma \in (0, 1)$. Chung et al. showed that, with probability tending to one as $N \to \infty$, this hybrid model generates a power-law graph with exponent β that satisfies the equation

$$\beta(1-\gamma) + \alpha\gamma(\beta-1) = 1 - \gamma\alpha^{\beta-1}. \tag{8.119}$$

This equation reduces to (8.118) when $\gamma = 1$. **Figure 8.2** displays a graph of this solution of (8.119), and also the solutions for $\gamma = 0.5, 0.7$, and 1.0.

Chung et al. claimed that the partial duplication process produces a power-law distribution. However, their proof does not show that a power law exists in this case

(Bebek et al., 2007). What Chung et al. failed to take into account was that the partial duplication process can create isolated nodes (i.e., nodes with degree 0). A node v' will be *isolated* if none of the edges of v is copied to v'. If a node is isolated, then it will always remain isolated throughout the entire duplication process. In fact, for $\alpha \leq \alpha^*$, the expected proportion of isolated nodes generated by this process increases through time, tending to a limit of 1, so that almost all nodes are isolated (Hermann and Pfaffelhuber, 2016). Furthermore, for all $k \geq 1$, the expected value of the degree distribution at time t converges to 0 as $t \to \infty$. Consequently, there is no limiting probability distribution for $\alpha > \alpha^*$. As a result, the partial duplication process cannot generate a power-law degree distribution. Although the non-isolated nodes comprise a single connected component, it is not known whether they exhibit a power law with exponent $\beta \leq 2$.

8.5.3 Citation Networks

A third type of copying model focused on citation networks, where references to cited papers are often copied without the author checking the original references (Krapivsky and Redner, 2005). As a result, the copying mechanism results in a higher frequency of references to more recent publications than to older publications.

This copying model starts by adding a node at each time step. The new node then randomly selects a "target" node amongst the existing nodes of the graph and links to it and to all neighboring nodes of the target node. This simple model, unlike the other copying models, does not include a probability mechanism that would determine to which nodes the links would go. Applying this copying model to the citation network of *Physical Review*, and treating the network as a directed graph, it was shown that the in-degree distribution is a power law with exponent $\beta = 2$, whilst a power law does not appear to be a good fit to the out-degree distribution, which may follow a Poisson distribution for large networks.

A more standard copying model is also mentioned in which a probability mechanism is introduced to determine whether the new node links only to the target node (with probability α) or only to each of the target node's neighbors (with probability $1 - \alpha$).

8.6 Further Reading

Books that have good discussions for the topics in this chapter include Durrett (2007), Kolaczyk (2009, Section 6.4), Newman (2010, Chapters 13 and 14), and Barrat, Barthélemy, and Vespignani (2013, Section 3.3).

Section 8.2. A book that describes the configuration model, including the excess-degree distribution, is Newman (2010, Chapter 13).

Section 8.2.6. Probability generating functions is a topic that has been part of the statistical literature for a long time. See, for example, Cramér (1946, p. 257), Kendall and Stuart (1963, Chapter 1), and Casella and Berger (1990, Chapter 2).

Section 8.4. See also the interesting chapter describing the development of preferential attachment, network growth, and the power laws by Barabási (2002, Chapter 7) (barabasi.com/f/633.pdf). See also Newman (2010, Chapter 14).

8.7 Exercises

Exercise 8.1 Consider the negative binomial distribution

$$p_k = \binom{k-1}{r-1} p^r (1-p)^{k-r}, \quad k = r, r+1, \ldots.$$

(a) Show this is a probability distribution.
(b) Using a probability generating function approach, find the mean and variance of this distribution.

Exercise 8.2 The geometric distribution is given by

$$p_k = p(1-p)^{k-1}, \quad k = 1, 2, \ldots.$$

(a) Show that the geometric distribution is a special case of the negative binomial distribution.
(b) Find the mean and variance of this distribution.

Exercise 8.3 Consider the power-law degree distribution having exponential cutoff

$$p_k = ck^{-\beta} e^{-\alpha k}, \quad k \geq 1,$$

where α, β, and c are constants. Find the mean and variance of this distribution.

Exercise 8.4 Define the *moment generating function* (mgf) of the random variable X as

$$M_X(s) = E\{e^{sX}\},$$

assuming that the expectation exists for s in some neighborhood of 0. Show that $E\{X^n\}$ is the nth derivative of the mgf evaluated at $s = 0$. Check this for $n = 1, 2$.

Exercise 8.5 For a gamma random variable X with density

$$p_{\alpha,\beta}(x) = \frac{1}{\Gamma(\alpha)\beta^\alpha} x^{\alpha-1} e^{-x/\beta},$$

where $0 < x < \infty$, $\alpha, \beta > 0$, and $\Gamma(\alpha)$ is the gamma function, find the mgf and its mean $E\{X\}$.

Exercise 8.6 For the binomial distribution, $X \sim \text{Bin}(n, p)$:

$$P\{X = x\} = \binom{n}{x} p^x (1-p)^{n-x},$$

where $x = 0, 1, 2, \ldots, n$ and $0 < p < 1$, find the mgf of X, and its mean and variance.

Exercise 8.7 Write a computer program to run the preferential attachment algorithm listed in **Algorithm 8.1**. Set $p = 0.25, 0.5, 0.75$. Run the algorithm for 10,000 iterations with each of these values of p. Draw the log–log plot of the degree distribution for each value of p. Do these degree distributions look like power-law distributions? For each case, estimate the power-law exponent β by linear regression. Compare results.

Exercise 8.8 Suppose the configuration model is defined so that $k_i = k$, $i = 1, 2, \ldots, N$. In other words, all nodes have the same degree k. Find (a) the degree distribution, (b) the probability generating functions $G_0(s)$ and $G_1(s)$, and (c) the excess-degree distribution.

Exercise 8.9 Prove Theorem 8.3.

Exercise 8.10 Write a computer program to run the random copying model listed in **Algorithm 8.2**. Apply it to a large network dataset of your choice. Set $m = 10, 15, 20$ and $\alpha = 0.1, 0.3, 0.5, 0.9$.

CHAPTER NINE

Network Sampling

When a network is too large to study completely, we sample from that network just as we would sample from any large population. The structure of network data, however, is more complicated than that of standard statistical data. The main question is, how can one sample from a network that has nodes and edges? Should we sample the nodes? Or should we sample the edges? The answers to these questions depend upon the complexity of the network. In this chapter, we examine various methods of sampling a network.

9.1 Introduction

So far, we have studied situations in which the primary interest was a network, $\mathcal{G} = (\mathcal{V}, \mathcal{E})$, with $N = |\mathcal{V}|$ nodes and $M = |\mathcal{E}|$ edges. The network structure of N nodes and M edges was viewed as the entire scenario that needed to be explained. The data collected from such a network could be viewed as a *census* of that network, in which every node and edge are completely recorded. For example, in social networks, we are only interested in the interactions between the 18 novice monks in Sampson's data and in the interactions betwen the 34 participants in Zachary's karate club data so that we can separate them into groups. We never considered these network data to be random samples from some much larger (possibly infinite) network of nodes and edges.

Network analysis takes on a completely different perspective when we regard the network in question as a sample having nodes and edges randomly drawn from a much larger network. In many situations, $\mathcal{G}$, which we will refer to as the "parent" or "underlying" network, may be too large to be stored in main memory and, therefore, because of its sheer size, cannot be studied in its entirety. In other situations, the observations that comprise the network may be too expensive to collect (e.g., experiments in biological networks), and so it may be appropriate to restrict the size of the network to a more manageable size. Even attempts at a network census may not capture all nodes and edges if the network has a complex structure and nodes are hard to identify and record.

9.1.1 Example: The Harvard College Influenza Study

Consider, for example, a social network in which it is important for public health officials to understand and predict the spread of infectious diseases. An example of this is the H1N1 influenza outbreak that occurred in late 2009 amongst Harvard College students. Well-connected individuals (i.e., those with high node degree or many contacts or friends) tend to be more exposed to such diseases than do those who are less well connected (i.e., those on the periphery of the network). So, it would pay to monitor these well-connected individuals in order to provide medical personnel with information regarding early detection of the disease. Reducing the rapid spread of the disease could then be accomplished by prophylactically vaccinating the high-degree individuals. However, this may be very difficult, if not impossible, because information regarding the complete knowledge of friendships within a large social network may not be readily available.

Thus, an approach in which the network is sampled could yield usable information about the potential spread of a disease. In this particular study, each Harvard College student provided the names of three friends, and, from those names, a random sample of 425 students was chosen for the study. These 425 students were then each asked to provide up to three friends. For each of these students, information on vaccination status, dorm of residence, and sports activities was recorded. They were also asked about flu symptoms or vaccinations received since the start of the study.

9.1.2 Why Sample?

The goal, then, is to create a more manageable sampled subnetwork, $\mathcal{G}^* = (\mathcal{V}^*, \mathcal{E}^*)$, say, of $\mathcal{G}$ that can be fitted into main memory or is financially less costly to deal with, and which can be viewed as a proxy for the entire network. We also try to ensure that $\mathcal{G}^*$ has as many properties of the parent network as possible. Thus, we wish to create a $\mathcal{G}^*$ that has $\mathcal{V}^* \subset \mathcal{V}$ and $\mathcal{E}^* \subset \mathcal{E}$, where $n = |\mathcal{V}^*| \ll N$ and $m = |\mathcal{E}^*| \ll M$, with the understanding that larger samples do a better job of approximating the parent network than do smaller samples.

For example, the World Wide Web is an enormous network, and it would be natural to create a smaller subnetwork so that we could carry out meaningful analyses of the data without encountering a storage problem. Network simulations are very important in studies of properties of networks and often it is more efficient to sample from the parent network and run the simulations from the subnetwork. Other networks are continuously getting bigger and bigger over time (e.g., streaming data that occur in online applications) and are now considered to be too huge to be studied as complete networks.

In communications networks, there is a sizeable literature focusing on extracting the most significant features of a very large network whilst reducing the size of the network. In an IP network, for example, nodes are routers or workstations and edges represent the ability of computer packets to move between such nodes. In this literature, preserving connectivity is the most desirable characteristic. Reducing network size is referred to as *semantic graph compression* (as opposed to *algorithmic graph compression*, in which the focus is on reducing the computational complexity of a graph algorithm). Many different compression schemes have been proposed, either for ranking nodes by their importance (e.g., through their degrees) or for combining similar nodes, which may have been duplicated for redundancy purposes, into a single node using a notion of similarity (see, e.g., Gilbert and Levchenko, 2004). We note that there is very little recognition that network sampling would be relevant in graph compression; an exception is Gile (2011).

So, what does it mean to sample a network? In statistical studies, we draw a random and representative sample from a large population, collect data taken from the sample as a way of understanding some population characteristic of interest, summarize the results, and then extrapolate those results to the characteristics of the population from which the sample was drawn. Does this approach extend to networks? How should we sample from the parent network in order to create a random and representative subnetwork? Should we randomly pick nodes from all those available in the network? How about picking random edges? What about sample size? How can we sample a network whose members are hard to reach? How do we evaluate the impact of treating the subnetwork as if it were the parent network? What can we say about the parent network when all we have is information about the subnetwork? Which nodes and edges should be regarded as "important"? And what sort of biases should we expect to see when sampling a network? All these questions are fundamental to the activity known as *network sampling*.

9.2 Objectives for Network Sampling

In traditional scientific sampling, a random sample, which is a subset of the population units, is required to satisfy two conditions: it has to be randomly selected and it has to be representative of the underlying population. If the underlying population has a particular structure, the sample, which operates as a proxy for the population, is expected to reflect that structure. Failure to uphold those two ideals will lead to bias in the sampling results.

Networks are examples of notoriously difficult populations from which to sample because of their particular structures. As a result, there are many different sampling methods that have been proposed for sampling from a network. Each of these sampling methods focuses upon a specific measure of representativeness that is judged to be appropriate for a given application.

Amongst the criteria that have been deemed important for a sample to emulate are the following graph-theoretic properties: *degree distribution, clustering coefficient, average path length between pairs of nodes, connectivity*, and *the power-law exponent*. Also, if *hubs* (nodes with very high degrees) are expected to be present in a network, then one would expect them to appear in the sample network. In each case, the sample is compared with the underlying population (when it is known) to see whether the sample possesses the same properties as does the underlying population. Note that these properties are characteristics of the parent network, and do not cover any consideration of whether or not the nodes and edges were randomly selected.

To formalize these ideas, let $\eta(\cdot)$ denote any of these graph properties (as a parameter) or as the distribution of a property (e.g., degree distribution). Then, $\mathcal{G}^*$ is declared to be a good sample from $\mathcal{G}$ if $\eta(\mathcal{G}^*) \approx \eta(\mathcal{G})$. We can put this into the form of an optimization problem. Find $\mathcal{G}^*$ such that

$$\mathcal{G}^* = \arg\min_{\mathcal{H} \subset \mathcal{G}} D(\eta(\mathcal{H}), \eta(\mathcal{G})), \tag{9.1}$$

where D is a distance function. A popular version of D used in social networks (see, e.g., Leskovec and Faloutsos, 2006; Ahmed, Neville, Kompella, 2012; Zhang, Kolaczyk, and Spencer, 2015) is the two-sample Kolmogorov–Smirnov D-statistic

$$D = \max_x |\widehat{F}(x) - \widehat{G}(x)|, \tag{9.2}$$

where $\widehat{F}$ and $\widehat{G}$ are the two sample cumulative distribution functions of the two sets of data, x runs over the range of the data, and $0 \leq D \leq 1$. Because problem (9.1) is intractable for most network properties, attention has instead been placed upon D in (9.2) as the primary way to compare a property of the sampled subnetwork with the same property of the parent network (Hübler et al., 2008).

9.3 Sampling of Nodes

Trying to provide an analogue for the random sampling of a population to a network would lead us to drawing a random sample of nodes from the network. The main assumption for node sampling (i.e., $V^* \subset V$) is that complete knowledge of the parent network has to be readily available for constructing the sample. Assuming that the nodes of the network are known may be reasonable in cases such as a telephone-calling network where the telephone company has complete records of its phone numbers (nodes), knows the total number of its customers, and may be able to extract all phone calls (edges) during a given period of time. For the researcher, this can be a computational problem as it requires random disk access to the complete set of nodes V and their neighbors.

9.3.1 Random-Node Sampling

For this sampling method (referred to as RNS), assuming the entire parent network is available, nodes are selected independently and uniformly at random, usually with replacement, from the parent network (i.e., V^* is a simple random sample drawn from V) and includes all of their mutual edges. Thus, $(v_i, v_j) \in \mathcal{E}^*$ iff, for some v_i, $v_j \in V^*$, the edge $(v_i, v_j) \in \mathcal{E}$. This sampling method is called *induced-subgraph sampling*. If $p(v)$ is the probability of choosing node v, then RNS says that

$$p(v) = \frac{1}{N}, \tag{9.3}$$

where N is the number of nodes in the parent network. When the nodes are selected for V^*, the sampled subnetwork is constructed to be the induced subnetwork over the nodes V^*. Put very simply, all edges in $\mathcal{E}$ that join nodes in $V^* \in \mathcal{G}$ are added to $\mathcal{E}^*$.

Criticisms. There is evidence that RNS does not accurately capture properties of networks having exact power-law degree distributions (Stumpf and Wiuf, 2005). However, for large enough degrees in that case, the degree distribution from induced subgraph sampling follows a power law and has the same exponent as the parent network (Stumpf and Wiuf, 2005). Also, the level of connectivity that exists in the parent network is unlikely to be preserved using RNS (Lee, Kim, and Jeong, 2006). The induced subnetwork derived by randomly sampling nodes could result in many isolated and disconnected nodes, which would not be a productive environment for any subsequent analysis. Moreover, RNS also fails to match the degree distribution, the clustering coefficient, and the average path length of the parent network.

9.3.2 Random-Node Sampling by Degree

This sampling method (referred to as RND) improves the performance of RNS. Nodes are selected randomly with probability proportional to their degrees, which is assumed known for every node in the parent network (Adamic et al., 2001). If $p(v)$ is the probability of choosing node v, then RND says that

$$p(v) = \frac{d_v}{2M},\qquad(9.4)$$

where d_v is the degree of node v and $M = |\mathcal{E}|$ is the number of edges. All mutual edges are included in the sample.

Biases and other issues. Because RND is biased towards higher-degree nodes, the degree distribution will be biased, the samples will be too dense, and the average degree will be higher than that for the parent network.

9.3.3 Random-Node Sampling by PageRank

In this sampling method (abbreviated here as RNP), a node is selected with probability proportional to its PageRank score, which is assumed known (Leskovec and Faloutsos, 2006). PageRank, which was developed by Sergey Brin and Larry Page (Brin and Page, 1998) for ranking the popularity of websites on the World Wide Web, looks at a webpage and declares its significance to be how many other webpages link to it.

Under the PageRank philosophy, links should not be treated as identical to each other; certain websites should be deemed more "important" than others. A node, therefore, is declared to be important if it is pointed to by its important neighbors. So, the significance of its neighbors is a major factor in a node's PageRank score, as is also the number of its neighbors, which reflects the strength of the network's connectivity.

PageRank. PageRank is defined as follows. Consider the Web as a directed network where nodes represent webpages and edges represent links between webpages. Brin and Page constructed a probability transition matrix that is stochastic and irreducible,[1] and yields a stationary distribution of a Markov chain whose state space is the set of all webpages.

Suppose N denotes the total number of webpages on the Web. Let $k > 0$ be the number of outgoing edges for the ith webpage. Let $\mathbf{P} = (p_{ij})$ be an $(N \times N)$-matrix of hyperlinks, where $p_{ij} = 1/k$ if the jth node is one of the outgoing edges, and $p_{ij} = 0$ otherwise. This definition of $\mathbf{P}$, however, does not take into account webpages without outgoing edges (called *dangling nodes*); rows of the matrix corresponding to dangling nodes will be rows that are completely made up of zeroes, so that $\mathbf{P}$ will not be a stochastic matrix. If the ith webpage has no outgoing edges, Brin and Page modified

---

[1]A matrix is *stochastic* if either each row or each column sums to one, *irreducible* if its graph shows that every node is reachable from every other node.

P to have $p_{ij} = 1/N$, for all j, so that its probability is spread out amongst all the webpages. In other words, the row vector $\mathbf{0}^\tau$ is replaced by the row vector $\frac{1}{N}\mathbf{1}_N^\tau$, where $\mathbf{1}_N$ is an N-vector of all 1s. Even though **P** is now a stochastic matrix, it does not guarantee the existence of a stationary vector of the Markov chain.

Towards this end, Brin and Page further adjusted **P** by forming

$$\widetilde{\mathbf{P}} = \alpha\mathbf{P} + (1-\alpha)\frac{1}{N}\mathbf{J}_N = \alpha\mathbf{P} + (1-\alpha)\mathbf{E}, \tag{9.5}$$

where $\mathbf{J}_N = \mathbf{1}_N\mathbf{1}_N^\tau$ is an $(N \times N)$-matrix each of whose elements is 1, and $\alpha \in (0,1)$ is the probability of not jumping to a random webpage. Note that α has been called the *damping factor* (Becchetti and Castillo, 2006) and $1 - \alpha$ has been termed the *teleportation probability*. Google, which was founded by Brin and Page, sets $\alpha = 0.85$.

The matrix $\widetilde{\mathbf{P}}$ is a convex combination of two stochastic matrices **P** and $\mathbf{E} = \frac{1}{N}\mathbf{J}_N$, and hence is stochastic, aperiodic, and irreducible. Thus, there exists a unique probability N-vector $\boldsymbol{\pi}$ such that

$$\boldsymbol{\pi}^\tau\widetilde{\mathbf{P}} = \boldsymbol{\pi}^\tau, \quad \boldsymbol{\pi}^\tau\mathbf{1}_N = 1. \tag{9.6}$$

From (9.5) and (9.6), and because $\mathbf{I}_N - \alpha\mathbf{P}$ is nonsingular and its inverse is non-negative (Berman and Plemmons, 1979), we have that

$$\boldsymbol{\pi}^\tau = \frac{1-\alpha}{N}\mathbf{1}_N^\tau(\mathbf{I}_N - \alpha\mathbf{P})^{-1}. \tag{9.7}$$

The vector $\boldsymbol{\pi}$ is called a *PageRank vector*, or *PageRank score*, and the ith element, π_i, of $\boldsymbol{\pi}$ is the PageRank score of the ith webpage. The PageRank score for node $v \in \mathcal{V}$ can be interpreted as the probability that a person randomly clicking on links (a so-called *web surfer*), jumping from webpage to webpage, would land on node v.

For an excellent review of the computational algorithms (e.g., the power method applied to (9.7)) used to evaluate the PageRank vector $\boldsymbol{\pi}$, see Langville and Meyer (2005).

9.3.4 Node Sampling by Metropolis Algorithm

Hübler et al. (2008) proposed a novel sampling method based upon a combination of the Metropolis algorithm for Markov chains and simulated annealing. Recall that a Markov chain is a stochastic process in which the probability of a future state given the past history of the process only depends upon the current state of the process. We have the following result.

Theorem 9.1 (Ergodic Theorem) *An ergodic Markov chain[2] converges to a stationary distribution iff any state can be reached from any other state in a finite number of moves.*

In Metropolis sampling, a sample S of n nodes is randomly drawn from the N nodes of the parent network $\mathcal{G} = (\mathcal{V}, \mathcal{E})$ according to the density $\pi(S)$ of the stationary

---

[2]A Markov chain, X_t, has the *ergodic property* if it is *irreducible* (i.e., every state is reachable from every other state), *aperiodic* (i.e., each state has period 1), and *positive recurrent* (i.e., there exists a stationary probability distribution for X_t).

distribution. The sampling process should produce a representative small subnet-work, where "representative" means that the subgraph will preserve certain properties of the parent graph.

Let $\{X_t, t = 0, 1, 2, \ldots\}$ denote a sequence of random variables, which need not be independent draws, such that, at each time $t \geq 0$, the next state X_{t+1} is sampled from a distribution that depends only on the current state X_t. Consider the transition from state S to state S'. The transition probability, $P\{X_{t+1} = S' | X_t = S\}$, can be partitioned into two terms: a *proposal distribution* $q(S'|S)$, which is the probability of proposing a move from state S to state S', and an *acceptance probability* $\alpha(S, S')$, which is the probability of accepting the move to state S'. If the move is rejected, the process returns to the current state (i.e., $X_{t+1} = X_t$) and attempts to move again. Thus,

$$P\{X_{t+1} = S' | X_t = S\} = q(S'|S)\alpha(S, S'), \quad S \neq S', \tag{9.8}$$

where

$$\alpha(S, S') = \min\left\{1, \frac{\pi(S')q(S|S')}{\pi(S)q(S'|S)}\right\} = \min\left\{1, \frac{\pi^*(S')q(S|S')}{\pi^*(S)q(S'|S)}\right\}, \tag{9.9}$$

where $\pi^*(S) \propto \pi(S)$ is the unnormalized density, and similarly for $\pi^*(S')$. If the proposal distribution is taken to be symmetric, that is, $q(S|S') = q(S'|S)$, then (9.9) reduces to $\alpha(S, S') = \min\{1, \pi^*(S')/\pi^*(S)\}$. The proposal distribution $q(\cdot|\cdot)$ can take any form and should be chosen to be easy to compute. From (9.9), we have the *detailed balance equation*,

$$\pi(S)P\{S'|S\} = \pi(S')P\{S|S'\}, \tag{9.10}$$

which, following integration wrt S, shows that if S is from $\pi(\cdot)$, then so will be S'. Thus, if a sample from the stationary distribution has been drawn, then all subsequent samples will be from that same distribution.

Let $\Delta_{\mathcal{G}, \eta}(S) = \Delta(\eta(S), \eta(\mathcal{G}))$ denote a measure of distance between $\eta(S)$ and $\eta(\mathcal{G})$, where $\eta(\mathcal{G})$ is some topological property of the network $\mathcal{G}$, and similarly for $\eta(S)$. Then, set

$$\pi_p^*(S) = \frac{1}{[\Delta_{\mathcal{G}, \eta}(S)]^p}, \tag{9.11}$$

for some large positive exponent $p \gg 0$. This has the effect of picking high-quality samples instead of selecting any of the more numerous, lower-quality samples.

One of the problems with this process concerns slow convergence: iterations of the Markov chain may get stuck in region heavily influenced by the starting distribution, or be trapped at a local maximum for long periods of time, which, in turn, can lead to erroneous results (Gelman, 1996; Gilks and Roberts, 1996).

Hübler et al. proposed avoiding this problem for optimization purposes by using simulated annealing to modify the stationary distribution. They introduced a parameter T (viewed as temperature) that is changed at each step in the process, so that the acceptance rate is increased at the beginning and, at the end, acceptance is made only if a move is deemed to be an improvement. So, the stationary distribution (9.11) becomes

$$\pi_{p, T}^*(S) = \frac{1}{[\Delta_{\mathcal{G}, \eta}(S)]^{p/T}}. \tag{9.12}$$

The idea behind that adjustment is that by raising the temperature (i.e., increasing the value of T), $\pi^*_{p,T}(S)$ will flatten out, so that the sampling process will be better able to search the support of $\pi^*_{p,T}(S)$, thereby skipping over local maxima. If we assume a symmetric proposal distribution, then (9.9) reduces to

$$\alpha(S, S') = \min \left\{ 1, \frac{\pi^*_{p,T}(S')}{\pi^*_{p,T}(S)} \right\} = \min \left\{ 1, \left[\frac{\Delta_{\mathcal{G},\eta}(S)}{\Delta_{\mathcal{G},\eta}(S')} \right]^{p/T} \right\}. \tag{9.13}$$

The authors recommend that the temperature parameter be reduced at each move according to the schedule $T_{t+1} = \gamma T_t$, where $0 < \gamma < 1$ and t is the tth step of the sampling procedure. They also recommend that p be taken to be $p_{\mathcal{G}} = 10 \cdot \frac{M}{N} \log_{10}(N)$, where N is the number of nodes and M is the number of edges in $\mathcal{G}$. The Metropolis algorithm then modifies node sampling by replacing some sampled nodes with other nodes. This strategy may provide a closer approximation of network properties and, hence, would be considered to be a reasonably faithful representation of the parent network.

The tests conducted by the authors indicated that their approach outperformed all previous approaches.

9.4 Sampling of Edges

The main assumption of edge sampling (i.e., $\mathcal{E}^* \subset \mathcal{E}$) is that complete knowledge of the parent network has to be available for determining the sample. It would be impossible to sample edges if the entire network of nodes and edges were not known to the researcher. Sampling of edges makes sense, for example, if the study were tracking the number of phone calls (where the nodes are the phone numbers) or tracking transactions made by financial institutions such as banks, which would be the nodes. In both these examples, the edges would be directed.

9.4.1 Random-Edge Sampling

In random-edge sampling (RES), the sample network consists of edges selected independently and uniformly at random (i.e., $\mathcal{E}^*$ is a simple random sample drawn from $\mathcal{E}$). When a particular edge is selected and added to $\mathcal{E}^*$, a partially induced subnetwork is obtained if both the end nodes of that chosen edge are included in $\mathcal{V}^*$. In other words, $v_i \in \mathcal{V}^*$ if, for some $v_j \in \mathcal{V}$, the edge $(v_i, v_j) \in \mathcal{E}^*$. It is partially induced because no edges are chosen additional to those selected by the sampling process. Thus,

$$\mathcal{E}^* = \{e_{ij} = (v_i, v_j) \in \mathcal{E} | v_i, v_j \in \mathcal{V}^*\}. \tag{9.14}$$

This method of sampling is called *incident-subgraph sampling*. If $p(e_{ij})$ is the probability that edge $e_{ij} = (v_i, v_j)$ is selected, then

$$p(e_{ij}) = \frac{1}{|\mathcal{E}|}, \tag{9.15}$$

where $|\mathcal{E}|$ is the number of edges in the set $\mathcal{E}$. A more detailed discussion may be found in Crane (2018, Chapter 9).

Biases and other issues. RES accurately approximates the average path length of the parent network. However, it fails to preserve most other network properties. For example, it is biased towards selecting high-degree nodes (which may be *hubs*), it underestimates the clustering coefficient, and does not preserve connectivity. Furthermore, because it chooses a subset of the edges (and, hence, of the nodes) of the network, the subnetwork $\mathcal{G}^*$ is biased downwards in approximating the degree distribution of $\mathcal{G}$. In other words, RES underestimates the node degrees of the network (Ahmed, Neville, and Kompella, 2014).

9.4.2 Random-Edge Sampling with Induced Subgraph

This sampling method (abbreviated to REI) improves the performance of RES by adding extra edges to the subnetwork (Ahmed, Neville, and Kompella, 2014). The sample subnetwork $\mathcal{G}^* = (\mathcal{V}^*, \mathcal{E}^*)$ consists of edges selected uniformly at random (using RES) plus a graph-induction step over the sampled nodes, namely, adding all other edges in $\mathcal{E}$ whose endpoints both exist in $\mathcal{V}^*$.

Biases and other issues. Graph induction helps to preserve the connectivity of $\mathcal{G}^*$ and increases the likelihood of triangles in $\mathcal{G}^*$, which, in turn, results in higher clustering coefficients and shorter average path length. It is also biased in selecting high-degree nodes, thereby preferring dense and highly clustered regions of the network, which improves the connectivity of $\mathcal{G}^*$. REI was found to outperform several other sampling techniques (RNS, RES, FFS; see Section 9.5.6) in matching the degree, path length, and clustering coefficient distributions of the original network $\mathcal{G}$.

9.5 Sampling by Network Exploration

The network-sampling methods of this section, which have also been called *topology-based sampling*, are used extensively in social science applications when knowledge of the network may not be available. This occurs especially when trying to sample *difficult-to-access populations*, where no sampling frame exists, and it is impossible to take a random sample. Examples include sampling those individuals diagnosed with HIV/AIDS, homeless persons, illegal drug users, or undocumented immigrants in a city. These individuals are hard to reach, and even if contact were established, it would be difficult to get them to respond to interviews, hard to keep track of them, and hard to maintain contact with them over time. Some attempts at sampling such populations have, however, been quite successful. That happens when researchers use sampling methods that take advantage of the social networks that link people together. However, such methods do not tend to work well when trying to sample large subpopulations.

9.5.1 Link-Trace Sampling

Link-trace sampling (LTS; Spreen, 1992) is also referred to as *crawling* or *online sampling*. It is based upon the assumption that difficult-to-access populations may be located through the social networks to which they belong. Examples of LTS include snowball sampling and respondent-driven sampling.

LTS starts with an integer k and an initial seed node $v \in \mathcal{V}$ from the parent network, so that $\mathcal{V}^* = \{v\}$. The next node is selected for inclusion into the sample from amongst the set of nodes directly connected to the node v already sampled (i.e., from the neighborhood of that node). The process is repeated in an iterative fashion until $|\mathcal{V}^*| = k$. The main question is how to choose a node from the neighborhood of a given node.

A particular application of this principle, when dealing with social networks, is to use a strategy such as breadth-first search (see Section 3.7.2), snowball sampling, random walk, or forest fire to choose individuals for inclusion in the sample. The choice of individuals usually is created by *chain-referral sampling*, whereby one individual already in the sample recommends another who would be appropriate to contact.

Biases and other issues. It is unlikely that the initial seed node would be randomly selected from the parent population, which would be one type of bias of the study. One of the issues of interest is how sensitive the results from the network sampling method are to the choice of seed node. Also, chain-referral samples tend to be biased towards those individuals who agree to participate and who will cooperate with the researcher. Selected individuals may try to protect friends from the study by not referring them to researchers, which would, therefore, bias the sample composition.

9.5.2 Snowball Sampling

Snowball sampling (SBS) is a special case of link-trace sampling and is the most popular sampling method for surveying rare, hidden, and difficult-to-access populations (Coleman, 1958). It was used, for example, in early studies of opiate addiction and marijuana smokers (Biernacki and Waldorf, 1981). It has also been used to study gangs, drug users, prostitution, slums, and the seriously ill.

In the absence of a sampling frame, snowball sampling is carried out in a sequential fashion by recruiting a small number of initial respondents (viewed as zero-wave individuals or seeds), usually identified through social media, interviewing them and asking for referrals to a next set of respondents. These first-wave responders are then interviewed and asked for a further set of responders, and so on, until a large enough sample is achieved. The parameters of this process, which need to be determined, are the number s of waves (depends upon how quickly the contacts dry up), and the number k of contacts (usually taken to be 3) to recruit at each stage. The image is like a snowball getting larger and larger as it rolls downhill, collecting more snow.

Thus, SBS adds nodes and edges from a seed node using a version of breadth-first search, but where only a fraction of a node's neighbors are explored (Goodman, 1961). The edges of this network are directed, showing the chain-referral process in which an individual member of a difficult-to-access population would recommend another individual from the same population for the survey.

Estimating the Size of a Difficult-to-Access Population

Perhaps the main interest in working with a difficult-to-access population is to estimate the size of that population. For example, Frank and Snijders (1994) used a one-wave snowball sample to estimate the total number of heroin users in the town of Groningen, Sweden. In their study, they recruited an initial sample of 34 persons, who were contacted through social assistance agencies, medical doctors, and by visiting known meeting points of heroin users (called *site sampling*). They justified the sampling process as being roughly representative and, although the participants in the initial sample were not randomly selected, they yielded results that were deemed to be close to what would be expected had the selection been made using probabilistic principles. The first wave totaled 237 persons after removing duplicates and those already recruited for the initial sample. Different statistical methods were then used to estimate the population size, including a design-based approach (regarding the network as fixed but unknown) and a model-based approach (assuming a probability framework for the initial sample).

Frank and Snijders set out the following design-based estimator of total sample size using moment estimators. Let N denote the total number of nodes and M denote the total number of edges in the network. Let $\mathbf{Y} = (Y_{ij})$ denote the adjacency matrix, where $Y_{ij} = 1$ if an edge exists between nodes v_i and v_j (i.e., if $(v_i, v_j) \in \mathcal{E}$), and 0 otherwise, $i, j = 1, 2, \ldots, N$. Let Z_i be an indicator variable that equals 1 if the node v_i were chosen for the initial sample, and 0 otherwise, where $P\{Z_i = 1\} = p_0$, $i = 1, 2, \ldots, N$. Let N_0 be the number of nodes in the initial sample, M_0 the number of edges in the initial sample, and M_1 the number of edges from nodes in the initial sample to nodes in the first wave. Then,

$$E\{N_0\} = E\left\{ \sum_i Z_i \right\} = Np_0, \tag{9.16}$$

$$E\{M_0\} = E\left\{ \sum_{i \neq j} Z_i Z_j Y_{ij} \right\} = (M - N)p_0^2, \tag{9.17}$$

$$E\{M_1\} = E\left\{ \sum_{i \neq j} Z_i (1 - Z_j) Y_{ij} \right\} = (M - N)p_0(1 - p_0). \tag{9.18}$$

To compute the moment estimators, we set each of the three quantities (9.16), (9.17), and (9.18) equal to their observed values n_0, m_0, and m_1, respectively. From (9.16), we have that $\widehat{N} = n_0/p_0$, and from (9.17) and (9.18), we have that $\widehat{p_0} = m_0/(m_0 + m_1)$. Plugging that back into the result for $\widehat{N}$, we get the estimated number of nodes in the network as

$$\widehat{N} = \frac{n_0}{m_0/(m_0 + m_1)} = \frac{n_0(m_0 + m_1)}{m_0}. \tag{9.19}$$

Thus, the estimated size of the network is found by multiplying the initial sample size, n_0, by a quantity that combines the number of edges in the initial sample and the number of edges from the initial sample to the first-wave sample.

In this study, $n_0 = 34$, $m_0 = 11$, and $m_1 = 237$. They adjusted the estimator (9.19) slightly so that $\widehat{N} = [n_0 m_0 + (n_0 - 1)m_1]/m_0$. The resulting estimate $\widehat{N} = 745$ was close to the police estimate of about 800 heroin addicts. The authors also derived a jackknife-based estimator of the variance of this estimator.

Biases and other issues. Although snowball sampling has proved to be popular, it does have its defects. First, as a nonprobability sampling method, SBS creates samples that are not randomly selected and may not even be representative of the underlying target population, thereby violating the fundamental rules of sampling. Statistical sampling issues are often ignored by researchers in favor of generating a large collection of individuals who show the requisite target behavior, such as drug abuse. Hence, the results from snowball sampling cannot be generalized to the underlying population. However, it may be the best that one can do to collect relevant data in a social network study of hard-to-reach individuals. This may not be a problem for some researchers. For example, in their article on the use of snowball sampling to study heroin addicts, Biernacki and Waldorf (1981) noted that

> *The purpose of the research was not to test a series of predetermined hypotheses on a representative sample which would allow for extrapolation to the whole population. Rather, the study's aim was to explore and analyze ... the social and psychological processes that worked to bring about a cessation of heroin addiction.*

Second, there will be various biases in the results due to the nonprobabilistic sampling method. One type of bias is that the individuals who constitute the seeds and also at each wave are often recruited as volunteers, and *volunteer bias* can exist. A good strategy for recruiting a seed would be to base the choice upon several indirect sources, such as online searches. Third, the samples will be biased towards high-degree individuals; those respondents, who have connections to many acquaintances, have a higher likelihood of being recruited to the study over respondents who have fewer connections. Fourth, it also suffers from *boundary bias* in that many peripheral nodes (i.e., those sampled at the last wave) will be missing a large number of neighbors (Lee, Kim, and Jeong, 2006). Fifth, there is no guarantee that the process will continue to gather respondents; indeed, studies have reported that only a few individuals were able to be recruited because the last-recruited respondent lacked any contacts.

9.5.3 Respondent-Driven Sampling

Respondent-driven sampling (RDS), introduced by Heckathorn (1997), extends snowball sampling by compensating for the fact that snowball sampling is carried out in a nonrandom way. It is also known as a *reweighted random walk*. Since its introduction, RDS has evolved into a number of different network sampling techniques. It is now not viewed as a single sampling method, but as a collection of methods that try to convert chain-referral sampling into an accurate sampling method.

It has been applied to the study of jazz musicians in New York City, San Francisco, New Orleans, and Detroit (Heckathorn and Jeffri, 2001; Gleiser and Danon, 2003), and to studies of aging artists and professional and semi-professional storytellers. It has also been used to survey disease prevalence (e.g., HIV/AIDS), sex workers, and drug users in difficult-to-access populations (Goel and Salganik, 2009; Gile, 2011). Another study used RDS to estimate the number of people who inject drugs in St. Petersburg, Russia, during 2012–2013, where the incidence and prevalence of HIV is highest among such people, many of whom are unaware of their HIV status (Crawford, Wu, and Heimer, 2018). The importance of RDS in addressing global public-health questions can be seen from its use by the Centers for Disease

Control and Prevention (CDC), which recently selected RDS for a 25-city study of injection-drug users, and Family Health International, a huge nonprofit organization specializing in global public health, which applied it in more than a dozen countries (Volz and Heckathorn, 2008).

RDS tries to create something that looks like a probability sample by weighting the sample using Markov chain theory and biased network theory. It starts with recruiting a modest number of initial respondents or seeds, and these seeds are asked to recruit other respondents directly. Respondents are rewarded financially for being interviewed, plus a reward for recruiting others and making sure they participate. They are also given a number of "coupons," which provide a unique ID number, the survey name, interview site address, date when the coupon can be redeemed, and other pertinent information for the project. These coupons are then used to recruit other people, who, in turn, become recruiters. All new recruits are similarly financially rewarded. Each newly selected set of respondents is referred to as a *wave*.

This process enables *referral chains* to be set up to penetrate social groups that are only accessible to group members. It is usually assumed that RDS is carried out with replacement, where "with replacement" means that an individual may be recruited into the sample more than a single time. It is, however, rare for this to happen in practice.

Sampling is terminated when either a targeted group is completely explored, or a prespecified minimum sample size is reached. In certain situations, sampling is continued until (or past the point when) an "equilibrium" in sample composition (e.g., males and females) is reached, which happens when the gender difference of one wave to the next is less than 2% (Wejnert, 2009). A web-based implementation of RDS has reported good results (Wejnert and Heckathorn, 2007).

A large part of the practice of RDS is to grow the referral chains (i.e., number of waves) so big that the sample network will become independent of the seeds from which it began. However, it has been reported that, in practice, the number of waves tends to be small, on the order of at most five, but more usually, fewer than 20 (Gile, 2011). Seeds can influence the rate at which networks grow and the speed at which sampling will occur. Long referral chains are initiated by respondents who should be well-motivated, enthusiastic, and committed to the goals of the study, and who are referred to as *super-seeds* (Gile and Handcock, 2010).

Horvitz–Thompson Estimator

We will need the following tool for estimating the mean of a population when the probability of inclusion may not be equal for each population unit. Suppose there are N individuals in the population. Let π_i denote the *inclusion probability* of the ith unit, which we assume to be known; this is the probability that the ith individual is included in the sample, and is the total of the probabilities of all samples that contain the ith unit. Note that $\sum_{i=1}^{N} \pi_i = n$.

Let X_i be a quantity of interest for the ith individual with observed value x_i. Then, the *Horvitz–Thompson estimator* (Horvitz and Thompson, 1952) of the population mean $\mu = \frac{1}{N} \sum_{i=1}^{N} X_i$ is

$$\widehat{\mu}_{HT} = \frac{1}{N} \sum_{i=1}^{N} \frac{X_i}{\pi_i} I_{[i \in S]}, \qquad (9.20)$$

where S represents the sample, $I_A = 1$ if A is true, and 0 otherwise, and we write $i \in S$ to mean $x_i \in S$. The estimator $\widehat{\mu}_{HT}$ is based upon an inverse weighting of the sample values, where the ith observation X_i is weighted by the reciprocal of its inclusion probability π_i. Treating the $x_1, x_2, \ldots, x_n$ as fixed, the expected value of $\widehat{\mu}_{HT}$ is given by the following:

$$
\begin{aligned}
\mathrm{E}\{\widehat{\mu}_{HT}\} &= \mathrm{E}\left\{ \frac{1}{N} \sum_{i=1}^{N} \frac{x_i}{\pi_i} I_{[i \in S]} \right\} \\
&= \frac{1}{N} \sum_{i=1}^{N} \frac{x_i}{\pi_i} \mathrm{E}\{I_{[i \in S]}\} \\
&= \frac{1}{N} \sum_{i=1}^{N} \frac{x_i}{\pi_i} \mathrm{P}\{I_{[i \in S]} = 1\} \\
&= \frac{1}{N} \sum_{i=1}^{N} \frac{x_i}{\pi_i} \pi_i \\
&= \mu,
\end{aligned}
\tag{9.21}
$$

so that $\widehat{\mu}_{HT}$ is an unbiased estimator of μ. It can be shown that the variance of $\widehat{\mu}_{HT}$ is

$$
\begin{aligned}
\mathrm{var}\{\widehat{\mu}_{HT}\} &= \frac{1}{N^2} \sum_{i=1}^{N} \sum_{j=1}^{N} \frac{x_i}{\pi_i} \frac{x_j}{\pi_j} (\pi_{ij} - \pi_i \pi_j) \\
&= \frac{1}{N^2} \sum_{i=1}^{N} \sum_{j=1}^{N} x_i x_j \left(\frac{\pi_{ij}}{\pi_i \pi_j} - 1 \right),
\end{aligned}
\tag{9.22}
$$

where $\pi_{ij} > 0$ is the joint inclusion probability that both the ith and jth individuals are included in the sample. An unbiased estimator of $\mathrm{var}\{\widehat{\mu}_{HT}\}$ is given by

$$
\begin{aligned}
\widehat{\mathrm{var}}\{\widehat{\mu}_{HT}\} &= \frac{1}{N^2} \sum_{i=1}^{N} \sum_{j=1}^{N} \frac{X_i}{\pi_i} \frac{X_j}{\pi_j} \left(\frac{\pi_{ij} - \pi_i \pi_j}{\pi_{ij}} \right) I_{[i \in S]} I_{[j \in S]} \\
&= \frac{1}{N^2} \sum_{i=1}^{N} \sum_{j=1}^{N} X_i X_j \left(\frac{1}{\pi_i \pi_j} - \frac{1}{\pi_{ij}} \right) I_{[i \in S]} I_{[j \in S]}.
\end{aligned}
\tag{9.23}
$$

See, for example, Särndal, Swensson, and Wretman (1992, Section 2.8) or Lohr (1999, Section 6.4.1).

RDS as MCMC

Goel and Salganik (2009) showed that RDS is equivalent to a Markov chain Monte Carlo (MCMC) importance-sampling process. We follow their development. In this scenario, MCMC is a first-order Markov chain operating on the state-space $\mathcal{V}$ of N nodes. This chain can be identified with a transition probability matrix $\mathbf{P} = (p_{ij})$, where

$$
p_{ij} = p(v_i, v_j) \geq 0, \quad \sum_{v_j \in \mathcal{V}} p(v_i, v_j) = 1,
\tag{9.24}
$$

that provides the probability of a transition from node v_i to node v_j. In other words, $p(v_i, v_j)$ is the probability that individual v_i recruits individual v_j. Now, if the chain is irreducible, there exists a unique stationary distribution, $\pi : \mathcal{V} \to \mathbb{R}$, such that

$$\sum_{v_i \in \mathcal{V}} \pi(v_i) p(v_i, v_j) = \pi(v_j), \quad j = 1, 2, \ldots, N, \tag{9.25}$$

or, in matrix notation,

$$\mathbf{P}\pi = \pi, \tag{9.26}$$

where $\pi = (\pi(v_1), \pi(v_2), \ldots, \pi(v_N))^\tau$ and $\pi^\tau \mathbf{1}_N = \sum_{i=1}^{N} \pi(v_i) = 1$.

Let $X_0, X_1, X_2, \ldots$ denote a realization of the Markov chain (i.e., participants in the study). If $X_0 \sim \pi$, then $X_i \sim \pi$, $i = 0, 1, 2, \ldots$. We use this setup to generate dependent samples from the distribution π. Thus, individual v_j has probability $P\{X_i = v_j\} = \pi(v_j)$ of being chosen, $j = 0, 1, 2, \ldots$. The transition matrix $\mathbf{P} = (p_{ij})$ corresponding to the first-order Markov chain has entries

$$p_{ij} = \begin{cases} P\{X_{t+1} = v_j | X_t = v_i\}, & \text{if } (v_i, v_j) \in \mathcal{E} \\ 0, & \text{otherwise} \end{cases}. \tag{9.27}$$

Given that $X_t = v_i$, then X_{t+1} is randomly and uniformly chosen from the neighbors of v_i. That is,

$$p_{ij} = P\{X_{t+1} = v_j | X_t = v_i\} = \frac{Y_{ij}}{d_{v_i}}, \tag{9.28}$$

where $Y_{ij} = I_{[(v_i, v_j) \in \mathcal{E}]}$ is the ijth entry in the adjacency matrix $\mathbf{Y}$, I_A is the indicator of event A (i.e., $I_A = 1$ if A is true, and 0 otherwise), and d_{v_i} is the degree of node $v_i \in \mathcal{V}$. If $\mathbf{D}^{-1} = \text{diag}\{1/d_{v_i}\}$ denotes the diagonal matrix of inverse degrees, then $\mathbf{P} = \mathbf{D}^{-1}\mathbf{Y}$.

Let $f : \mathcal{V} \to \mathbb{R}$ be any function that assigns a characteristic (i.e., an individual trait) to a node. We wish to estimate the population mean,

$$\mu_f = \frac{1}{N} \sum_{i=1}^{N} f(v_i) \tag{9.29}$$

of f. Unfortunately, the sample mean with sample size n, $\frac{1}{n} \sum_{i=0}^{n-1} f(X_i)$, is not an unbiased estimator of the mean μ_f. Instead, it is an unbiased estimator of $E_\pi\{f\} = \sum_{i=1}^{N} f(v_i)\pi(v_i)$. However, we can correct the estimator by noting that

$$E_\pi \left\{ \frac{f(X_i)}{N \cdot \pi(X_i)} \right\} = \sum_{i=1}^{N} \frac{f(v_i)}{N \cdot \pi(v_i)} \pi(v_i) = \frac{1}{N} \sum_{i=1}^{N} f(v_i). \tag{9.30}$$

So, the weighted sample mean,

$$\widehat{\mu}_f = \frac{1}{n} \sum_{i=0}^{n-1} \frac{f(X_i)}{N \cdot \pi(X_i)} \tag{9.31}$$

is an unbiased estimator of the mean, μ_f, of f: that is, $E_\pi\{\widehat{\mu}_f\} = \mu_f$. This is the Horwitz–Thompson estimator.

If $D \subseteq \mathcal{V}$ are those individuals that have a specific trait (e.g., infected with a disease), then we set $f(v_i) = 1$ if $v_i \in D$, and 0 otherwise. The estimator (9.31) reduces to

$$\widehat{\mu}_{\text{trait}} = \frac{1}{n} \sum_{i=0}^{n-1} \frac{I_{[X_i \in D]}}{N \cdot \pi(X_i)}, \qquad (9.32)$$

where $I_A = 1$ if A is true, and 0 otherwise. This yields an estimate of the proportion of individuals in the population with the trait.

Because N is unknown in practice, we use an unbiased estimator of N:

$$\widehat{N} = \sum_{i=0}^{n-1} \frac{1}{\pi(X_i)}. \qquad (9.33)$$

This yields an alternative estimator for μ_f given by

$$\widehat{\mu}_f^* = \frac{\sum_{i=0}^{n-1} \frac{f(X_i)}{\pi(X_i)}}{\sum_{i=0}^{n-1} \frac{1}{\pi(X_i)}}, \qquad (9.34)$$

which estimates the prevalence of the trait in the population. The estimator $\widehat{\mu}_f^*$ is asymptotically unbiased for μ_f. The quantities $\{1/\pi(X_i)\}$ are called the *importance weights*. If all importance weights were the same (i.e., $\pi(X_i) = \pi$), then (9.34) reduces to the mean, $\widehat{\mu}_f^* = \frac{1}{n} \sum_{i=0}^{n-1} f(X_i)$.

Assume that the network is connected and, hence, irreducible. We apply a symmetric weight function $W(x, y)$ to the edge joining nodes x and y, where $W(x, y) = W(y, x)$. Typically, researchers set $W(x, y) = 1$, which corresponds to the assumption that participants recruit their contacts uniformly at random, and that all contacts approached agree to participate. This assumption would fail in most studies because of selection bias: recruiters decide whom to enlist, and recruits decide whether to participate or not. Let $A \subseteq \mathcal{V}$. Then, the weight of A is given by

$$W_A = \sum_{x \in A} \sum_{y \in \mathcal{V}} W(x, y). \qquad (9.35)$$

If x is a single node (i.e., $A = \{x\}$), we write W_x, which is the *self-reported* personal degree of individual x (Salganik and Heckathorn, 2004).

The extent to which people recall how many of their acquaintences are members of a certain subpopulation was studied by Zheng, Salganik, and Gelman (2006), who analyzed response data on the variation of counts when respondents were asked questions such as: How many males do you know in state or federal prison?; How many people do you know named X? (where X is a specified male or female name); How many twins do you know?; How many homeless people do you know?; How many people do you know who have AIDS?; and so on.[3] The answers to these questions can then be used to estimate the size of the respondent's social network. For example, if the ith respondent gave the count a_{ik} to the question regarding

---

[3]Gelman and Romano (2010) have written an hilarious version of this article entitled *"How many zombies do you know? Using indirect survey methods to measure alien attacks and outbreaks of the undead,"* which was published on Scienceblogs.com. It can be found at www.stat.columbia.edu/~gelman/research/published/zombies.pdf or arxiv.org/pdf/1003.6087.pdf. See also andrewgelman.com/2015/05/16/apology-to-george-a-romero/.

subpopulation k, and if the average network size of subpopulation k is b_k, then a preliminary estimate of the size of the ith respondent's social network is $(a_{ik}/b_k) \times N_{\text{pop}}$, where N_{pop} is the population total.

Now, think of RDS as a random walk on the weighted network defined by the transition probability

$$p(x, y) = \frac{W(x, y)}{W_x}. \tag{9.36}$$

Recall that $p(x, y)$ is the probability that individual x recruits individual y. Then, the random walk has the unique stationary distribution $\pi(x) = W_x/W_y$. So, the importance-sampling estimator (9.34) of μ_f reduces to the *Volz–Heckathorn estimator* (VH; Volz and Heckathorn, 2008),

$$\widehat{\mu}_f^{VH} = \frac{\sum_{i=0}^{n-1} \frac{f(X_i)}{W_{X_i}}}{\sum_{i=0}^{n-1} \frac{1}{W_{X_i}}}, \tag{9.37}$$

where W_{X_i} is the *self-reported* degree of individual X_i. Assuming that a single individual is recruited randomly from a list of possible contacts at each wave of the chain, the VH estimator (9.37) has the large-sample property that

$$\sqrt{n}(\widehat{\mu}_f^{VH} - \mu_f) \to \mathcal{N}(0, \sigma_f^2), \quad \text{as } n \to \infty. \tag{9.38}$$

Thus, $\widehat{\mu}_f^{VH}$ is an asymptotically unbiased, importance-sample estimator of μ_f. The variance, σ_f^2, however, is difficult to estimate because it depends upon the variance of f and the dependencies caused by the Markov chain.

If f were defined as the indicator function of whether v_i has the trait, then (9.37) reduces to

$$\widehat{p}_{\text{trait}}^{VH} = \frac{\sum_{i=0}^{n-1} \frac{I_{\{X_i \in D\}}}{W_{X_i}}}{\sum_{i=0}^{n-1} \frac{1}{W_{X_i}}}, \tag{9.39}$$

where D is the set of individuals who have the given trait. The estimator (9.39) is the ratio of two Horvitz–Thompson estimators, of which the numerator is an unbiased estimator of the population total and the denominator is an unbiased estimator of the population size.

Variance Estimation

A few articles have touched upon the estimation of the variance of the RDS estimator. See, for example, Verdery et al. (2015). This is a difficult problem because successive observations obtained through RDS are *not* independent and may even violate the first-order Markov (FOM) assumption used in RDS. If the FOM assumption were violated (and there is evidence that it frequently is for real network data), possibly due to the sample consisting of a mixture of members of different communities, then the accuracy of the estimated variance of RDS estimates may be biased downwards. There is also evidence (Goel and Salganik, 2010) that RDS mean estimates may exhibit very high sampling variance relative to simple random sampling, even when assumptions hold. It has been suggested that RDS sample sizes should be at least twice that of a comparable simple random sample.

Several estimators of sampling variance have been proposed. The most commonly used are the Salganik estimator (Salganik, 2006), which uses a bootstrap procedure, and the Volz–Heckathorn estimator (Volz and Heckathorn, 2008) given by (9.37). These two variance estimation strategies attempt to deal with dependencies generated by successive (or near-successive) referred individuals. Salganik's bootstrap procedure, for example, modifies the usual bootstrap procedure, where bootstrap samples are selected by random sampling with replacement from the original sample, by incorporating information on how the people were recruited, thereby preserving some of the dependencies in the data.

Few studies of these variance estimators have appeared so far. However, in real network data studies (Verdery et al., 2015), both variance estimators were found to be poor estimates of the true variance and exhibited substantial downward bias, and the magnitude of this bias turned out to be inconsistent from network to network. The assessment of Verdery et al. is that current techniques of RDS variance estimation should be viewed as "wildly inaccurate," and that further research on this topic needs to be undertaken. We refer the interested reader to those references for further details.

Biases and other issues. Since the RDS technique was introduced, research has centered on scenarios in which various types of bias would be present. An RDS sample may be biased towards high-degree individuals because it is more likely that an individual with a large number of links to other individual members of the target population will be included in the sample than someone with few links to population members. Bias may also be present if the initial seed were not chosen randomly and if the underlying population had a tendency towards social homophily, in which people tend to be friends with others like themselves.

So far, most estimation procedures for RDS assume that the social network in the population is undirected, which does not reflect reality because relationships in social networks tend to be directed, and those relationships may or may not be reciprocal. If a person P_1 were asked to list his/her friends, he/she may list person P_2. However, when person P_2 is asked to list his/her friends, P_1 may or may not be on P_2's list. Neglecting to take into account directed edges could introduce substantial bias and variance into the RDS estimator.

Other issues encountered in using RDS in real-world surveys include failure in reaching the target sample size, slow recruitment due to loss of coupons by participants or due to negligence from effects of drugs, and lack of funds by participants to obtain transportation to the study center.

9.5.4 Random-Walk Sampling

This sampling method (referred to as RWS) was originally used to study structure in urban networks (Spreen, 1992). The RWS process is carried out as follows. A random walk, $\{X_t\}$, is simulated on the network, starting at a randomly selected node in $\mathcal{V}$, where X_t is a random variable that denotes the node selected at the tth step. A sequence of individuals is selected by linking one individual to another, and so on. Each person in the network who is known by another person in the network has a nonzero chance of being chosen for the sample. Each step involves randomly choosing a member of the current participant's social network. The sample consists of the selected nodes together with the edges, which are visited by a random walker.

For a given node v, let $\mathcal{N}(v)$ be the neighborhood of v, and let $d_v = |\mathcal{N}(v)|$ be the degree of v. Then, RWS may be considered as a finite Markov chain with transition matrix

$$\mathbf{P} = (p_{ij}), \quad v_i, v_j \in \mathcal{V}, \tag{9.40}$$

where

$$p_{ij} = \mathrm{P}\{X_{t+1} = v_j | X_t = v_i\} = \begin{cases} \frac{1}{d_{v_i}}, & \text{if } v_j \in \mathcal{N}(v_i) \\ 0, & \text{otherwise} \end{cases} \tag{9.41}$$

is the probability of selecting node v_j at the $(t+1)$st step if the node at the tth step is v_i. The stationary distribution $\pi = (\pi_j)$ of this Markov chain is given by (Lovász, 1993)

$$\pi_j = \frac{d_{v_j}}{2M}, \quad v_j \in \mathcal{V}, \tag{9.42}$$

where $M = |\mathcal{E}|$ is the number of edges in the network $\mathcal{G}$. The stationary distribution is unique if the network is connected.

Biases and other issues. Random-walk sampling has been found to outperform snow-ball sampling. It is, however, a rather slow sampling procedure and can be expensive to implement when viewed by staff time and investment. RWS has to be carried out with complete knowledge of the network, and it is not unusual for the random walk to get trapped inside a subnetwork (i.e., a local neighborhood) whose characteristics differ from those of the parent network. To avoid this scenario, it is common to ignore the first set of moves (called the *burn-in period*) or to employ multiple independent random walks each starting at a different node of the network, leading to *frontier sampling* (Ribeiro and Towsley, 2010). RWS is biased towards selecting nodes with high degree and has been found not to match the degree distribution of the underlying network.

9.5.5 Random-Walk Sampling with Random Jumps

This is a variant of RWS, which also needs complete network information. It is designed to avoid getting stuck in some local neighborhood of the network. The sampling procedure (which is abbreviated to RWJ) is similar to random-walk sampling but it incorporates a jump to a random node with probability p (Ribeiro and Towsley, 2010). The literature recommends $p = 0.15$.

RWJ starts at a random node $v \in \mathcal{V}$ with neighborhood $\mathcal{N}(v)$. Then, a random walk is carried out as follows: with probability $1 - p$, an outgoing edge $(v, v') \in \mathcal{E}$ is selected at random and the walker moves to the node $v' \in \mathcal{N}(v)$ in the neighborhood of the current node, and with probability p, the walker is teleported to a random node $\sigma \in \mathcal{V}$ and the random walk is restarted. The Markov chain corresponding to RWJ has transition matrix

$$\mathbf{P} = (p_{ij}), \quad v_i, v_j \in \mathcal{V}, \tag{9.43}$$

where

$$p_{ij} = \begin{cases} \frac{p}{N} + \frac{1-p}{d_{v_i}}, & \text{if } d_{v_i} > 0 \\ \frac{1}{N}, & \text{if } d_{v_i} = 0 \end{cases}. \tag{9.44}$$

The process stops when the required number of nodes has been visited. Jumps can be carried out using probability proportional to PageRank or probability proportional to node degrees.

Biases and other issues. If the network were directed, there may exist a node with in-degree zero, so that the node cannot be reached by a sampling process such as RWS; this would then prevent estimation of various network characteristics. However, an advantage of RWJ is that that node can be reached by RWJ if it is chosen for the random jump.

If jumping to random nodes is not acted upon, then nodes will be chosen in the same way as is RWS, and so RWJ will still be biased towards high-degree nodes. Although RWJ was originally proposed for estimating the out-degree distributions of a large directed online social network, empirical evidence shows that estimating the in-degree distribution is much more difficult, requiring knowledge of most directed edges in the network if the edges are hidden. However, it does preserve the clustering coefficient.

9.5.6 Forest-Fire Sampling

Forest-fire sampling (FFS) is a probabilistic version of a breadth-first search (Leskovec and Faloutsos, 2006). A broad neighborhood of a randomly selected seed node is retrieved from a partial breadth-first search, in which only a fraction of neighbors is followed for each node.

The algorithm starts by picking a node at random and adds it to the sample. A random proportion of the node's outgoing edges are then "burned," and those edges, along with the incident nodes, are added to the sample. The number of edges sampled at each step is drawn from a geometric distribution with mean $p/(1 - p)$, where the authors recommend $p = 0.7$. The process is repeated recursively for each burned neighbor until no further node is selected to be burned. Then, a new random node is chosen to continue the process until the desired sample size is obtained.

Biases and other issues. The forest-fire sampling algorithm has been shown to have excellent overall performance. However, it fails to match path length and clustering coefficient of the parent network, and it underestimates the power-law exponent due to its bias towards high-degree nodes.

9.6 Further Reading

Network sampling is discussed in the books by Kolaczyk (2009, Chapter 5), Newman (2010, Section 3.7), Thompson (2012, Chapter 15), Chaudhuri (2014), and Kolaczyk (2017, Chapter 3).

Section 9.1.1. For information on the Harvard College influenza study, see Christakis and Fowler (2010).

Section 9.5. An excellent review of sampling of drug-abuse populations is given by Taylor and Griffiths (2005). See also Semaan, Lauby, and Liebman (2002), and Spreen and Coumans (2003). A discussion of the difficulties in sampling lesbian, gay, and bisexual populations is given by Meyer and Wilson (2009).

Section 9.5.3. There are a number of excellent books on Markov chain Monte Carlo. They include Neal (1993), Berg (2004), Robert and Casella (2004), and Gamerman and Lopes (2006). We highly recommend the edited volumes by Brooks et al. (2011) and Gilks, Richardson, and Spiegelhalter (1996). Books that discuss MCMC include Gelman et al. (1995, Chapter 11), Carlin and Louis (2000, Chapter 5), and MacKay (2003, Chapter 29). Survey articles on MCMC include Besag et al. (1995), Cowels and Carlin (1996), Andrieu et al. (2003), and Diaconis (2009).

9.7 Exercises

Exercise 9.1 Draw a random sample of size $n = 200$ from a Gaussian distribution with mean 100 and standard deviation 0.7. Draw another random sample (independent of the first) of size 100 from a Gaussian distribution with mean 80 and standard deviation 0.3. Display the empirical cumulative distribution functions of each sample on the same graph. Does it look like they are different? Compute the two-sample Kolmogorov–Smirnov statistic. Based upon your result or otherwise, which do you think has a more important influence on the value of this statistic: the difference in sample sizes, means, or standard deviations?

Exercise 9.2 For the Horvitz–Thompson estimator, let π_{ij} be the probability that the ith and jth items in the population are both in the sample. Show that (a) $\sum_{j \neq i}^{N} \pi_{ij} = (n-1)\pi_i$ and (b) $\sum_{i=1}^{N} \sum_{j>i} \pi_{ij} = \frac{1}{2} n(n-1)$.

Exercise 9.3 Show that the variance of the Horvitz–Thompson estimator $\widehat{\mu}_{HT}$ in (9.20) is given by (9.22).

Exercise 9.4 Show that the estimator $\widehat{\text{var}}\{\widehat{\mu}_{HT}\}$ as given by (9.23) is unbiased for $\text{var}\{\widehat{\mu}_{HT}\}$.

Exercise 9.5 Show that the weighted sample mean $\widehat{\mu}_f$ given in (9.31) is unbiased for the population mean μ_f in (9.29).

Exercise 9.6 In the discussion of the PageRank statistic (see Section 9.3.3), we have to invert the matrix $\mathbf{I}_N - \alpha \mathbf{P}$. Show that $\mathbf{I}_N - \alpha \mathbf{P}$ is nonsingular and that its inverse is non-negative.

Exercise 9.7 Let S denote the set of sampled nodes and let $N(S)$ be the set of neighbors of S. Let the expansion factor $X(S)$ be defined by (9.45). Show that the impact on $X(S)$ depends upon whether $|N(\{v\}) - (N(S)) \cup S|$ is larger (or smaller) than $X(S)$.

CHAPTER TEN

Parametric Network Models

In this chapter, we introduce a number of parametric statistical models that have been used to model network data. The social network literature has named them p_1, p_2, and p^* models, the last of which has also been referred to as an ERGM (exponential random graph model). These models originated in the 1980s and 1990s as new statistical approaches were developed for analyzing social network data.

10.1 Introduction

The main problem faced by researchers was how to define a statistical approach (e.g., inference using parametric probability models) that would show whether or not a social network consisting of personal relationships differed significantly from chance. Little research on this problem had appeared in the social network literature before 1980, although much had been written on analyzing social network data.

A breakthrough occurred when Holland and Leinhardt (1981) changed the focus from independent edges to independent *dyads*.

Definition 10.1 A *dyad* is the pair $D_{ij} = (Y_{ij}, Y_{ji})$, where $Y_{ij} = 1$ if there is an edge between nodes v_i and v_j, and 0 otherwise, $i, j = 1, 2, \ldots, N$. The possible values for D_{ij} are $(1, 1), (1, 0), (0, 1)$, or $(0, 0)$.

Probability distributions, such as p_1 and p_2, are defined over independent dyads.

In the p_2 and p^* models, the network is used as the dependent variable and covariates appear as explanatory variables. The challenge is how to use the covariates to explain the network structure.

Recall that $\mathbf{Y} = (Y_{ij})$ is the random adjacency matrix corresponding to the graph. We assume $Y_{ii} = 0$ (i.e., there are no self-loops). The Y_{ij} can also represent edge weights if the graph is weighted, and for undirected graphs, $\mathbf{Y}$ is a symmetric matrix with $Y_{ij} = Y_{ji}$.

10.2 Exponential Families

There have been many articles in the network literature that assume parametric models for network data. The variables of a random network are often assumed (or known) to be jointly distributed as a member of an exponential family. Indeed, exponential families cover a wide variety of continuous univariate and multivariate distributions, including the multivariate Gaussian and Wishart distributions, as well as every discrete probability distribution. Because we will refer to exponential families in this chapter and in other parts of the book, we provide here a brief introduction to exponential families. See Brown (1986) for further details.

We have the following definition of an exponential family:

Definition 10.2 An *exponential family* is a family of parametric distributions

$$\mathcal{P} = \{P_\theta, \theta \in \mathbb{R}^k\} \tag{10.1}$$

whose density functions can be represented in *canonical form* as

$$p_\theta(\mathbf{x}) = e^{\theta^\tau \mathbf{t}(\mathbf{x}) - A(\theta)} h(\mathbf{x}), \tag{10.2}$$

where $\mathbf{x} = (x_1, \dots, x_r)^\tau$ is a point in a sample space that could be a discrete set, or $\mathbb{R}^r$, or a product of them. The functions $h(\mathbf{x})$, $\mathbf{t}(\mathbf{x}) = (t_1(\mathbf{x}), \dots, t_k(\mathbf{x}))^\tau$, and $A(\theta)$ are known functions, where $\theta = (\theta_1, \dots, \theta_k)^\tau$ is a parameter vector and k is the *dimension* of the exponential family.

Definition 10.3 The function $A(\theta)$ is called the *normalization function* or the *log-partition function*. The derivatives of $A(\theta)$ are the cumulants of $\mathbf{t}(\mathbf{x})$; for example,

$$A'(\theta) = \mathrm{E}_\theta[\mathbf{t}(\mathbf{X})], \quad A''(\theta) = \mathrm{cov}_\theta[\mathbf{t}(\mathbf{X})]. \tag{10.3}$$

Hence, $A(\theta)$ is also known as the *cumulant generating function*. Because any covariance matrix is non-negative definite, it follows that $A(\theta)$ is a convex function of θ.

For identifiability purposes, we assume that any two probability distributions P_θ, $P_{\theta'} \in \mathcal{P}$ are equal iff $\theta = \theta'$. Note that choices of θ and $\mathbf{t}(\mathbf{x})$ are not unique: we can, for example, multiply each θ_i in (10.2) by a constant and then divide the corresponding $t_i(\mathbf{x})$ by the same constant without changing their product $\theta^\tau \mathbf{t}(\mathbf{x})$.

Definition 10.4 An exponential family is said to have *order k* if $p_\theta(\mathbf{x})$ cannot be represented by a natural parameter vector θ with fewer than k components.

Definition 10.5 The *natural parameter space* is given by

$$\Theta = \left\{ \theta \in \mathbb{R}^k : \int e^{\theta^\tau \mathbf{t}(\mathbf{x}) - A(\theta)} h(\mathbf{x}) d\mathbf{x} < \infty \right\}, \tag{10.4}$$

and the vector θ is referred to as the *natural parameter vector*.

Because the density $p_\theta(\mathbf{x})$ integrates (or sums) to one,

$$e^{A(\theta)} = \int e^{\theta^\tau t(\mathbf{x})} h(\mathbf{x}) d\mathbf{x} < \infty \tag{10.5}$$

is the normalization constant. In the discrete case, the integral in Θ is replaced by a sum.

Definition 10.6 If, for every $\theta \in \Theta$, there exists a probability distribution $P_\theta \in \mathcal{P}$ with density of the form (10.2), then $\mathcal{P}$ is said to be a *full exponential family*.

Definition 10.7 If Θ is also an open subset of $\mathbb{R}^k$, then $\mathcal{P}$ is said to be a *regular exponential family*, in which case, Θ is also a convex subset of $\mathbb{R}^k$.

Definition 10.8 If $\mathbf{t}(\mathbf{x})$ and $\boldsymbol{\theta}$ do not satisfy linear constraints, the family $\mathcal{P}$ is said to be *minimally represented* (or just *minimal*).

Definition 10.9 If $\mathcal{P}$ is minimal and if Θ contains a k-dimensional rectangle, then the family $\mathcal{P}$ is said to have *full rank* (i.e., rank = k), and the function $\mathbf{t}(\mathbf{x})$ is a k-dimensional *minimal sufficient statistic* for $\mathcal{P}$. If $\mathcal{P}$ has full rank, then the exponential family is said to be *linear*.

A single-parameter exponential family has been characterized by Efron (1975) as a "straight line through the space of possible probability distributions" having no curvature. Efron's results have since been extended to the multiparameter case (Reeds, 1975; Madsen, 1979).

These definitions assume implicitly that the number of θs is equal to k; however, in certain situations, this may not hold: the number of θs (and, hence, the dimension of Θ) may be strictly smaller than k. This idea leads to the following definition:

Definition 10.10 Let $\mathcal{P} = \{P_{\boldsymbol{\theta}}, \boldsymbol{\theta} \in \Theta\}$ be a regular exponential family of order k. If $\Theta_0 \subseteq \mathbb{R}^q$ is a (smooth) q-dimensional submanifold of $\Theta \subseteq \mathbb{R}^k$, $q < k$, then the subfamily $\mathcal{P}_0 = \{P_{\boldsymbol{\theta}}, \boldsymbol{\theta} \in \Theta_0\}$ of $\mathcal{P}$ is said to be a *curved exponential family* of order q.

Thus, for curved exponential families, the minimal sufficient statistic $\mathbf{t}(\mathbf{x})$ has higher dimensionality than that of the natural parameter space Θ. In other words, we have k sufficient statistics but only $q < k$ parameters of interest. Note that for curved exponential families, the natural parameter space Θ_0 is not a convex subset of $\mathbb{R}^q$.

10.3 The p_1 Model

One of the first statistical models for a directed binary network was the so-called p_1 model (Holland and Leinhardt, 1981).

10.3.1 Dyads

Consider the $(N \times N)$-matrix $\mathbf{D} = (D_{ij})$, where $D_{ij} = (Y_{ij}, Y_{ji})$, $i < j$, is called a *dyad* (or *pair*). Define the following probabilities:

$$m_{ij} = \mathrm{P}\{D_{ij} = (1,1)\} = \mathrm{P}\{Y_{ij} = 1, Y_{ji} = 1\}, \quad i < j, \tag{10.6}$$

$$a_{ij} = \mathrm{P}\{D_{ij} = (1,0)\} = \mathrm{P}\{Y_{ij} = 1, Y_{ji} = 0\}, \quad i \neq j, \tag{10.7}$$

$$a_{ji} = \mathrm{P}\{D_{ij} = (0,1)\} = \mathrm{P}\{Y_{ij} = 0, Y_{ji} = 1\}, \quad i \neq j, \tag{10.8}$$

$$n_{ij} = \mathrm{P}\{D_{ij} = (0,0)\} = \mathrm{P}\{Y_{ij} = 0, Y_{ji} = 0\}, \quad i < j, \tag{10.9}$$

where $m_{ij} + a_{ij} + a_{ji} + n_{ij} = 1$ for all $i < j$. These probabilities represent states of the dyad D_{ij}: m_{ij} is the probability of a mutual or reciprocated pair, a_{ij} and a_{ji} are the probabilities of an asymmetric or nonreciprocated pair, and n_{ij} is the probability of a null pair.

Background. The word "dyad" has been used since the late nineteenth century in the sociological literature to study social relations, such as the strength, direction, and duration of such relations. In the political science and international relations literature, it was used to assess which pairs of countries were dangerous, or politically relevant, or jointly democratic, and it was used to study the ebb and flow of international politics and economics, and the general principles behind the dynamics of conflicts. It has also been used in work on investments, trade, and in civil war studies, and since the 1950s as a way of formalizing relations amongst tensors in the physics literature. See Dorff and Ward (2012) for details and references.

10.3.2 Probability of a Directed Graph

The primary assumption of p_1 models is that the dyads $\{D_{ij}\}$ are statistically independent. This assumption enables us to estimate the model parameters and it allows efficient ways of performing simulations. The probability of a given directed graph $\mathbf{y}$ is

$$
\begin{aligned}
P\{\mathbf{Y} = \mathbf{y}\} &= \prod_{i,j} P\{Y_{ij} = y_{ij}, Y_{ji} = y_{ji}\} \\
&= \prod_{i<j} m_{ij}^{y_{ij}y_{ji}} \cdot \prod_{i\neq j} a_{ij}^{y_{ij}(1-y_{ji})} \cdot \prod_{i<j} n_{ij}^{(1-y_{ij})(1-y_{ji})} \\
&= \frac{1}{Z} \exp\left\{ \sum_{i<j} \rho_{ij} y_{ij} y_{ji} + \sum_{i\neq j} \theta_{ij} y_{ij} \right\},
\end{aligned}
\tag{10.10}
$$

where

$$
\rho_{ij} = \log_e\left\{ \frac{m_{ij}n_{ij}}{a_{ij}a_{ji}} \right\}, \quad \theta_{ij} = \log_e\left\{ \frac{a_{ij}}{n_{ij}} \right\},
\tag{10.11}
$$

and Z is the normalizing constant. In the social networks literature, the parameter ρ_{ij} has been called the *reciprocation effect*. Although the independence condition for the dyads appears to be restrictive, it has been shown to be satisfied by many types of social networks.

Both ρ_{ij} and θ_{ij} can be viewed as log-odds ratios. Thus, ρ_{ij} is the log-odds ratio in the 2×2 contingency table associated with the dyad (Y_{ij}, Y_{ji}) and measures the relative increase of odds of $Y_{ij} = 1$ given $Y_{ji} = 1$ (i.e., reciprocity); similarly, θ_{ij} is the log-odds of Y_{ij} given $Y_{ji} = 0$, measuring the relative increase of odds of $Y_{ij} = 1$ given $Y_{ji} = 0$ (i.e., the likelihood of an asymmetric dyad).

The distribution (10.10) is a member of the exponential family: the parameters are $\{\rho_{ij}\}$ and $\{\theta_{ij}\}$, the statistics are $\{y_{ij}\}$ and $\{y_{ij}y_{ji}\}$, and the normalizing constant is $Z = \sum_{i<j} \log n_{ij}$.

10.3.3 Special Cases

The problem with the probability (10.10) is that it contains too many parameters to be statistically useful. For example, standard asymptotic arguments fail as the number of parameters increases as the number of nodes, N, increases. One way to overcome this problem is to impose any of the following conditions:

- $\rho_{ij} = \rho$, *homogeneity or constant reciprocation effect.*
- $\rho_{ij} = 0$, *no reciprocation effect.*
- $\rho_{ij} = \rho + \rho_i + \rho_j$, *edge-dependent reciprocation effect.*

The first two conditions were introduced by Holland and Leinhardt (1981), whilst the third was studied by Fienberg and Wasserman (1981a,b). In their original development of the p_1 model, Holland and Leinhardt proposed to set $\rho_{ij} = \rho$ and

$$\theta_{ij} = \theta + \alpha_i + \beta_j, \ i \neq j. \tag{10.12}$$

For identifiability purposes, we impose the side conditions $\alpha_+ = \beta_+ = 0$, where $\alpha_+ = \sum_i \alpha_i$ and $\beta_+ = \sum_j \beta_j$.

10.3.4 Model Parameters

The parameters $\{\rho_{ij}\}, \theta, \{\alpha_i\}$, and $\{\beta_j\}$ are interpreted in the literature on social networks as follows:

- ρ_{ij}: *overall mutuality* or *reciprocity* parameter (the tendency towards a reciprocal relationship).
- θ: *global density* parameter (the overall tendency of the network to have edges).
- α_i: *individual expansiveness, gregariousness,* or *productivity* parameter (the tendency to choose others).
- β_j: *attractiveness* or *individual popularity* parameter (the tendency to be chosen by others).

Define the N-vectors

$$\boldsymbol{\alpha} = (\alpha_1, \ldots, \alpha_N)^\tau, \ \boldsymbol{\beta} = (\beta_1, \ldots, \beta_N)^\tau. \tag{10.13}$$

To simplify this model, certain of these parameters may be set equal to zero.

Set $\rho_{ij} = \rho$. Then the probability of a dyad (Y_{ij}, Y_{ji}) is given by

$$P\{Y_{ij} = y_{ij}, Y_{ji} = y_{ji}\}$$
$$= \frac{1}{Z_{ij}} \exp\{\rho y_{ij} y_{ji} + y_{ij}(\theta + \alpha_i + \beta_j) + y_{ji}(\theta + \alpha_j + \beta_i)\}, \tag{10.14}$$

where $y_{ij}, y_{ji} \in \{0,1\}$ and

$$Z_{ij} = 1 + e^{\theta + \alpha_i + \beta_j} + e^{\theta + \alpha_j + \beta_i} + e^{\rho + 2\theta + \alpha_i + \alpha_j + \beta_i + \beta_j} \tag{10.15}$$

is the normalizing constant. If we set $(y_{ij}, y_{ji}) = (1,1), (1,0), (0,1), (0,0)$, we can express the dyad probabilities, m_{ij}, a_{ij}, a_{ji}, and n_{ij}, in terms of the parameters:

$$m_{ij} = \frac{1}{Z_{ij}} \exp\{\rho + 2\theta + \alpha_i + \alpha_j + \beta_i + \beta_j\}, \tag{10.16}$$

Table 10.1 Numerators of the probabilities of dyadic outcomes $D_{ij} = (Y_{ij}, Y_{ji})$ for the p_1 model. To convert the entries into probabilities, divide each entry by Z_{ij}

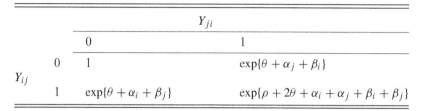

		Y_{ji}	
		0	1
Y_{ij}	0	1	$\exp\{\theta + \alpha_j + \beta_i\}$
	1	$\exp\{\theta + \alpha_i + \beta_j\}$	$\exp\{\rho + 2\theta + \alpha_i + \alpha_j + \beta_i + \beta_j\}$

$$a_{ij} = \frac{1}{Z_{ij}} \exp\{\theta + \alpha_i + \beta_j\}, \tag{10.17}$$

$$a_{ji} = \frac{1}{Z_{ij}} \exp\{\theta + \alpha_j + \beta_i\}, \tag{10.18}$$

$$n_{ij} = \frac{1}{Z_{ij}}. \tag{10.19}$$

See **Table 10.1**. These probabilities are often rewritten to incorporate the normalizing constant Z_{ij}. Set $\lambda_{ij} = -\log(Z_{ij})$. Then the above probabilities are given by

$$\log(m_{ij}) = \lambda_{ij} + \rho + 2\theta + \alpha_i + \alpha_j + \beta_i + \beta_j, \tag{10.20}$$

$$\log(a_{ij}) = \lambda_{ij} + \theta + \alpha_i + \beta_j, \tag{10.21}$$

$$\log(a_{ji}) = \lambda_{ij} + \theta + \alpha_j + \beta_i, \tag{10.22}$$

$$\log(n_{ij}) = \lambda_{ij}. \tag{10.23}$$

The *model parameters* are $\rho, \theta, \{\alpha_i\}, \{\beta_j\}$, and $\{\lambda_{ij}\}$. Of especial interest is the case in which $\alpha_i = \beta_j = 0$, for each i and j, and $\rho = 0$. This special case is essentially the Erdős–Rényi random graph model for directed graphs in which every directed edge has an equal chance of occurring.

Under the conditions $\rho_{ij} = \rho$ and $\theta_{ij} = \theta + \alpha_i + \beta_j$, the joint likelihood (10.14) reduces to

$$p_1(\mathbf{y}) = P\{\mathbf{Y} = \mathbf{y}\} = \frac{1}{Z} \exp\left\{\rho m + \theta y_{++} + \sum_i \alpha_i y_{i+} + \sum_j \beta_j y_{+j}\right\}, \tag{10.24}$$

where $Z = Z(\rho, \theta, \boldsymbol{\alpha}, \boldsymbol{\beta})$ is the normalizing constant.

10.3.5 Sufficient Statistics

The sufficient statistics are:

- $y_{++} = \sum_{i,j} y_{ij}$, the total number of edges in the graph.
- $y_{i+} = \sum_j y_{ij}$, the number of edges sent from the ith node (the "out-degree" of the ith node).
- $y_{+j} = \sum_i y_{ij}$, the number of edges received by the jth node (the "in-degree" of the jth node).
- $m = \sum_{i<j} y_{ij} y_{ji}$, the number of mutual (or reciprocated) edges.

The minimal sufficient statistics are $m, y_{i+}, y_{+j}, i, j = 1, 2, \ldots, N$.

10.3.6 Undirected Graphs

The probability density (10.24) was obtained for a directed graph. To obtain an analogous result for an undirected graph, we note that the adjacency matrix $\mathbf{Y}$ will be symmetric (i.e., $y_{ij} = y_{ji}$), which simplifies certain terms of (10.24). For example, $m = \sum_{i<j} y_{ij} y_{ji} = \sum_{i<j} y_{ij}^2 = \sum_{i<j} y_{ij} = y_{++}$ and $y_{i+} = y_{+i}$. The resulting probability density function is given by

$$p_1(\mathbf{y}) = \mathrm{P}\{\mathbf{Y} = \mathbf{y}\} = \frac{1}{Z} \exp \left\{ \rho m + \sum_i \alpha_i y_{i+} \right\}, \qquad (10.25)$$

where we replaced $\rho + \theta$ by ρ and $\alpha_i + \beta_i$ by α_i. The obvious side-condition to impose is $\alpha_+ = 0$.

10.3.7 ML Estimation of Model Parameters

The goal is to estimate the parameters of the p_1 distribution. The most popular estimation method for this problem is maximum likelihood (ML). From (10.10), the likelihood function is given by

$$l(\mathbf{p}) = \prod_{i<j} m_{ij}^{y_{ij} y_{ji}} a_{ij}^{y_{ij}(1-y_{ji})} a_{ji}^{(1-y_{ij})y_{ji}} n_{ij}^{(1-y_{ij})(1-y_{ji})}, \qquad (10.26)$$

where $\mathbf{p} = (\{m_{ij}\}, \{a_{ij}\}, \{a_{ji}\}, \{n_{ij}\})^\tau$, and the ML estimator of $\mathbf{p}$ is

$$\widehat{\mathbf{p}} = \arg \max_{\mathbf{p}} l(\mathbf{p}). \qquad (10.27)$$

From $\widehat{\mathbf{p}}$, we can estimate the parameter vector

$$\boldsymbol{\omega} = (\rho, \theta, \boldsymbol{\alpha}^\tau, \boldsymbol{\beta}^\tau)^\tau, \qquad (10.28)$$

where we impose the side condition that $\alpha_+ = \beta_+ = 0$. The normalization constants, $\{\lambda_{ij}, i < j\}$, do not need to be estimated, although they can be obtained from $\boldsymbol{\omega}$. Although a number of estimation methods have been proposed, few of them have provided any formal supporting theoretical results.

10.3.8 Computational Algorithms for ML Estimation

Generalized Iterated Scaling Algorithm
We can write the dyad probability (10.14) in the following way. Let

$$\mathbf{d}_{ij} = \begin{pmatrix} y_{ij}(1 - y_{ji}) \\ (1 - y_{ij})y_{ji} \\ y_{ij} y_{ji} \end{pmatrix} = \begin{pmatrix} d_{1ij} \\ d_{2ij} \\ d_{3ij} \end{pmatrix} \qquad (10.29)$$

and

$$\boldsymbol{\xi}_{ij} = \begin{pmatrix} \alpha_i + \beta_j + \theta \\ \alpha_j + \beta_i + \theta \\ \xi_{1ij} + \xi_{2ij} + \rho \end{pmatrix} = \begin{pmatrix} \xi_{1ij} \\ \xi_{2ij} \\ \xi_{3ij} \end{pmatrix}. \qquad (10.30)$$

Then

$$\xi_{ij}^{\tau} \mathbf{d}_{ij} = y_{ij}(\alpha_i + \beta_j + \theta) + y_{ji}(\alpha_j + \beta_i + \theta) + y_{ij} y_{ji} \rho. \tag{10.31}$$

So, the numerator in (10.14) can be expressed as $\exp\{\xi_{ij}^{\tau} \mathbf{d}_{ij}\}$. We can, therefore, write

$$P\{\mathbf{Y} = \mathbf{y}\} = \exp\{\xi^{\tau} \mathbf{d} - A(\xi)\}, \tag{10.32}$$

where

$$A(\xi) = \sum_{i<j} \log\{1 + e^{\xi_{1ij}} + e^{\xi_{2ij}} + e^{\xi_{2ij}}\} \tag{10.33}$$

is the log-normalization factor. Because the likelihood function (10.32) for the p_1 model is an exponential family, we can find ML estimates of these parameters by equating the minimal sufficient statistics to their expected values. This approach by Holland and Leinhardt (1981) led to a *generalized iterated scaling (GIS) algorithm*. Although the GIS algorithm is known to converge, there is no information regarding the speed of that convergence. Indeed, there is empirical evidence that convergence in some cases can be very slow.

Iterative Proportional Fitting Algorithm (IPF)

Following the Holland and Leinhard article, several attempts were made to improve the performance of the p_1 model by proposing extensions to the GIS algorithm. Because all effects in the p_1 model are treated as fixed effects, the p_1 model can be represented as a four-way contingency table and is a member of the class of log-linear models that arise from a product-multinomial sampling scheme (Fienberg and Wasserman, 1981a). To be more specific, let

$$W_{ijkl} = \begin{cases} 1, & \text{if } D_{ij} = (Y_{ij}, Y_{ji}) = (k, l) \\ 0, & \text{otherwise} \end{cases}. \tag{10.34}$$

where $k, l \in \{0, 1\}$. Thus, we have transformed each dyad into a 2×2 table (i.e., Y_{ij} vs. Y_{ji}), in which one cell equals 1 (i.e., when $Y_{ij} = k$ and $Y_{ji} = l$) and the other three cells are 0. If we bring together the 2×2 tables from all $\binom{N}{2}$ dyads, this yields an $N \times N \times 2 \times 2$ cross-classified four-way array $\mathbf{W} = (W_{ijkl})$ with "structural" zeroes (i.e., no self-loops) along the diagonal of the $N \times N$ subtable. Let $\mathbf{w}$ denote a realization of $\mathbf{W}$. It was shown that fitting the p_1 model to the $\mathbf{y}$ data with $\rho_{ij} = \rho$ is equivalent to fitting the log-linear model [12][13][14][23][24][34] to the $\mathbf{w}$ data, where the model contains all main effects and 2-factor interactions, but no 3-factor or 4-factor interactions. If we set $\rho_{ij} = \rho + \rho_i + \rho_j$, where the $\{\rho_{ij}\}$ are normalized to sum to zero, the p_1 model corresponds to fitting the log-linear model [12][134][234]. Thus, we can apply the *iterative proportional fitting (IPF) algorithm* to $\mathbf{w}$ to obtain ML estimates of the parameters in (10.28).

Iterative Generalized Least-Squares Algorithm

A different algorithm was proposed by van Duijn, Snijders, and Zijlstra (2004) in which the dyads are modeled as a generalized linear model. The probability of a dyad (Y_{ij}, Y_{ji}) having the value (y_{ij}, y_{ji}) in the p_1 model can be expressed as the product of two probabilities: the unconditional probability of Y_{ij}, and the conditional probability of Y_{ji} given Y_{ij}. Thus,

$$
\begin{aligned}
P\{Y_{ij} &= y_{ij}, Y_{ji} = y_{ji}\} \\
&= P\{Y_{ij} = y_{ij}\}P\{Y_{ji} = y_{ji}|Y_{ij} = y_{ij}\} \\
&= \frac{1}{Z_{1ij}}[\exp\{y_{ij}\theta_{ij}\} + \exp\{y_{ij}(\rho + \theta_{ij}) + \theta_{ji}\}] \times \frac{1}{Z_{2ji}}\exp\{y_{ji}(\rho y_{ij} + \theta_{ji})\},
\end{aligned}
\tag{10.35}
$$

where

$$
Z_{1ij} = Z_{ij}, \quad Z_{2ji} = 1 + e^{\theta_{ji} + \rho y_{ij}}
\tag{10.36}
$$

are normalization constants, and $\rho_{ij} = \rho$ and $\theta_{ij} = \theta + \alpha_i + \beta_j$. Now, let the dyad pair be modeled as follows:

$$
Y_{ij} = E\{Y_{ij}\} + \epsilon_{1ij},
\tag{10.37}
$$

$$
Y_{ji} = E\{Y_{ji}|Y_{ij}\} + \epsilon_{2ji}.
\tag{10.38}
$$

Taking expectations of (10.37) and (10.38), we have that $E\{\epsilon_{1ij}\} = E\{\epsilon_{2ji}\} = 0$. Also, $\operatorname{var}\{\epsilon_{1ij}\} = \operatorname{var}\{Y_{ij}\}$ and $\operatorname{var}\{\epsilon_{2ji}\} = \operatorname{var}\{Y_{ji}|Y_{ij}\}$. Setting $y_{ij} = 1$ in the first term of (10.35) yields

$$
E\{Y_{ij}\} = P\{Y_{ij} = 1\} = \frac{1}{Z_{1ij}}[\exp\{\theta_{ij}\} + \exp\{\theta_{ij} + \theta_{ji} + \rho\}],
\tag{10.39}
$$

and setting $y_{ji} = 1$ in the second term of (10.35):

$$
E\{Y_{ji}|Y_{ij} = y_{ij}\} = P\{Y_{ji} = 1|Y_{ij} = y_{ij}\} = \frac{1}{Z_{2ji}}\exp\{\theta_{ji} + y_{ij}\rho\}.
\tag{10.40}
$$

So, $E\{Y_{ij}\}$ is a function of $\boldsymbol{\mu}_{1ij} = (\theta, \rho, \alpha_i, \alpha_j, \beta_i, \beta_j)^\tau$, and $E\{Y_{ji}|Y_{ij} = y_{ij}\}$ is a function of $\boldsymbol{\mu}_{2ji} = (\theta, \rho, \alpha_j, \beta_i)^\tau$. Denote $E\{Y_{ij}\}$ by $F_1(\boldsymbol{\mu}_{1ij})$, and $E\{Y_{ji}|Y_{ij} = y_{ij}\}$ by $F_2(\boldsymbol{\mu}_{2ji}, y_{ij})$. Dyad components are either 0 or 1, and so their variances are

$$
\operatorname{var}\{Y_{ij}\} = F_1(\boldsymbol{\mu}_{1ij})(1 - F_1(\boldsymbol{\mu}_{1ij})),
\tag{10.41}
$$

$$
\operatorname{var}\{Y_{ji}|Y_{ij} = y_{ij}\} = F_2(\boldsymbol{\mu}_{2ji}, y_{ij})(1 - F_2(\boldsymbol{\mu}_{2ji}, y_{ij})).
\tag{10.42}
$$

Thus, ML estimates of $\boldsymbol{\mu} = (\theta, \rho, \boldsymbol{\alpha}^\tau, \boldsymbol{\beta}^\tau)^\tau$ can be obtained using the *iterative generalized least squares (IGLS) algorithm* (see, e.g., McCullagh and Nelder, 1989, Section 2.5), where linearization of the nonlinear link functions $F_1(\boldsymbol{\mu}_{1ij})$ and $F_2(\boldsymbol{\mu}_{2ji}, y_{ij})$ using the first-order term in a Taylor-series expansion requires an additional step in the optimization process. Further details can be found in van Duijn, Snijders, and Zijlstra (2004).

10.3.9 Inferential Difficulties

Much work has been carried out with the goal of obtaining conditions for the existence of an ML estimator of the model parameters $\boldsymbol{\omega}$ and goodness-of-fit testing of p_1 models (see Haberman, 1981). The algebraic and geometric properties of the p_1 model have been studied in a series of articles by Petrović, Rinaldo, and Fienberg (2010), Rinaldo, Petrović, and Fienberg (2010), and Fienberg, Petrović, and

Rinaldo (2011). Some of these results are obtained based upon the toric parameterization of the p_1 model, and involve algebraic methods such as toric ideals, polytopes, facial and co-facial sets, and Markov bases. Even for small N (e.g., $N = 4$), they show that, for many of the different graphs, the ML estimator does not exist.

In particular, the above references noted the following formidable statistical problems that have contributed to a poor understanding of the p_1 model:

1. In practice, all we have is a single observed network, and a dyad D_{ij} can be observed in only one of its four possible states. So, a dyad is a random vector in $\mathbb{R}^4$ having the multinomial distribution, $D_{ij} \sim \mathcal{M}(1, \mathbf{p}_{ij})$, with size 1 and class probability $\mathbf{p}_{ij}$, where

$$\mathbf{p}_{ij} = (m_{ij}, a_{ij}, a_{ji}, n_{ij})^\tau \qquad (10.43)$$

is an unknown vector. Also, the multinomial vectors for the dyads are mutually independent. Thus, the sparse statistical information available for the network severely limits the estimation of the parameters of the p_1 distribution.

2. Estimating the parameters of p_1, especially ρ, can be an extremely challenging problem. A possible approach is through computing conditional ML estimates. But even that turns out to be very complicated because large summations need to be computed over sets which are hard to enumerate.

3. The number of network parameters increases as the number of nodes increases. This is very different from standard statistical models where the number of parameters remains fixed and independent of sample size. As a result, the usual asymptotic distribution theory for ML estimates and chi-squared goodness-of-fit statistics is not applicable.

The articles by Petrović, Rinaldo, and Fienberg do not discuss whether or not the ML estimates obtained via the Holland–Leinhardt and van Duijn–Snijders–Zijlstra algorithms satisfy the conditions for the existence of ML estimators.

10.4 The p_2 Model

The p_2 model (Van Duijn, 1995; Lazega and van Duijn, 1997; van Duijn, Snijders, and Zijlstra, 2004; Zijlstra, van Duijn, and Snijders, 2009), which builds upon the p_1 model, also assumes dyad independence. The network is modeled as the dependent variable and covariates are incorporated into the model in such a way that the network structure is kept intact. Recall that covariates were not included in the p_1 model. This formulation leads to a version of a logistic regression model adapted to network data.

In the p_2 model, the varying parameters are characterized as random effects in a multilevel model. *Random effects* are unobservable random variables that are drawn independently from the same underlying distribution and are used to explain the variability of the observations. In contrast, the p_1 model considers the $\{\alpha_i\}$ and $\{\beta_j\}$ parameters as *fixed effects* because they are specifically chosen for analysis and are constant across observations (and, hence, do not get modeled). *Multilevel* (or *hierarchical*) *models* (see, e.g., Gelman and Hill, 2007) are usually viewed as random-effects models in which either the observations are independently drawn from a sequence of nested levels or the parameters of the model are controlled by the hyperparameters of a higher-level model.

10.4.1 Example: Friendship Network of Lawyers

The p_2 model was first used to study lawyers working for Spencer, Grace, & Robbins, a New England corporate law firm, during 1988–1991 (Lazega and van Duijn, 1997; van Duijn, Snijders, and Zijlstra, 2004). The 71 lawyers in the firm consisted of 36 partners and 35 salaried associates. The firm is a relatively decentralized organization, traditional rather than bureaucratic, that was formed following a merger.

All lawyers in the firm were interviewed and each was handed a list of all the members of the firm. Three networks were studied: advice ("To whom do you go for basic professional advice?"), collaboration ("Check the names of those with whom you have worked, on at least one case"), and friendship ("Check the names of those you socialize with outside work") networks. For example, the friendship network of the 35 associates contained 595 dyadic relationships, each of which is interpreted as a binary directed relational variable, and there are 62 mutual and 58 asymmetric dyads present within the 1190 potential relations.[1]

Seven covariates are used to study the three networks: (1) status (partner or associate); (2) office location (the firm has three offices, in Boston, Hartford, and Providence); (3) practice specialty (half the firm are litigators, while everyone else is considered to be a "corporate" lawyer); (4) seniority (three levels for partners and five levels for associates, depending upon when the lawyer joined the firm); (5) gender; (6) age (26–53 years); and (7) law school attended (Harvard or Yale University, University of Connecticut, and other). These covariates were chosen to formalize the structure of the law firm.

10.4.2 Example: Bullying in Schools

The p_2 model was also used to study the practice of bullying in schools in the Netherlands, with the goals of identifying the bullies and their victims, and evaluating intervention programs to prevent bullying (Veenstra et al., 2007; Zijlstra, Veenstra, and van Duijn, 2008). Longitudinal survey data were available on pre-adolescents (mean age of 11 years old), in which the development of mental health and social development was followed from pre-adolescence into adulthood.[2]

The data structure was hierarchical with three levels: 54 classes (Level 3), consisting of 918 students (Level 2), with 7–33 students per class, were asked to report about their relationships with classmates. The students were given a list of all classmates and asked to nominate them on bullying and victimization. The number of nominations they could make was unlimited, and the questions were asked at the dyadic level. This yielded bidirectional information of each student and resulted in 13,606 dyadic relations (Level 1). The dependent variable is the network on self-reported bullying (i.e., "Whom do you bully?" and "By whom are you bullied?"). Covariates included

---

[1]The data are available in two files: `Lazega.lawfirmdata` (3 (71 × 71)-matrices, one for each type of network studied) and `Lazega.attinfo` (a (71 × 7)-matrix of covariate information on each of the 71 lawyers in the firm). The variables are coded in the UCINET DL format.

[2]The bullying study is part of the TRAILS (TRacking Adolescents' Individual Lives Survey) long-term multidisciplinary study of the psychological, social, and physical development of adolescents and young adults. The dataset can be obtained by requesting it from www.trails.nl. Thanks to René R. Veenstra for this information.

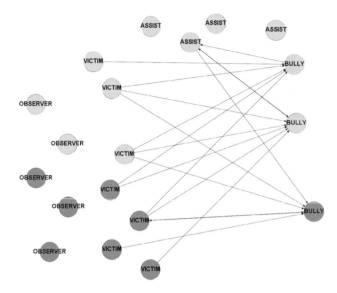

Figure 10.1 Network of bullying in a school. Nodes represent students in a 6th-grade middle-school class, blue for boys, pink for girls. A directed edge goes from node v_i to node v_j if student i identifies student j as a bully. The "bully" is the main perpetrator of the violence, while the "assist" is someone who helps the bully. The node labeled "victim" identifies a student that only has edges emanating from the node, and "observer" is a node without edges. Adapted from Huitsing and Veenstra (2012).

gender (divided up into four types of bullying: boy–boy, boy–girl, girl–boy, and girl–girl as a reference group), percentage of boys in a class, a teacher rating of each student in terms of aggressiveness and vulnerability, and a dyadic covariate on dislike relations (i.e., "Whom do you not like at all?").

A specific example of a network of bullying in a 6th-grade middle-school class in the Netherlands is displayed in **Figure 10.1** (adapted from Figure 1 in Huitsing and Veenstra, 2012). There were 19 children (11 boys, 8 girls) in this class. After parental consent was obtained for all children, a questionnaire was completed by each student at school under trained supervision. Examples of different types of bullying were explained, as well as examples of what is not considered bullying. As well as the questions mentioned above, other questions such as "Who starts when you are victimized?," "Which classmates assist the bully when you are victimized?," "Which classmates are usually present when you are victimized?," and "Which classmates defend you when you are victimized?" The results from the survey showed that there were three "ringleader" bullies, who supported each other and were also assisted by four other students.

10.4.3 Model Definitions

We first define the following matrices of node-specific covariates. Let the $(N \times r)$-matrix $\mathcal{X}_1$ consist of the values of a set of r covariates that express the attractiveness (or popularity) of each of N nodes, and let the $(N \times s)$-matrix $\mathcal{X}_2$ consist of the values of a set of s covariates that express the productivity (or sociability) of the same N nodes. The two sets of variables can overlap with each other.

Let the r-vector $\boldsymbol{\gamma}_1$ and the s-vector $\boldsymbol{\gamma}_2$ be two vectors of regression coefficient parameters. Then, we model the two N-vectors, $\boldsymbol{\alpha}$ and $\boldsymbol{\beta}$, from the p_1 model (see (10.13)) as follows:

$$\boldsymbol{\alpha} = \mathcal{X}_1\boldsymbol{\gamma}_1 + \boldsymbol{\xi}, \tag{10.44}$$

$$\boldsymbol{\beta} = \mathcal{X}_2\boldsymbol{\gamma}_2 + \boldsymbol{\eta}, \tag{10.45}$$

where the N-vectors $\boldsymbol{\xi} = (\xi_1, \ldots, \xi_N)^\tau$ and $\boldsymbol{\eta} = (\eta_1, \ldots, \eta_N)^\tau$ represent random error terms. The terms ξ_i and η_j represent the unexplained parts of the sender effect of node i and the receiver effect of node j, respectively, given the node-dependent covariates. We can combine these two equations into a single equation:

$$\begin{pmatrix} \boldsymbol{\alpha} \\ \boldsymbol{\beta} \end{pmatrix} = \begin{pmatrix} \mathcal{X}_1 & \mathbf{0} \\ \mathbf{0} & \mathcal{X}_2 \end{pmatrix} \begin{pmatrix} \boldsymbol{\gamma}_1 \\ \boldsymbol{\gamma}_2 \end{pmatrix} + \begin{pmatrix} \boldsymbol{\xi} \\ \boldsymbol{\eta} \end{pmatrix}, \tag{10.46}$$

where the random errors are assumed to be distributed as iid bivariate Gaussian:

$$\mathbf{C}_i = \begin{pmatrix} \xi_i \\ \eta_i \end{pmatrix} \sim \mathcal{N}_2(\mathbf{0}, \Sigma), \quad i = 1, 2, \ldots, N, \tag{10.47}$$

and where

$$\Sigma = \begin{pmatrix} \sigma_\xi^2 & \sigma_{\xi\eta} \\ \sigma_{\xi\eta} & \sigma_\eta^2 \end{pmatrix}. \tag{10.48}$$

In other words:

$$\mathrm{var}\{\xi_i\} = \sigma_\xi^2, \quad i = 1, 2, \ldots, N,$$
$$\mathrm{var}\{\eta_j\} = \sigma_\eta^2, \quad j = 1, 2, \ldots, N,$$
$$\mathrm{cov}\{\xi_i, \eta_i\} = \sigma_{\xi\eta}, \quad i = 1, 2, \ldots, N,$$
$$\mathrm{cov}\{\xi_i, \xi_j\} = \mathrm{cov}\{\eta_i, \eta_j\} = \mathrm{cov}\{\xi_i, \eta_j\} = 0, \quad \text{if } i \neq j.$$

An alternative way of writing this is

$$\begin{pmatrix} \boldsymbol{\xi} \\ \boldsymbol{\eta} \end{pmatrix} \sim \mathcal{N}_{2N}\left(\begin{pmatrix} \mathbf{0} \\ \mathbf{0} \end{pmatrix}, \begin{pmatrix} \Sigma_{\xi\xi} & \Sigma_{\xi\eta} \\ \Sigma_{\eta\xi} & \Sigma_{\eta\eta} \end{pmatrix} \right), \tag{10.49}$$

and $\Sigma_{\xi\xi} = \mathrm{diag}\{\sigma_\xi^2\}$, $\Sigma_{\eta\eta} = \mathrm{diag}\{\sigma_\eta^2\}$, and $\Sigma_{\xi\eta} = \mathrm{diag}\{\sigma_{\xi\eta}\} = \Sigma_{\eta\xi}^\tau$, where $\mathrm{diag}\{\sigma\}$ is the $(N \times N)$ diagonal matrix with all diagonal entries equal to σ and all other entries zero. Furthermore, θ and ρ are allowed to vary over dyads, and so are again represented by θ_{ij} and ρ_{ij}, respectively, where ρ is assumed to be constant within dyads: $\rho_{ij} = \rho_{ji}$.

The next assumption of the p_2 model relates to the parameters θ_{ij} and ρ_{ij} from the p_1 model (see (10.11)). Define the $(k_1 \times N(N-1))$-matrix $\mathcal{W}_1$ and the $(k_2 \times N(N-1))$-matrix $\mathcal{W}_2$ so that they contain dyad-specific covariates. Let $\mathbf{W}_{1ij}$ denote the column of $\mathcal{W}_1$ that corresponds to the directed $i \rightarrow j$ relation and let $\mathbf{W}_{2ij}$ denote the column of $\mathcal{W}_2$ that corresponds to the directed $i \rightarrow j$ relation, $i < j$. The assumption says that θ_{ij} and ρ_{ij} are linearly related to $\mathbf{W}_{1ij}$ and $\mathbf{W}_{2ij}$, respectively, so that[3]

$$\theta_{ij} = \theta + \boldsymbol{\delta}_1^\tau \mathbf{W}_{1ij}, \tag{10.50}$$

$$\rho_{ij} = \rho + \boldsymbol{\delta}_2^\tau \mathbf{W}_{2ij}, \tag{10.51}$$

---

[3]These expressions are slightly different from those that appear in van Duijn, Snijders, and Zijlstra (2004, p. 6) because all of our vectors are column vectors.

where the k_1-vector $\boldsymbol{\delta}_1$ and the k_2-vector $\boldsymbol{\delta}_2$ are vectors of regression coefficients. Reciprocity is assumed to be constant within dyads, so that $\rho_{ij} = \rho_{ji}$. Moreover, for the matrix $\mathcal{W}_2$, we have that $\mathbf{W}_{2ij} = \mathbf{W}_{2ji}$. The unknown fixed parameters are collected into the k-vector of parameters

$$\boldsymbol{\omega} = (\theta, \rho, \boldsymbol{\gamma}_1^\tau, \boldsymbol{\gamma}_2^\tau, \boldsymbol{\delta}_1^\tau, \boldsymbol{\delta}_2^\tau)^\tau. \tag{10.52}$$

10.4.4 Estimation of the p_2 Model

The p_2 model is formed by adding covariates to the p_1 model. Hence, even though there is some similarity between the two models, the p_2 model is more complicated, as is the problem of estimating its parameters.

Iterative Generalized Least-Squares Algorithm

The p_2 model contains "fixed" regression coefficients, $\boldsymbol{\gamma}_1, \boldsymbol{\gamma}_1, \boldsymbol{\delta}_1, \boldsymbol{\delta}_2$, and random effects, $\boldsymbol{\xi}$ and $\boldsymbol{\eta}$. Thus, we can formulate the p_2 model as a generalized linear mixed (GLM) model for data having a crossed structure. Thus, the GLM model can be estimated by the *iterative generalized least-squares (IGLS) algorithm* in a similar way as was proposed for the p_1 model (van Duijn, Snijders, and Zijlstra, 2004).

The IGLS algorithm consists of three steps. In the first step, the random effects are set to zero, the "fixed" effects vector is estimated through a linearization approximation to the likelihood function (by using the first-order term in a Taylor-series expansion around the current estimates of the fixed parameters), and then an iterative generalized least-squares procedure is applied; in the second step, a regression model of the random effects is formed with parameter vector $(\sigma_\xi^2, \sigma_\eta^2, \sigma_{\xi\eta})^\tau$, and the parameters of the random effects are estimated by generalized least-squares; finally, the various quantities are updated.

The three steps in the estimation process are computationally intensive, especially the second step in which large matrices need to be inverted. Furthermore, it has been shown that IGLS-type algorithms applied in similar models often yield biased estimates for both random and fixed parameters (Rodriguez and Goldman, 1995), and we can expect the IGLS algorithm for the p_2 model to do the same.

Markov Chain Monte Carlo

MCMC is used most often within a Bayesian context, where prior distributions, likelihood function, and posterior distribution provide the core components of a particular view of statistical inference. Because of the computational and statistical difficulties of the frequentist IGLS procedure, Zijlstra, van Duijn, and Snijders (2009) proposed a Bayesian view of network estimation for the p_2 model.

Assuming the parameters Σ in (10.48) and $\boldsymbol{\omega} = (\omega_i)$ in (10.52) are a priori independent, the p_2 distribution can be factored as follows:

$$P\{\mathbf{Y}, \mathbf{C}, \Sigma, \boldsymbol{\omega}\} = P_Y\{\mathbf{Y}|\mathbf{C}, \boldsymbol{\omega}\} P_C\{\mathbf{C}|\Sigma\} P_\Sigma\{\Sigma\} P_\omega\{\boldsymbol{\omega}\}, \tag{10.53}$$

where

$$\mathbf{C} = (\mathbf{C}_1^\tau, \dots, \mathbf{C}_N^\tau)^\tau, \quad \mathbf{C}_i = (\xi_i, \eta_i)^\tau, \ i = 1, 2, \dots, N, \tag{10.54}$$

is a $2N$-vector of random effects. Priors are placed on the random-effects parameters (prior = product, over all N nodes, of bivariate Gaussians) and on the covariance matrix Σ of the random effects (prior = inverse Wishart with identity covariance matrix), on the fixed-effects parameters $\omega_i \in \{\boldsymbol{\gamma}_1^\tau, \boldsymbol{\gamma}_2^\tau, \boldsymbol{\delta}_1^\tau, \boldsymbol{\delta}_2^\tau\}$ (prior = multivariate Gaussian with zero mean and diagonal covariance matrix), and on the parameters ρ and θ (prior = Gaussian with zero mean).

We need to obtain the conditional distribution $P\{\boldsymbol{\omega}|\mathbf{Y}, \mathbf{C}, \Sigma\}$. However, we cannot sample directly from this distribution. In principle, we could do this by deriving two other conditional distributions, $P\{\mathbf{C}|\mathbf{Y}, \boldsymbol{\omega}, \Sigma\}$ and $P\{\Sigma|\mathbf{C}\}$. From $P\{\mathbf{C}|\mathbf{Y}, \boldsymbol{\omega}, \Sigma\}$ we obtain $\mathbf{C}$, which is used in $P\{\Sigma|\mathbf{C}\}$ to find Σ, and then $\mathbf{C}$ and Σ are used to find $P\{\boldsymbol{\omega}|\mathbf{Y}, \mathbf{C}, \Sigma\}$. The problem is that we cannot sample directly from $P\{\mathbf{C}|\mathbf{Y}, \boldsymbol{\omega}, \Sigma\}$.

We can, however, sample from the conditional distribution $P\{\Sigma|\mathbf{C}\}$. From Bayes's theorem, this conditional distribution can be factored as

$$P\{\Sigma|\mathbf{C}\} \propto P_C\{\mathbf{C}|\Sigma\}P_\Sigma\{\Sigma\}. \tag{10.55}$$

Thus, the conditional distribution of Σ is proportional to the density of the random effects times the inverse-Wishart prior for Σ. Hence, $P\{\Sigma|\mathbf{C}\}$ has an inverse-Wishart distribution. If $\mathbf{A} \sim \mathcal{W}(\nu, \Sigma)$, then $\mathbf{A}^{-1} \sim \mathcal{W}^{-1}(\nu, \Sigma^{-1})$ (Anderson, 1984, Section 7.7.1). Thus, we can sample from the conditional distribution of Σ by inverting a draw from the Wishart distribution for the conditional distribution of Σ^{-1}. This is accomplished using a computationally efficient algorithm by Odell and Feiveson (1966).

In order to sample from $P\{\mathbf{C}|\mathbf{Y}, \boldsymbol{\omega}, \Sigma\}$ and $P\{\boldsymbol{\omega}|\mathbf{Y}, \mathbf{C}, \Sigma\}$, we apply the Metropolis–Hastings algorithm (Metropolis et al., 1953; Hastings, 1970). Examples from a large class of sampling algorithms, which are popularly referred to as Markov chain Monte Carlo (MCMC), have been proposed based upon the Metropolis–Hastings algorithm.

Consider sampling the random effects $\mathbf{C}_i = (\xi_i, \eta_i)^\tau$. Denote a proposal by $\mathbf{C}_i^* = (\xi_i^*, \eta_i^*)^\tau$ and the current value of $\mathbf{C}_i$ by $\mathbf{C}_i^{(t-1)} = (\xi_i^{(t-1)}, \eta_i^{(t-1)})^\tau$. Let $\mathbf{C}_i^{(0)}$ be the appropriate starting values. Let $h(\mathbf{C}_i)$ be a proposal-generating function for $\mathbf{C}_i$. For $t = 1, 2, \ldots$, compute the ratio

$$a = \frac{P\{\mathbf{C}_i^*|\mathbf{Y}, \boldsymbol{\omega}, \Sigma\}h(\mathbf{C}_i^{(t-1)}|\mathbf{C}_i^*)}{P\{\mathbf{C}_i^{(t-1)}|\mathbf{Y}, \boldsymbol{\omega}, \Sigma\}h(\mathbf{C}_i^*|\mathbf{C}_i^{(t-1)})}. \tag{10.56}$$

If $a \geq 1$, accept the proposal and set $\mathbf{C}_i^{(t)} = \mathbf{C}_i^*$. Otherwise, accept the proposal with probability a or reject it with probability $1 - a$. If the proposal is rejected, retain the current value by setting $\mathbf{C}_i^{(t)} = \mathbf{C}_i^{(t-1)}$.

Similarly, consider sampling the fixed parameters $\boldsymbol{\omega}$. Denote a proposal by $\boldsymbol{\omega}^*$ and the current value of $\boldsymbol{\omega}$ by $\boldsymbol{\omega}^{(t-1)}$. Let $\boldsymbol{\omega}^{(0)}$ be the appropriate starting values. Let $g(\boldsymbol{\omega})$ be a proposal-generating function for $\boldsymbol{\omega}$. For $t = 1, 2, \ldots$, compute the ratio

$$b = \frac{P\{\boldsymbol{\omega}^*|\mathbf{Y}, \mathbf{C}, \Sigma\}g(\boldsymbol{\omega}^{(t-1)}|\boldsymbol{\omega}^*)}{P\{\boldsymbol{\omega}^{(t-1)}|\mathbf{Y}, \mathbf{C}, \Sigma\}g(\boldsymbol{\omega}^*|\boldsymbol{\omega}^{(t-1)})}. \tag{10.57}$$

If $b \geq 1$, accept the proposal and set $\boldsymbol{\omega}^{(t)} = \boldsymbol{\omega}^*$. Otherwise, accept the proposal with probability b or reject it with probability $1 - b$. If the proposal is rejected, retain the current value by setting $\boldsymbol{\omega}^{(t)} = \boldsymbol{\omega}^{(t-1)}$.

A number of MCMC algorithms that incorporate specific h and g functions were proposed by Zijlstra, van Duijn, and Snijders (2009), including a random walk algorithm with Metropolis steps and an independence chain sampler with Metropolis–Hastings steps. They suggest approximating the h and g functions by multivariate Gaussians. For both types of algorithms, the covariance matrices of the Gaussian approximations to the conditional distributions, Ψ_ω for the fixed parameters ω and Ψ_C for the C_i, all i, play a major role. However, identifying appropriate matrices Ψ_ω and Ψ_C turns out to be the hardest part of running these algorithms.

Empirical assessments. Simulations of the two algorithms provide the following comparisons. MCMC tends to show rapid convergence to the posterior means, although the standard errors of those posterior means tend to be underestimated and slow to converge. When the IGLS and MCMC algorithms are compared, the results tend to be mixed. In some situations, the following can occur: (1) an IGLS algorithm can fail to converge and its estimates can be biased; (2) MCMC can overestimate the variances of the random effects; and (3) the MCMC algorithms become increasingly inefficient and more time-consuming as the complexity of the model increases.

10.5 Markov Random Graphs

The p_1 and p_2 models focus on dyadic relations, where the dyads $\{D_{ij}\}$ are assumed to be statistically independent. These models cannot capture links between nodes that have a triangular configuration (or, indeed, anything more complicated). In social networks, this is similar to assuming that people form links to other people independently of their other social connections. This is a very strong independence assumption, and it is too unrealistic to be useful either theoretically or practically. Fortunately, it can be dispensed with because of the development of more general models (e.g., p^*) that relax the independence assumption.

10.5.1 Dependence Graphs

A specific type of dyad dependence is provided by the *dependence graph*,

$$\mathcal{D} = \mathcal{D}(\mathcal{G}) = (\mathcal{V}_\mathcal{D}, \mathcal{E}_\mathcal{D}), \tag{10.58}$$

which specifies a dependence structure between the edges. The dependence graph assumes a Markov relationship between dyads for an undirected graph $\mathcal{G}$ in which edges, or subsets of edges, are "conditionally dependent" given the values of all other links in the graph (Frank and Strauss, 1986).

> **Definition 10.11** Two edges are said to be *conditionally dependent* if the conditional probability that the edges are present, given all other edges in the network, is *not* equal to the product of their marginal conditional probabilities.

A version of the dependence graph was also developed for directed graphs.

Let (v_i, v_j) represent the edge between nodes v_i and v_j. In social networks, (v_i, v_j) is often referred to as a *couple*. The nodes of the dependence graph $\mathcal{D}$ are the $\binom{|\mathcal{V}|}{2}$ possible edges,

$$\mathcal{V}_\mathcal{D} = \{(v_i, v_j) : v_i, v_j \in \mathcal{V}, i \neq j\}, \tag{10.59}$$

from the undirected random graph $\mathcal{G} = (\mathcal{V}, \mathcal{E})$, where $|\mathcal{V}|$ is the number of nodes in $\mathcal{G}$. Let Y_{ij} be a random variable that equals 1 or 0 according as the edge (v_i, v_j) is present or absent, respectively, in $\mathcal{G}$, and let $Y_{ij,kl}^c$ represent all edges of $\mathcal{G}$ *except* (v_i, v_j) *and* (v_k, v_l).

Definition 10.12 (Dawid, 1979) If X and Y are *independent*, we write $X \perp\!\!\!\perp Y$.

Definition 10.13 (Dawid, 1979) If X and Y are *conditionally independent given* Z, we write $X \perp\!\!\!\perp Y | Z$.

Note that $X \not\!\perp\!\!\!\perp Y | Z$ denotes that X and Y are *not conditionally independent given* Z. Then, the edges,

$$\mathcal{E}_{\mathcal{D}} = \{((v_i, v_j), (v_k, v_l)) : Y_{ij} \not\!\perp\!\!\!\perp Y_{kl} | Y_{ij,kl}^c\}, \qquad (10.60)$$

of $\mathcal{D}$ are pairs of edges of $\mathcal{G}$ that are dependent, conditional upon the remaining edges of $\mathcal{G}$. So, the edges of a dependence graph link all pairs of conditionally dependent edges. Note that because conditional dependence between pairs of edges is not uniquely defined, different versions of a dependence graph may be possible.

10.5.2 Markov Random Graphs

Definition 10.14 An undirected graph $\mathcal{G}$ is a *Markov random graph* (or has *Markov dependence*) if the dependence graph $\mathcal{D}$ contains no edge between disjoint edges (v_i, v_j) and (v_k, v_l).

Thus, the random variables Y_{ij} and Y_{kl} are independent, conditional upon the values of all other edges (i.e., $Y_{ij,kl}^c$), iff they do not share a node in $\mathcal{D}$; that is, iff $\{v_i, v_j\} \cap \{v_k, v_l\} = \emptyset$ (Besag, 1974).

Definition 10.15 Two edges are dependent, conditional upon the values of all other edges in the network, iff they have a node in common.

In terms of social interactions, this condition can be interpreted as meaning that one's social activity can be influenced only by one's friends, not by other persons who are not one's friends. If we apply this Markov assumption to random directed graphs, we obtain *Markov random directed graphs*, which are generalizations of Markov random fields that have been applied to spatial interaction data (Besag, 1974).

In a Markov graph, the cliques (i.e., complete subgraphs of the dependence graph in which every pair of nodes is linked by an undirected edge) are referred to as *edges*, *stars*, and *triangles*. In **Figure 10.2**, we see that undirected edges link two nodes, stars can be 2-stars, 3-stars, etc., depending upon how many other nodes are linked by undirected edges to the node in question.

Definition 10.16 *Triangles* are three nodes (a "triad") linked together by undirected edges. A *star* is an "in-star" if the node in question receives edges from other nodes or an "out-star" if the node sends out edges to other nodes. In the case of a directed graph, a "mixed-star" is anything else. A triangle is *cyclic* if the triad is (Y_{ij}, Y_{jk}, Y_{ki}).

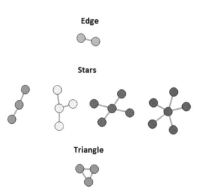

Figure 10.2 An edge, (2-, 3-, 4-, and 5-) stars, and a triangle for the Markov random graph model, the triad model, and the exponential random graph model.

10.5.3 Triad Model

A simple type of undirected Markov random graph that incorporates edges, stars, and triangles is the *triad model* (a special case of the p^* model).

Definition 10.17 (The Triad Model) The triad model has probability distribution

$$p^*(\mathbf{y}) = P(\mathbf{Y} = \mathbf{y}) = \frac{1}{Z(\boldsymbol{\omega})} \exp\left\{\rho L(\mathbf{y}) + \sigma S(\mathbf{y}) + \theta T(\mathbf{y})\right\}, \tag{10.61}$$

where $\boldsymbol{\omega} = (\rho, \sigma, \theta)^{\tau}$ is the vector of parameters in the model and $Z(\boldsymbol{\omega})$ is the normalizing constant.

The sufficient statistics are:

- $L(\mathbf{Y}) = \sum_{i,j} Y_{ij} = Y_{++}$, the number of edges ($v_i \leftrightarrow v_j$).
- $S(\mathbf{Y}) = \sum_{i \neq j \neq k} Y_{ij} Y_{ik}$, the number of 2-stars ($v_i \leftrightarrow v_j, v_i \leftrightarrow v_k$).
- $T(\mathbf{Y}) = \sum_{i \neq j \neq k} Y_{ij} Y_{jk} Y_{ki}$, the number of triangles ($v_i \leftrightarrow v_j, v_j \leftrightarrow v_k, v_k \leftrightarrow v_i$).

The edges are contained within the 2-stars, and the 2-stars are contained within the triangles. The parameters ρ, σ, and θ are usually unknown; ρ is referred to as *overall density* (general level of connectedness), and σ and θ correspond to *clustering* and *transitivity* ("friends of my friends are my friends"), respectively, of the edges.

We have the following special cases of the triad model (10.61).

Definition 10.18 (The Clustering Model) When $\rho = \theta = 0$, the model only depends upon the number of 2-stars,

$$P(\mathbf{Y} = \mathbf{y}) = \frac{1}{Z(\sigma)} \exp\{\sigma S(\mathbf{y})\}, \quad -\infty < \sigma < \infty, \tag{10.62}$$

and the resulting model is known as the *clustering model*.

Definition 10.19 (The Erdős–Rényi Model) When $\sigma = \theta = 0$, we have the *Erdős–Rényi model*,

$$P(\mathbf{Y} = \mathbf{y}) = \frac{1}{Z(\rho)} \exp\{\rho L(\mathbf{y})\}, \quad -\infty < \rho < \infty, \tag{10.63}$$

where the edges are iid with $Y_{ij} = 1$ (present) with probability $p = e^\rho/(1 + e^\rho)$, and $Y_{ij} = 0$ (absent) with probability $1 - p = 1/(1 + e^\rho)$, and the normalizing constant is $Z(\rho) = (1 + e^\rho)^m$, where $m = \binom{|\mathcal{V}|}{2}$.

Although the Erdős–Rényi model (like the p_1 model) is unrealistic for most types of real networks, it has often been used as a null model against which other models can be compared.

10.6 Exponential Random Graph Models

The *exponential random graph model (ERGM)* was originally created for spatial interaction models (Besag, 1975). It contains as a special case the Markov random graph model and incorporates arbitrary statistics and higher-order terms into the form of the exponential family. The ERGM has also been extended to directed graphs (Frank, 1991).

Definition 10.19 (The ERGM) The ERGM (which is also referred to as the p^* model in the social networks literature) has probability distribution

$$p^*(\mathbf{y}) = P(\mathbf{Y} = \mathbf{y}) = \frac{1}{Z(\boldsymbol{\omega})} \exp\left\{\rho L(\mathbf{y}) + \sum_{i=2}^{d} \sigma_i S_i(\mathbf{y}) + \theta T(\mathbf{y})\right\}. \tag{10.64}$$

10.6.1 Parameters and Sufficient Statistics

If we collect all $d+1$ variables into a column vector $\mathbf{u} = (u_i)$ and all $d+1$ parameters into a column vector $\boldsymbol{\omega} = (\omega_i)$, then we can write (10.64) as follows:

$$p^*(\mathbf{y}) = P(\mathbf{Y} = \mathbf{y}) = \exp\left\{\sum_i \omega_i u_i(\mathbf{y}) - A(\boldsymbol{\omega})\right\} \tag{10.65}$$

$$= \exp\left\{\boldsymbol{\omega}^\tau \mathbf{u}(\mathbf{y}) - A(\boldsymbol{\omega})\right\}, \tag{10.66}$$

where

$$A(\boldsymbol{\omega}) = \log Z(\boldsymbol{\omega}) = \sum_{\mathbf{y}} \exp\{\boldsymbol{\omega}^\tau \mathbf{u}(\mathbf{y})\} \tag{10.67}$$

is the normalizing constant defined to ensure that $p^*(\mathbf{y})$ sums to one. The vectors of sufficient statistics and parameters are given by

$$\mathbf{u}(\mathbf{Y}) = (L(\mathbf{Y}), S_2(\mathbf{Y}), \dots, S_d(\mathbf{Y}), T(\mathbf{Y}))^\tau, \tag{10.68}$$

$$\boldsymbol{\omega} = (\rho, \sigma_2, \dots, \sigma_d, \theta)^\tau, \tag{10.69}$$

respectively (Wasserman and Patterson, 1996), where

- $L(\mathbf{Y}) = S_1(\mathbf{Y}) = \sum_{i,j} Y_{ij} = Y_{++}$, the number of edges (or 1-stars),
- $S_2(\mathbf{Y}) = \sum_{i \neq j \neq k} Y_{ij} Y_{ik}$, the number of 2-stars,
- $S_3(\mathbf{Y}) = \sum_{i \neq j \neq k \neq l} Y_{ij} Y_{ik} Y_{il}$, the number of 3-stars,

and so on, in the network $\mathbf{Y}$, and

- $T(\mathbf{Y}) = \sum_{i \neq j \neq k} Y_{ij} Y_{jk} Y_{ki}$, the number of triangles (cyclic triads) in $\mathbf{Y}$.

The value of d determines the *degree distribution* of **y**. Stars can be subdivided into in-stars, out-stars, and mixed-stars depending upon the orientations of the arrows.

10.6.2 Evaluating the Normalization Constant

The most complicated problem of using an ML approach to fit an ERGM to network data is that we have to evaluate the normalizing constant $Z(\boldsymbol{\omega})$ (or, equivalently, $A(\boldsymbol{\omega})$). For most networks, evaluating $Z(\boldsymbol{\omega})$ is usually intractable (except for very small networks) and difficult to differentiate analytically, which makes estimation and inference difficult, if not impossible. Moreover, the sum in (10.67) is taken over the entire sample space of all possible undirected graphs on N nodes, which amounts to $\exp\{\binom{N}{2}\log 2\} = 2^{N(N-1)/2}$ graphs, a huge number for any realistic N. For example, $\binom{7}{2} = 21$ and $2^{21} = 2,097,152$, while $\binom{9}{2} = 36$ and $2^{36} = 68,719,476,736$. The overwhelming size of the sample space makes evaluating the likelihood function for a given $\boldsymbol{\omega}$ (or even maximizing over $\boldsymbol{\omega}$) impossible. As a result, any attempt to compute ML estimates of network parameters directly will usually be unsuccessful.

Some progress in approximating the normalization constant for ERGMs in the case of dense graphs (i.e., when the number of nodes, N, is comparable to the square root of the number of edges) has been made by Chatterjee and Diaconis (2013), who used a large-deviation approach. Many previous attempts to approximate the normalization constant have been made, but they are probably useful only in the case of graphs with small numbers of nodes.

10.6.3 Logit Models

Attempts to derive estimates of the parameters of the ERGM probability distribution have to navigate past the problem of evaluating the normalization constant. One method is to form an expression, such as a ratio of ERGM probabilities, in which the normalization constant drops out (Strauss and Ikeda, 1990).

Let $\mathbf{Y} = \{Y_{kl} : k,l = 1, 2, \ldots, N, \ k \neq l\}$ denote the $(N \times N)$ adjacency matrix of the graph $\mathcal{G}$. Let $\mathbf{Y}_{ij}^{+} = \{Y_{kl} : Y_{ij} = 1\}$ denote the adjacency matrix in which the edge (v_i, v_j) is forced to be present and $\mathbf{Y}_{ij}^{-} = \{Y_{kl} : Y_{ij} = 0\}$ denote the adjacency matrix in which the edge (v_i, v_j) is forced to be absent. Furthermore, let $\mathbf{Y}_{ij}^{c} = \{Y_{kl} : (v_k, v_l) \neq (v_i, v_j)\}$ denote the complementary relation for the edge (v_i, v_j). Then, the conditional probability[4]

$$P\{Y_{ij} = 1 | \mathbf{Y}_{ij}^{c} = \mathbf{y}_{ij}^{c}\} = \frac{P\{\mathbf{Y} = \mathbf{y}_{ij}^{+}\}}{P\{\mathbf{Y} = \mathbf{y}_{ij}^{+}\} + P\{\mathbf{Y} = \mathbf{y}_{ij}^{-}\}}$$

$$= \frac{\exp\{\boldsymbol{\omega}^{\tau}\mathbf{u}(\mathbf{y}_{ij}^{+})\}}{\exp\{\boldsymbol{\omega}^{\tau}\mathbf{u}(\mathbf{y}_{ij}^{+})\} + \exp\{\boldsymbol{\omega}^{\tau}\mathbf{u}(\mathbf{y}_{ij}^{-})\}} \tag{10.70}$$

does not involve the normalization constant $Z(\boldsymbol{\omega})$.

---

[4]Note that $\mathbf{y}_{ij}^{c}$, $\mathbf{y}_{ij}^{+}$, and $\mathbf{y}_{ij}^{-}$ are each $(N \times N)$-matrices.

Now, consider the odds ratio of the edge (v_i, v_j) being present versus that it is absent:

$$\frac{P\{Y_{ij} = 1 | \mathbf{Y}_{ij}^c = \mathbf{y}_{ij}^c\}}{P\{Y_{ij} = 0 | \mathbf{Y}_{ij}^c = \mathbf{y}_{ij}^c\}} = \frac{\exp\{\boldsymbol{\omega}^\tau \mathbf{u}(\mathbf{y}_{ij}^+)\}}{\exp\{\boldsymbol{\omega}^\tau \mathbf{u}(\mathbf{y}_{ij}^-)\}} = \exp\{\boldsymbol{\omega}^\tau [\mathbf{u}(\mathbf{y}_{ij}^+) - \mathbf{u}(\mathbf{y}_{ij}^-)]\}. \qquad (10.71)$$

Taking logs of both sides yields the *logit model*,

$$\log\left(\frac{P\{Y_{ij} = 1 | \mathbf{Y}_{ij}^c = \mathbf{y}_{ij}^c\}}{P\{Y_{ij} = 0 | \mathbf{Y}_{ij}^c = \mathbf{y}_{ij}^c\}}\right) = \text{logit}(P\{Y_{ij} = 1 | \mathbf{Y}_{ij}^c = \mathbf{y}_{ij}^c\})$$

$$= \boldsymbol{\omega}^\tau \Delta(\mathbf{y}_{ij}), \qquad (10.72)$$

where $\text{logit}(p) = \log\{p/(1-p)\}$ and

$$\Delta(\mathbf{y}_{ij}) = \mathbf{u}(\mathbf{y}_{ij}^+) - \mathbf{u}(\mathbf{y}_{ij}^-) \qquad (10.73)$$

represents the differences in the vector $\mathbf{u}$ of sufficient statistics when y_{ij} changes from 1 to 0. The vector $\Delta(\mathbf{y}_{ij})$ is called the vector of *change statistics*.

For the p_1 directed-graph model, for example, its defining equation (10.24) can be expressed in logit form as

$$\text{logit}(P\{Y_{ij} = 1 | \mathbf{Y}_{ij}^c = \mathbf{y}_{ij}^c\}) = \rho y_{ji} + \theta + \alpha_i + \beta_j. \qquad (10.74)$$

When the graph is undirected, then the logit model version of (10.25) is

$$\text{logit}(P\{Y_{ij} = 1 | \mathbf{Y}_{ij}^c = \mathbf{y}_{ij}^c\}) = \rho + \alpha_i + \alpha_j, \qquad (10.75)$$

where $\sum_i \alpha_i = 0$ for identifiability. For the p^* model (10.64), the logit form is given by

$$\text{logit}(P\{Y_{ij} = 1 | \mathcal{Y}_{ij}^c = \mathbf{y}_{ij}^c\}) = \rho + \sigma \Delta S + \theta \Delta T, \qquad (10.76)$$

where ΔS and ΔT represent the changes in S and T, respectively, when y_{ij} changes value from 1 to 0. See **Exercise 10.2**.

10.6.4 Pseudo-likelihood Function

The pseudo-likelihood method was introduced by Besag (1975), who studied spatially dependent variables in a network structure. To avoid having to compute the normalization constant, he approximated the joint probability of a set of variables by the product of conditional probabilities of each variable given the values of all other variables. This method has since been extended in several directions.

Definition 10.20 (Strauss and Ikeda, 1990) The *pseudo-likelihood function* is defined as

$$\text{PL}(\boldsymbol{\omega}) = \prod_{i \neq j} P\{Y_{ij} = y_{ij} | \mathbf{Y}_{ij}^c = \mathbf{y}_{ij}^c\}. \qquad (10.77)$$

This can be considered to be a local approximation to the likelihood function. We use the following notation for convenience. Let

$$P_{ij} = P\{Y_{ij} = 1 | \mathbf{Y}_{ij}^c = \mathbf{y}_{ij}^c\}, \quad Q_{ij} = 1 - P_{ij} = P\{Y_{ij} = 0 | \mathbf{Y}_{ij}^c = \mathbf{y}_{ij}^c\}. \qquad (10.78)$$

Then the pseudo-likelihood function becomes

$$\text{PL}(\boldsymbol{\omega}) = \prod_{i \neq j} P_{ij}^{y_{ij}} Q_{ij}^{1 - y_{ij}}. \tag{10.79}$$

Definition 10.21 The *maximum pseudo-likelihood (MPL) estimator* of the p^*-model parameters is the value of $\boldsymbol{\omega}$ that maximizes $\text{PL}(\boldsymbol{\omega})$:

$$\boldsymbol{\omega}_{\text{MPL}} = \arg\max_{\omega} \text{PL}(\boldsymbol{\omega}). \tag{10.80}$$

Because the conditional probability P_{ij} does not involve the normalizing constant $Z(\boldsymbol{\omega})$, $\text{PL}(\boldsymbol{\omega})$ is easier to maximize wrt $\boldsymbol{\omega}$ than the likelihood function. If the conditional probability does not depend upon $\mathbf{Y}_{ij}^c$, then the MPL estimate will coincide with the ML estimate. If the dependencies between elements of $\mathbf{Y}$ are weak, then the MPL estimate should yield a close approximation to the ML estimate. In general, however, we should not expect that to be the case, and the two types of estimates should differ.

Maximizing $\text{PL}(\boldsymbol{\omega})$ wrt $\boldsymbol{\omega}$ is equivalent to an ML fit of a logistic regression to the model (10.76) for independent observations $\{y_{ij}\}$ (Strauss and Ikeda, 1990). To see this, let ω_k be the kth parameter in $\boldsymbol{\omega}$. The pseudo-likelihood estimators are found by differentiating the log-pseudo-likelihood function

$$\log \text{PL}(\boldsymbol{\omega}) = \sum_{i \neq j} [y_{ij} \log P_{ij} + (1 - y_{ij}) \log Q_{ij}] \tag{10.81}$$

wrt ω_k and setting the result equal to zero. The MPL estimators are solutions to the equations

$$\frac{\partial}{\partial \omega_k} \log \text{PL}(\boldsymbol{\omega}) = \sum_{i,j} \left\{ \frac{y_{ij}}{P_{ij}} \frac{\partial P_{ij}}{\partial \omega_k} + \frac{1 - y_{ij}}{Q_{ij}} \left(-\frac{\partial P_{ij}}{\partial \omega_k} \right) \right\} = 0, \ k = 1, 2, \ldots. \tag{10.82}$$

These equations simplify to

$$\sum_{i,j} \frac{1}{P_{ij} Q_{ij}} (y_{ij} - P_{ij}) \frac{\partial P_{ij}}{\partial \omega_k} = 0, \ k = 1, 2, \ldots, \tag{10.83}$$

which can be written as

$$\frac{\partial}{\partial \omega_k} \sum_{i,j} w_{ij} (y_{ij} - P_{ij})^2 = 0, \ k = 1, 2, \ldots, \tag{10.84}$$

where the $w_{ij} = 1/(P_{ij} Q_{ij})$ are treated as constants in the differentiation. If the $\{y_{ij}\}$ were conditionally independent (which they are not), then these equations would look like a special case of ML equations for the generalized linear model (see, e.g., McCullagh and Nelder, 1989, Section 2.5). Hence, setting the derivatives of the pseudo-likelihood function equal to zero (while adopting some simplifying assumptions) yields equations that are identical to those derived from a logistic regression. These equations can be solved using *iteratively reweighted least squares* (although, for some models, there are alternative methods available), with the weights $\{w_{ij}\}$ recomputed at each step from the current value of $\boldsymbol{\omega}$.

Some theoretical results for the MPL approach have appeared. Under certain conditions, if the MPL estimator exists, it is unique (Handcock, 2003). Furthermore, the MPL estimator has been shown to be consistent and asymptotically Gaussian for particular types of Markov random field models (Comets and Janžura, 1998).

10.6.5 Inferential Difficulties

The pseudo-likelihood approach is widely used and is regarded as a practical alternative to ML. Unfortunately, it has many flaws. Furthermore, we have very little information about its statistical properties. There are a number of serious problems that have been identified with the MPL method, based upon theoretical considerations, simulations, and empirical results.

Dyads are typically not independent. We have seen that the pseudo-likelihood for an ERGM is equivalent to the likelihood for an appropriate logistic regression. Then, the parameters of the ERGM can be estimated by using a standard logistic regression computer program. Logistic regression models assume that the observations (dyads, in this case) are independent. Such an assumption may hold, for example, in biological networks that have very few or even no triangles (Saul and Filkov, 2007).

However, in general, *dyads are not independent*. The dependency amongst the dyads derives from the Markov assumption of the model. As a result, the sufficient statistics are not independent (even though they are assumed to be independent in the pseudo-likelihood function). The PL approach can also produce estimates with infinite values (which is not a particularly useful trait) even when the PL function converges. Consequently, when dependence is strong but ignored, the MPL method has been described as "unreliable," "arbitrary," "completely unreasonable," and "misleading"; in particular, it causes parameter estimates to be biased.

Degeneracy. Parameter estimation in ERGMs is also complicated by the unusual behavior of the likelihood function (Handcock, 2003; Rinaldo, Fienberg, and Zhou, 2009; Chatterjee and Diaconis, 2013). As a rule, one expects a probability distribution to place most of its mass on outcomes that are highly likely to occur in practice. *Degeneracy* (or *near degeneracy*) occurs when a disproportionately large amount of the probability mass is placed on only a few possible outcomes, leading to a bimodal or possibly multimodel distribution.

In the present context, a random graph model is said to be *degenerate* if the model places almost all of its probability mass on two graph configurations, such as an empty graph (i.e., $Y_{ij} = 0$, for all i, j) and/or a completely connected graph (i.e., $Y_{ij} = 1$, for all i, j), neither of which can be viewed as realistic, interesting, or useful. Degeneracy, therefore, acts as a warning sign that the proposed model is such a bad fit to the observed data that the data could never have been observed under that model. Many popular graph models have been shown to be degenerate.

Degeneracy can also result in problems in fitting ERGMs. For certain values of the coefficients ω of the sufficient statistics, the model transitions from a very sparse graph to a very dense graph, without passing through the intermediate stages.[5] The instantaneous transition between empty and full graphs at certain values of ω can induce "instability" problems for the MPL method (in the sense of Breiman (1996),

---

[5]This effect is reminiscent of the sudden appearance of a phase transition in random graph models. See the discussion in Section 4.5.

---

who defined *instability* for model selection). Stability in a random graph model occurs when small perturbations in the values of certain parameters lead to small perturbations in the probabilistic structure of the model. Instability, on the other hand, means that very similar parameter values can lead to quite different graph structures (Handcock, 2003). At a critical value of ω when the empty graph instantaneously transitions to the full graph, any simulation-based algorithm (e.g., MCMC) for random graph models can fail to converge, especially as the number of nodes increases.

Standard errors. Because the dyads in the model are correlated (and possibly strongly so), the standard errors of parameter estimates obtained through logistic regression and interpreted as if they were reasonable estimates of variability of MPL estimates are probably too small (Snijders, 2002). The best one can say about these estimated standard errors is that they may be an approximation to truth.

10.6.6 Markov Chain Monte Carlo Methods

Difficulties in applying the MPL method to the estimation of the ERGM parameters led to simulation-based procedures, such as *MCMC*.

MCMC methods consist of the *Metropolis–Hastings algorithms* (Metropolis et al., 1953; Hastings, 1970; Peskun, 1973) and the *Gibbs sampler* (Geman and Geman, 1984), which is almost a special case of the Metropolis–Hastings algorithm. These algorithms entered the statistics lexicon at the same time (during the 1990s) as computational facilities became faster, more efficient, and more local, software became readily available (e.g., BUGS[6] in 1991), and a wide range of applications (e.g., spatial statistics, Bayesian inference, computer vision, speech processing, computational biology, robotics, agriculture) demonstrated that the MCMC method could indeed be practical. Following Gelfand and Smith (1990), Geyer and Thompson (1992), and Tierney (1994), the MCMC method was recognized as a real breakthough in statistical computing, and the whole field exploded with books, articles, workshops, and conferences.

In this particular application, the MCMC method samples from the ERGM by using the Metropolis (or Metropolis–Hastings) algorithm. The *Metropolis algorithm* runs as follows. We generate a sequence of adjacency matrices (corresponding to their graphs),

$$\mathbf{y}^{(0)}, \mathbf{y}^{(1)}, \mathbf{y}^{(2)}, \ldots, \mathbf{y}^{(b-1)}, \mathbf{y}^{(b)}, \tag{10.85}$$

as follows. The algorithm is initialized by setting $\mathbf{y}^{(0)}$ to be the starting values of the Markov chain, which can be taken to be the MPLE values. At the bth iteration ($b = 1$, $2, \ldots$), the algorithm makes a random choice of whether to remain at $\mathbf{y}^{(b-1)}$ for an additional step or to transition to a proposed future state $\mathbf{y}^*$. The state $\mathbf{y}^*$ is identical to $\mathbf{y}^{(b-1)}$, except that a pair of nodes, v_i, v_j, is selected at random and the edge between them, $y_{ij}^{(b-1)}$, is set to $1 - y_{ij}^{(b-1)}$ (i.e., a 0 is switched to a 1 or a 1 is switched to a 0).[7] We compute the odds ratio of the proposed and current states,

---

[6]Bayesian updating with Gibbs sampling. See Spiegelhalter, Thomas, and Best (1999).
[7]In the social networks literature, this is often referred to as a "toggle."

$$r_M = \frac{P\{Y = y^*\}}{P\{Y = y^{(b-1)}\}}. \tag{10.86}$$

If $r_M \geq 1$, we accept the proposed state and set $y^{(b)} = y^*$ as the next state in the Markov chain. Otherwise, we accept the proposed state with probability r_M or reject it with probability $1 - r_M$. If the proposed state is rejected, we set $y^{(b)} = y^{(b-1)}$. At each iteration step, a pair of nodes is selected at random, the edge between them is toggled, and a determination is made regarding the next state in the chain. These iterations continue until the joint distribution of $y^{(b)}$ stabilizes. Note that the normalizing constants in the ratio r_M cancel out, which makes this method computationally attractive.

Hastings (1970) studied an important generalization of the Metropolis algorithm by introducing a *proposal distribution*, $q(y_1|y_2)$, which is defined as the probability that $Y = y_1$ given that $Y = y_2$. The generalization multiplies r_M by the ratio $q(y^{(b-1)}|y^*)/q(y^*|y^{(b-1)})$, so that

$$r_{MH} = \frac{P\{Y = y^*\}q(y^{(b-1)}|y^*)}{P\{Y = y^{(b-1)}\}q(y^*|y^{(b-1)})}. \tag{10.87}$$

In the presence of symmetry, $q(y^*|y^{(b-1)}) = q(y^{(b-1)}|y^*)$, and the *Metropolis–Hastings algorithm* reduces to the Metropolis algorithm. The success (or failure) of the Metropolis–Hastings algorithm depends upon the choice of proposal distribution q. It is also usual to run the Markov chain for a long time so that its starting position is forgotten, and then throw away the first m samples (referred to as "burn-in" samples), where m is large, assuming that the chain has not yet reached equilibrium. How long the burn-in period should be is still open to discussion (see, e.g., Jones and Hobert, 2001), although 1–2% of a run is customary. It has been suggested that high autocorrelation (say, greater than 0.4) in the Markov chain can be reduced by subsampling every Δ iterations and throwing the rest away (Geyer, 1991, 1992); often the best strategy is to take $\Delta = 1$ (Geyer and Thompson, 1992).

MCMC methods are now generally preferred over MPL methods. If dependence is not an issue, MPL is faster, more efficient, requires no iterations, and the MPLE is the same as the MLE. If dependence is an issue, but is ignored, then the MPLE will typically be unreliable and will fail to be close to the MLE. Simulation studies have even suggested that the MPL estimator is inconsistent (Corander, Dahmström, and Dahmström, 1998, 2002). Thus, MCMC, despite being computationally intensive and impracticable for large networks, is usually the estimation method of choice because of its reliability.

MCMC has been used to obtain an approximate MLE. If the likelihood function is intractable, we first approximate it by generating samples from a distribution with known parameter values; then, the maximum of that approximation, which we refer to as the MCMC-MLE, can be used to estimate the MLE (Geyer, 1991; Geyer and Thompson, 1992). Recall the form of the ERGM (10.66):

$$P_\omega\{Y = y\} = \exp\{\omega^\tau u(y) - A(\omega)\}, \tag{10.88}$$

where $u(y)$ is a vector of statistics, ω is a vector of parameters, and $A(\omega)$ is the normalization constant given by (10.67). Fix ω^0 as an arbitrary value of ω and consider the difference $A(\omega) - A(\omega^0)$. Then,

$$e^{A(\boldsymbol{\omega})-A(\boldsymbol{\omega}^0)} = \frac{\sum_{\mathbf{y}} e^{\boldsymbol{\omega}^\tau \mathbf{u}(\mathbf{y})}}{e^{A(\boldsymbol{\omega}^0)}}$$

$$= \sum_{\mathbf{y}} e^{\boldsymbol{\omega}^\tau \mathbf{u}(\mathbf{y})} \cdot e^{-(\boldsymbol{\omega}^0)^\tau \mathbf{u}(\mathbf{y})} \cdot \frac{e^{(\boldsymbol{\omega}^0)^\tau \mathbf{u}(\mathbf{y})}}{e^{A(\boldsymbol{\omega}^0)}}$$

$$= \sum_{\mathbf{y}} e^{(\boldsymbol{\omega}-\boldsymbol{\omega}^0)^\tau \mathbf{u}(\mathbf{y})} \cdot \frac{e^{(\boldsymbol{\omega}^0)^\tau \mathbf{u}(\mathbf{y})}}{e^{A(\boldsymbol{\omega}^0)}}$$

$$= \mathrm{E}_{\boldsymbol{\omega}^0}\{e^{(\boldsymbol{\omega}-\boldsymbol{\omega}^0)^\tau \mathbf{u}(\mathbf{Y})}\}. \tag{10.89}$$

So, we can estimate $e^{A(\boldsymbol{\omega})-A(\boldsymbol{\omega}^0)}$ by the sample mean

$$\frac{1}{m} \sum_{i=1}^{m} e^{(\boldsymbol{\omega}-\boldsymbol{\omega}^0)^\tau \mathbf{u}(\mathbf{Y}_i)}, \tag{10.90}$$

where $\mathbf{Y}_1, \mathbf{Y}_2, \ldots, \mathbf{Y}_m$ is a sample of m random graphs each drawn from the probability distribution with parameter vector $\boldsymbol{\omega}^0$. We can generate such a sample using MCMC.

Let $\mathbf{y}_a$ denote a single graph of $\mathbf{Y}$, and let $\ell(\boldsymbol{\omega}) = \boldsymbol{\omega}^\tau \mathbf{u}(\mathbf{y}_a) - A(\boldsymbol{\omega})$ denote the log-likelihood function for the ERGM (10.88) computed from $\mathbf{y}_a$. The log-likelihood ratio is

$$\lambda(\boldsymbol{\omega}, \boldsymbol{\omega}^0) = \ell(\boldsymbol{\omega}) - \ell(\boldsymbol{\omega}^0)$$

$$= (\boldsymbol{\omega} - \boldsymbol{\omega}^0)^\tau \mathbf{u}(\mathbf{y}_a) - \log \mathrm{E}_{\boldsymbol{\omega}^0}\{e^{(\boldsymbol{\omega}-\boldsymbol{\omega}^0)^\tau \mathbf{u}(\mathbf{Y})}\}. \tag{10.91}$$

The first term on the rhs is a simple function of the sufficient statistics and, hence, is easy to compute. The second term involves an expectation under the probability distribution (10.88) with $\boldsymbol{\omega} = \boldsymbol{\omega}^0$, and is more difficult to compute. However, we can estimate $\lambda(\boldsymbol{\omega}, \boldsymbol{\omega}^0)$ by

$$\widehat{\lambda}_m(\boldsymbol{\omega}, \boldsymbol{\omega}^0) = (\boldsymbol{\omega} - \boldsymbol{\omega}^0)^\tau \mathbf{u}(\mathbf{y}_a) - \log \left\{ \frac{1}{m} \sum_{i=1}^{m} e^{(\boldsymbol{\omega}-\boldsymbol{\omega}^0)^\tau \mathbf{u}(\mathbf{Y}_i)} \right\}, \tag{10.92}$$

where $\mathbf{Y}_1, \mathbf{Y}_2, \ldots, \mathbf{Y}_m$ constitute an MCMC sample, drawn from a probability distribution with parameter $\boldsymbol{\omega}^0$.

Next, we apply *the Strong Law of Large Numbers (SLLN) for Markov chains*. First, we need the following definitions:

Definition 10.22 A Markov chain is called *irreducible* if, starting at any state of the Markov chain, there is a positive probability that the chain is able to access all other states, and is called *aperiodic* if the chain does not get trapped in cycles.

Definition 10.23 The distribution of a Markov chain is called *stationary* (or *invariant*) if it does not depend upon the origin of time.

The SLLN states that

Theorem 10.1 (SLLN) *If $\{X_i\}$ is an irreducible and aperiodic Markov chain with stationary distribution π, then for any integrable, real-valued function f, the SLLN holds; that is, $\frac{1}{n}\sum_{i=1}^{n} f(X_i) \to \mathrm{E}_\pi\{f(X)\}$, a.s. $[\pi]$, as $n \to \infty$.*

The nice thing about MCMC algorithms is that the Markov chains so formed are irreducible and aperiodic, and the target distribution is stationary; hence, the assumptions of the SLLN for Markov chains are satisfied. It follows that

$$\widehat{\lambda}_m(\boldsymbol{\omega}, \boldsymbol{\omega}^0) \to \lambda(\boldsymbol{\omega}, \boldsymbol{\omega}^0), \quad a.s., \quad \text{as } m \to \infty. \tag{10.93}$$

Thus, given the sample size m, maximizing $\widehat{\lambda}_m(\boldsymbol{\omega}, \boldsymbol{\omega}^0)$ wrt $\boldsymbol{\omega}$ will yield an approximation, $\widehat{\boldsymbol{\omega}}_m$, to the MLE, $\widehat{\boldsymbol{\omega}}$. We refer to $\widehat{\boldsymbol{\omega}}_m$ as the MCMC-MLE of $\boldsymbol{\omega}$. The MCMC-MLE will be different depending upon the chosen $\mathbf{y}_a$. Iterating this method by drawing fresh samples at the current MCMC-MLE, and then re-estimating, yields the unique maximum of the likelihood function (Hunter and Handcock, 2006). Uniqueness is due to the concavity of the likelihood function for an ERGM.

To simulate a Markov chain with stationary distribution given by (10.88), we can use the results of Section 10.7.3 in which

$$\frac{P_{\boldsymbol{\omega}}\{Y_{ij} = 1 | \mathbf{Y}_{ij}^c = \mathbf{y}_{ij}^c\}}{P_{\boldsymbol{\omega}}\{Y_{ij} = 0 | \mathbf{Y}_{ij}^c = \mathbf{y}_{ij}^c\}} = \exp\{\boldsymbol{\omega}^\tau \Delta(\mathbf{y}_{ij})\} \tag{10.94}$$

is easily computed for the next step of the chain. Recall that the vector of *change statistics*, $\Delta(y_{ij}) = \mathbf{u}(\mathbf{y}_{ij}^+) - \mathbf{u}(\mathbf{y}_{ij}^-)$, represents the effect on the summary statistics of a randomly chosen single dyad y_{ij} being toggled either from 0 to 1 ($\mathbf{u}(\mathbf{y}_{ij}^+)$) or from 1 to 0 ($\mathbf{u}(\mathbf{y}_{ij}^-)$), while the rest of the graph remains unchanged.

The error in using the MCMC-MLE as an approximation to the MLE depends upon the accuracy of the estimate of the likelihood function. It was found that the greater the distance between $\boldsymbol{\omega}$ and the starting parameter vector, $\boldsymbol{\omega}^0$, the less reliable is the accuracy of the likelihood estimate (Geyer and Thompson, 1992; Hummel, Hunter, and Handcock, 2012). In fact, a maximum step size of $\| \boldsymbol{\omega} - \boldsymbol{\omega}^0 \| \leq \delta$ was recommended (Geyer and Thompson, 1992), where $\delta = \sqrt{2}$ was judged to be "about the right size," but no more informative guidelines were given. When $\boldsymbol{\omega}$ is not near $\boldsymbol{\omega}^0$, a larger number of samples is needed to compensate for the lack of precision (the starting point, $\boldsymbol{\omega}^0$, for the Markov chain is usually taken to be the MPLE, which can be far away from the MLE) and to improve the chance of convergence.

10.7 Latent Space Models

Consider a network consisting of N nodes, $v_1, v_2, \ldots, v_N$, which can be represented by an $(N \times N)$ adjacency matrix $\mathbf{Y} = (Y_{ij})$, where $Y_{ij} = 1$ if nodes v_i and v_j are connected by an edge, and $Y_{ij} = 0$ otherwise. *Latent space models* (Hoff, Raftery, and Handcock, 2002) posit that each node, v_i, say, can be viewed as corresponding to a point (or *position*) $\mathbf{z}_i$ in some space. The points $\mathbf{z}_1, \mathbf{z}_2, \ldots, \mathbf{z}_N$ are assumed to be independent and exist in a k-dimensional *latent space* (or *social space*, as referred to by those working in social networks) $\mathbb{R}^k$, where we think of the space as low dimensional, $k = 2$, or maybe 3.

While nodes correspond to points, how should we think of an edge, (v_i, v_j), between the pair of nodes v_i and v_j in this latent space? A key feature of the latent space model is the notion of distance, $d_{ij} = d(\mathbf{z}_i, \mathbf{z}_j)$, between pairs of points, $\mathbf{z}_i$ and $\mathbf{z}_j$. Such a distance determines whether or not an edge is present between v_i

and v_j (i.e., whether $Y_{ij} = 1$ or 0). Essentially, the smaller the distance, the higher the probability that an edge is present, while the larger the distance, the smaller the probability of there being such an edge.

The latent space model is a conditional probability model. By this we mean that the presence or absence of an edge, (v_i, v_j), between a pair of nodes, v_i and v_j, say, is stochastically independent of all other edges in the network, conditional on the unobserved positions, $\mathbf{z}_i$ and $\mathbf{z}_j$, in latent space. This representation can provide useful insight into the structure of the network, which also allows a visual and probabilistic interpretation.

Examples of datasets that have used latent space models (see, e.g., Hoff, Raftery, and Handcock, 2002), include Sampson's monk data (see Section 2.5.4), marriage data of Florentine families (see **Exercise 2.4**), and neural data from the roundworm *Caenhorhabditis elegans* (see Section 2.6.2).

Suppose that edge information is available through r dyad-specific covariates. Recall that a dyad is a pair $D_{ij} = (Y_{ij}, Y_{ji})$ with $i < j$. We denote the r-vector of covariates corresponding to D_{ij} by $\mathbf{X}_{ij} = (X_{1,ij}, \ldots, X_{r,ij})^\tau$. The collection of all covariates is given by $\mathbf{X} = \{\mathbf{X}_{ij}\}$. Then,

$$P\{\mathbf{Y}|\mathbf{Z}, \mathbf{X}, \boldsymbol{\beta}\} = \prod_{i \neq j} P\{y_{ij}|\mathbf{z}_i, \mathbf{z}_j, \mathbf{x}_{ji}, \boldsymbol{\beta}\}, \tag{10.95}$$

where $\mathbf{Z} = (\mathbf{z}_1, \ldots, \mathbf{z}_N)$ is a $(k \times N)$-matrix of positions to be determined and $\boldsymbol{\beta} = (\boldsymbol{\beta}_0^\tau, \beta_1)^\tau$ is an $(r + 1)$-vector of parameters to be estimated, where $\boldsymbol{\beta}_0^\tau = (\beta_{01}, \ldots, \beta_{0r})^\tau$. The next question is how should we model the probability $P\{y_{ij}|\mathbf{z}_i, \mathbf{z}_j, \mathbf{x}_{ji}, \boldsymbol{\beta}\}$?

Hoff, Raftery, and Handcock (2002) proposed two methods based upon Euclidean latent space – a distance model and a projection model – for estimating $P\{Y_{ij}|\mathbf{z}_i, \mathbf{z}_j, \mathbf{x}_{ji}, \boldsymbol{\beta}\}$. The two models are similar, both weighting the covariate values by a vector $\boldsymbol{\beta}$, but they use different ways to express the notion of distance between nodes.

10.7.1 Distance Model

For the latent space model, the probability of an edge (i.e., $Y_{ij} = 1$) is assumed to depend upon the distance, $d_{ij} = d(\mathbf{z}_i, \mathbf{z}_j)$, between the two points, $\mathbf{z}_i$ and $\mathbf{z}_j$, and on covariates, $\mathbf{x}_{ij}$, specific to each node. The distance d_{ij} between points $\mathbf{z}_i$ and $\mathbf{z}_j$ has to satisfy the following conditions:

- $d_{ii} = 0$, for all i.
- $d_{ij} = d_{ji} > 0$, for all $i \neq j$.
- $d_{ij} \leq d_{ik} + d_{kj}$, for all i, j, k.

The most widely used distance is Euclidean distance, d_{ij}, where

$$\begin{aligned} d_{ij}^2 = \parallel \mathbf{z}_i - \mathbf{z}_j \parallel^2 &= (\mathbf{z}_i - \mathbf{z}_j)^\tau (\mathbf{z}_i - \mathbf{z}_j) \\ &= \parallel \mathbf{z}_i \parallel^2 + \parallel \mathbf{z}_j \parallel^2 - 2\mathbf{z}_i^\tau \mathbf{z}_j, \end{aligned} \tag{10.96}$$

where $\parallel \mathbf{z}_i \parallel^2 = \mathbf{z}_i^\tau \mathbf{z}_i$.

We assume that the distribution of the $\{\mathbf{z}_i\}$ is given by

$$\mathbf{z}_i \sim f_k(\mathbf{z}|\boldsymbol{\phi}), \quad \mathbf{z}_i \in \mathbb{R}^k, \tag{10.97}$$

where $\boldsymbol{\phi}$ is a k-vector of parameters that characterize the distribution of the latent vectors in $\mathbb{R}^k$. In their treatment of the distance model, Hoff et al. assume that $f_k(\mathbf{z}|\boldsymbol{\phi})$ is bivariate standard Gaussian, $\mathcal{N}_2(\mathbf{0}, \mathbf{I}_2)$. If the network were viewed as having community structure, Handcock, Raftery, and Tantrum (2007) take $f_k(\mathbf{z}|\boldsymbol{\phi})$ to be a mixture of Q multivariate Gaussian distributions,

$$\mathbf{z}_i \sim \sum_{q=1}^{Q} \lambda_q \mathcal{N}_k(\boldsymbol{\mu}_q, \sigma_q^2 \mathbf{I}_k), \quad i = 1, 2, \ldots, N, \tag{10.98}$$

where λ_q is the probability that a node is a member of the qth community, $\lambda_q \geq 0$, $q = 1, 2, \ldots, Q$, and $\sum_{q=1}^{Q} \lambda_q = 1$, and the means, $\{\boldsymbol{\mu}_q\}$, and variances, $\{\sigma_q^2\}$, are different for the different communities.

For binary data $\mathbf{Y} = (Y_{ij})$, where

$$Y_{ij} \sim \text{Bernoulli}(p_{ij}), \quad i \neq j, \tag{10.99}$$

are independent, $p_{ij} = P\{Y_{ij} = 1\}$, and $\boldsymbol{\eta} = (\eta_{ij})$, where

$$\eta_{ij} = \log\left(\frac{p_{ij}}{1 - p_{ij}}\right), \tag{10.100}$$

the logistic regression model is usually the model of choice. Thus, if $P\{Y_{ij} = 1|\mathbf{z}_i, \mathbf{z}_j, \mathbf{x}_{ij}, \boldsymbol{\beta}\}$ is the probability of an edge, then its logit is

$$\begin{aligned}
\eta_{ij} &= \text{logit}(P\{Y_{ij} = 1|\mathbf{z}_i, \mathbf{z}_j, \mathbf{x}_{ij}, \boldsymbol{\beta}\}) \\
&= \boldsymbol{\beta}_0^{\tau} \mathbf{x}_{ij} - \beta_1 d_{ij} \\
&= \sum_{l=1}^{r} \beta_{0l} x_{l,ij} - \beta_1 d_{ij},
\end{aligned} \tag{10.101}$$

where $\mathbf{x}_{ij} = (x_{1,ij}, \ldots, x_{r,ij})^{\tau}$ and $\text{logit}(p) = \log\{p/(1-p)\}$. The points $\{\mathbf{z}_i\}$ are scaled so that $\{N^{-1} \sum_{i=1}^{N} \| z_i \|^2\}^{1/2} = 1$. This distance model satisfies the conditions of *reciprocity* and *transitivity*. Reciprocity means that if an edge (v_i, v_j) exists, then with high probability, so does the edge (v_j, v_i). Transitivity means that if the edges (v_i, v_j) and (v_j, v_k) exist (so that the distances d_{ij} and d_{jk} will not be large), then, with high probability, the edge (v_i, v_k) will also exist. This distance model is most appropriate for undirected networks or for directed networks with a strong reciprocity feature.

Estimation. Estimation for the distance model is given by the following process. The likelihood function is

$$L(\mathbf{Y}|\boldsymbol{\eta}) = \prod_{i=1}^{N} \prod_{j \neq i} p_{ij}^{y_{ij}} (1 - p_{ij})^{1 - y_{ij}}. \tag{10.102}$$

The log-likelihood function is

$$\ell(\mathbf{Y}|\boldsymbol{\eta}) = \sum_{i=1}^{N} \sum_{j \neq i} \{y_{ij} \log p_{ij} + (1 - y_{ij}) \log(1 - p_{ij})\}$$

$$= \sum_{i=1}^{N} \sum_{j \neq i} \{y_{ij}[\log p_{ij} - \log(1 - p_{ij})] + \log(1 - p_{ij})\}$$

$$= \sum_{i=1}^{N} \sum_{j \neq i} \left\{ y_{ij} \log \left(\frac{p_{ij}}{1 - p_{ij}} \right) + \log(1 - p_{ij}) \right\}. \tag{10.103}$$

Now, $p_{ij} = \frac{e^{\eta_{ij}}}{1 + e^{\eta_{ij}}}$ and

$$\log(1 - p_{ij}) = \log \left(1 - \frac{e^{\eta_{ij}}}{1 + e^{\eta_{ij}}} \right) = \log \left(\frac{1}{1 + e^{\eta_{ij}}} \right) = -\log(1 + e^{\eta_{ij}}). \tag{10.104}$$

So, the log-likelihood function is

$$\ell(\mathbf{Y}|\boldsymbol{\eta}) = \sum_{i=1}^{N} \ell_i, \tag{10.105}$$

where

$$\ell_i = \sum_{j \neq i} \{y_{ij} \eta_{ij} - \log(1 + e^{\eta_{ij}})\}. \tag{10.106}$$

Maximum-likelihood estimates can be derived by obtaining a set of distances that maximize the likelihood (or log-likelihood). This turns out to be a very difficult problem for linear optimization.

A suggestion by Hoff et al. (and also by Handcock et al.) is to consider a Bayesian approach to the estimation problem whereby, given prior information on $\boldsymbol{\beta}$ and $\mathbf{Z}$, we sample from the posterior distribution. So, we apply a nonlinear optimization algorithm, where starting values for the latent positions $\{z_{ij}\}$ can be derived from multidimensional scaling (MDS) methods (see, e.g., Izenman, 2013, Chapter 13) followed by a Metropolis–Hastings updating algorithm.

Problems arise from this computational approach, however, because of the invariance properties of the distance model. The derived positions are invariant under rotation, reflection, and translation, so that the resulting distances, and also the model itself, do not change from these operations. Consequently, there is an infinite number of latent positions that are equivalent, in the sense that they yield an identical log-likelihood. This means that there is a set of equivalence classes of latent positions in which the class [$\mathbf{Z}$] is invariant under rotation, reflection, and translation. For a given class of latent positions [$\mathbf{Z}$], we carry out a Procrustes transformation of $\mathbf{Z}$,

$$\mathbf{Z}^* = \arg \min_{\mathbf{TZ}} \text{tr}\{(\mathbf{Z}_0 - \mathbf{TZ})^{\tau}(\mathbf{Z}_0 - \mathbf{TZ})\}, \tag{10.107}$$

where $\mathbf{Z}_0$ represents a fixed set of "reference" positions (such as those obtained from MDS) and $\mathbf{T}$ is a transformation matrix that ranges over the set of rotations, reflections, and translations. Thus, $\mathbf{Z}^*$ is that member of the class $[\mathbf{Z}]$ that is closest to $\mathbf{Z}_0$. The least-squares solution is given by

$$\mathbf{Z}^* = \mathbf{Z}_0 \mathbf{Z}^\tau (\mathbf{Z}\mathbf{Z}_0^\tau \mathbf{Z}_0 \mathbf{Z}^\tau)^{-1/2} \mathbf{Z}, \tag{10.108}$$

and an iterative algorithm for computing $\mathbf{Z}^*$ that involves MCMC sampling from the posterior distibution is provided by Hoff, Raftery, and Handcock (2002).

10.7.2 Projection Model

The second type of latent model described by Hoff et al. is the projection model in which the latent points $\{\mathbf{z}_i\}$ are embedded onto a k-dimensional hypersphere having unit radius. If $k = 2$, for example, the points appear on the unit circle. Distance between two points $\mathbf{z}_i$ and $\mathbf{z}_j$ in this latent space (i.e., on the hypersphere) is measured by angular distance,

$$d(\mathbf{z}_i, \mathbf{z}_j) = \frac{\mathbf{z}_i^\tau \mathbf{z}_j}{\| \mathbf{z}_i \| \cdot \| \mathbf{z}_j \|}. \tag{10.109}$$

Thus, assuming that the vectors $\mathbf{z}_i$ and $\mathbf{z}_j$ are normalized to have unit length, the dot product $\mathbf{z}_i^\tau \mathbf{z}_j$ is used as a similarity measure between the two points $\mathbf{z}_i$ and $\mathbf{z}_j$. The measure (10.109) is not a true distance, but has been used often to describe the distance between two points on a hypersphere.

The existence of an edge between two nodes, therefore, depends upon the angle between the corresponding points on the hypersphere (or circle in the $k = 2$ case). In other words, it is likely that there is an edge between v_i and v_j if the angle between $\mathbf{z}_i$ and $\mathbf{z}_j$ is acute (i.e., if $\mathbf{z}_i^\tau \mathbf{z}_j > 0$), neutral if the angle is a right angle, and likely to be no edge if the angle is obtuse (i.e., if $\mathbf{z}_i^\tau \mathbf{z}_j < 0$).

Again, using the logistic regression model, we have that the probability of an edge between the nodes v_i and v_j depends upon the magnitude of

$$\begin{aligned} \eta_{ij} &= \text{logit } \mathrm{P}\{Y_{ij} = 1 | \mathbf{z}_i, \mathbf{z}_j, \mathbf{x}_{ij}, \boldsymbol{\theta}\} \\ &= \boldsymbol{\theta}^\tau \mathbf{x}_{ij} - \frac{\mathbf{z}_i^\tau \mathbf{z}_j}{\| \mathbf{z}_j \|}. \end{aligned} \tag{10.110}$$

Several extensions and generalizations to the form of η_{ij} and to alternative spaces have been proposed by Smith, Asta, and Calder (2019).

Estimation. Estimation for the projection model follows the same computational algorithm as proposed for the distance model. The main difference is that the log-likelihood of the projection model is invariant under rotation and reflection (but not translation), and so the matrix $\mathbf{T}$ ranges over the set of rotations and reflections.

10.8 Further Reading

Section 10.2. Although the basic idea behind exponential families of distributions dates back to the period 1934–1936 with articles by Fisher (1934), Darmois (1935), Koopman (1936), and Pitman (1936), reference to "exponential families" of distributions first appeared around 1950. Exponential families are described by Kass and Vos (1997) using the notions of differential geometry. Exponential families have long been part of statistical theory; see the books by Lehmann (1959, Section 2.7), Barndorff-Nielsen (1978), Casella and Berger (1990, Section 3.3), and the monograph by Brown (1986).

Curved exponential families were introduced and developed by Efron (1975, 1978) as a way of describing the curvature of a statistical problem; see also Amari (1982), Kass (1989), Lauritzen (1996, Appendix D), and Wainwright and Jordan (2008).

Sections 10.4.4 and 10.6.6. We recommend the book on MCMC edited by Gilks, Richardson, and Spiegelhalter (1996). Surveys of MCMC are given by Andrieu et al. (2003) and Robert and Casella (2011). Special issues of *Statistical Science* on MCMC are November 2001 [**16**(4)] and February 2004 [**19**(1)].

There are many excellent books on stochastic processes that include sections or chapters on Markov chains. These include Lamperti (1977), Doob (1990), Ross (1996), Durrett (2012), Karlin and Taylor (2012), and Parzen (2015). An excellent book on Markov chains that uses a constructive approach is Freedman (1971).

Section 10.6. A detailed account of ERGMs for social networks is given by Lusher, Kosinen, and Robins (2013).

Section 10.6.3. There are many excellent books dealing with logistic models, contingency tables, and log-linear models. These include Haberman (1974), Bishop, Fienberg, and Holland (1975), Agresti (1990), Christensen (1997), and Lloyd (1999). The book by McCullagh and Nelder (1989) on generalized linear models includes material (in Chapter 6) on log-linear models.

10.9 Exercises

Exercise 10.1 For a network dataset of your choosing, fit an ERGM to those data by using an appropriate software package. Give a full report on your results.

Exercise 10.2 Let X be a random variable and let $M_X(s) = \mathrm{E}\{e^{sX}\}$ denote its *moment generating function*. Show that

(a) $M_X(s)$ is continuous and nondecreasing on $[0, 1]$.

(b) $M_X(s)$ is strictly convex on $[0, 1]$ if $\mathrm{P}\{X \geq 2\} > 0$.

Exercise 10.3 Let X follow the binomial distribution $\mathrm{Bin}(p, N)$, where $\mathrm{P}\{X = x\} = \binom{n}{x} p^x q^{n-x}$, $p \in (0, 1)$, $q = 1 - p$. Show that its moment generating function is $M_X(s) = (q + pe^s)^N$, and find its first two moments. Show that the binomial distribution is a member of an exponential family.

Exercise 10.4 Let X follow the Poisson distribution $\mathcal{P}(\lambda)$, where $\mathrm{P}\{X = x\} = \lambda^x e^{-x}/x!$, $x = 0, 1, 2, \ldots$, $\lambda > 0$. Show that its moment generating function is $M_X(s) = e^{\lambda(e^s - 1)}$, and find its first two moments. Show that the Poisson distribution is a member of an exponential family.

Exercise 10.5 Suppose $X_1, X_2, \ldots, X_N$ are independent random variables, where X_i is distributed as Poisson $\mathcal{P}(\lambda_i)$. Show that the distribution of their sum, $\sum_{i=1}^{N} X_i$, is also Poisson.

Exercise 10.6 The random variable X has the logarithmic-series distribution with parameter $0 < p < 1$ if

$$P\{X = x\} = \frac{-(1-p)^x}{x \log(p)},$$

where $x = 1, 2, \ldots$.

(a) Show that these probabilities sum to one.
(b) Find the mean and variance of X.
(c) Find the moment generating function of X.
(d) Is the logarithmic-series distribution a member of an exponential family?

Exercise 10.7 Let X follow the power-series distribution, which has the form

$$P\{X = x\} = \frac{a(x)\theta^x}{C(\theta)},$$

where $x = 0, 1, 2, \ldots$, $a(x) \geq 0$, and $\theta > 0$.

(a) Is the power-series distribution a member of an exponential family?
(b) Find the moment generating function of X.
(c) Show that the Poisson distribution is a special case of the power-series distribution.
(d) What is the relationship (if any) between the logarithmic-series distribution and the power-series distribution?

Exercise 10.8 Let $T_X(s) = \log(M_X(s))$. Show that

(a) $\frac{d}{ds}T_X(s)|_{s=0} = E\{X\}$.
(b) $\frac{d^2}{ds^2}T_X(s)|_{s=0} = \text{var}\{X\}$.

Exercise 10.9 Show that the logit form of the p^* model is given by (10.82).

CHAPTER ELEVEN

Graph Partitioning: I. Graph Cuts

In this and the next three chapters, we study the problem of *graph partitioning*, which deals with the question of how to divide up the nodes into homogeneous groups (if they exist). This topic is currently receiving the greater part of research activity within the network science community. In this chapter, we describe various methods for graph partitioning using *graph cuts*, including binary cuts, ratio cuts, normalized cuts, and multiway cuts.

11.1 Introduction

The problem of partitioning the nodes of a graph $\mathcal{G} = (\mathcal{V}, \mathcal{E})$ into smaller subsets of nodes is similar to the problem of partitioning multivariate data points into homogeneous clusters (see, e.g., Izenman, 2013, Chapter 12). In some applications, data clusters may be disjoint subsets of the dataset, whilst in other applications, the clusters overlap in some way, thereby making classification problems more difficult. In an analogous situation with graph partitioning, the goal is to separate the nodes into disjoint clusters. However, for certain networks, that may not be possible because clusters may overlap; this topic will be the focus of Chapter 14.

The graph-partitioning problem has been found to be important in a number of different fields and disciplines, such as very large-scale integration (VLSI) design (Kahng et al., 2011), computer vision, and image segmentation (Cai, Wu, and Chung, 2006). Graph partitioning has proved to be vital in applications such as partitioning of electrical circuits (Schweikert and Kernighan, 1972), speech separation (Bach and Jordan, 2006), and detection of a brain tumor in MRI images (Padole and Chaudhari, 2012).

Graph partitioning is especially important for monitoring power grids, where the goal is to split the system so as to avoid cascading failures leading to catastrophic blackouts. In road networks, nodes can be intersections and edges can be road segments, and it is desired to partition the roads for efficient routing purposes (Delling and Werneck, 2013).

Another major application of graph partitioning occurs in parallel computing, where the available processors have to be divided up into separate clusters. The goal is to create an efficient and balanced computing environment so that data can be assigned evenly to the various clusters with minimum communication between clusters.

In many of these applications, the dimensions are huge, which increases the computational requirements and provides the researcher with substantial challenges.

11.1.1 Example: Legislative Redistricting

A recent application of the graph-partitioning process is to legislative redistricting (Duchin, 2018). See also Fifield et al. (2020). The goal is to find an optimal partition of a state into a number of congressional districts subject to the following criteria. Each district in the partition should be:

1. *Geographically compact.* Because there is no uniformly agreed-upon definition of compactness, it is typically assessed by eyeballing the result.
2. *Contiguous.* Each district in the partition has to be connected.
3. *Transitable.* From the *Federalist Papers*: "The natural limit of a democracy is that distance from the central point which will just permit the most remote citizen to assemble as often as their public functions demand."
4. *Respectful of community boundaries.*
5. *Districts should have equal populations.*

The collection of geographical units of a state (e.g., counties, townships, or voting precincts) can be viewed as the set $\mathcal{V}$ of nodes in a graph $\mathcal{G} = (\mathcal{V}, \mathcal{E})$ and the links between neighboring nodes constitute the elements of the set $\mathcal{E}$ of edges of the graph. Districts are then disjoint connected subgraphs of $\mathcal{G}$, where by "connected" (see criterion 2) we mean that any pair of nodes in a district is linked through a succession of nodes and edges.

Redistricting can result in districts having bizarre shapes, a feature that has been labeled as "partisan gerrymandering."[1] There have, therefore, been attempts to create automated computational algorithms for legislative redistricting. Knowing the geographical distribution of voters can also be an important aspect of any redistricting process. So, drawing district boundaries can be seen as equivalent to a graph-cut problem, where the set $\mathcal{V}$ of nodes is to be partitioned into K blocks (i.e., districts) and each node has to be a member of one and only one block.[2]

In this chapter, we describe various methods for graph partitioning using *graph cuts*. First, we discuss various algorithms for carrying out binary cuts of a graph (i.e., partitioning the set of nodes $\mathcal{V}$ into two disjoint subsets $\mathcal{V}_1$ and $\mathcal{V}_2$, so that $\mathcal{V} = \mathcal{V}_1 \cup \mathcal{V}_2$). We describe the Kernighan–Lin algorithm and two popular graph-cutting algorithms, ratio cuts and normalized cuts, and their solutions via a spectral approach and subsequent approximations. Then, we discuss multiway cuts, where the nodes of a graph are partitioned into more than two disjoint subsets.

---

[1]Gerrymandering is named after Elbridge Gerry, a signer of the Declaration of Independence, governor of Massachusetts in 1812, and vice-president to James Madison. His fame is now due to his rearranging a district in the North Shore of Boston into an unnatural shape that favored his party over a rival party.

[2]For example, prior to 2000, the Commonwealth of Pennsylvania had 21 congressional districts; after the 2000 Census, a population decrease reapportioned Pennsylvania into 19 congressional districts; after the 2010 Census, the number of congressional districts was reduced to $K = 18$. The 2020 Census reduced it even further to 17 districts.

11.2 Binary Cuts

Let $\mathcal{G} = (\mathcal{V}, \mathcal{E})$ denote an undirected graph or network, where $\mathcal{V}$ is the set of N nodes of the network and $\mathcal{E}$ is the set of edges between pairs of nodes. Let $\mathbf{W} = (w_{ij})$ be a symmetric weight matrix (i.e., $w_{ij} = w_{ji}$, where $w_{ij} \geq 0$ for $(v_i, v_j) \in \mathcal{E}$ and $w_{ij} = 0$ otherwise) that provides the weights on the edges of the network, where w_{ij} represents the degree of similarity between node v_i and node v_j. The matrix $\mathbf{W}$ is sometimes referred to as a *connectivity* or *similarity matrix*.

In this chapter, $\mathcal{V}_1$ and $\mathcal{V}_2$ are two disjoint subsets of the nodes in $\mathcal{V}$ that satisfy

- $\mathcal{V} = \mathcal{V}_1 \cup \mathcal{V}_2$,
- $\mathcal{V}_1 \cap \mathcal{V}_2 = \emptyset$,

where $\mathcal{V}_i$ consists of $N_i = |\mathcal{V}_i|$ nodes, $i = 1, 2$, and $|\mathcal{V}| = N = N_1 + N_2$. Partitioning $\mathcal{V}$ into two disjoint subsets is often called *bipartitioning* or *bisectioning* of $\mathcal{V}$. Optimal partitioning of $\mathcal{V}$ involves the specification of an objective function and its minimization (or maximization).

11.2.1 Minimum Cuts

Amongst the various objective functions used in graph partitioning, the simplest is the "cut" of a partition.

Definition 11.1 Any edge that joins a node from one subset $\mathcal{V}_1$ to a node in the other subset $\mathcal{V}_2$ is called a *cut edge*. The number of cut edges is called the *cut size*.

Definition 11.2 The *cut value* between $\mathcal{V}_1$ and $\mathcal{V}_2$ is the sum of the weights of the cut edges joining the two subsets:

$$\text{cut}\{\mathcal{V}_1, \mathcal{V}_2\} = \sum_{u \in \mathcal{V}_1} \sum_{v \in \mathcal{V}_2} w_{uv}. \tag{11.1}$$

Definition 11.3 The *minimum cut* (or *mincut*) of $\mathcal{V}$ is the bipartition of $\mathcal{V}$ that minimizes the cut value.

11.2.2 Example: Parallel Computing

The partitioning problem is a major issue in parallel computing. It is not unusual today for researchers to have personal computers with two or more processors, so that computations can be spread across a number of processors by assigning to each processor a subset of the computations. Thus, computations on a two-processor computer can be simplified by treating the circuit as a graph with gates viewed as nodes of the graph and an electrical connection joining two gates viewed as an edge, and then assigning half the nodes to one processor and half to the other.

So, how should we partition the network nodes into two equal-size subsets to yield a balanced workload? Furthermore, how should we arrange the computations so that as little of the time as possible is taken up transmitting values back and forth from one processor to the other? It becomes a problem when this type of cross-processor communication slows down the computations on parallel computers. It is, therefore, important to try to minimize communication between processors, which corresponds to minimizing the cut size.

11.2.3 Kernighan–Lin Algorithm

The most popular algorithm for the bipartitioning of V into two equal-size disjoint subsets is the heuristic *Kernighan–Lin algorithm*[3] (Kernighan and Lin, 1970). The algorithm was originally motivated by showing how it could be used to partition electronic circuits when circuits had to be split evenly between two printed circuit boards, and where the number of connections were to be minimized to reduce costs and increase efficiency.

Let $N = 2m$ be the number of nodes in the network. The Kernighan–Lin algorithm produces two sets V_1 and V_2 such that

$$|V_1| = |V_2| = m, \quad V_1 \cup V_2 = V, \quad V_1 \cap V_2 = \emptyset, \tag{11.2}$$

while minimizing cut$\{V_1, V_2\}$. The algorithm begins with an initial partition of V into V_1 and V_2, where $|V_1| = |V_2| = m$ and $V_1 \cap V_2 = \emptyset$. At the conclusion of the algorithm, the partition $\mathcal{P} = \{V_1, V_2\}$ is turned into an "optimal" partition, $\mathcal{P}^* = \{V_1^*, V_2^*\}$. This is accomplished by swapping a subset $A \subseteq V_1$ with a subset $B \subseteq V_2$ so that $|A| = |B|$, $A = V_1 \cap V_2^*$, and $B = V_1^* \cap V_2$. The optimal partition is then given by

$$V_1^* = (V_1 \backslash A) \cup B, \quad V_2^* = (V_2 \backslash B) \cup A. \tag{11.3}$$

The problem is to identify A and B.

In order to generate A and B, Kernighan and Lin introduced the two ideas of external cost and internal cost. The *external cost*, E_a, of a node $a \in V_1$ is the sum of the weights of the edges between a and the nodes that are not members of V_1, while the *internal cost*, I_a, of $a \in V_1$ is the sum of the weights of the edges between a and nodes that are also in V_1. In other words,

$$E_a = \sum_{v \in V_2} w_{av}, \quad I_a = \sum_{v \in V_1} w_{av}, \quad a \in V_1. \tag{11.4}$$

For a given node $a \in V_1$, let D_a denote the difference between its external cost and the internal cost,

$$D_a = E_a - I_a. \tag{11.5}$$

So, D_a represents the benefit of moving node a from V_1 to V_2. Similarly, to preserve the balance between V_1 and V_2, a node b is moved from V_2 to V_1 whenever a node a is moved from V_1 to V_2. We can write the external cost as

$$E_a = w_{ab} + \sum_{v \in V_2, v \neq b} w_{av}, \tag{11.6}$$

---

[3] Brian W. Kernighan, who worked at Ball Laboratories, coined the name Unix, contributed to the development of the Unix operating system, and coauthored popular manuals on Unix (Kernighan and Pike, 1984) and the *C* programming language (Kernighan and Ritchie, 1978).

so that

$$D_a = w_{ab} + \sum_{v \in V_2, v \neq b} w_{av} - I_a. \tag{11.7}$$

Similarly, for $b \in V_2$:

$$D_b = w_{ab} + \sum_{u \in V_1, u \neq a} w_{bu} - I_b. \tag{11.8}$$

So, moving a from V_1 to V_2 reduces the cost by $D_a - w_{ab}$ and moving b from V_2 to V_1 reduces the cost by $D_b - w_{ab}$. The total cost reduction (or "gain") due to this swap is, therefore,

$$g_{ab} = D_a + D_b - 2w_{ab}. \tag{11.9}$$

Next, we update the D-values from this swap. Let $x \in V_1 \backslash \{a\}$. If $b \in V_2$ moves to V_1, the internal cost of x increases by w_{xb}, while if $a \in V_1$ moves to V_2, the internal cost of x decreases by w_{xa}. So, the new internal cost of x is $I'_x = I_x - w_{xa} + w_{xb}$. Similarly, the new external cost of x is $E'_x = E_x + w_{xa} - w_{xb}$. A similar argument holds for $y \in V_2 \backslash \{b\}$. Putting these results of a swap together, the new D-values are $D'_x = E'_x - I'_x$ and $D'_y = E'_y - I'_y$. That is,

$$D'_x = D_x + 2w_{xa} - 2w_{xb}, \quad \text{for all } x \in V_1 \backslash \{a\}, \tag{11.10}$$

$$D'_y = D_y + 2w_{yb} - 2w_{ya}, \quad \text{for all } y \in V_2 \backslash \{b\}. \tag{11.11}$$

Next, we use a greedy procedure to identify the subsets A and B.

Compute g_{ab} for all $a \in V_1$ and $b \in V_2$. Find the pair (a_1, b_1) that yields the largest gain, g_1. Lock a_1 and b_1 in place. This results in the subsets $X = V_1 \backslash \{a_1\}$ and $Y = V_2 \backslash \{b_1\}$. Update the D-values of the remaining "free" elements not yet locked. Compute the gains. Find a second pair of nodes, a_2 and b_2, that yield the largest gain, g_2. Lock a_2 and b_2 in place. The total gain from swapping a_1 with b_1, followed by swapping a_2 with b_2, is $G_2 = g_1 + g_2$. Continue this swapping process, so that after $k \leq m$ swaps, $A = \{a_1, a_2, \ldots, a_k\}$ and $B = \{b_1, b_2, \ldots, b_k\}$, and the total gain is $G_k = \sum_{i=1}^{k} g_i$. The process terminates when the current partition cannot be improved any further. This occurs when there is no k such that $G_k > 0$.

The resulting partition of V into $V'_1 \cup V'_2$ with $V'_1 = (V_1 \backslash A) \cup B$ and $V'_2 = (V_2 \backslash B) \cup A$, where $V'_1 \cap V'_2 = \emptyset$ and $|V'_1| = |V'_2| = m$, constitutes a single iteration of the Kernighan–Lin algorithm. The entire process is repeated, this time starting with V'_1 and V'_2. The iterations stop when no further improvements can be obtained; that is, when $G_k \leq 0$.

The algorithm is very sensitive to the initial partition of V, where prior information can be helpful. As for algorithmic complexity, computing external and internal costs has a complexity of $\mathcal{O}(N^2)$ and finding pairs of nodes to swap in a single iteration has complexity $\mathcal{O}(N^2 \log N)$. Thus, the complexity of the algorithm is $\mathcal{O}(N^2 \log N)$, which shows that the algorithm is computationally intensive and can be slow if the initial partition is poor.

There have been several variations on the Kernighan–Lin algorithm. These include adapting the algorithm for unequal-sized subsets and multiway partitions, and the Fiduccia–Mattheyses algorithm.

Table 11.1 Adjacency matrix for example of Kernighan–Lin algorithm

	1	2	3	4	5	6
1	0	1	1	1	0	0
2	1	0	0	0	0	0
3	1	0	0	0	0	0
4	1	0	0	0	1	1
5	0	0	0	1	0	1
6	0	0	0	1	1	0

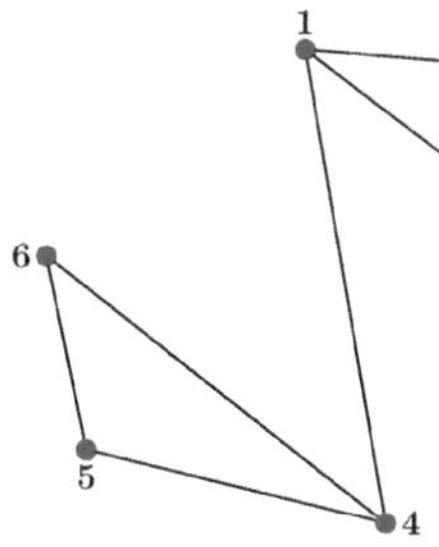

Figure 11.1 Worked Example 11.1. Network graph of the adjacency matrix in Table 11.1. There are six nodes and six edges.

Worked Example 11.1

Consider the graph displayed in **Figure 11.1** and the adjacency matrix of that graph in **Table 11.1**. The six nodes are $\mathcal{V} = \{v_1, v_2, v_3, v_4, v_5, v_6\}$, which we abbreviate to $\{1,2,3,4,5,6\}$. Suppose we start the Kernighan–Lin algorithm by partitioning $\mathcal{V}$ into the two disjoint subsets

$$\mathcal{V}_1 = \{2,5,6\}, \quad \mathcal{V}_2 = \{1,3,4\}. \tag{11.12}$$

We assume that the weights w_{uv} are specified by the adjacency matrix; that is, that $w_{uv} = 1$ if there is an edge joining nodes u and v, and $w_{uv} = 0$ otherwise.

First, we compute the D-values. For example, node 1, which is in $\mathcal{V}_2$, is connected to nodes 3 and 4 in $\mathcal{V}_2$, and so $I_1 = \sum_{v \in \mathcal{V}_2} w_{1v} = 2$; furthermore, node 1 is connected to node 2 in $\mathcal{V}_1$, and so $E_1 = \sum_{v \in \mathcal{V}_1} w_{1v} = 1$. Thus,

$$\begin{aligned}
D_1 &= E_1 - I_1 = 1 - 2 = -1, \\
D_2 &= E_2 - I_2 = 1 - 0 = 1, \\
D_3 &= E_3 - I_3 = 0 - 1 = -1, \\
D_4 &= E_4 - I_4 = 2 - 1 = 1, \\
D_5 &= E_5 - I_5 = 1 - 1 = 0, \\
D_6 &= E_6 - I_6 = 1 - 1 = 0.
\end{aligned}$$

Next, we compute the gains for all possible swaps:

$$\begin{aligned}
g_{21} &= D_2 + D_1 - 2w_{21} = 1 + (-1) - 2(1) = -2, \\
g_{23} &= D_2 + D_3 - 2w_{23} = 1 + (-1) - 2(0) = 0, \\
g_{24} &= D_2 + D_4 - 2w_{24} = 1 + 1 - 2(0) = 2,
\end{aligned}$$

$$g_{51} = D_5 + D_1 - 2w_{51} = 0 + (-1) - 2(0) = -1,$$
$$g_{53} = D_5 + D_3 - 2w_{53} = 0 + (-1) - 2(0) = -1,$$
$$g_{54} = D_5 + D_4 - 2w_{54} = 0 + 1 - 2(1) = -1,$$
$$g_{61} = D_6 + D_1 - 2w_{61} = 0 + (-1) - 2(0) = -1,$$
$$g_{63} = D_6 + D_3 - 2w_{63} = 0 + (-1) - 2(0) = -1,$$
$$g_{64} = D_6 + D_4 - 2w_{64} = 0 + 1 - 2(1) = -1.$$

The largest g-value is g_{24}. Thus, $(a_1, b_1) = (2,4)$ and the maximum gain is $g_{24} = g_1 = 2$. Lock nodes 4 (in V_1) and 2 (in V_2) in place, and write [4] and [2]. We update the subsets as follows:

$$X = V_1 \setminus \{2\} = \{5, 6\},$$
$$Y = V_2 \setminus \{4\} = \{1, 3\}.$$

Both X and Y are not empty. The two subsets are now $V_1^{(1)} = \{[4], 5, 6\}$ and $V_2^{(1)} = \{1, [2], 3\}$.

Now, we update the D-values of the nodes. The new D-values are

$$D_1' = D_1 + 2w_{14} - 2w_{12} = (-1) + 2(1) - 2(1) = -1,$$
$$D_3' = D_3 + 2w_{34} - 2w_{32} = (-1) + 2(0) - 2(0) = -1,$$
$$D_5' = D_5 + 2w_{52} - 2w_{54} = 0 + 2(0) - 2(1) = -2,$$
$$D_6' = D_6 + 2w_{62} - 2w_{64} = 0 + 2(0) - 2(1) = -2.$$

We next update the gains:

$$g_{51} = D_5' + D_1' - 2w_{51} = -2 + (-1) - 2(0) = -3,$$
$$g_{61} = D_6' + D_1' - 2w_{61} = -2 + (-1) - 2(0) = -3,$$
$$g_{53} = D_5' + D_3' - 2w_{53} = -2 + (-1) - 2(0) = -3,$$
$$g_{63} = D_6' + D_3' - 2w_{63} = -2 + (-1) - 2(0) = -3.$$

All four gains are equal and negative. So, the iterations stop because no further improvements can be made by swapping nodes.

Now, we determine the value of k. We have

$$G_1 = g_1 = 2,$$
$$G_2 = g_1 + g_2 = 2 + (-3) = -1.$$

The k with the largest G_k is $k = 1$. This yields the final partition:

$$V_1^* = \{1, 2, 3\}, \quad V_2^* = \{4, 5, 6\}. \tag{11.13}$$

11.2.4 Ratio Cuts

Let V_1 and V_2 be two subsets with V_i consisting of $N_i = |V_i|$ nodes, $i = 1, 2$, and $N = N_1 + N_2$.

Definition 11.4 The *ratio-cut objective function* for two disjoint subsets, V_1 and V_2, is defined as

$$\text{RatioCut}(V_1, V_2) = \text{cut}\{V_1, V_2\} \left\{ \frac{1}{N_1} + \frac{1}{N_2} \right\} = N \cdot \frac{\text{cut}\{V_1, V_2\}}{N_1 N_2}. \tag{11.14}$$

Because $N = |\mathcal{V}|$ is fixed and, therefore, can be dropped from further consideration, the ratio-cut optimization problem seeks to find $\mathcal{V}_1$ (and, hence, $\mathcal{V}_2$) to minimize

$$\frac{\text{cut}\{\mathcal{V}_1, \mathcal{V}_2\}}{N_1 N_2}. \tag{11.15}$$

The RatioCut criterion allows subset sizes to differ, but it is still biased towards finding subsets of the same size. The problem of finding a partition of $\mathcal{V}$ into two equal-sized subsets by minimizing RatioCut is an $\mathcal{NP}$-hard problem.

Applications

The RatioCut criterion has been used in the following problems.

- *VLSI computer-aided design.* In this application, the huge number of components that make up today's VLSI circuits are partitioned into interconnected subcircuits or "blocks," which, after being connected, operate as the original circuit. A *block* can be a collection of transistors or gates, or a complete circuit. The goals of circuit partitioning are to reduce the number of blocks and minimize the number of inter-block signals. Application of the RatioCut criterion to circuit partitioning has proved to be very successful and efficient, and has enabled huge cost savings to take place in solving digital circuit layout problems (Wei and Cheng, 1991; Hagen and Kahng, 1992).

- *Air-traffic control system.* In this application, airspace is partitioned into disjoint regions for the dual purposes of safety and efficiency. An air-traffic controller supervises traffic in a limited area, called an *air-traffic sector*. Controllers are qualified for working on a set of sectors, called a *functional airspace block*. Controllers only know air-traffic sectors on which they are qualified, and it is unusual for controllers to know other sectors. The system is represented by a graph $\mathcal{G} = (\mathcal{V}, \mathcal{E})$, where nodes represent air-traffic sectors and edges represent aircraft flow between sectors. Each edge (v_i, v_j) has a weight w_{ij}, which is the number of aircraft that fly from v_i to v_j, and vice versa. Each node has a weight, which is the sum of the weights of connected edges plus aircraft departures and landings if the sector is connected with an airport. The problem is to construct blocks of sectors to maximize the flow of aircraft within blocks, whilst minimizing aircraft flow between blocks. Bichot and Durand (2006) studied the RatioCut method for partitioning European airspace (759 nodes, 3165 edges) into blocks.

A Spectral Approach

We follow the development of von Luxburg (2007). Define the N-vector $\mathbf{s} = (s_1, \ldots, s_N)^\tau \in \mathbb{R}^N$ so that

$$s_i = \begin{cases} \sqrt{N_2/N_1}, & \text{if } v_i \in \mathcal{V}_1 \\ -\sqrt{N_1/N_2}, & \text{if } v_i \in \mathcal{V}_2 \end{cases}. \tag{11.16}$$

Then, the *unnormalized graph Laplacian*[4] is defined as

$$\mathbf{L} = \mathbf{D} - \mathbf{Y}, \tag{11.17}$$

---

[4]See Chapter 13 for more details on graph Laplacians.

where $\mathbf{D} = (D_{ij})$ and

$$D_{ii} = \sum_j Y_{ij} \tag{11.18}$$

is the diagonal degree matrix formed from $\mathbf{Y}$. If $\mathbf{1}_N$ is the N-vector of 1s, then $\mathbf{D} = \mathbf{Y}\mathbf{1}_N$. We have that

$\mathbf{s}^\tau \mathbf{L}\mathbf{s}$

$$= \frac{1}{2} \sum_{i,j=1}^n w_{ij}(s_i - s_j)^2$$

$$= \frac{1}{2} \sum_{i \in \mathcal{V}_1, j \in \mathcal{V}_2} w_{ij} \left(\sqrt{\frac{N_2}{N_1}} + \sqrt{\frac{N_1}{N_2}} \right)^2 + \frac{1}{2} \sum_{i \in \mathcal{V}_2, j \in \mathcal{V}_1} w_{ij} \left(-\sqrt{\frac{N_2}{N_1}} - \sqrt{\frac{N_1}{N_2}} \right)^2$$

$$= \text{cut}\{\mathcal{V}_1, \mathcal{V}_2\} \left(\frac{N_2}{N_1} + \frac{N_1}{N_2} + 2 \right)$$

$$= N \cdot \text{RatioCut}\{\mathcal{V}_1, \mathcal{V}_2\}. \tag{11.19}$$

So, finding a vector $\mathbf{s}$ to minimize the quadratic form $\mathbf{s}^\tau \mathbf{L}\mathbf{s}$ is equivalent to finding subsets $\mathcal{V}_1$ and $\mathcal{V}_2$ to minimize RatioCut($\mathcal{V}_1, \mathcal{V}_2$). Furthermore,

$$\mathbf{s}^\tau \mathbf{1}_N = \sum_{i=1}^N s_i$$

$$= \sum_{i \in \mathcal{V}_1} \sqrt{\frac{N_2}{N_1}} - \sum_{i \in \mathcal{V}_2} \sqrt{\frac{N_1}{N_2}}$$

$$= N_1 \sqrt{\frac{N_2}{N_1}} - N_2 \sqrt{\frac{N_1}{N_2}} = 0, \tag{11.20}$$

$$\| \mathbf{s} \|^2 = \sum_{i=1}^N s_i^2 = N_1 \cdot \frac{N_2}{N_1} + N_2 \cdot \frac{N_1}{N_2} = N. \tag{11.21}$$

So, our optimization problem boils down to

$$\min_{\mathcal{V}_1} \mathbf{s}^\tau \mathbf{L}\mathbf{s} \quad \text{subject to} \quad \mathbf{s} \perp \mathbf{1}_N, \quad \| \mathbf{s} \| = \sqrt{N}. \tag{11.22}$$

Because the entries of $\mathbf{s}$ can have only one of two possible values, this is a discrete optimization problem that is $\mathcal{NP}$-hard.

Relaxation

One way to resolve this difficulty is to abandon the discreteness requirement and instead allow s_i to take arbitrary values in $\mathbb{R}$; that is,

$$\min_{\mathbf{s} \in \mathbb{R}^N} \mathbf{s}^\tau \mathbf{L}\mathbf{s} \quad \text{subject to} \quad \mathbf{s} \perp \mathbf{1}_N, \quad \| \mathbf{s} \| = \sqrt{N}. \tag{11.23}$$

This can be solved using the *Rayleigh–Ritz theorem*.[5]

[5]Named after Lord Rayleigh and Waltham Ritz.

Theorem 11.1 (Rayleigh–Ritz) *Let $\mathbf{A}$ be a symmetric $(N \times N)$-matrix with eigenvalues $\lambda_1 \leq \lambda_2 \leq \cdots \leq \lambda_N$ and eigenvector $\mathbf{v}_i$ associated with λ_i, $i = 1, 2, \ldots, N$. Then,*

$$\max_{\mathbf{x} \neq \mathbf{0}} \frac{\mathbf{x}^\tau \mathbf{A} \mathbf{x}}{\mathbf{x}^\tau \mathbf{x}} = \lambda_N, \tag{11.24}$$

with maximum attained at $\mathbf{x} = \mathbf{v}_N$, and, for $k = 1, 2, \ldots, N - 1$,

$$\max_{\mathbf{x} \neq \mathbf{0},\, \mathbf{x} \in \{\mathbf{v}_{N-k+1}, \ldots, \mathbf{v}_N\}^\perp} \frac{\mathbf{x}^\tau \mathbf{A} \mathbf{x}}{\mathbf{x}^\tau \mathbf{x}} = \lambda_{N-k}, \tag{11.25}$$

with maximum attained at $\mathbf{x} = \mathbf{v}_{N-k}$. Similarly,

$$\min_{\mathbf{x} \neq \mathbf{0}} \frac{\mathbf{x}^\tau \mathbf{A} \mathbf{x}}{\mathbf{x}^\tau \mathbf{x}} = \lambda_1, \tag{11.26}$$

with minimum attained at $\mathbf{x} = \mathbf{v}_1$, and, for $i = 2, 3, \ldots, N$,

$$\min_{\mathbf{x} \neq \mathbf{0},\, \mathbf{x} \in \{\mathbf{v}_1, \ldots, \mathbf{v}_{i-1}\}^\perp} \frac{\mathbf{x}^\tau \mathbf{A} \mathbf{x}}{\mathbf{x}^\tau \mathbf{x}} = \lambda_i, \tag{11.27}$$

with minimum attained at $\mathbf{x} = \mathbf{v}_i$.

The solution to the optimization problem (11.23) is to take $\mathbf{s}$ to be the eigenvector corresponding to the second-smallest eigenvalue of $\mathbf{L}$. (The smallest eigenvalue of $\mathbf{L}$ is 0 with eigenvector $\mathbf{1}_N$.)

Rounding

This solution now has to be turned back into an indicator variable. The result is the following:

$$\begin{array}{ll} v_i \in \mathcal{V}_1, & \text{if } s_i \geq 0 \\ v_i \in \mathcal{V}_2, & \text{if } s_i < 0 \end{array}. \tag{11.28}$$

11.2.5 Normalized Cuts

The *normalized cut* (Shi and Malik, 2000) aims to minimize the connections emerging from each subset to the rest of the nodes relative to the connections within the subset.

Definition 11.5 The *normalized-cut objective function* is given by

$$\begin{aligned} \text{Ncut}\{\mathcal{V}_1, \mathcal{V}_2\} &= \frac{\text{cut}\{\mathcal{V}_1, \mathcal{V}_2\}}{\text{cut}\{\mathcal{V}_1, \mathcal{V}\}} + \frac{\text{cut}\{\mathcal{V}_2, \mathcal{V}_1\}}{\text{cut}\{\mathcal{V}_2, \mathcal{V}\}} \\ &= \text{cut}\{\mathcal{V}_1, \mathcal{V}_2\} \left\{ \frac{1}{\text{cut}\{\mathcal{V}_1, \mathcal{V}\}} + \frac{1}{\text{cut}\{\mathcal{V}_2, \mathcal{V}\}} \right\} \\ &= L \cdot \frac{\text{cut}\{\mathcal{V}_1, \mathcal{V}_2\}}{\kappa_1 \kappa_2}, \end{aligned} \tag{11.29}$$

where $\kappa_r = \text{cut}\{\mathcal{V}_r, \mathcal{V}\}$ is the total number of edges joining nodes in $\mathcal{V}_r$ to all nodes in the network, $r = 1, 2$, and $L = \kappa_1 + \kappa_2$.

We often see cut$\{\mathcal{V}_r, \mathcal{V}\}$ written either as assoc$\{\mathcal{V}_r, \mathcal{V}\}$ or as vol$\{\mathcal{V}_r\}$, the volume of $\mathcal{V}_r$. The goal is to find the cut that minimizes Ncut.

We define the constant h_{Ncut} as follows:

$$h_{\text{Ncut}} = \min_{\mathcal{V}_1} \ \text{Ncut}\{\mathcal{V}_1, \mathcal{V}_2\}. \qquad (11.30)$$

The Ncut value is minimized when cut$\{\mathcal{V}_1, \mathcal{V}\} = $ cut$\{\mathcal{V}_2, \mathcal{V}\}$ (i.e., when $\kappa_1 = \kappa_2$). It tends to be large for a cut that separates out isolated nodes (such as mincut produces) because the cut value will be a high percentage of the total connection from $\mathcal{V}_1$ to all nodes in the graph. As a result, the Ncut function penalizes unbalanced partitions (such as mincut produces).

Unfortunately, minimizing Ncut is an $\mathcal{NP}$-hard problem. So, we pursue a strategy of manipulating the problem into a more useful problem (which is still $\mathcal{NP}$-hard) and then relaxing the problem to get it into a continuous form, which can then be solved exactly.

The Cheeger Constant

The Ncut measure is closely related to the Cheeger constant[6] in spectral geometry. We can express Ncut as follows:

$$\text{Ncut}\{\mathcal{V}_1, \mathcal{V}_2\} = \frac{\text{cut}\{\mathcal{V}_1, \mathcal{V}_2\}}{\min\{\text{cut}\{\mathcal{V}_1, \mathcal{V}\}, \text{cut}\{\mathcal{V}_2, \mathcal{V}\}\}}$$

$$+ \frac{\text{cut}\{\mathcal{V}_1, \mathcal{V}_2\}}{\max\{\text{cut}\{\mathcal{V}_1, \mathcal{V}\}, \text{cut}\{\mathcal{V}_2, \mathcal{V}\}\}}$$

$$= h_{\mathcal{G}}(\mathcal{V}_1) + \frac{\text{cut}\{\mathcal{V}_1, \mathcal{V}_2\}}{\max\{\text{cut}\{\mathcal{V}_1, \mathcal{V}\}, \text{cut}\{\mathcal{V}_2, \mathcal{V}\}\}}, \qquad (11.31)$$

where

$$h_{\mathcal{G}}(\mathcal{V}_1) = \frac{\text{cut}\{\mathcal{V}_1, \mathcal{V}_2\}}{\min\{\text{cut}\{\mathcal{V}_1, \mathcal{V}\}, \text{cut}\{\mathcal{V}_2, \mathcal{V}\}\}} \qquad (11.32)$$

is referred to as the *Cheeger ratio* of $\mathcal{V}_1 \subset V$. The *Cheeger constant* h_G of the graph $\mathcal{G}$ is defined as the fewest number of edges that would have to be cut away from the graph $\mathcal{G}$ so that the graph formed from $\mathcal{V}_1$ and its edges would be separated from the rest of the graph; that is,

$$h_{\mathcal{G}} = \min_{\mathcal{V}_1} \ h_{\mathcal{G}}(\mathcal{V}_1). \qquad (11.33)$$

So, Ncut is a version of the Cheeger constant. In fact, the relationship can be expressed in terms of the inequality

$$\frac{h_{\text{Ncut}}}{2} \leq h_{\mathcal{G}} \leq h_{\text{Ncut}}. \qquad (11.34)$$

The upper bound, $h_{\mathcal{G}} \leq h_{\text{Ncut}}$, is derived by minimizing (11.31) over choice of $\mathcal{V}_1$. For the lower bound, we have $h_{\mathcal{G}}(\mathcal{V}_1) \geq \frac{1}{2}\text{Ncut}\{\mathcal{V}_1, \mathcal{V}_2\}$, whence, $h_{\mathcal{G}} \geq h_{\text{Ncut}}/2$. However, like the Ncut problem, finding $h_{\mathcal{G}}$ is nontrivial and is an $\mathcal{NP}$-hard problem.

---

[6]The constant, ratio, and inequality are each named after Jeffrey L. Cheeger. See Cheeger (1970).

In light of the fact that finding $h_{\mathcal{G}}$ is an $\mathcal{NP}$-hard problem, an important result in graph theory is the *Cheeger inequality*, which is used to approximate the Cheeger constant and is given by

$$\frac{h_{\mathcal{G}}}{2} \leq \lambda_2 \leq 2h_{\mathcal{G}}, \tag{11.35}$$

where λ_2 is the second-smallest eigenvalue of the normalized Laplacian $\mathbf{L}^{sym} = \mathbf{D}^{-1/2}\mathbf{Y}\mathbf{D}^{-1/2}$. For any graph, $\lambda_1 = 0$. The proof of (11.35) is long and can be found in Chung (1997, 2007).

Applications

The Ncut measure has proved to be the most popular method for graph partitioning in a number of different areas and disciplines. The minimum of the Ncut measure is identical to the *conductance* of a graph, which measures how well-connected the graph is. It is defined as the minimum conductance over all possible cuts of the graph. If the conductance of a graph is small, it implies that there is a "bottleneck" in the graph (i.e., there are two distinct subsets of nodes that are not well-connected to each other), while a high value implies that the graph is well-connected. Conductance has been studied at length in graph theory, machine learning, and in problems involving the "mixing rate" of Markov chains (Sinclair, 1992); finding the conductance of a graph is well-known to be an $\mathcal{NP}$-hard problem.

The Ncuts measure has also been applied, for example, to the following types of problems.

- *Detection of brain tumor.* Brain tumors can be benign or malignant and their location influences the type of symptoms that can occur. Diagnostic information is obtained through an MRI (magnetic resonance image), where pixels represent nodes in a weighted undirected graph. The Ncut algorithm is applied directly to the MRI, which computes the exact tumor area, its centroid, eccentricity, and bounding box, which are used to estimate the shape of the tumor region. MRIs are computed from different planes and the Ncut algorithm is applied to each of them, and the tumor shape is then determined (Padole and Chaudhari, 2012).
- *Image segmentation.* In this application, the pixels of an image are modeled as nodes in an undirected graph that are then partitioned using Ncut into disjoint subsets, a process that can be extremely computationally intensive (see, e.g., Cai, Wu, and Chung, 2006). Many examples can be found in the biomedical imaging literature (Carballido-Gamio, Belongie, and Majumdar, 2004; van den Heuvel, Mandl, and Pol, 2008).
- *Network tomography.* In this application, the interest is in a network's internal characteristics, and *network mapping*, which is the study of the Internet's physical characteristics. One of the most annoying aspects of modern-day communications is the existence (and persistence) of spam e-mails (estimated to be over 80% of e-mails). The hidden economic costs of having to deal with spam are astronomical. The NSA has monitored spamming activity since 2004, and has used spectral clustering and the Ncut measure to help identify the social networks of spammers (National Security Agency, 2010).

- *Web clustering.* In this application, the Ncut measure is used to cluster millions of documents on the World Wide Web by incorporating distinct topic areas, textual similarity, co-citation information, and link structures into a single similarity metric (see, e.g., He et al., 2002).

The biggest challenge in carrying out any of these applications is to come up with computational methods for solving separation problems efficiently. In image segmentation problems, we see hundreds of thousands of pixels resulting in very large adjacency and weighted adjacency matrices that cannot be stored in main memory. For images of 256×256 pixels, for example, the adjacency matrices contain 65,536 pixels, and so the corresponding adjacency matrix is huge, of size $65,536 \times 65,536$. In speech separation problems (Bach and Jordan, 2006), 4 seconds of speech sampled at 5.5 kHz provides us with 22,000 samples, and so the corresponding adjacency matrices have size $22,000 \times 22,000$.

Spectral Segmentation

We ignore the constant L in (11.29) because it plays no role in the optimization procedure. Following Newman (2013), we define the N-vector $\mathbf{s} = (s_1, \ldots, s_N)^{\tau} \in \mathbb{R}^N$ so that

$$s_i = \begin{cases} \sqrt{\kappa_2/\kappa_1}, & \text{if } v_i \in \mathcal{V}_1 \\ -\sqrt{\kappa_1/\kappa_2}, & \text{if } v_i \in \mathcal{V}_2 \end{cases}, \quad i = 1, 2, \ldots, N, \tag{11.36}$$

where κ_r is the total number of degrees in the rth group ($r = 1, 2$). Now,

$$\mathbf{k}^{\tau}\mathbf{s} = \sum_i k_i s_i$$

$$= \sqrt{\frac{\kappa_2}{\kappa_1}} \sum_{i \in \mathcal{V}_1} k_i - \sqrt{\frac{\kappa_1}{\kappa_2}} \sum_{i \in \mathcal{V}_2} k_i$$

$$= \sqrt{\kappa_2 \kappa_1} - \sqrt{\kappa_1 \kappa_2} = 0. \tag{11.37}$$

Furthermore,

$$\mathbf{s}^{\tau}\mathbf{D}\mathbf{s} = \sum_i k_i s_i^2 = \frac{\kappa_2}{\kappa_1} \sum_{i \in \mathcal{V}_1} k_i + \frac{\kappa_1}{\kappa_2} \sum_{i \in \mathcal{V}_2} k_i = \kappa_2 + \kappa_1 = L. \tag{11.38}$$

Note that

$$s_i + \sqrt{\frac{\kappa_1}{\kappa_2}} = \frac{L}{\sqrt{\kappa_1 \kappa_2}} \delta_{c_i, 1}, \quad s_i - \sqrt{\frac{\kappa_2}{\kappa_1}} = -\frac{L}{\sqrt{\kappa_1 \kappa_2}} \delta_{c_i, 2}. \tag{11.39}$$

Thus,

$$\sum_{i, j} Y_{ij} \left(s_i + \sqrt{\frac{\kappa_1}{\kappa_2}} \right) \left(s_i - \sqrt{\frac{\kappa_2}{\kappa_1}} \right) = -\frac{L^2}{\kappa_1 \kappa_2} \sum_{i, j} Y_{ij} \delta_{c_i, 1} \delta_{c_i, 2}$$

$$= -L^2 \frac{m_{12}}{\kappa_1 \kappa_2}, \tag{11.40}$$

where

$$m_{12} = \sum_{i,j} Y_{ij}\delta_{c_i,1}\delta_{c_j,2} = \text{cut}\{\mathcal{V}_1,\mathcal{V}_2\} \qquad (11.41)$$

is the total number of edges between subsets $\mathcal{V}_1$ and $\mathcal{V}_2$.

We can express this result in matrix terms. First, note that

$$\mathbf{k} = \mathbf{Y}\mathbf{1}_N, \quad \mathbf{1}_N^\tau \mathbf{Y}\mathbf{1}_N = L. \qquad (11.42)$$

The lhs of (11.40) is

$$\left(\mathbf{s} + \sqrt{\frac{\kappa_1}{\kappa_2}}\mathbf{1}_N\right)^\tau \mathbf{Y}\left(\mathbf{s} - \sqrt{\frac{\kappa_2}{\kappa_1}}\mathbf{1}_N\right) = \mathbf{s}^\tau \mathbf{Y}\mathbf{s} - L. \qquad (11.43)$$

In other words,

$$\text{Ncut}\{\mathcal{V}_1,\mathcal{V}_2\} = \frac{\text{cut}\{\mathcal{V}_1,\mathcal{V}_2\}}{\kappa_1\kappa_2} = \frac{L - \mathbf{s}^\tau \mathbf{Y}\mathbf{s}}{L^2}. \qquad (11.44)$$

So, finding $\mathcal{V}_1$ and $\mathcal{V}_2$ to minimize $\text{Ncut}\{\mathcal{V}_1,\mathcal{V}_2\}$ is equivalent to finding $\mathbf{s}$ to maximize $\mathbf{s}^\tau \mathbf{Y}\mathbf{s}$, subject to $\mathbf{s}$ being defined by (10.36). Because the elements of $\mathbf{s}$ are each discrete-valued with two possible values, this is a discrete optimization problem, which is $\mathcal{NP}$-hard.

Relaxation. As an approximation to the solution, we relax the requirement that the s_i have to be two-valued and, instead, allow them to take any real values, subject to the constraints (11.27).

Let λ and μ be Lagrangian multipliers. Set up the following optimization function:

$$f(\mathbf{s},\lambda,\mu) = \frac{1}{2}\mathbf{s}^\tau \mathbf{Y}\mathbf{s} - \lambda(\mathbf{k}^\tau \mathbf{s}) - \frac{1}{2}\mu(\mathbf{s}^\tau \mathbf{D}\mathbf{s} - L). \qquad (11.45)$$

Differentiate wrt $\mathbf{s}$ and set the result equal to zero to get

$$\mathbf{Y}\mathbf{s} = \mu \mathbf{D}\mathbf{s} + \lambda \mathbf{k}. \qquad (11.46)$$

Premultiply by $\mathbf{1}_N^\tau$:

$$(\mathbf{1}_N^\tau \mathbf{Y})\mathbf{s} = \mu(\mathbf{1}_N^\tau \mathbf{D})\mathbf{s} + \lambda \mathbf{1}_N^\tau \mathbf{k}. \qquad (11.47)$$

Using the fact that $\mathbf{1}_N^\tau \mathbf{Y} = \mathbf{1}_N^\tau \mathbf{D} = \mathbf{k}^\tau$, we have that

$$\mathbf{k}^\tau \mathbf{s} = \mu \mathbf{k}^\tau \mathbf{s} + \lambda L, \qquad (11.48)$$

and because $\mathbf{k}^\tau \mathbf{s} = 0$, this shows that $\lambda = 0$. So, (11.46) becomes

$$\mathbf{Y}\mathbf{s} = \mu \mathbf{D}\mathbf{s}. \qquad (11.49)$$

Thus, $\mathbf{s}$ is an eigenvector corresponding to the eigenvalue μ of the generalized eigenequation of $\mathbf{Y}$ wrt $\mathbf{D}$. Premultiplying (11.49) by $\mathbf{s}^\tau$ gives $\mathbf{s}^\tau \mathbf{Y}\mathbf{s} = \mu \mathbf{s}^\tau \mathbf{D}\mathbf{s}$, and (11.44) becomes

$$\text{Ncut}\{V_1,V_2\} = \frac{L - \mu \mathbf{s}^\tau \mathbf{D}\mathbf{s}}{L^2} = \frac{1 - \mu}{L}, \qquad (11.50)$$

where we used (11.38). This is minimized by maximizing the value of μ. However, the largest eigenvalue μ_1 of $\mathbf{Y}$ wrt $\mathbf{D}$ has eigenvector $\mathbf{1}_N$, which fails to satisfy (11.38), and so cannot be used. Instead, we take the eigenvector $\mathbf{s}_2$ corresponding to the second-largest eigenvalue μ_2 as our solution to the relaxed problem.

Note that we can express the generalized eigenequation (11.49) in simpler terms. Let $\mathbf{v} = \mathbf{D}^{1/2}\mathbf{s}$. Then, the generalized eigenequation can be written as

$$(\mathbf{D}^{-1/2}\mathbf{YD}^{-1/2})\mathbf{v} = \mu\mathbf{v}, \tag{11.51}$$

which means that we compute the second-largest eigenvalue μ_2 and associated eigenvector $\mathbf{v}_2$ of the symmetric matrix $\mathbf{D}^{-1/2}\mathbf{YD}^{-1/2}$, and then transform to $\mathbf{s}_2 = \mathbf{D}^{-1/2}\mathbf{v}_2$.

Rounding. The problem with this solution, however, is that rounding the entries of the eigenvector will not reflect the values of the s_i as given by (11.36). In fact, the values of s_i are not constant but depend upon the configurations of $\mathcal{V}_1$ and $\mathcal{V}_2$. In applications, we often want to maintain a balanced partition by assigning half the number of nodes in the network to each of the two subsets; that is, κ_1 and κ_2 should be roughly equal, so that $s_i = \pm 1, i = 1, 2, \ldots, N$. So, we round the entries of the eigenvector to these values, and then assign the node v_i to $\mathcal{V}_1$ or to $\mathcal{V}_2$ according as the sign of the ith entry in the eigenvector is positive or negative, respectively.

Probabilistic Interpretations

There is a random walk interpretation of the normalized-cuts method (Meilă and Shi, 2001). The Ncuts solution to partitioning $\mathcal{V}$ is the generalized eigenvector $\mathbf{s}$ associated with the second-largest generalized eigenvalue of the equation $\mathbf{Ys} = \mu\mathbf{Ds}$, which we can also write as $\mathbf{Ps} = \mu\mathbf{s}$, where $\mathbf{P} = \mathbf{D}^{-1}\mathbf{Y}$. Each row of the matrix $\mathbf{P}$ sums to one, and so $\mathbf{P}$ is a stochastic matrix. In fact, $\mathbf{P} = (P_{ij})$ is the transition matrix of a Markov random walk, where P_{ij} is the probability of moving from node v_i to node v_j, given that we are currently at node v_i. It follows that if $(\mu, \mathbf{s})$ is an eigenvalue–eigenvector pair of $\mathbf{P}$, where the eigenvalues of $\mathbf{P}$ lie between -1 and $+1$, and $\mathbf{P} = \mathbf{D}^{-1}\mathbf{Y}$, then $(\mu, \mathbf{s})$ also is the generalized eigenvalue–eigenvector pair of $\mathbf{Ys} = \mu\mathbf{Ds}$. In other words,

Theorem 11.2 *The solution of the spectral problem corresponding to a relaxation of the Ncut algorithm is equivalent to computing the eigenvalues and eigenvectors of the stochastic matrix* $\mathbf{P}$.

Meilă and Shi also studied the Ncut multiway case and gave conditions under which the eigenvectors associated with the largest K (out of N) eigenvalues of $\mathbf{P}$ would be piecewise constant, which would help partition $\mathcal{V}$, with the subsets having roughly an equal number of nodes. The conditions essentially boil down to $\mathbf{P}$ being a block-stochastic matrix.

Another probabilistic interpretation of Ncut is given by Nadler and Galun (2006) and Gavish and Nadler (2013). They consider the relationship between the Ncut measure of a given partition (interpreted as a Markov random walk on a weighted graph) and the eigenvalues that correspond to local random walks on each of these partitions. They show that the Ncut measure has a number of serious limitations:

- Ncut is not a suitable measure for the quality of clustering.
- For sparsely connected subsets, the Ncut measure essentially considers only the time it takes for a random walk to exit a given subset, without taking into consideration the internal structure of the subsets.
- The Ncut measure will fail when attempting to cluster large networks that contain structures of widely differing sizes and density.

Hyperplane Clustering Method

Another view of the spectral theory of normalized cuts has been given by Rahimi and Recht (2004). They assume that each node of a network represents a vector-valued observation; the problem is then one of classifying each observation into one of two possible classes, thereby also partitioning the set of nodes into two subsets. In other words, given an undirected network $\mathcal{G} = (\mathcal{V}, \mathcal{E})$, if the node $v_i \in \mathcal{V}$ represents the point $\mathbf{x}_i \in \mathbb{R}^r$, $i = 1, 2, \ldots, N$, then the goal of this approach is to classify each $\mathbf{x}_i$ (and, hence, also v_i) into one of two possible classes.

One way of dealing with the binary classification problem is through a kernel-based approach, such as support vector machines (see, e.g., Izenman, 2013, Chapter 11). If it is not possible to separate the points in *input space* $\mathbb{R}^r$, it may be possible instead to do it by a nonlinear transformation into a high-dimensional *feature space* $\mathcal{H}$ (which could be infinite dimensional), and then pass a hyperplane through the points in $\mathcal{H}$ so that points on one side of the hyperplane are classified into one group while points on the other side of the hyperplane are classified into the other group. It is generally assumed that $\mathcal{H}$ is an $N_\mathcal{H}$-dimensional reproducing-kernel Hilbert space of real-valued functions on $\mathbb{R}$ with inner product $\langle \cdot, \cdot \rangle$ and norm $\| \cdot \|$.

Let $\mathbf{x}_i \in \mathbb{R}^r$ be the data point corresponding to node $v_i \in V$, $i = 1, 2, \ldots, N$. Let $\Phi : \mathbb{R}^r \to \mathcal{H}$ be a nonlinear mapping of the form

$$\Phi(\mathbf{x}_i) = (\phi_1(\mathbf{x}_i), \ldots, \phi_{N_\mathcal{H}}(\mathbf{x}_i))^\tau \in \mathcal{H}, \quad i = 1, 2, \ldots, N, \tag{11.52}$$

and, in an obvious abuse of notation, let

$$\Phi = (\Phi(\mathbf{x}_1), \ldots, \Phi(\mathbf{x}_N)) \tag{11.53}$$

denote the $(N_\mathcal{H} \times N)$-matrix whose columns are formed from (11.52). Let $K(\mathbf{x}, \mathbf{y})$ be a *Mercer kernel* (i.e., a positive-definite function of two vectors in $\mathbb{R}^r$). A popular example of a kernel function K is the Gaussian radial basis function (RBF) given by

$$K(\mathbf{x}, \mathbf{y}) = \exp \left\{ -\frac{\| \mathbf{x} - \mathbf{y} \|^2}{2\sigma^2} \right\}, \tag{11.54}$$

where $\sigma > 0$ is a scale parameter. Note that for the Gaussian RBF, $\| \Phi(\mathbf{x}) \|^2 = K(\mathbf{x}, \mathbf{x}) = 1$, which means that the points in input space are mapped onto a sphere in feature space $\mathcal{H}$.

We now show that the matrices $\mathbf{W}$ and $\mathbf{D}$ of normalized cuts both have geometric interpretions in feature space. Recall that $\mathbf{W} = (w_{ij})$ is a symmetric weight matrix, where w_{ij} is the weight assigned to the edge between node v_i and node v_j. The relaxed version of Ncuts (with $\mathbf{Y}$ replaced by the more general weight matrix $\mathbf{W}$) involves computing the eigenvector associated with the second-largest eigenvalue of the matrix $\mathbf{D}^{-1/2} \mathbf{W} \mathbf{D}^{-1/2}$. Set

$$w_{ij} = K(\mathbf{x}_i, \mathbf{x}_j) = \langle \Phi(\mathbf{x}_i), \Phi(\mathbf{x}_j) \rangle = \Phi(\mathbf{x}_i)^\tau \Phi(\mathbf{x}_j), \quad i, j = 1, 2, \ldots, N. \tag{11.55}$$

The $(N \times N)$-matrix $\mathbf{W} = (w_{ij}) = (K(\mathbf{x}_i, \mathbf{x}_j)) = \Phi^\tau \Phi$ is the *Gram matrix* of K wrt $\mathbf{x}_1, \ldots, \mathbf{x}_N$. We next express the diagonal entries of the matrix $\mathbf{D} = \text{diag}\{D_{ii}\}$ as follows:

$$D_{ii} = \sum_{j=1}^N K(\mathbf{x}_i, \mathbf{x}_j) = \Phi(\mathbf{x}_i)^\tau \sum_{j=1}^N \Phi(\mathbf{x}_j) = N\Phi(\mathbf{x}_i)^\tau \overline{\Phi}, \quad i = 1, 2, \ldots, N, \tag{11.56}$$

where $\overline{\Phi} = \frac{1}{n} \sum_{j=1}^{N} \Phi(\mathbf{x}_j)$. Let θ_i be the angle in feature space between the vector $\Phi(\mathbf{x}_i)$ and the mean vector $\overline{\Phi}$. Then,

$$\cos(\theta_i) = \frac{\Phi(\mathbf{x}_i)^{\tau} \overline{\Phi}}{\| \overline{\Phi} \|}. \tag{11.57}$$

Thus,

$$D_{ii} = N \| \overline{\Phi} \| \cos(\theta_i) \tag{11.58}$$

can be viewed as the distance between the point $\Phi(\mathbf{x}_i)$ and the average point $\overline{\Phi}$.

Consider now an $N_{\mathcal{H}}$-dimensional point, $\Phi(\mathbf{x})$, in feature space $\mathcal{H}$. A hyperplane in $\mathcal{H}$ that passes through the origin is given by $\{\Phi(\mathbf{x})|\Phi(\mathbf{x})^{\tau}\boldsymbol{\beta} = 0\}$, where $\boldsymbol{\beta}$ is a $N_{\mathcal{H}}$-vector of weights with norm $\| \boldsymbol{\beta} \|$. So, the distance from the point $\Phi(\mathbf{x})$ to the hyperplane is $\Phi(\mathbf{x})^{\tau}\boldsymbol{\beta}$. The class of the point depends upon which side of the hyperplane the point falls. Thus, the class y ($= \pm 1$) is given by the sign of the distance, namely, $y = \text{sign}\{\Phi(\mathbf{x})^{\tau}\boldsymbol{\beta}\}$.

We define a simple measure of the "gap" between the two classes in feature space as the average squared distance between the N points and the hyperplane:

$$M(\boldsymbol{\beta}) = \frac{1}{N} \sum_{i=1}^{n} (\Phi(\mathbf{x}_i)^{\tau}\boldsymbol{\beta})^2 = \frac{1}{N}\boldsymbol{\beta}^{\tau}\Phi\Phi^{\tau}\boldsymbol{\beta}, \tag{11.59}$$

where we assume that every point in M is assigned equal weight. We want the hyperplane to pass through the dataset and we would also like the classes to contain roughly an equal number of points. So, we add a "balancing" constraint that the average signed distance to the hyperplane is to be zero; that is, $\frac{1}{N} \sum_{i=1}^{n} \Phi(\mathbf{x}_i)^{\tau}\boldsymbol{\beta} = 0$, or $\overline{\Phi}^{\tau}\boldsymbol{\beta} = 0$.

We seek to maximize this gap measure subject to the norm constraint and the balancing constraint:

$$\boldsymbol{\beta}^* = \arg\max_{\boldsymbol{\beta}} \; M(\boldsymbol{\beta}) \tag{11.60}$$

$$\text{subject to } \| \boldsymbol{\beta} \| = 1, \; \overline{\Phi}^{\tau}\boldsymbol{\beta} = 0. \tag{11.61}$$

From the last constraint, we need to find a $\boldsymbol{\beta}$ that is orthogonal to $\overline{\Phi}$. So, define the $(N_{\mathcal{H}} \times N_{\mathcal{H}})$-matrix

$$\mathbf{Z} = \mathbf{I}_{N_{\mathcal{H}}} - \frac{\overline{\Phi} \, \overline{\Phi}^{\tau}}{\| \overline{\Phi} \|^2}. \tag{11.62}$$

The matrix $\mathbf{Z}$ is symmetric and idempotent (i.e., $\mathbf{Z}^2 = \mathbf{Z}$). Hence, $\mathbf{Z}$ is a projection matrix that spans the space of vectors orthogonal to $\overline{\Phi}$. We now force $\boldsymbol{\beta}$ to lie in the span of $\mathbf{Z}$ by premultiplying $\boldsymbol{\beta}$ by $\mathbf{Z}$ in the optimization problem. This yields

$$\max_{\boldsymbol{\beta}} \; \boldsymbol{\beta}^{\tau}\mathbf{Z}^{\tau}\Phi\Phi^{\tau}\mathbf{Z}\boldsymbol{\beta} \tag{11.63}$$

$$\text{subject to } \boldsymbol{\beta}^{\tau}\mathbf{Z}^{\tau}\mathbf{Z}\boldsymbol{\beta} = 1. \tag{11.64}$$

Note that the balancing constraint, $\overline{\Phi}^{\tau}\boldsymbol{\beta} = 0$, is no longer needed in the optimization problem. Using a Lagrangian multiplier λ, differentiating wrt $\boldsymbol{\beta}$, and setting the result equal to zero, the maximum is attained by the eigenvector $\boldsymbol{\beta}_1$ associated with the largest eigenvalue: λ_1, of the following generalized eigenequation:

$$\mathbf{Z}^\tau \Phi \Phi^\tau \mathbf{Z} \boldsymbol{\beta} = \lambda \mathbf{Z} \boldsymbol{\beta}, \tag{11.65}$$

where we used the symmetry and idempotency properties of $\mathbf{Z}$.

We can simplify this result. Premultiply this equation by Φ^τ and then let $\boldsymbol{\alpha} = \Phi^\tau \mathbf{Z} \boldsymbol{\beta}$. Then (11.65) becomes

$$\Phi^\tau \mathbf{Z} \Phi \boldsymbol{\alpha} = \lambda \boldsymbol{\alpha}. \tag{11.66}$$

So, the solution of the constrained optimization problem is given by the eigenvector $\boldsymbol{\alpha}_1$ associated with the largest eigenvalue λ_1 of $\Phi^\tau \mathbf{Z} \Phi$. But, from (11.62),

$$\Phi^\tau \mathbf{Z} \Phi = \Phi^\tau \left(\mathbf{I}_{N_{\mathcal{H}}} - \frac{\overline{\Phi}\, \overline{\Phi}^\tau}{\| \overline{\Phi} \|^2} \right) \Phi. \tag{11.67}$$

Now, $\Phi^\tau \Phi = \mathbf{W}$ and

$$
\begin{aligned}
N \Phi^\tau \overline{\Phi} &= \frac{1}{N} \sum_{i=1}^{N} (\Phi(\mathbf{x}_1), \dots, \Phi(\mathbf{x}_n))^\tau \Phi(\mathbf{x}_i) \\
&= \left(\sum_i K(\mathbf{x}_1, \mathbf{x}_i), \dots, \sum_i K(\mathbf{x}_N, \mathbf{x}_i) \right)^\tau \\
&= (D_{11}, \dots, D_{NN})^\tau \\
&= \mathbf{W} \mathbf{1}_N.
\end{aligned}
\tag{11.68}
$$

Similarly, $N \overline{\Phi}^\tau \Phi = \mathbf{1}_N^\tau \mathbf{W}$. So, $N^2 \Phi^\tau \overline{\Phi}\, \overline{\Phi}^\tau \Phi = \mathbf{W} \mathbf{1}_N \mathbf{1}_N^\tau \mathbf{W}$. Also,

$$
\begin{aligned}
N^2 \| \overline{\Phi} \|^2 &= \sum_i \sum_j \Phi(\mathbf{x}_i)^\tau \Phi(\mathbf{x}_j) \\
&= \sum_i D_{ii} \\
&= \mathbf{1}_N^\tau \mathbf{W} \mathbf{1}_N.
\end{aligned}
\tag{11.69}
$$

Putting it all together, we compute the eigenvector $\boldsymbol{\alpha}_1$ associated with the largest eigenvalue λ_1 of the $(N \times N)$-matrix

$$\Phi^\tau \mathbf{Z} \Phi = \mathbf{W} - \frac{\mathbf{W} \mathbf{1}_N \mathbf{1}_N^\tau \mathbf{W}}{\mathbf{1}_N^\tau \mathbf{W} \mathbf{1}_N}. \tag{11.70}$$

Set $\boldsymbol{\beta}_1 = \mathbf{Z} \Phi \boldsymbol{\alpha}_1$. Then,

$$
\begin{aligned}
\mathbf{Z} \Phi \Phi^\tau \mathbf{Z} \boldsymbol{\beta}_1 &= \mathbf{Z} \Phi \Phi^\tau \mathbf{Z} (\mathbf{Z} \Phi \boldsymbol{\alpha}_1) \\
&= \mathbf{Z} \Phi \Phi^\tau \mathbf{Z} (\Phi \boldsymbol{\alpha}_1) \\
&= \lambda_1 \mathbf{Z} (\mathbf{Z} \Phi \boldsymbol{\alpha}_1) \\
&= \lambda_1 \mathbf{Z} \boldsymbol{\beta}_1,
\end{aligned}
\tag{11.71}
$$

and so λ_1 and $\boldsymbol{\beta}_1$ satisfy the eigenequation (11.65). From this solution, the N data points are classified according to the sign of $\boldsymbol{\alpha}_1$:

$$\widehat{\mathbf{y}} = \text{sign}\{\Phi^\tau \boldsymbol{\beta}_1\} = \text{sign}\{\Phi^\tau \mathbf{Z}\Phi\boldsymbol{\alpha}_1\} = \text{sign}\{\boldsymbol{\alpha}_1\}. \tag{11.72}$$

So, we do not have to compute $\boldsymbol{\beta}_1$.

In examples of their method, Rahimi and Recht show that while this solution appears to be insensitive to a single outlier in the data, it may, however, be sensitive to several outliers.

11.3 Multiway Cuts

A graph is described as $\mathcal{G} = (\mathcal{V}, \mathcal{E})$, where $\mathcal{V} = \{v_i\}$ is the set of nodes and $\mathcal{E}$ is the set of edges.

Definition 11.6 A *partition* of $\mathcal{V}$ into K parts (or "clusters") is defined as

$$\mathcal{P}_K = \{\mathcal{V}_1, \mathcal{V}_2, \ldots, \mathcal{V}_K\}, \tag{11.73}$$

where $\mathcal{P}_K$ satisfies

- $\mathcal{V} = \mathcal{V}_1 \cup \cdots \cup \mathcal{V}_K$,
- $\mathcal{V}_i \cap \mathcal{V}_j = \emptyset$, all $i \neq j$.

Many graph-partitioning algorithms that seek to obtain a partition $\mathcal{P}_K$ into K disjoint clusters do so using a strategy of iterative binary cuts. First, partition the graph into two equal-sized ("balanced") clusters; then, partition each of those clusters into two balanced clusters; and so on. Because this procedure is neither computationally efficient nor particularly reliable, we generalize the binary-cut algorithms to partitioning the graph into K clusters directly.

Recall the following definitions. A node v_i is a *neighbor* of node v_j if the edge $(v_i, v_j) \in \mathcal{E}$. The number of neighbors of a node v is the *degree*, D_v, of that node. If a node $v_i \in \mathcal{V}_k$ has a neighbor $v_j \in \mathcal{V} \backslash \mathcal{V}_k$, then v_i is called a *boundary node* and the edge (v_i, v_j) is called a *cut edge*. We are often interested in the set

$$E_{k\ell} = \{(v_i, v_j) \in \mathcal{E} : v_i \in \mathcal{V}_k, v_j \in \mathcal{V}_\ell, k \neq \ell\} \tag{11.74}$$

of all cut edges between distinct subsets $\mathcal{V}_k$ and $\mathcal{V}_\ell$ of $\mathcal{V}$, $k, \ell = 1, 2, \ldots, K$. In many applications, a graph partition coincides with a clustering of the nodes into an unknown number, K, of subsets.

Definition 11.7 For a partition of $\mathcal{V}$ into K subsets of nodes, we define the *cut objective function* as

$$\text{cut}\{\mathcal{P}_K\} = \sum_{i<j} \text{cut}\{\mathcal{V}_i, \mathcal{V}_j\} = \frac{1}{2} \sum_{i=1}^{K} \text{cut}\{\mathcal{V}_i, \mathcal{V} \backslash \mathcal{V}_i\}. \tag{11.75}$$

The $\frac{1}{2}$ factor is needed so that we do not count every edge twice when dealing with undirected networks.

Definition 11.8 The *minimum-cut* solution is given by the partition $\mathcal{P}_K$ that yields the minimum of cut$\{\mathcal{P}_K\}$.

The minimum-cut solution, however, does not usually produce reasonable partitions. What happens in practice is that it tends to separate small sets of isolated nodes from the rest of the graph. This occurs because the cut value increases with an increasing number of cut edges that cross from one subset to the other.

Let $N_i = |\mathcal{V}_i|$ be the number of nodes in the ith subset, $i = 1, 2, \ldots, K$, so that $\sum_{i=1}^{K} N_i = N$.

Definition 11.9 Generalizations of the ratio-cut and normalized-cut objective functions are given by the following criteria:

$$\text{RatioCut}\{\mathcal{P}_K\} = \sum_{i=1}^{K} \frac{\text{cut}\{\mathcal{V}_i, \mathcal{V}\backslash\mathcal{V}_i\}}{N_i}, \tag{11.76}$$

$$\text{Ncut}\{\mathcal{P}_K\} = \sum_{i=1}^{K} \frac{\text{cut}\{\mathcal{V}_i, \mathcal{V}\backslash\mathcal{V}_i\}}{\text{cut}\{\mathcal{V}_i, \mathcal{V}\}}. \tag{11.77}$$

In each case, the problem is to find $\mathcal{P}_K = \{\mathcal{V}_1, \mathcal{V}_2, \ldots, \mathcal{V}_K\}$ such that the appropriate objective function is minimized.

Because the denominator of each term in Ncut can be written as

$$\text{cut}\{\mathcal{V}_i, \mathcal{V}\} = \text{cut}\{\mathcal{V}_i, \mathcal{V}_i\} + \text{cut}\{\mathcal{V}_i, \mathcal{V}\backslash\mathcal{V}_i\}, \tag{11.78}$$

the Ncut objective function is small if, for all i, the cut size within the ith subset, cut$\{\mathcal{V}_i, \mathcal{V}_i\}$, is greater than the cut size between the ith subset and the other nodes, cut$\{\mathcal{V}_i, \mathcal{V}\backslash\mathcal{V}_i\}$. Hence, the Ncut objective function penalizes unbalanced partitions, unlike the RatioCut objective function, which does not (Bach and Jordan, 2006). If we require that the partition be "balanced," with subsets of roughly equal numbers of nodes, and that some "cut" objective criterion be minimized, then these problems are $\mathcal{NP}$-complete problems (Wagner and Wagner, 1993).

11.3.1 Ratio Cuts

Define K indicator N-vectors, $\mathbf{h}_j = (h_{1j}, \ldots, h_{Nj})^\tau$, $j = 1, 2, \ldots, K$, where

$$h_{ij} = \begin{cases} 1/\sqrt{N_j}, & \text{if } v_i \in \mathcal{V}_j \\ 0, & \text{otherwise} \end{cases}, \quad i = 1, 2, \ldots, N, \quad j = 1, 2, \ldots, K. \tag{11.79}$$

Arrange these indicator vectors as columns of the $(N \times K)$-matrix

$$\mathbf{H} = (\mathbf{h}_1, \ldots, \mathbf{h}_K), \tag{11.80}$$

where $\mathbf{H}^\tau\mathbf{H} = \mathbf{I}_K$. Recall that $\mathbf{L} = \mathbf{Y} - \mathbf{D}$ is given by (11.17). It follows that

$$\mathbf{h}_i^\tau\mathbf{L}\mathbf{h}_i = \frac{\text{cut}\{\mathcal{V}_i, \mathcal{V}\backslash\mathcal{V}_i\}}{N_i} \tag{11.81}$$

and

$$\mathbf{h}_i^\tau\mathbf{L}\mathbf{h}_i = (\mathbf{H}^\tau\mathbf{L}\mathbf{H})_{ii}. \tag{11.82}$$

Hence

$$\text{RatioCut}\{\mathcal{P}_K\} = \sum_{i=1}^{K} \mathbf{h}_i^{\tau} \mathbf{L} \mathbf{h}_i = \sum_{i=1}^{K} (\mathbf{H}^{\tau} \mathbf{L} \mathbf{H})_{ii} = \text{tr}\{\mathbf{H}^{\tau} \mathbf{L} \mathbf{H}\}. \tag{11.83}$$

The problem of minimizing $\text{RatioCut}\{\mathcal{P}_K\}$ can be expressed as follows:

$$\min_{\mathcal{P}_K} \text{tr}\{\mathbf{H}^{\tau} \mathbf{L} \mathbf{H}\} \text{ subject to } \mathbf{H}^{\tau} \mathbf{H} = \mathbf{I}_K. \tag{11.84}$$

This minimization problem is an $\mathcal{NP}$-hard problem. Relaxing the problem, we permit $\mathbf{H}$ to take arbitrary real values:

$$\min_{\mathbf{H} \in \mathbb{R}^{N \times K}} \text{tr}\{\mathbf{H}^{\tau} \mathbf{L} \mathbf{H}\} \text{ subject to } \mathbf{H}^{\tau} \mathbf{H} = \mathbf{I}_K. \tag{11.85}$$

As before for the binary case, we use the Rayleigh–Ritz theorem to provide the solution of this constrained trace-minimization problem. Thus, the columns of $\mathbf{H}$ are given by the first K eigenvectors of $\mathbf{L}$. The matrix $\mathbf{H}$ is the same matrix $\mathbf{U}$ of the unnormalized spectral clustering algorithm. The next step, which is to translate the real-valued solution to a discrete partition, uses the kmeans algorithm applied to the rows of $\mathbf{H}$. The solution is then the general unnormalized spectral clustering algorithm.

There is no guarantee that the approximate solution to the RatioCut problem is close to the exact solution (von Luxburg, 2007). Simple examples (with $k = 2$) have been studied by Guattery and Miller (1998), in which there is a large discrepancy between the two solutions, even to the point of yielding completely different partitions.

11.3.2 Normalized Cuts

As with RatioCut, we define an $(N \times K)$-matrix $\mathbf{H} = (\mathbf{h}_1, \ldots, \mathbf{h}_K)$ of indicator vectors, where $\mathbf{h}_j = (h_{1j}, \ldots, h_{Nj})^{\tau}$, $j = 1, 2, \ldots, K$, but, in this case, we set

$$h_{ij} = \begin{cases} 1/\sqrt{\text{vol}(\mathcal{V}_j)}, & \text{if } v_i \in \mathcal{V}_j \\ 0, & \text{otherwise} \end{cases}, \quad i = 1, 2, \ldots, N, \quad j = 1, 2, \ldots, K, \tag{11.86}$$

where $\text{vol}(A) = \sum_{i \in A} D_{ii}$ is the *size* of the subset $A \subset \mathcal{V}$ as measured by the sum of the degrees of all the nodes in A. As before, $\mathbf{H}^{\tau} \mathbf{H} = \mathbf{I}_K$. Also, we have that

$$\mathbf{h}_i^{\tau} \mathbf{D} \mathbf{h}_i = 1, \quad \mathbf{H}_i^{\tau} \mathbf{L} \mathbf{h}_i = \frac{\text{cut}\{\mathcal{V}_i, \mathcal{V} \backslash \mathcal{V}_i\}}{\text{vol}\{\mathcal{V}_i\}}. \tag{11.87}$$

The problem of finding a partition $\mathcal{P}_K$ to minimize $\text{Ncut}\{\mathcal{P}_K\}$ can, therefore, be expressed as

$$\min_{\mathcal{P}_K} \text{tr}\{\mathbf{H}^{\tau} \mathbf{L} \mathbf{H}\} \text{ subject to } \mathbf{H}^{\tau} \mathbf{D} \mathbf{H} = \mathbf{I}_K. \tag{11.88}$$

This minimization problem is an $\mathcal{NP}$-complete problem. Relaxing the discreteness of the minimization problem, we set $\mathbf{T} = \mathbf{D}^{1/2} \mathbf{H}$, so that (11.88) can be written as

$$\min_{\mathbf{T} \in \mathbb{R}^{N \times K}} \text{tr}\{\mathbf{T}^{\tau} \mathbf{D}^{1/2} \mathbf{L} \mathbf{D}^{1/2} \mathbf{T}\} \text{ subject to } \mathbf{T}^{\tau} \mathbf{T} = \mathbf{I}_K. \tag{11.89}$$

This is another constrained trace-minimization problem. The solution $\mathbf{T}$ has columns that are the first K eigenvectors of the matrix $\mathbf{L}^{rw}$, or the first K eigenvectors of the generalized eigenproblem $\mathbf{L}^{rw}\mathbf{u} = \lambda\mathbf{Du}$. This corresponds to the normalized spectral clustering algorithm of Section 13.3.2.

11.4 Further Reading

Books on graph partitioning include the edited volume of Bichot and Siarry (2011). Chapters or sections on graph partitioning, graph cuts, and cut distance in books include Kolaczyk (2009, Section 4.3.3), Newman (2010, Chapter 11), and Lovász (2012, Chapter 8).

11.5 Exercises

Exercise 11.1 Given the graph in **Figure 11.1**, and the interim partition $A = \{1,2,6\}$ and $B = \{3,4,5\}$ of $\mathcal{V}$, show that the Kernighan–Lin algorithm yields the final partition $\mathcal{V}_1 = \{1,2,3\}$ and $\mathcal{V}_2 = \{4,5,6\}$. Show that this partition cannot be improved any further.

Exercise 11.2 Write a computer program for the RatioCut process for K classes using the description in Section 11.3.1. Apply it to the partition of a large network dataset of your choice. Comment on your results.

Exercise 11.3 Let $\mathcal{G}$ denote a graph and let $\mathbf{L} = \mathbf{D} - \mathbf{Y}$ denote the unnormalized graph Laplacian, where $\mathbf{Y}$ is the adjacency matrix of $\mathcal{G}$ and $\mathbf{D} = \mathbf{Y}\mathbf{1}_N$ is the diagonal matrix formed from $\mathbf{Y}$. Suppose N is an eigenvalue of $\mathbf{L}$. Show that the value 0 can occur at most once as an eigenvalue of $\mathbf{L}$.

Exercise 11.4 Pick a cut that divides Sampson's monk data (Section 2.5.4) into two groups. Find the normalized cut (Ncut) and RatioCut for that cut. Try different cuts to optimize Ncut and RatioCut. Compare these two types of graph cuts for those data.

Exercise 11.5 Pick a cut that divides Zachary's karate club data (Section 2.5.5) into two groups. Find the normalized cut (Ncut) and RatioCut for that cut. Try different cuts to optimize Ncut and RatioCut. Compare these two types of graph cuts for those data.

CHAPTER TWELVE

Graph Partitioning: II. Community Detection

Much of what is studied in network analysis derives from the observation that nodes in a graph G often tend to form "cohesive groups" or, as they are called in social networks, *communities*, *clusters*, or *modules*. This chapter describes statistical methods for identifying such communities using the stochastic blockmodel and an approach based upon the concept of modularity.

12.1 Introduction

One can loosely think of cohesive groups or communities in network data as various equivalence classes, in which the nodes within such a community connect with each other more often than they connect with the nodes in other communities. Thus, nodes within the same community tend to be similar to each other and share common features, while nodes from different communities tend to have little in common with each other. Sometimes, these communities are identified by giving them names. We have the following examples.

- In *social networks*, the communities may be associated with specific social factions (e.g., nations, cities, villages, or families, with individuals perhaps defined through covariates such as age, race, gender, socioeconomic status, geographical location, or political party affiliation).
- In *protein–protein interaction (PPI) networks*, communities may be associated with the many protein groupings associated with the Gene Ontology (GO) project[1] (Airoldi et al., 2008; Fu, Song, and Xing, 2009; Daudin, 2011).
- In *computer networks*, online communities may be associated with peer-to-peer (P2P) file-sharing groups that form webpages to discuss different topics of interest.
- In *terrorist networks*, communities may be associated with terrorist cells.

In many situations, little or nothing will be known about the various communities, and it may be natural to try to identify communities though covariates. However, it

---

[1]The Gene Ontology project, which is a major bioinformatics initiative to annotate genes, gene products, and sequences in a wide variety of organisms, is part of the *Gene Ontology Consortium*, and has its website at www.geneontology.org.

is more likely that communities will be recognized not by using such knowledge, but instead by using only the observed relationships between nodes.

The unsupervised learning problem of determining which nodes of a network are members of which groups is often referred to as "community discovery" or "community detection" (Newman, 2004a; Fortunato, 2010; Zhao, Levina, and Zhu, 2012). This problem has much in common with that of clustering multivariate data into an unknown number of distinct groups (see, e.g., Izenman, 2013, Chapter 12). Because partitioning even a relatively small dataset in an exhaustive search for all possible clusters quickly becomes computationally intractable, many different clustering algorithms have been proposed to discover the best arrangement of clusters, where "best" is defined in terms of some optimality criterion. The same concerns also hold true for clustering nodes within a network graph.

Another approach to this problem is through model-based methods, such as the stochastic blockmodel, which we describe in this chapter.

12.2 Equivalence Concepts

The main problem in modeling and analyzing this type of network structure in social networks has been how to define the notion of "similarity" or "equivalence" when nodes are assigned to their respective equivalence classes.

12.2.1 Structural Equivalence

An early attempt to define similarities amongst nodes and amongst edges used a formal algebraic approach (Lorrain and White, 1971). Given a directed graph G, this approach describes a partitioning of the nodes into disjoint groups (or "blocks") of *structurally equivalent* nodes, defined as follows:

Definition 12.1 Two nodes are said to be *structurally equivalent* iff both relate to every other node in G in the same way.

In a binary network, this means that the two nodes have identical links to every other node in the network. Thus, within the set of structurally equivalent nodes, any node can be substituted for any other node because the patterns of links for each node are identical.

At the heart of the concept of structural equivalence was a need to identify the relationships that each individual has with all other individuals, as well as the various types of such relationships. An argument made against structural equivalence was that it did not admit that individuals from different populations or groups may perform the same roles and that, therefore, they should be classified as structurally equivalent.

Consider the following simple example that illustrates the problems of this definition that arise due to the presence of hierarchical relationships. Although teachers within a single school are considered to be structurally equivalent, teachers in different schools are not. According to this definition, principals (or headmasters) of different schools (within the same school district and having the same superintendent above them) who perform the same roles within their respective schools, are also not considered to be structurally equivalent.

Because nodes hardly ever exhibit exact structural equivalence (especially in larger networks), this definition of structural equivalence was felt by many to be too restrictive for practical use. The pros and cons of this definition of equivalence became a source of controversy in the sociology literature.

12.2.2 Alternative Deterministic Equivalences

Disaffection with the notion of structural equivalence led to an alternative version called *regular equivalence* (Sailer, 1978; White and Reitz, 1983; Everett and Borgatti, 1994):

Definition 12.2 Two nodes are said to be *regularly equivalent* iff they have the same pattern of links with members of other sets of nodes that are also regularly equivalent.

This is a recursive definition: regular equivalence classes of nodes consist of nodes that have similar links to any of the members of other regular equivalence classes; it does not mention links to particular other nodes. Nodes that are structurally equivalent are also regularly equivalent, but the converse does not hold.

Consider again the teacher example. Two teachers are equivalent because each has a certain pattern of links, for example, to their own relatives. Yet, because a teacher does not have any links to the relatives of other teachers, they are not structurally equivalent. Teachers have no links with the superintendent, but each has a link with a principal/headmaster. The pattern of links of each teacher is, therefore, identical to that of teachers in other schools. Each principal/headmaster has a link to the superintendent and also a link to each teacher in their respective school. The superintendent has a link to each principal/headmaster, but not to the teachers. Thus, in this example, there are three regular-equivalent groups: superintendent, principals/headmasters, and teachers. So, regular equivalence provides a way of identifying "social roles" in the network.

Other alternatives to the principle of structural equivalence in social networks include *approximate structural equivalence* (Sailer, 1978) and *role equivalence* (Winship and Mandel, 1983).

12.2.3 Stochastic Equivalence

Because of the heated debate surrounding the notion of structural equivalence (see, e.g., White, Boorman, and Breiger, 1976; Faust, 1988; Wasserman and Faust, 1994), "structural equivalence" was generalized to "stochastic equivalence" by adding a probability component (Feinberg and Wasserman, 1981a; Holland, Laskey, and Leinhardt, 1983; Fienberg, Meyer, and Wasserman, 1985):

Definition 12.3 Two nodes are *stochastically equivalent* iff the nodes are "exchangeable" with respect to the probability distribution.[2]

---

[2]For a discussion of the notion of "exchangeable," see Section 15.3.

To be more specific, let $P\{v_i R v_j\}$ denote the probability of observing a connection of type R between the ith node v_i and the jth node v_j.

- In an undirected graph, let $v_i \leftrightarrow v_j$ denote the event that nodes v_i and v_j are connected.
- If, on the other hand, the network graph is directed, let $v_i \rightarrow v_j$ denote the event that node v_i is sending an edge to node v_j, and let $v_i \leftarrow v_j$ denote the reverse event, that v_i is receiving an edge from v_j.

We have the following two definitions:

Definition 12.4 Let $\mathcal{G}$ be an undirected network graph. Then, two nodes, v_i and v_j, are *stochastically equivalent* iff they are equally likely to connect to a common third node, v_k. That is,

$$P\{v_i \leftrightarrow v_k\} = P\{v_j \leftrightarrow v_k\}, \quad \text{for every } k. \tag{12.1}$$

Definition 12.5 Let $\mathcal{G}$ be a directed network graph. Then, two nodes, v_i and v_j, are *stochastically equivalent* iff the following two probability statements hold:

$$P\{v_i \rightarrow v_k\} = P\{v_j \rightarrow v_k\}, \quad \text{for every } k, \tag{12.2}$$
$$P\{v_i \leftarrow v_k\} = P\{v_j \leftarrow v_k\}, \quad \text{for every } k. \tag{12.3}$$

We say that v_i and v_j are *stochastically equivalent senders* iff (12.2) holds, and are *stochastically equivalent receivers* iff (12.3) holds.

See Rohe and Yu (2012). Structural equivalence implies stochastic equivalence, but the converse statement does not hold.

12.3 Deterministic Blockmodels

12.3.1 Blockmodels

The idea of structural equivalence led to the introduction of a *blockmodel* (White, Boorman, and Breiger, 1976), which became an important concept in social network analysis. The basic idea of a blockmodel[3] is to take a large, possibly confusing network and reduce it to a smaller, more readily interpretable configuration.

In a blockmodel, each of the N nodes of $\mathcal{G} = (\mathcal{V}, \mathcal{E})$, $\mathcal{E} \subseteq \mathcal{V} \times \mathcal{V}$, is assumed to be a member of exactly one of K prespecified blocks (i.e., distinct groups or communities) independent of the other nodes, where K is assumed to be known. Blockmodels are used primarily in social networks, where a node is called an *actor*, and a block, which is often called a *position*, a *class*, or a *color*, is usually viewed within a specific social structure. Thus, reflecting the notion of structural equivalence, *two individuals occupy the same position in a social structure if and only if they are related to the same individuals in the same way* (Winship and Mandel, 1983).

In a blockmodel, the edges of $\mathcal{G}$ are deterministic in that each edge is either present or absent with probability 1 or 0, respectively. Let $\mathbf{Y} = (Y_{ij})$ be the $(N \times N)$ *adjacency matrix* (also known as a *sociomatrix*) associated with the graph $\mathcal{G}$, where $Y_{ij} = 1$ if nodes $v_i, v_j \in \mathcal{V}$ are joined by an edge, and $Y_{ij} = 0$ otherwise. The blockmodel

---

[3]Some authors write blockmodel as two words "block model."

algorithm partitions the N nodes into blocks by simultaneously permuting the rows and columns of the adjacency matrix to expose patterns in the entries. The partition is created so that the nodes assigned to the same block are deemed to be approximately structurally equivalent. The blocks are interpreted as positions. This entire approach is deterministic and depends upon algorithms for the optimal partitioning of the nodes of a graph.

Image Matrix

Following a partition of the $(N \times N)$ adjacency matrix, we form the $(K \times K)$ *image matrix*, $K < N$, in which each entry in the image matrix is associated with a corresponding block of nodes in the adjacency matrix. Because the blocks are formed using the principle of structural equivalence, the image matrix has entries that are either 0 or 1. A 1 indicates that there is a link between the row block and the column block, while a 0 indicates that no such link exists.

Different proposals have been put forward for determining the image matrix from the permuted adjacency matrix. First, there is a *fat fit*, in which a block with all 1s is coded as 1 and a block with all 0s is coded as 0. Then, there is a *lean fit*, which has two versions: in a *zeroblock*, if the block has all 0s, then it is coded as 0, while otherwise, it is coded as 1; in a *oneblock*, if the block has all 1s, it is coded as 1, while otherwise, it is coded as 0.

However, in social network data, the entries in the blockmodel are not so clearcut: it does not always follow that each block contains either all 1s or all 0s. Often, a block will contain a mixture of 1s and 0s, and a decision will have to be made as to whether to assign a 1 or a 0 to that entry in the image matrix as an indicator of the entries in the block. One way is to assign either a 1 or a 0 according to whether a majority (or some specified proportion) of the nodes in the block have that value.

More commonly, one applies an α-*density criterion* (Arabie, Boorman, and Levitt, 1978), where the threshold α is the observed proportion of 1s in the entire blockmodel (ignoring diagonal entries). If the observed proportion of 1s in any block exceeds α, the block is coded as a 1 in the image matrix, whereas if the observed proportion of 1s is at most α, the block is coded as a 0 in the image matrix.

Reduced Graph

From the image matrix, we draw a *reduced graph* whose K nodes correspond to the blocks and whose directed edges correspond to the 1s in the image matrix, where the edge goes from the row block to the column block. A 1 along the diagonal of the image matrix implies that the arrow goes from that node back to that node in a loop.

12.3.2 Example: Sampson's Monk Data

To use Sampson's data as binary sociomatrices, we set all nonzero entries to be 1.[4] The adjacency matrix of the "like" responses for the third time point (T_4) is given in **Table 3.1** and the network graph corresponding to that adjacency matrix is displayed in **Figure 2.10**.

---

[4]Some authors retain only the highest choice (3) for analysis of the monk data and remove the other two choices, while other articles retain the top two choices (2 and 3); in either case, the retained choices were coded as 1 and the rest as 0.

Table 12.1 Blockmodel (permuted adjacency matrix) for Sampson's monk data on the "like" responses at the third time point (T_4). The four blocks, YT = {1, 2, 7, 12, 14, 15, 16}, LO = {4, 5, 6, 9, 11}, O = {3, 17, 18}, I = {8, 10, 13}, are defined by Sampson's analysis

#	1	2	7	12	14	15	16	4	5	6	9	11	3	17	18	8	10	13
1	0	0	0	1	1	0	0	0	0	0	0	0	1	0	0	0	0	0
2	1	0	1	1	0	0	0	0	0	0	0	0	0	0	0	0	0	0
7	0	1	0	1	0	0	1	0	0	0	0	0	0	0	0	0	0	0
12	1	1	1	0	0	0	0	0	0	0	0	0	0	0	0	0	0	0
14	1	0	0	1	0	1	0	0	0	0	0	0	0	0	0	0	0	0
15	0	1	1	1	0	0	0	0	0	0	0	0	0	0	0	0	0	0
16	0	1	1	0	0	1	0	0	0	0	0	0	0	0	0	0	0	0
4	0	0	0	0	0	0	0	0	1	1	0	0	0	0	0	0	1	0
5	0	0	0	0	0	0	0	1	0	0	1	1	0	0	0	0	0	0
6	0	0	0	0	0	0	0	1	1	0	1	0	0	0	0	0	0	0
9	0	0	0	0	0	0	0	0	1	0	0	1	0	0	0	1	0	0
11	0	0	0	0	1	0	0	0	1	0	0	0	0	0	0	1	0	0
3	1	0	0	0	0	0	0	0	0	0	0	0	0	1	1	0	0	1
17	0	1	0	0	0	0	0	0	0	0	0	0	1	0	1	0	0	0
18	0	1	0	0	0	0	0	0	0	0	0	0	1	1	0	0	0	0
8	0	0	0	0	0	0	0	1	0	1	1	0	0	0	0	0	0	0
10	0	0	0	0	0	0	0	1	1	0	1	0	0	0	0	0	0	1
13	0	0	1	0	0	0	0	0	1	0	0	0	0	0	1	0	0	0

The blockmatrix (permuted adjacency matrix) derived from Sampson's analysis is given in **Table 12.1**, where there are four lean-fit blocks and one fat-fit block. Because there are several blocks with a mixture of 0s and 1s, it is necessary to decide whether to assign a 1 or a 0 to those blocks. We computed $\alpha = 56/306 = 0.183$, or 18.3%. The percentages for each block are given in the left panel of **Table 12.2**. Any block whose density of 1s is larger than 0.183 would be assigned a 1, and 0 otherwise; the resulting image matrix is given in the right panel of **Table 12.2**. The reduced graph derived from the image matrix is displayed in **Figure 12.1**. We see that the YT block and the O block are detached from the rest of the network, and the LO and I blocks, although detached from the YT and O blocks, are linked together in both directions.

Other analysts have leaned towards three (rather than four) blocks, with the interstitials (or waverers) being split up. Although Armand (13) was classified by Sampson as a member of the Interstitials, his three "likes" were to members of other blocks; as a result, he was placed by other analysts into either the Loyal Opposition (Breiger, Boorman, and Arabie, 1975) or the Outcasts (White, Boorman, and Breiger, 1976; Airoldi et al., 2008). The other members of the Interstitials, Victor (8) and Ramuald (10), have been placed with the Loyal Opposition (White, Boorman, and Breiger, 1976).

If we calculate the image matrix for three blocks, the reduced graph will consist of three isolated blocks with links only to themselves and no links to one another (see **Exercise 12.1**).

Table 12.2 Left panel: The proportion (and percentage) of 1s in each block of the blockmodel (permuted adjacency matrix) for Sampson's monk data on the "like" responses at the third time point (T_4). Note that diagonal entries are not included in the counts. Right panel: The image matrix derived from the blockmodel, where the proportion of 1s in a block has to be greater than 18.3% for a 1 to be assigned to that block

	YT	LO	O	I
YT	20/42 (47.62)	0/25 (0)	1/21 (4.76)	0/21 (0)
LO	1/35 (2.86)	10/20 (50)	0/15 (0)	3/15 (20)
O	3/21 (14.29)	0/15 (0)	6/6 (100)	1/9 (11.11)
I	1/21 (4.76)	7/15 (46.67)	1/9 (11.11)	1/6 (16.67)

	YT	LO	O	I
YT	1	0	0	0
LO	0	1	0	1
O	0	0	1	0
I	0	1	0	0

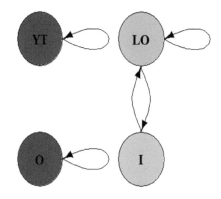

Figure 12.1 Sampson's monk network. The four nodes and the edges form a graphical representation of the image matrix in Table 12.4. The nodes represent the groups of Young Turks (YT), Loyal Opposition (LO), Outcasts (O), and Interstitials (I).

12.4 Stochastic Blockmodels

Let $\mathcal{G} = (\mathcal{V}, \mathcal{E})$ be a binary random graph, where $\mathcal{V} = \{v_1, \ldots, v_N\}$ are the nodes and $|\mathcal{V}| = N$ is the number of nodes. Let $\mathbf{Y} = (Y_{ij})$ be an $(N \times N)$ *random adjacency matrix*, where

$$Y_{ij} = \begin{cases} 1, & \text{if nodes } v_i, v_j \in \mathcal{V} \text{ are joined by an edge (i.e., if } (v_i, v_j) \in \mathcal{E}) \\ 0, & \text{otherwise} \end{cases} \quad (12.4)$$

The total number of degrees in the graph is given by $\sum_{i,j} Y_{ij}$. The rows and columns of $\mathbf{Y}$ are the same set of nodes, which are ordered in the same way. If $\mathcal{G}$ is an undirected graph, the edges will be unordered pairs, so that $\mathbf{Y}$ will be symmetric. If $\mathcal{G}$ is a directed graph, then the edges will be ordered pairs and $\mathbf{Y}$ will generally not be symmetric. Assume also that, for both cases, there are no self-loops (i.e., $Y_{ii} = 0$, for all $v_i \in \mathcal{V}$).

12.4.1 Defining Stochastic Blockmodels

The *stochastic blockmodel* (Fienberg and Wasserman, 1981; Holland, Laskey, and Leinhardt, 1983; Wang and Wong, 1987), also referred to as the *planted partition model* in theoretical computer science and the *inhomogeneous random graph model* in mathematics, was introduced for social networks by grafting probability concepts onto the deterministic blockmodel. This was accomplished by incorporating a $(K \times K)$-matrix $\mathbf{P} = (p_{ij})$ of *connection probabilities* into the model having K *blocks* (or *latent classes*), where K is much smaller than N. The entry $p_{ij} \in [0, 1]$ is the probability that an edge exists between a node in block B_i and a node in block B_j, for $i, j \in \{1, 2, \ldots, K\}$. For identifiability reasons, it is assumed that no two rows of $\mathbf{P}$ are the same and that no two columns of $\mathbf{P}$ are the same. As with the deterministic blockmodel, each node is assumed to be a member of one and only one of the K blocks. We assume that K does not depend upon N.

Random assignments of nodes into blocks are carried out using a random *block membership function*, $\phi : \{v_1, v_2, \ldots, v_N\} \to \{B_1, B_2, \ldots, B_K\}$, where the ith block is denoted by B_i, and K is known. Thus, $\phi(v_i) = B_j$ means that node v_i is a member of block B_j; we will also express this idea by writing $v_i \in B_j$. The function ϕ is unobserved and has to be estimated. We have the following definition:

> **Definition 12.6** In a *stochastic blockmodel*, two nodes are members of the same block iff they are stochastically equivalent.

Membership of a node in any one of these blocks is based upon covariate values corresponding to that node. The nodes within a block are assumed to be stochastically equivalent, by which we mean that different nodes in the same block will have the same probability of linking to all other nodes in the graph $\mathcal{G}$.

Stochastic blockmodels are defined by using indicator variables to denote membership of a node in one of the K groups. For the ith node in $\mathcal{G}$ (i.e., $v_i \in \mathcal{V}$), let $\mathbf{Z}_i$ be a K-vector of unobserved latent variables, where

$$\mathbf{Z}_i = (Z_{1i}, \ldots, Z_{Ki})^\tau, \quad i = 1, 2, \ldots, N, \tag{12.5}$$

where each Z_{ki} is an indicator variable,

$$Z_{ki} = \begin{cases} 1, & \text{if } v_i \in B_k \\ 0, & \text{otherwise} \end{cases}, \quad k = 1, 2, \ldots, K. \tag{12.6}$$

The $(K \times N)$-matrix,

$$\mathbf{Z} = (\mathbf{Z}_1, \ldots, \mathbf{Z}_N), \tag{12.7}$$

has a single 1 in each column and at least one 1 in each row, all the other entries being zeroes. The matrix $\mathbf{Z} \in \{0, 1\}^{K \times N}$ is often called the *block membership matrix* and is analogous to the setup used in one-way analysis of variance. Then,

$$Z_{+i} = \sum_{k=1}^{K} Z_{ki} = 1, \quad i = 1, 2, \ldots, N. \tag{12.8}$$

The $\mathbf{Z}_1, \ldots, \mathbf{Z}_N$ are iid random K-vectors, drawn from the multinomial distribution,

$$\mathbf{Z}_i \stackrel{iid}{\sim} \mathcal{M}(1, \boldsymbol{\pi}), \quad i = 1, 2, \ldots, N, \tag{12.9}$$

where $\boldsymbol{\pi} = (\pi_1, \ldots, \pi_K)^\tau$ denotes the *block probability vector*. Thus,

$$P\{v_i \in B_k\} = P\{Z_{ki} = 1\} = \pi_k, \quad k = 1, 2, \ldots, K, \tag{12.10}$$

$$\sum_{k=1}^{K} \pi_k = 1. \tag{12.11}$$

Given that $v_i \in B_a$ and $v_j \in B_b$, the random edge Y_{ij} between v_i and v_j is drawn from a Bernoulli distribution with probability $p_{ab} \in [0, 1]$,

$$Y_{ij}|(Z_{ai}Z_{bj} = 1) \sim \text{Bernoulli}(p_{ab}), \quad i \neq j, \tag{12.12}$$

where

$$p_{ab} = P\{Y_{ij} = 1|Z_{ai} = 1, Z_{bj} = 1\} = P\{Y_{ij} = 1|Z_{ai}Z_{bj} = 1\}. \tag{12.13}$$

If $a \neq b$, p_{ab} represents the expected number of links between block a and block b, while if $a = b$, it represents twice the number of links within block a. Let $\mathbf{P} = (p_{ab})$ be the $(K \times K)$-matrix of connection probabilities between blocks. Thus, conditional on the block identities of the nodes, $\{\mathbf{Z}_1, \ldots, \mathbf{Z}_N\}$, the probability that nodes $v_i \in B_a$ and $v_j \in B_b$ are connected through an edge is determined only by their block memberships. We shall find it convenient to describe this as follows. Let $\mathbf{c} = (c_1, \ldots, c_N)^\tau = (\phi(v_1), \ldots, \phi(v_N))^\tau$ denote the block labels for each of the N nodes. Instead of writing "$Z_{ai} = 1$" to say that node v_i is a member of block B_a, we will write the equivalent statement "$c_i = B_a$." So, $c_i = \phi(v_i)$ is the block membership for node v_i.

The edges $\{Y_{ij}\}$ are assumed to be independently Bernoulli generated as

$$Y_{ij}|\mathbf{c} \sim \text{Bernoulli}(p_{c_i c_j}), \quad i, j = 1, 2, \ldots, N, \tag{12.14}$$

where $p_{c_i c_j}$ is the probability that $Y_{ij} = 1$ given the node labels c_i and c_j of nodes v_i and v_j, respectively. Thus,

$$E\{Y_{ij}|\mathbf{c}\} = p_{c_i c_j}, \quad i, j = 1, 2, \ldots, N. \tag{12.15}$$

Theorem 12.1 *Given ϕ (or $\mathbf{c}$), the probability of observing adjacency matrix $\mathbf{Y}$ is*

$$P\{\mathbf{Y}|\mathbf{c}\} = \prod_{i<j} P\{Y_{ij}|\mathbf{c}\} \tag{12.16}$$

$$= \prod_{i=1}^{N} \prod_{j=i+1}^{N} p_{c_i c_j}^{Y_{ij}} (1 - p_{c_i c_j})^{1-Y_{ij}}. \tag{12.17}$$

The product is taken over $i < j$ because the graph is assumed to be undirected and without self-loops.

Definition 12.7 A stochastic blockmodel is defined by the K-vector of block probabilities $\boldsymbol{\pi} = (\pi_1, \ldots, \pi_K)^\tau \in (0, 1)^K$ and the $(K \times K)$ symmetric matrix $\mathbf{P} = (p_{c_i c_j})$ of probabilities.

Given $\mathbf{Y}$, N, and K, the parameters of this model are, therefore, $(\mathbf{P}, \boldsymbol{\pi})$. In general, these parameters will be unknown and will have to be estimated.

12.4.2 ML Estimation for Stochastic Blockmodels

We follow the approach of Karrer and Newman (2011a), who make an assumption that has little practical consequence other than that it simplifies the technical arguments. We assume that, for large N, the Bernoulli probability (12.14) can be approximated by a Poisson probability:

$$Y_{ij}|\mathbf{c} \sim \mathcal{P}(p_{c_i c_j}), \quad i < j, \tag{12.18}$$

$$\tfrac{1}{2}Y_{ii}|\mathbf{c} \sim \mathcal{P}(\tfrac{1}{2}p_{c_i c_i}). \tag{12.19}$$

This approximation is reasonable if the Bernoulli distribution has a small mean and if the network is sparse, as are most real networks (Zhao, Levina, and Zhu, 2012).

Substituting (12.18) and (12.19) into (12.17), the complete-data likelihood of $\mathbf{Y}$ is

$$P\{\mathbf{Y}|\mathbf{c}\} = \prod_{i<j} \frac{p_{c_i c_j}^{Y_{ij}} \exp\{-p_{c_i c_j}\}}{Y_{ij}!} \times \prod_i \frac{(\tfrac{1}{2}p_{c_i c_i})^{Y_{ii}/2} \exp\{-\tfrac{1}{2}p_{c_i c_i}\}}{(Y_{ii}/2)!}$$

$$= C \cdot \prod_i \prod_{i<j} \left(p_{c_i c_j}^*\right)^{Y_{ij}^*} \exp\left\{-p_{c_i c_j}^*\right\}, \tag{12.20}$$

where $Y_{ij}^* = \left(1 - \tfrac{1}{2}\delta_{ij}\right)Y_{ij}$, $p_{c_i c_j}^* = \left(1 - \tfrac{1}{2}\delta_{ij}\right)p_{c_i c_j}$, $\delta_{ij} = 1$ if $i = j$, and 0 otherwise, and

$$C = \frac{1}{\prod_{i<j} Y_{ij}! \cdot \prod_i 2^{Y_{ii}/2}(Y_{ii}/2)!} \tag{12.21}$$

is the normalizing constant.

Next, we rewrite (12.20) by taking the products over blocks instead of over nodes. By expanding the products in (12.20) and collecting terms so that nodes that are members of the same block are grouped together (see **Exercise 12.2**), we have the following result.

Theorem 12.2 *The complete-data likelihood of observing the adjacency matrix* $\mathbf{Y}$ *is given by*

$$P\{\mathbf{Y}|\mathbf{c}\} = C \cdot \prod_{r=1}^{K} \prod_{s=1}^{K} p_{rs}^{m_{rs}/2} \exp\left\{-\tfrac{1}{2}N_r N_s p_{rs}\right\}, \tag{12.22}$$

where $p_{rs} = P\{Y_{ij} = 1|c_i = r, c_j = s\} = E\{Y_{ij}|c_i = r, c_j = s\} = p_{sr}$,

$$m_{rs} = \sum_{i=1}^{N} \sum_{j=1}^{N} Y_{ij}\delta_{c_i,r}\delta_{c_j,s} \tag{12.23}$$

is the total number of edges between blocks r *and* s, $N_r = \sum_{i=1}^{N} \delta_{c_i,r}$ *is the number of nodes in the* rth *block, and* C *is given by* (12.21).

Maximizing (12.22) is equivalent to maximizing the logarithm of the expression (ignoring the constant term C and other constants)

$$\log P\{\mathbf{Y}|\mathbf{c}\} = \sum_{r,s} (m_{rs} \log p_{rs} - N_r N_s p_{rs}), \tag{12.24}$$

wrt the symmetric $(K \times K)$-matrix of probabilities, $\mathbf{P} = (p_{rs})$. Differentiating $\log \mathbf{P}$ wrt p_{rs}, and setting the result equal to zero, yields the following equation:

$$\frac{\partial \log \mathbf{P}}{\partial p_{rs}} = \frac{m_{rs}}{p_{rs}} - N_r N_s = 0, \tag{12.25}$$

from which we have the result.

Theorem 12.3 *The MLE of* p_{rs} *is*

$$\widehat{p}_{rs} = \frac{m_{rs}}{N_r N_s}. \tag{12.26}$$

Substituting $\widehat{p}_{rs}$ for p_{rs} in (12.24) and removing irrelevant constants, yields the next result.

Theorem 12.4 *The stochastic blockmodel likelihood criterion is given by*

$$\mathcal{Q}_{SBM} = \sum_{r=1}^{K} \sum_{s=1}^{K} m_{rs} \log \left(\frac{m_{rs}}{N_r N_s} \right). \tag{12.27}$$

This criterion is computed for all possible partitions of the network to find the partition that yields its maximum value. Zhao, Levina, and Zhu (2012) found that, in practice, the Poisson-based criterion $\mathcal{Q}_{SBM}$ produced solutions identical with the (correct) Bernoulli likelihood.

Properties. Several articles have studied maximum-likelihood (ML) estimation of the parameters in the stochastic blockmodel. These include Snijders and Nowicki (1997), who proposed an EM-type algorithm for networks with fewer than 20 nodes; Nowicki and Snijders (2001), who described a Bayesian approach for networks with at most 200 nodes using Gibbs sampling and a computer program BLOCKS with a manual given in Snijders and Nowicki (2007); Celisse, Daudin, and Pierre (2012) and Bickel et al. (2013), both of whom studied asymptotic properties of the model, such as consistency and asymptotic normality of the MLEs; and Karrer and Newman (2011a), who proposed adjusting the model to take into account the degree distributions of the nodes.

Conditions under which the parameters in an SBM are identifiable were discussed by Allman, Matias, and Rhodes (2009, 2011) and by Celisse, Daudin, and Pierre (2012). Proofs of consistency of the ML estimates (for K fixed and $N \rightarrow \infty$) have been given by Snijders and Nowicki (1997) for $K = 2$ (using a Bayesian approach with MCMC methods), and extended to general fixed K by Celisse, Daudin, and Pierre (2012) and by Bickel et al. (2013), who also provided a proof of the asymptotic normality of ML estimates. Bickel and Chen (2009) used a profile likelihood approach to prove consistency for SBM estimates.

Computational problems. Two major problems, however, have been identified with the ML method. First, as with many applications of the ML method, the iterative process may get trapped within one of the multiple local maxima of the likelihood function. One way to deal with this computational problem is to run the process several times from different starting points, each time optimizing the likelihood function using a label-switching algorithm (Stephens, 2000; Bickel and Chen, 2009). Second, the log-likelihood (12.17) of the observed network data (ignoring the constant term),

$$\log P\{\mathbf{Y}|\mathbf{c}\} = \sum_i \sum_{i \le j} [Y_{ij} \log(p_{c_i c_j}) + (1 - Y_{ij}) \log(1 - p_{c_i c_j})], \tag{12.28}$$

is so complicated, involving a search over K^N terms in order to assign N nodes to K communities, that its maximization is computationally infeasible for large networks; that is, optimization over all possible label assignments is an $\mathcal{NP}$-hard problem (Bickel et al., 2013). As a result, the theoretical results on consistency are viewed as having little practical use. This difficulty means that the MLE is computable only for very small networks, probably up to about 20 nodes (Celisse, Daudin, and Pierre, 2012).

12.4.3 Approximating MLEs

As we just noted, the biggest problem with MLE for SBMs is its computational limits. To overcome these limits, one might try to use the EM (expectation-maximization) algorithm (Dempster, Laird, and Rubin, 1977). Unfortunately, the EM algorithm uses the conditional distribution $P\{\cdot|\mathbf{Y}\}$, which is also computationally intractable. Several alternative procedures have been suggested. We outline some of these alternatives.

1. Variational Approximation

One approach that has been proposed is to use the variational approximation method (Daudin, Picard, and Robin, 2008). This method is claimed to be efficient enough that it can deal with networks of up to 3000 nodes (Celisse, Daudin, and Pierre, 2012). In a review article, Daudin (2011) goes even further by arguing that the variational approximation method should be able to deal with at most 5000 nodes, which would cover most biological and ecological networks, and many social networks. See also Ormerod and Wand (2010).

Variational approximation methods have their origins in the calculus of variations. Variational calculus is used for optimizing a functional over a class of functions on which that functional depends. Approximations may be needed for cases in which constraints are imposed upon the class of functions. Variational approximations for statistical models are now being used as a competitor to MCMC methods, especially for Bayesian inference problems, when the size of the problem increases and MCMC begins to exhibit computational difficulties.

The variational approach (Jordan et al., 1999; Jaakkola, 2000) attacks this problem by finding a lower bound on the log-likelihood, and then maximizes that lower bound. This approach provides an approximation to the ML estimate. See also Bishop (2006, Sections 9.4 and 10.1).

Consider the following lower bound on the log-likelihood $\log P\{\mathbf{Y}|\mathbf{c}\}$:

$$\mathcal{J}(q_{\mathbf{Y}}) = \log P\{\mathbf{Y}|\mathbf{c}\} - KL(q_{\mathbf{Y}}(\cdot) \| P\{\cdot|\mathbf{Y}\}), \tag{12.29}$$

where

$$KL(q \| p) = E_q\{\log q\} - E_q\{\log p\} = \sum q \log \left\{ \frac{q}{p} \right\} \tag{12.30}$$

is the Kullback–Leibler divergence between q and p (which is always non-negative), $P\{\mathbf{c}|\mathbf{Y}\}$ is the true conditional distribution of $\mathbf{c}$ given $\mathbf{Y}$, and $q_{\mathbf{Y}}$ is an approximation to this conditional distribution. Note that $q_{\mathbf{Y}}$ depends upon the data $\mathbf{Y}$.

Now, $\mathcal{J}(q_{\mathbf{Y}}) = \log P\{\mathbf{Y}|\mathbf{c}\}$ iff $KL(q_{\mathbf{Y}}(\cdot) \| P\{\cdot|\mathbf{Y}\}) = 0$, which will hold iff $q_{\mathbf{Y}}(\cdot) = P\{\cdot|\mathbf{Y}\}$. Unfortunately, as we have already noted, $P\{\cdot|\mathbf{Y}\}$ is intractable and, hence, cannot be computed. So, the plan is to find a $q_{\mathbf{Y}}$ in some given class $\mathcal{Q}$ of distributions that is "best" (in a KL sense) at approximating $P\{\cdot|\mathbf{Y}\}$.

One way of proceeding is to be more specific about the class $\mathcal{Q}$. We limit the class $\mathcal{Q}$ to a family of "tractable" distributions q for which the lower bound can be evaluated, and then optimize within this class to find the $q = q_{\mathbf{Y}}$ that most closely approximates $P\{\cdot|\mathbf{Y}\}$. The variational approximation $q_{\mathbf{Y}}$ to $P\{\cdot|\mathbf{Y}\}$ is found by minimizing the KL term in (12.29):

$$q_{\mathbf{Y}} = \arg\min_{q \in \mathcal{Q}} KL(q \| P\{\cdot|\mathbf{Y}\}), \tag{12.31}$$

assuming such a minimum exists. We choose $\mathcal{Q}$ by assuming that it contains all those q that can be factorized into a product of independent marginal distributions. From (12.9), Daudin, Picard, and Robin (2008) assumed that $\mathbf{Z}_i = (Z_{1i}, \ldots, Z_{Ki})^\tau$ has the multinomial distribution $\mathcal{M}(1, \boldsymbol{\pi}_i)$ with density function $q(\mathbf{z}_i; \boldsymbol{\pi}_i)$ and parameter vector $\boldsymbol{\pi}_i = (\pi_{1i}, \ldots, \pi_{Ki})^\tau$, $\pi_{ki} \in [0, 1]$, where $\sum_{k=1}^{K} \pi_{ki} = 1$, $i = 1, 2, \ldots, N$. They then assumed that $q_{\mathbf{Y}}$ could be expressed as the product of independent multinomial distributions:

$$q_{\mathbf{Y}}(\mathbf{z}) = \prod_{i=1}^{N} q(\mathbf{z}_i; \boldsymbol{\pi}_i) = \prod_{i=1}^{N} \prod_{k=1}^{K} \pi_{ki}^{z_{ki}}. \tag{12.32}$$

This type of factorization, which is commonly referred to in statistical physics as a *mean-field approximation* (see, e.g., Peterson and Anderson, 1987), allows us to optimize each marginal $q(\mathbf{z}_i; \boldsymbol{\pi}_i)$ wrt $\boldsymbol{\pi}_i$, separately, one at a time.

The next step uses the EM algorithm, which alternates between two optimization problems (**Algorithm 12.1**).

Properties. Celisse, Daudin, and Pierre (2012) showed that variational estimators and MLEs are asymptotically equivalent and both are consistent. Bickel et al. (2013) showed that the variational method enjoys the same asymptotic normality properties as the ordinary likelihood, but can be computed at a faster speed.

Algorithm 12.1 *Variational EM algorithm for the SBM*

---

1. *Initialize* $\mathbf{Z}$ or the model parameters $\theta = (\mathbf{P}, \{\boldsymbol{\pi}_i\})$, where $\boldsymbol{\pi}_i$ is the block probability K-vector for the ith node, $i = 1, 2, \ldots, N$, and $\mathbf{P} = (p_{ab})$ is the $(K \times K)$ connectivity matrix.
2. *vE-step:* Hold all model parameters θ fixed at their current values and maximize $\mathcal{J}(q_{\mathbf{Y}})$ wrt an approximation $q_{\mathbf{Y}}$ of the conditional distribution.
3. *vM-step:* Hold the approximation $q_{\mathbf{Y}}$ fixed and maximize $\mathcal{J}(q_{\mathbf{Y}})$ wrt the model parameters θ.
4. *Iterate* between the E-step and the M-step until convergence.

Estimation of the variational parameters depends strongly upon the initialization for those parameters because the KL divergence may contain many local minima (Salter-Townshend and Murphy, 2012). The usual recommendation, therefore, is to start the algorithm from different values of the variational parameters and see whether the algorithm converges to different minima from the different initial values.

Applications. There have been several applications of the variational approximation to the SBM. We mention two of these applications.

(i) Biological networks. Daudin, Picard, and Robin (2008) applied this variational method to the biological network of *Escherichia coli*, in which the nodes are chemical reactions and a pair of nodes are connected by an edge if a compound produced by the first node is a part of the second node (or vice versa). The network data consist of 605 nodes and 1782 edges (including 12 nodes with degree 0, 44 with degree 1, and 150 with degree 2). The authors considered their algorithm to be the only method then currently available for handling such a large network. They fitted the variational algorithm with $K = 21$ classes to the data.

(ii) Financial networks. The variational EM approximation approach to estimating an SBM was used by De la Concha, Martinez-Jaramillo, and Carmona (2017) in a first-ever attempt at using SBMs to model the structural complexity and level of interconnectedness in a banking system. They studied various types of multilayer market interactions. In this case study, there were seven layers: transactions in the securities market, repurchase agreements markets, payment system flows, interbank deposits and loans, cross-holding of securities, foreign exchange transactions, and derivatives transactions. Eleven years of daily data on the Mexican banking system were made available as time series (January 2005 to December 2015). The banking network was described by a time sequence of adjacency matrices. Although the numbers of nodes and edges of each of those adjacency matrices are not explicitly stated in the article, the graphs each suggest that there are 44 banks (nodes) in the banking network. SBM analysis suggested that the banks should be categorized into $K = 3$ groups. Computations were carried out using the `blockmodels` package in R.

2. Belief Propagation

A Bayesian approach is typically a computationally intensive affair and, hence, is not used much for community detection, except possibly for networks of moderate size. Using an MCMC algorithm (e.g., Gibbs sampling) to compute network estimates within a Bayesian framework is much slower than using variational methods, and is only helpful when dealing with only a hundred or so nodes.

Decelle et al. (2013) present the problem of determining the topology of an SBM through the lens of statistical physics, using terms such as Boltzmann distribution, generalized Potts model, Hamiltonian, local magnetic field, Gibbs state, spin glasses, and thermodynamic limit to motivate the model and to describe its structure. Given a network graph $\mathcal{G}$, which is generated by an SBM, they study the problem of estimating K and the matrix $\mathbf{P}$, as well as the number of nodes, N_k, within the kth group. They show that there are three regions relating to the ranges of parameter values: one region where inference is impossible; a second region where inference is possible, but exponentially difficult; and a third region where a belief propagation algorithm carries out computations to infer the parameter values in an optimal manner, and in linear time. The boundaries between these regions correspond to phase transitions in various models of statistical physics. By incorporating a belief propagation (BP) algorithm into the Bayesian approach, they are able to construct an iterative algorithm that has good properties for networks of moderate size. In the limit, as $N \to \infty$, their algorithm either produces the exact parameter values, or reports that inference is impossible or algorithmically difficult.

The authors report that, if the values of the parameters are known, their BP algorithm, which is used to determine which nodes are assigned to which group and the probability that each node belongs to each group, is much faster than an equivalent MCMC algorithm. To estimate the parameters, if they are unknown, an EM algorithm is applied with the BP algorithm used for the E-step. Like all iterative algorithms, BP performance success depends largely on using good starting values. No asymptotic results are available yet.

3. Pseudo-Likelihood

The basic idea of pseudo-likelihood (PL) is that it simplifies the likelihood by ignoring some of the dependency structure in the data. Amini et al. (2013) use a PL algorithm to analyze the SBM. In this context, they ignore the symmetry of the adjacency matrix $\mathbf{Y}$. Next, they apply "block compression" by partitioning the set of nodes, $\mathcal{V}$, into blocks, and then compute the likelihood of row totals within these blocks. This provides a fast approximation to the SBM likelihood, which, in turn, can be used to analyze a network of "tens of millions of nodes."

Although the joint likelihood of $\mathbf{Y}$ and $\mathbf{c}$ can be maximized through the EM algorithm, the E-step is an $\mathcal{NP}$-hard task because it involves a maximization over all possible label assignments. To overcome this problem, the authors set up a "partition vector" $\mathbf{e} = (e_1, \ldots, e_N)^\tau$, where $e_i \in \{1, 2, \ldots, K\}$. The vector $\mathbf{e}$, which a user selects as an initial approximation to $\mathbf{c}$, assigns each of the N nodes to one of K blocks (communities), where the value of K (and the assignments) reflects either prior knowledge or an educated guess.

Consider e_j. If $e_j = k$, the indicator variable $I_{[e_j = k]} = 1$; otherwise, it equals 0. Now, add over all N entries in the ith row of $\mathbf{Y}$ that correspond with nodes that $\mathbf{e}$ has assigned to the kth block:

$$b_{ik} = \sum_{j=1}^{N} Y_{ij} I_{[e_j = k]}, \quad i = 1, 2, \ldots, N, \quad k = 1, 2, \ldots, K. \tag{12.33}$$

Let $\mathbf{b}_i = (b_{i1}, b_{i2}, \ldots, b_{iK})^\tau$. Now, define the $(K \times K)$ "confusion" matrix $\mathbf{R} = (R_{ka})$, where

$$R_{ka} = \frac{1}{N} \sum_{i=1}^{N} I_{[e_i = k, c_i = a]}, \quad k, a = 1, 2, \ldots, K. \tag{12.34}$$

So, R_{ka} is the average (over all N nodes) number of nodes that are initially classified into the kth block but which really belong to the ath block. Denote the kth row of $\mathbf{R}$ by $\mathbf{R}_{k.}^\tau$ and the ℓth column by $\mathbf{R}_{.\ell}$. Set $\mathbf{\Lambda} = (\lambda_{\ell k})$, where $\lambda_{\ell k} = N \mathbf{R}_{k.}^\tau \mathbf{R}_{.\ell}$. Let $\lambda_\ell = \sum_{k=1}^{K} \lambda_{\ell k}$.

The authors next observe that for the ith node, conditional on $\mathbf{c}$ with $c_i = \ell$, the entries of $\mathbf{b}_i$ are mutually independent. Furthermore, b_{ik}, which is a sum of independent Bernoulli variables, is approximately Poisson with parameter $\lambda_{\ell k}$. Thus, assuming $\boldsymbol{\Lambda}$ has no identical rows, $\mathbf{b}_i$ consists of approximately Poisson entries.

Putting everything together, we have the following result.

Theorem 12.5 *Ignoring any dependence amongst the Poisson entries in $\{\mathbf{b}_i\}$ and defining the $\{c_i\}$ as latent variables, the pseudo-log-likelihood is given by*

$$\ell_{\mathrm{PL}}(\boldsymbol{\pi},\boldsymbol{\Lambda};\{\mathbf{b}_i\}) = \sum_{i=1}^{N} \log \left(\sum_{\ell=1}^{K} \pi_\ell e^{-\lambda_\ell} \prod_{k=1}^{K} \lambda_{\ell k}^{b_{ik}} \right), \tag{12.35}$$

where, from (12.9), $\boldsymbol{\pi} = (\pi_1,\ldots,\pi_K)^\tau$ and the constant is ignored. Maximizing $\ell_{\mathrm{PL}}(\boldsymbol{\pi},\boldsymbol{\Lambda};\{\mathbf{b}_i\})$ over $\boldsymbol{\pi}$ and $\boldsymbol{\Lambda}$ yields the maximum pseudo-log-likelihood estimate of $(\boldsymbol{\pi},\boldsymbol{\Lambda})$.

Maximizing (12.35) wrt $\boldsymbol{\pi}$ and $\boldsymbol{\Lambda}$ can be accomplished through the EM algorithm applied to mixture models. See Amini et al. (2013) for details. The partition vector $\mathbf{e}$ is updated at each EM iteration until convergence, so that its entries become the most likely block label for each node. The EM algorithm is known to converge to a local maximum (under certain identifiability conditions). Simulations indicate that the algorithm converges quickly, especially if $\mathbf{e}$ is a good initial value.

Properties. There is, however, a major drawback to the pseudo-likelihood method. When a network has hubs or a high variability in degree size of the nodes, the pseudo-likelihood method has been found to fail, instead dividing the nodes by degree size, low-degree nodes being placed into one block and high-degree nodes being placed into the other block. In an empirical study, Amini et al. showed that such an incorrect assignment occurred for the political blogs network data (see Section 13.3.5). They overcame this problem by creating the conditional pseudo-likelihood method, in which the pseudo-likelihood is conditioned on the observed node degrees.

They tackle the question of consistency for the pseudo-likelihood estimates, both of node-label assignment and of parameter estimation, for SBMs. They take the SBM to have $K = 2$ blocks, which are balanced in the sense that each block contains $N/2$ nodes. In both directed and undirected cases, they show that, under certain regularity conditions, the estimates are uniformly consistent. Empirical study of a number of variations on the theme of pseudo-likelihood shows that the estimation algorithm works well even when starting values are non-informative. They also found that pseudo-likelihood was faster than belief propagation but about the same on accuracy for community detection problems.

4. Semidefinite Programming

To overcome the computational difficulties in estimating an SBM, this semidefinite programming (SDP) relaxation method was first proposed by Abbe, Bandeira, and Hall (2016), who studied the case of $K = 2$ communities (the "planted bisection" model). Their work was generalized to K communities by Amini and Levina (2018). The authors work with a balanced model, where "balance" is taken to mean that the sizes of the different communities are equal, each containing N/K nodes, where N is the number of nodes, K is the number of communities, and N/K is assumed to be an integer.

The balanced planted partition model, denoted by $PP^{\text{bal}}(p,q)$, depends upon two parameters, p and q, where $p > q$. In this model, any two nodes in the same community are connected by an edge wpr p, while two nodes in different communities are connected wpr q. The model assumes that the $(K \times K)$ symmetric matrix $\mathbf{P}$ showing probabilities of connections between communities has the form

$$\mathbf{P} = q\mathbf{J}_K + (p - q)\mathbf{I}_K, \tag{12.36}$$

where $\mathbf{J}_K = \mathbf{1}_K \mathbf{1}_K^{\tau}$ is the $(K \times K)$-matrix of 1s, $\mathbf{1}_K$ is the K-vector of all 1s, and $\mathbf{I}_K$ is the $(K \times K)$ identity matrix. In other words, $\mathbf{P}$ is assumed to have p on each of the diagonal entries and q as off-diagonal entries.

The MLE of the SBM is obtained by varying the parameter space of the optimization problem. First, Amini and Levina re-express the log-likelihood. Let $\mathbf{M} = \mathbf{Z}^{\tau}\mathbf{PZ}$, where the $(K \times N)$-matrix $\mathbf{Z}$ is given in (12.7). So, $\mathbf{M} = (m_{ij})$ is a block-constant, rank-K, $(N \times N)$-matrix. For any function f and matrix $\mathbf{M}$, define f applied to the ijth entry of $\mathbf{M}$ as $(f \circ \mathbf{M})_{ij} = f(m_{ij})$. Thus,

$$f \circ \mathbf{M} = f \circ (\mathbf{Z}^{\tau}\mathbf{PZ}) = \mathbf{Z}^{\tau}(f \circ \mathbf{P})\mathbf{Z}. \tag{12.37}$$

From (12.17), the log-likelihood of the SBM is

$$\ell(\mathbf{Z}, \mathbf{P}) = \sum_{i<j}[Y_{ij} \log m_{ij} + (1 - Y_{ij}) \log(1 - m_{ij})]$$
$$= \sum_{i<j}[Y_{ij}(f \circ \mathbf{M})_{ij} + (g \circ \mathbf{M})_{ij}], \tag{12.38}$$

where we set $f(x) = \log(\frac{x}{1-x})$ and $g(x) = \log(1-x)$. Now, we define $\langle \mathbf{A}, \mathbf{B} \rangle = \text{tr}(\mathbf{AB})$ for symmetric matrices $\mathbf{A}$ and $\mathbf{B}$. Then, using (12.37), we can rewrite (12.38) as

$$2\ell(\mathbf{Z}, \mathbf{P}) = \langle \mathbf{Y}, f \circ \mathbf{M} \rangle_0 + \langle \mathbf{J}_N, g \circ \mathbf{M} \rangle_0$$
$$= \langle \mathbf{Y}, \mathbf{Z}^{\tau}(f \circ \mathbf{P})\mathbf{Z} \rangle_0 + \langle \mathbf{J}_N, \mathbf{Z}^{\tau}(g \circ \mathbf{P})\mathbf{Z} \rangle_0, \tag{12.39}$$

where $\langle \mathbf{A}, \mathbf{B} \rangle_0 = \langle \mathbf{A}, \mathbf{B} \rangle - \sum_i A_{ii} B_{ii}$. Note that for the PP-model, $f \circ \mathbf{P} = f(q)\mathbf{J}_K + (f(p) - f(q))\mathbf{I}_K$ and $g \circ \mathbf{P} = g(q)\mathbf{J}_K + (g(p) - g(q))\mathbf{I}_K$, and $\mathbf{Z}^{\tau}\mathbf{J}_K\mathbf{Z} = \mathbf{J}_N$. Expanding (12.39), we have that

$$2\ell(\mathbf{Z}, \mathbf{P}) = (f(p) - f(q))\langle \mathbf{Y}, \mathbf{Z}^{\tau}\mathbf{Z} \rangle_0 + (g(p) - g(q))\langle \mathbf{J}_N, \mathbf{Z}^{\tau}\mathbf{Z} \rangle_0 + C, \tag{12.40}$$

where C is a constant independent of $\mathbf{Z}$. Now, $p > q$ implies that $f(p) > f(q)$ and $g(p) < g(q)$. Thus,

$$\frac{2\ell(\mathbf{Z}, \mathbf{P})}{f(p) - f(q)} = \langle \mathbf{Y}, \mathbf{Z}^{\tau}\mathbf{Z} \rangle - \lambda \langle \mathbf{J}_N, \mathbf{Z}^{\tau}\mathbf{Z} \rangle + C', \tag{12.41}$$

where

$$\lambda = \frac{g(q) - g(p)}{f(p) - f(q)} > 0, \tag{12.42}$$

and C' is another constant independent of $\mathbf{Z}$. Because $(\mathbf{Z}^{\tau}\mathbf{Z})_{ii} = 1$, for all i, $\langle \cdot, \cdot \rangle_0$ was replaced by $\langle \cdot, \cdot \rangle$ in (12.41).

Next, Amini and Levina introduce a relaxation process so that the log-likelihood can be maximized wrt $\mathbf{Z}$. The estimator of $\mathbf{Z}$ is given by

$$\widehat{\mathbf{Z}} = \arg \max_{\mathbf{Z} \in \mathcal{Z}} \{\langle \mathbf{Y}, \mathbf{Z}^{\tau}\mathbf{Z} \rangle - \lambda \langle \mathbf{J}_N, \mathbf{Z}^{\tau}\mathbf{Z} \rangle\}. \tag{12.43}$$

Now, let $\mathbf{X} = \mathbf{Z}^\tau \mathbf{Z}$. Each $X_{ij} = 1$ if v_i and v_j are members of the same community, and $X_{ij} = 0$ otherwise. So, (12.43) becomes

$$\widehat{X} = \arg\max_{\mathbf{X} \in \mathcal{X}} \{ \langle \mathbf{Y}, \mathbf{X} \rangle - \lambda \langle \mathbf{J}_N, \mathbf{X} \rangle \}. \tag{12.44}$$

Note that while $\mathbf{Z}$ is maximized over the subspace $\mathcal{Z}$, $\mathbf{X}$ is maximized over the subspace $\mathcal{X}$. The type of relaxation process depends upon the form of $\mathcal{X}$.

The relaxation proposed by Amini and Levina is to take $\mathbf{Z} = \mathbf{A}\mathbf{Z}_0\mathbf{B}$, where $\mathbf{A}$ and $\mathbf{B}$ are permutation matrices. Under the balanced planted partition model, $\mathbf{Z}_0 = \mathbf{1}_{N/K}^\tau \otimes \mathbf{I}_K$, where $\otimes$ is the Kronecker product.[5] The $(K \times N)$-matrix $\mathbf{Z}_0$ assigns nodes consecutively to communities $1, 2, \ldots, K$. The corresponding $(N \times N)$-matrix $\mathbf{X}$ has the form $\mathbf{X} = \mathbf{B}^\tau \mathbf{X}_0 \mathbf{B}$, where $\mathbf{X}_0 = \mathbf{J}_{N/K} \otimes \mathbf{I}_K$ is a block-diagonal matrix with each diagonal block equal to $\mathbf{J}_{N/K}$, and, again, $\mathbf{B}$ is a permutation matrix. They then show that the term $\lambda \langle \mathbf{J}_N, \mathbf{X} \rangle$ can be dropped from (12.44) because $\mathbf{X}\mathbf{1}_N = (N/K)\mathbf{1}_N$. This leads us to the following proposal:

Definition 12.8 (Amini and Levina, 2018) *The proposed relaxation result for the MLE of the SBM is*

$$\widehat{\mathbf{X}} = \arg\max_{\mathbf{X}} \langle \mathbf{Y}, \mathbf{X} \rangle = \arg\max_{\mathbf{X}} \sum_{i,j} Y_{ij} X_{ij}, \tag{12.45}$$

subject to certain restrictions on $\mathbf{X}$, including non-egativity ($\mathbf{X} \geq \mathbf{0}$), semidefiniteness ($\mathbf{X} \succeq \mathbf{0}$), a diagonal restriction ($\mathrm{diag}(\mathbf{X}) = \mathbf{1}_N$), and a size restriction ($\mathbf{X}\mathbf{1}_N = (N/K)\mathbf{1}_N$).

See also Chen and Xu (2016) for similar results.

Properties. Under certain regularity conditions, Amini and Levina establish consistency results for this estimation technique and variations of this technique. They also show, via simulations, that this method outperforms spectral methods for fitting SBMs with a large number of blocks. The high computational cost of SDP relaxations is expected to be reduced by future developments in optimization research.

12.4.4 p_1 Blockmodels

The theory of statistical inference for a stochastic blockmodel can be developed by considering the p_1 distribution for dyads given in Section 10.3 (Wang and Wong, 1987). To do this, we insert an interaction term into θ_{ij} in (10.12) to account for the block structure. Let

$$d_{ijk\ell} = \begin{cases} 1, & \text{if node } v_i \in B_k \text{ and node } v_j \in B_\ell \\ 0, & \text{otherwise} \end{cases}, \tag{12.46}$$

$k, \ell = 1, 2, \ldots, K$. There are, therefore, K^2 blocks. Now, we assign to each $B_k \times B_\ell$ block a parameter $\gamma_{k\ell}$, where we assume (for identifiability purposes) that $\gamma_{k+} = \gamma_{+\ell} = 0$, $k, \ell = 1, 2, \ldots, K$. Let $\boldsymbol{\gamma} = (\gamma_{k\ell})$ be the K^2-vector of block parameters. The interaction model is given by

---

[5]Note that we take the *left* Kronecker product, $\mathbf{A} \otimes \mathbf{B} = (\mathbf{A}B_{jk})$.

$$\theta_{ij} = \theta + \alpha_i + \beta_j + \sum_k \sum_\ell \gamma_{k\ell} d_{ijk\ell}. \tag{12.47}$$

Substituting this form of θ_{ij} into (10.24) and setting $\rho_{ij} = \rho$ for $i < j$ yields the following result.

Theorem 12.6 *The probability distribution for the p_1 blockmodel is given by*

$$p_1(\mathbf{y}|\boldsymbol{\gamma}) = P\{\mathbf{Y} = \mathbf{y}|\boldsymbol{\gamma}\}$$

$$= C \exp\left\{ \rho m + \theta y_{++} + \sum_{k,\ell} \gamma_{k\ell} y_{++}(B_k \times B_\ell) + \sum_i \alpha_i y_{i+} + \sum_j \beta_j y_{+j} \right\}, \tag{12.48}$$

where $y_{++}(B_k \times B_\ell)$ is the observed total in the $B_k \times B_\ell$ block, and

$$C = C(\rho, \theta, \{\gamma_{k\ell}\}, \{\alpha_i\}, \{\beta_j\}) = \prod_{i<j} c_{ij} \tag{12.49}$$

is the normalizing constant, where

$$c_{ij} = (1 + e^{\theta_{ij}} + e^{\theta_{ji}} + e^{\rho + \theta_{ij} + \theta_{ji}})^{-1}. \tag{12.50}$$

There are $(K-1)^2$ independent block parameters in (12.48). For example, Sampson's monk data has $K = 4$, and so there are $(4-1)^2 = 9$ free block parameters in the full model. The distribution (12.48) is a member of the exponential family with parameters $\rho, \theta, \boldsymbol{\gamma}, \boldsymbol{\alpha}$, and $\boldsymbol{\beta}$, where $\boldsymbol{\alpha} = (\alpha_1, \ldots, \alpha_K)^\tau$ and $\boldsymbol{\beta} = (\beta_1, \ldots, \beta_K)^\tau$.

If we wished to consider a smaller p_1 model with fewer parameters than the full model, we could restrict the block parameters in $\boldsymbol{\gamma}$ to a more manageable number. So, suppose we reduce $\boldsymbol{\gamma}$ to $\gamma_1, \ldots, \gamma_h$, where $h \leq (K-1)^2$, by setting U_s to be the union of all those blocks that have the same block parameter γ_s, $s = 1, 2, \ldots, h$. Then, we can write (12.48) as

$$p_1(\mathbf{y}|\boldsymbol{\gamma}) = C \exp\left\{ \rho m + \theta y_{++} + \sum_s \gamma_s y_{++}(U_s) + \sum_i \alpha_i y_{i+} + \sum_j \beta_j y_{+j} \right\}, \tag{12.51}$$

where $y_{++}(U_s)$ denotes the observed total number of edges in U_s, $s = 1, 2, \ldots, h$.

Maximum-Likelihood Estimation

The unknown parameters in the reduced p_1-blockmodel distribution (12.51) are $\boldsymbol{\pi} = (\rho, \theta, \boldsymbol{\gamma}^\tau, \boldsymbol{\alpha}^\tau, \boldsymbol{\beta}^\tau)^\tau$, and these can be estimated by maximum likelihood. We denote the ML estimates of these parameters by

$$\widehat{\boldsymbol{\pi}} = (\widehat{\rho}, \widehat{\theta}, \widehat{\boldsymbol{\gamma}}^\tau, \widehat{\boldsymbol{\alpha}}^\tau, \widehat{\boldsymbol{\beta}}^\tau)^\tau, \tag{12.52}$$

where $\widehat{\boldsymbol{\gamma}} = (\widehat{\gamma}_{kl})$, $\widehat{\boldsymbol{\alpha}} = (\widehat{\alpha}_1, \ldots, \widehat{\alpha}_K)^\tau$, and $\widehat{\boldsymbol{\beta}} = (\widehat{\beta}_1, \ldots, \widehat{\beta}_K)^\tau$. Because we are dealing with a member of the exponential family, we can derive the MLEs of these parameters by equating the sufficient statistics to their expected values. The sufficient statistics of the reduced p_1-blockmodel distribution (12.51) are $M = \sum_{i<j} Y_{ij} Y_{ji}$, $\{Y_{++}(U_s)\}$, $\{Y_{i+}\}$, and $\{Y_{+j}\}$. Thus, the likelihood equations are

$$M = E_\pi\{M\} = \widehat{m}_{++}, \tag{12.53}$$
$$Y_{++}(U_s) = E_\pi\{Y_{++}(U_s)\} = \widehat{m}_{++}(U_s) + \widehat{y}_{++}(U_s), \quad s = 1, 2, \ldots, h, \tag{12.54}$$
$$Y_{i+} = E_\pi\{Y_{i+}\} = \widehat{m}_{i+} + \widehat{y}_{i+}, \quad i = 1, 2, \ldots, N, \tag{12.55}$$
$$Y_{+j} = E_\pi\{Y_{+j}\} = \widehat{m}_{+j} + \widehat{y}_{+j}, \quad j = 1, 2, \ldots, N, \tag{12.56}$$

where

$$m_{ij} = (1/c_{ij})e^{\theta_{ij} + \theta_{ji}}, \quad i < j, \tag{12.57}$$
$$y_{ij} = (1/c_{ij})e^{\theta_{ij}}, \quad i \neq j. \tag{12.58}$$

Here, c_{ij} is given by (12.50) and θ_{ij} by (12.47). The likelihood equations for the p_1 model can be obtained from the likelihood equations for the p_1 blockmodel by removing the blocking equations (12.54) from the estimation process.

The likelihood equations (12.53)–(12.56) can be solved using the generalized iterative scaling (GIS) algorithm (Darroch and Ratcliff, 1972). Although the GIS algorithm converges to the ML estimates, it is known to have an "excruciatingly slow" convergence rate relative to the standard iterated scaling algorithm (Fienberg and Wasserman, 1981a). The blockmodel of Wang and Wong (1987) cannot be formulated as an ordinary log-linear model, and so the standard iterative scaling algorithm is not applicable. Wang and Wong choose to use a slightly modified version of the GIS algorithm to find the ML estimates. When the full model contains a large number of blocks, the resulting large number of blocking parameters that need to be estimated will slow down the convergence rate of the algorithm; this should be a good reason for using the reduced model with a small number (possibly one) of blocking parameters. No studies have yet been carried out on the convergence rate of this algorithm.

Wang and Wong compute the ML estimates for Sampson's monk data and show that with only a single blocking parameter (i.e., $h = 1$), the p_1 blockmodel yields a much better fit than the p_1 model, with the sum of absolute residuals reduced by almost 40% by adding a blocking term to the p_1 model.

12.4.5 Degree-Corrected Stochastic Blockmodels

Although the stochastic blockmodel is a great idea, it fails as an approximation to real-world networks. First, it lacks flexibility in understanding heterogeneous network structures that one typically finds in network data. Second, every node located within a community is treated by the model as homogeneous or stochastically equivalent. Third, it tends to place nodes with low degrees in different blocks than nodes with high degrees, and it tries hard not to place nodes with very different degrees into the same block. Fourth, it does not provide for the possibility of "hubs" or central nodes that possess high degrees and which have been shown to be common in real networks. There have been several attempts to "repair" the stochastic blockmodel so that it can take account of these points (e.g., p^* models and p_1 models), but, for one reason or another, most of these have been found to be less than satisfying as networks expand and become more complex.

One generalization of the stochastic blockmodel that has been more successful in allowing heterogeneity of degree within blocks is the *degree-corrected stochastic blockmodel* (Karrer and Newman, 2011a). In this approach, the heterogeneity in the degree structure of the nodes is accounted for by inserting a "degree parameter" $\theta_i > 0$ (associated with node v_i) into the model to represent the probability that node v_i is able to link to neighboring nodes in the network, $i = 1, 2, \ldots, N$. We will see that the degree-corrected stochastic blockmodel allows nodes of different degrees to be clustered together.

Let $\boldsymbol{\theta} = (\theta_1, \ldots, \theta_N)^\tau$ and $\Theta = \mathrm{diag}\{\theta_1, \ldots, \theta_N\}$. The change to the usual stochastic blockmodel is to replace (12.18) and (12.19) by

$$\mathrm{E}\{Y_{ij} | \mathbf{c}, \boldsymbol{\theta}\} = \theta_i \theta_j p_{c_i c_j}, \ i \neq j, \tag{12.59}$$

$$\mathrm{E}\{\tfrac{1}{2} Y_{ii} | \mathbf{c}, \boldsymbol{\theta}\} = \tfrac{1}{2} \theta_i^2 p_{c_i c_i}, \tag{12.60}$$

where $\theta_i \theta_j p_{c_i c_j} \in [0, 1]$. In this formulation, we allow self-loops, each of which gets counted twice (i.e., $Y_{ii} = 2$); hence, the reason for the multiplier of $\frac{1}{2}$ in (12.60). We can express this in matrix notation as (Qin and Rohe, 2013)

$$\mathrm{E}\{\mathbf{Y} | \mathbf{c}, \boldsymbol{\theta}\} = \Theta \mathbf{Z}^\tau \mathbf{P} \mathbf{Z} \Theta, \tag{12.61}$$

where $\mathbf{Z} = (Z_{ai}) \in \{0, 1\}^{K \times N}$ is the block-membership matrix (i.e., $Z_{ai} = 1$ iff node v_i is a member of block a, and 0 otherwise) and $\mathbf{P} = (p_{c_i c_j})$ is a $(K \times K)$ symmetric, positive-definite matrix that identifies relationships between and within blocks. For identifiability purposes, the degree parameters are constrained by Karrer and Newman to satisfy the following condition:

$$\sum_{i=1}^{N} \theta_i \delta_{c_i, r} = 1, \ r = 1, 2, \ldots, K, \tag{12.62}$$

where $\delta_{a,b}$ is the Kronecker delta, which equals 1 if $a = b$, and 0 otherwise. This constraint says that within each block, the θs sum to 1. Note that the standard stochastic blockmodel is a special case of the degree-corrected stochastic blockmodel where the θs are equal.

Maximum-Likelihood Estimation

Following Karrer and Newman, we again assume that the Bernoulli probability is approximated by a Poisson probability:

$$Y_{ij} | (\mathbf{c}, \boldsymbol{\theta}) \sim \mathcal{P}(\theta_i \theta_j p_{c_i c_j}), \ i < j, \tag{12.63}$$

$$\tfrac{1}{2} Y_{ii} | (\mathbf{c}, \boldsymbol{\theta}) \sim \mathcal{P}(\tfrac{1}{2} \theta_i^2 p_{c_i c_i}). \tag{12.64}$$

This approximation is reasonable if the Bernoulli distribution has a small mean and if the network is sparse, as are most real networks (Zhao, Levina, and Zhu, 2012). Putting it all together, the complete-data likelihood of $\mathbf{Y}$ is

$$P\{\mathbf{Y} | \mathbf{c}, \boldsymbol{\theta}\} = \prod_{i < j} \frac{(\theta_i \theta_j p_{c_i c_j})^{Y_{ij}} \exp\{-\theta_i \theta_j p_{c_i c_j}\}}{Y_{ij}!} \times \prod_i \frac{(\tfrac{1}{2} \theta_i^2 p_{c_i c_i})^{Y_{ii}/2} \exp\{-\tfrac{1}{2} \theta_i^2 p_{c_i c_i}\}}{(Y_{ii}/2)!}$$

$$= C \cdot \prod_i \theta_i^{k_i} \cdot \prod_{i \leq j} p_{c_i c_j}^{Y_{ij}^*} \exp\{-\theta_{ij} p_{c_i c_j}\}, \tag{12.65}$$

where $Y_{ij}^* = (1 - \frac{1}{2}\delta_{ij})Y_{ij}$, $\theta_{ij} = (1 - \frac{1}{2}\delta_{ij})\theta_i\theta_j$, $\delta_{ij} = 1$ if $i = j$, and 0 otherwise, $k_i = \sum_{j=1}^N Y_{ij}$ is the degree of node v_i, and

$$C = \frac{1}{\prod_{i<j} Y_{ij}! \cdot \prod_i 2^{Y_{ii}/2}(Y_{ii}/2)!} \tag{12.66}$$

is the normalizing constant. After some algebra, (12.65) can be written as

$$P\{\mathbf{Y}|\mathbf{c},\boldsymbol{\theta}\} = C \cdot \prod_i \theta_i^{k_i} \cdot \prod_{r,s} w_{rs}^{m_{rs}/2} \exp\left\{-\tfrac{1}{2}w_{rs}\right\}, \tag{12.67}$$

where $w_{rs} = E\{Y_{ij}|c_i = r, c_j = s\}$ and

$$m_{rs} = \sum_{i,j=1}^N Y_{ij}\delta_{c_i,r}\delta_{c_j,s} \tag{12.68}$$

is the total number of edges between blocks r and s.

Maximizing this expression (subject to the constraint (12.62)) is equivalent to maximizing the logarithm of the expression (ignoring the constant term C)

$$\log P\{\mathbf{Y}|\mathbf{c},\boldsymbol{\theta}\} = 2\sum_{i=1}^N k_i \log \theta_i + \sum_{r,s}(m_{rs}\log w_{rs} - w_{rs}) - \lambda\left(\sum_i \theta_i\delta_{c_i,r} - 1\right), \tag{12.69}$$

wrt $\boldsymbol{\theta} = (\theta_i)$ and $\mathbf{W} = (w_{rs})$, where λ is a Lagrangian multiplier. Differentiating $\log P$ wrt θ_i, w_{rs}, and λ, and setting each result equal to zero, yields the following equations:

$$\frac{\partial \log P}{\partial w_{rs}} = \frac{m_{rs}}{w_{rs}} - 1 = 0, \tag{12.70}$$

$$\frac{\partial \log P}{\partial \theta_i} = \frac{2k_i}{\theta_i} - \lambda\delta_{c_i,r} = 0, \tag{12.71}$$

$$\frac{\partial \log P}{\partial \lambda} = \sum_i \theta_i\delta_{c_i,r} - 1 = 0. \tag{12.72}$$

From (12.70), the MLE of w_{rs} is

$$\widehat{w}_{rs} = m_{rs}. \tag{12.73}$$

The solution of (12.71) yields $\widehat{\theta}_i = 2k_i/(\lambda\delta_{c_i,r})$. From (12.72), we have the constraint (12.62). Substituting $\widehat{\theta}_i$ for θ_i in the constraint, we have that $\widehat{\lambda} = 2\sum_i k_i$. Substituting $\widehat{\lambda}$ for λ in $\widehat{\theta}_i$ and solving, we have the following maximum-likelihood estimator of θ_i:

Theorem 12.7 (Karrer and Newman, 2011a) *The MLE for the degree parameter θ_i is given by*

$$\widehat{\theta}_i = \frac{k_i}{\kappa_{c_i}}, \tag{12.74}$$

where k_i is the degree of node v_i and

$$\kappa_r = \sum_i k_i \delta_{c_i,r} = \sum_{s=1}^{K} m_{rs} \qquad (12.75)$$

is the total of the degrees of all nodes in community r.

The maximum of the likelihood function is given by

$$\log P = 2 \sum_{i=1}^{n} k_i \log \widehat{\theta}_i + \sum_{r,s} (m_{rs} \log \widehat{w}_{rs} - \widehat{w}_{rs})$$

$$= 2 \sum_{i=1}^{n} k_i \log(k_1/\kappa_{c_i}) + \sum_{r,s} (m_{rs} \log m_{rs} - m_{rs})$$

$$= 2 \sum_i k_i \log(k_i/\kappa_{c_i}) + \sum_{r,s} m_{rs} \log m_{rs} - 2m, \qquad (12.76)$$

where $m = \frac{1}{2} \sum_{r,s} m_{rs}$. Consider the first term in (12.76):

$$2 \sum_i k_i \log(k_i/\kappa_{c_i}) = 2 \sum_i k_i \log k_i - 2 \sum_i k_i \log \kappa_{c_i}$$

$$= 2 \sum_i k_i \log k_i - 2 \sum_i k_i \left(\sum_r \delta_{c_i,r} \log \kappa_r \right)$$

$$= 2 \sum_i k_i \log k_i - 2 \sum_r \left[\left(\sum_i k_i \delta_{c_i,r} \right) \log \kappa_r \right]$$

$$= 2 \sum_i k_i \log k_i - \sum_r \kappa_r \log \kappa_r - \sum_s \kappa_s \log \kappa_s$$

$$= 2 \sum_i k_i \log k_i - \sum_{r,s} m_{rs} \log(\kappa_r \kappa_s). \qquad (12.77)$$

Substituting (12.77) into (12.76) and omitting the constant terms $2 \sum_i k_i \log k_i$ and $2m$ yields the following criterion:

Definition 12.9 The *Karrer–Newman likelihood criterion* is defined by

$$Q_{KN} = \sum_{r,s} m_{rs} \log \left(\frac{m_{rs}}{\kappa_r \kappa_s} \right). \qquad (12.78)$$

If we compare (12.78) with (12.27), we see that the criteria are very similar, except N_r is replaced by κ_r.

The maximum of Q_{KN} is computed over all possible partitions of the network. It turns out that if one moves a node from one community to another, the resulting change in the log-likelihood can be computed very quickly. With that in mind, an efficient algorithm similar to the Kernighan–Lin algorithm can be computed (Karrer

---

Algorithm 12.2 *Karrer–Newman algorithm*

---

1. Start by randomly partitioning the nodes of the network into an initial set of K communities.
2. Move a node sequentially from a current community to each of the other communities and compute the value of the objective function due to each move. The node is ultimately moved to the community that maximizes the objective function. Note that each node may be moved only once in the same round. It is not important to keep the community sizes constant.
3. After moving each node, the states through which the process passed are examined and the one with the highest objective score is used as the starting configuration for the next round of the process.
4. Stop when a complete iteration fails to increase the objective function.

---

and Newman, 2011a); see **Algorithm 12.2**. This algorithm is best implemented by starting at a number of different initial partitions and then taking the best result over all runs.

Karrer and Newman applied this algorithm to the karate club network (Section 2.5.5) and an undirected version of the political blogs network (Section 13.3.5), both having $K = 2$ communities. They found that the standard SBM algorithm failed to partition the network into the two known groups; instead, it placed high-degree nodes into one group and low-degree nodes into the other. The degree-corrected SBM partitioned the network into the correct two groups, except for a single misclassified node that fell on the boundary of the two groups. Similar findings were observed for the political blogs network, where the division occurred between the two political factions.

Bayesian Approach

A nonparametric Bayesian approach to the degree-corrected SBM (Herlau, Schmidt, and Mørup, 2013) uses the Poisson probabilities (12.63) and (12.64) for the edge variable Y_{ij}, and priors on the model are given by the Chinese Restaurant Process as a prior on the node partitions, a Dirichlet prior distribution on the θ_i (reparameterized to make them scale invariant across communities of different sizes), a conjugate Gamma distribution as a prior on p_{ij}, and non-informative priors on each of the hyperparameters. The resulting model, called an *infinite-degree-corrected SBM*, was used to analyze simulated data networks and seven real-data networks; it was found to be successful in estimating the number of communities in the network and, using Markov chain Monte Carlo (MCMC) algorithms, was able to predict the presence of edges in the network.

In the Bayesian framework, MCMC methods have been developed, but they only work for networks with a few hundred nodes.

12.5 Modularity

A popular measure of the adequacy of a specific partition of a network is "modularity," a quality function that has values ranging from -1 to $+1$ (Newman and Girvan, 2004; Newman, 2006). See also Newman (2010, Section 7.13).

12.5.1 Defining Modularity

Definition 12.10 *Modularity* is defined as the number (or fraction) of edges that fall within the given communities minus the expected such number (or fraction) if edges are randomly distributed in a network with the same number of nodes and where each node retains its original degree.

To formalize this definition of modularity, let κ_r be the total number of degrees for those nodes in the rth community; see (12.75). Let

$$L = \sum_{r=1}^{K} \kappa_r = \sum_{i,j=1}^{N} Y_{ij} \left(\sum_{r,s=1}^{K} \delta_{c_i,r} \delta_{c_j,s} \right) = \sum_{i,j=1}^{N} Y_{ij} = 2m \qquad (12.79)$$

be the total number of degrees in the network, where $\delta_{c_i,r} = 1$ if $c_i = r$, and 0 otherwise. The fraction of edges that join two nodes within the same community is

$$\frac{\sum_{i,j} Y_{ij} \delta_{c_i,c_j}}{\sum_{i,j} Y_{ij}} = \frac{1}{L} \sum_{i,j} Y_{ij} \delta_{c_i,c_j}, \qquad (12.80)$$

where $\delta_{c_i,c_j} = 1$ if $c_i = c_j$, and 0 otherwise.

Definition 12.11 The general formulation of modularity is given by

$$Q = \sum_{i,j} \left(\frac{Y_{ij}}{L} - P_{ij} \right) \delta_{c_i,c_j} = \frac{1}{L} \sum_{i,j} (Y_{ij} - L \cdot P_{ij}) \delta_{c_i,c_j}, \qquad (12.81)$$

where P_{ij} is the probability of an edge between nodes v_i and v_j under a null model of random edge placement.

Thus, $L \cdot P_{ij}$ can be viewed as the expected number of edges randomly placed between nodes v_i and v_j in the null model. Because dividing the sum in (12.81) by L does not affect the algorithm, we do not include L as the divisor in the criterion.

12.5.2 Erdős–Rényi Modularity

For the standard (i.e., non-degree-corrected) stochastic blockmodel, the null case, referred to as the *Erdős–Rényi random graph* (Zhao, Levina, and Zhu, 2012), assumes that there are no communities in the network (i.e., $K = 1$) and that edges are randomly placed in the network of nodes with constant probability $P_{ij} = p$. We estimate p by $\widehat{p} = L/N^2$. Substituting this value into (12.81) and noting that the only nonzero terms in Q are those in which $c_i = c_j = r$, say, we group the contributions together into communities and re-express in terms of communities. Thus, we have that

$$\sum_{i,j}(Y_{ij} - L \cdot P_{ij})\delta_{c_i,c_j} = \sum_{i,j}\left(Y_{ij} - \frac{L^2}{N^2}\right)\delta_{c_i,c_j}$$

$$= \sum_r \sum_{i,j}\left(Y_{ij} - \frac{L^2}{N^2}\right)\delta_{c_i,r}\delta_{c_j,r}$$

$$= \sum_r \left(\sum_{ij} Y_{ij}\delta_{c_i,r}\delta_{c_j,r} - \frac{L^2}{N^2}\left(\sum_i \delta_{c_i,r}\right)^2\right),$$

$$(12.82)$$

where we used the fact that $\delta_{c_i,c_j} = \sum_r \delta_{c_i,r}\delta_{c_j,r}$.

Definition 12.12 The *Erdős–Rényi modularity* is given by

$$Q_{ER} = \sum_{r=1}^{K}\left(m_{rr} - \frac{L^2 N_r^2}{N^2}\right),$$

$$(12.83)$$

where N_r is the number of nodes in community r and m_{rs} is the total number of edges between communities r and s (see (12.68)).

12.5.3 Newman–Girvan Modularity

Now consider the degree-corrected stochastic blockmodel. In the null case, we randomly generate edges between pairs of nodes assuming no communities (i.e., $K = 1$). Because k_i is the degree of node v_i and L is the total number of degrees, then $p_i = k_i/L$ is the probability of a random edge emanating from node v_i. The probability of placing a random edge between nodes v_i and v_j is $P_{ij} = p_i p_j = k_i k_j/L^2$ (because edges are placed independently of each other). It follows that if edges are placed at random, the expected number of edges between nodes v_i and v_j is $L \cdot P_{ij} = k_i k_j/L$. Substituting this into (12.81) and noting that the only nonzero terms derive from pairs of nodes that are members of the same community, we group these contributions together (by setting $c_i = c_j = r$, say) and rewrite the sum over node pairs in terms of a sum over the communities,

$$\sum_{i,j}(Y_{ij} - L \cdot P_{ij})\delta_{c_i,c_j} = \sum_{i,j=1}^{N}\left(Y_{ij} - \frac{k_i k_j}{L}\right)\delta_{c_i,c_j}$$

$$= \sum_r \left(\sum_{i,j} Y_{ij}\delta_{c_i,r}\delta_{c_j,r} - \frac{1}{L}\left(\sum_i k_i \delta_{c_i,r}\right)^2\right).$$

$$(12.84)$$

Definition 12.13 (Bickel and Chen, 2009) The *Newman–Girvan modularity* is defined as

$$Q_{NG} = \sum_{r=1}^{K}\left(m_{rr} - \frac{\kappa_r^2}{L}\right).$$

$$(12.85)$$

Table 12.4 Community-detection criteria. SBM is stochastic blockmodel and DC-SBM is degree-corrected stochastic blockmodel. Adapted from Zhao, Levina, and Zhu (2012, Table 1)

	SBM	DC-SBM
Likelihood	$Q_{SBM} = \sum_{r,s} m_{rs} \log\left(\frac{m_{rs}}{N_r N_s}\right)$	$Q_{KN} = \sum_{r,s} m_{rs} \log\left(\frac{m_{rs}}{\kappa_r \kappa_s}\right)$
Modularity	$Q_{ER} = \sum_r \left(m_{rr} - \frac{L^2 N_r^2}{N^2}\right)$	$Q_{NG} = \sum_r \left(m_{rr} - \frac{\kappa_r^2}{L}\right)$

See **Table 12.4**. The algorithm searches for that partition of nodes to maximize Q_{NG}. The quantity Q_{NG} measures the number of edges in the network that connect nodes within the same community minus the expected value of the same quantity in a network with the same community partitions but with random edges between the nodes. If network data correspond to the null scenario, we would see $Q_{NG} = 0$. Any nonzero value of Q_{NG} suggests a deviation from randomness. Strong community structure, beyond that expected from the null model, would be indicated by a value of Q_{NG} near 1. Although the criterion $Q_{NG} < 1$, it generally does not attain the value 1 and, indeed, its maximum can be much less than 1. Newman and Girvan report that, typically, $Q_{NG} \in [0.3, 0.7]$, and higher values than 0.7 are very unusual. Newman (2010, p. 225) presents a normalized version of Q_{NG} whose maximum is 1, but notes that it is rarely used.

12.5.4 Optimal Modularity: Spectral Methods

We have seen that there has been a lot of interest in recent years concerning the problem of detecting communities in networks. One method of determining how to partition nodes in a network into communities is to carry out an exhaustive search of all possible partitions and identify that partition that is viewed as the "best" because it optimizes some criterion. The criterion of choice has been that of modularity Q, and the goal is to discover that partition that maximizes Q. However, such a program of exhaustive search to maximize Q is an $\mathcal{NP}$-hard problem; it would be incredibly costly to run, especially when analyzing a very large network composed of millions of nodes.

As a result, a number of different types of approximation algorithms for maximizing Q have been proposed to find the "best" partition of the network into communities. Modularity is a general concept for measuring the strength of a network that has been partitioned into "modules" (or communities). Optimizing the modularity criterion has been shown to be an effective way of learning how to partition network data (Newman, 2006).

In this section, we describe spectral methods for community detection based upon eigenvalues and eigenvectors of certain matrices that represent network structure. We follow the development of Newman (2013a).

Two Communities

Consider, first, the case of two communities (i.e., $K = 2$). Define

$$s_i = \begin{cases} +1, & \text{if node } v_i \text{ belongs to community 1} \\ -1, & \text{if node } v_i \text{ belongs to community 2} \end{cases}. \tag{12.86}$$

From the entries of $\mathbf{s} = (s_1, \ldots, s_N)^\tau$, we construct the following indicator variable:

$$\frac{1}{2}(s_i s_j + 1) = \begin{cases} 1, & \text{if nodes } v_i \text{ and } v_j \text{ belong to the same community} \\ 0, & \text{otherwise} \end{cases}. \tag{12.87}$$

We can now rewrite the modularity $\mathcal{Q}_{NG}$ as

$$\mathcal{Q}_{NG} = \frac{1}{2} \sum_{i,j} \left(Y_{ij} - \frac{k_i k_j}{L} \right) (s_i s_j + 1) = \frac{1}{2} \sum_{i,j} B_{ij} s_i s_j = \frac{1}{2} \mathbf{s}^\tau \mathbf{B} \mathbf{s}, \tag{12.88}$$

where

$$\mathbf{B} = \mathbf{Y} - \frac{\mathbf{k}\mathbf{k}^\tau}{L}, \tag{12.89}$$

$\mathbf{s} = (s_1, \ldots, s_N)^\tau$, and $\mathbf{k} = (k_1, \ldots, k_N)^\tau$. Thus,

$$B_{ij} = Y_{ij} - \frac{k_i k_j}{L}, \quad i, j = 1, 2, \ldots, N. \tag{12.90}$$

Definition 12.14 The $(N \times N)$-matrix $\mathbf{B} = (B_{ij})$ is real and symmetric and is referred to as the *modularity matrix*. Because $\sum_i B_{ij} = \sum_j B_{ij} = 0$, it follows that $\mathbf{B}$ has an eigenvalue 0 with associated eigenvector $\mathbf{1}_N = (1, 1, \ldots, 1)^\tau$.

The problem, therefore, is, for a given modularity matrix $\mathbf{B}$, to find an N-vector $\mathbf{s}$ that will maximize the quadratic form $\mathcal{Q}_{NG}$ subject to $s_i \in \{-1, +1\}, i = 1, 2, \ldots, N$. This is an integer quadratic programming problem that is $\mathcal{NP}$-complete (Brandes et al., 2008). The difficulty here is that the s_i are discrete-valued, which makes finding the solution computationally hard. Accordingly, several approximation methods have been proposed to overcome this computational problem. We describe a method we have seen before, known as "relaxation and rounding."

From (12.86), we write $\mathbf{s}$ as a linear combination of the normalized eigenvectors $\mathbf{u}_i$ of $\mathbf{B}$. Thus,

$$\mathbf{s} = \sum_{i=1}^N \alpha_i \mathbf{u}_i, \quad \alpha_i = \mathbf{u}_i^\tau \mathbf{s}. \tag{12.91}$$

Let β_i be the eigenvalue corresponding to the eigenvector $\mathbf{u}_i$, and assume the eigenvalues of $\mathbf{B}$ are arranged in decreasing order of magnitude; i.e., that $\beta_1 \geq \beta_2 \geq \cdots \geq \beta_n$. Substituting this expression into (12.88) and dropping the factor $\frac{1}{2}$ yields

$$\begin{aligned} \mathcal{Q}_{NG} &= \left(\sum_i \alpha_i \mathbf{u}_i \right)^\tau \mathbf{B} \left(\sum_j \alpha_j \mathbf{u}_j \right) \\ &= \sum_i \sum_j \alpha_i \alpha_j \beta_i (\mathbf{u}_i^\tau \mathbf{u}_j) \\ &= \sum_i \alpha_i^2 \beta_i \\ &= \sum_i (\mathbf{u}_i^\tau \mathbf{s})^2 \beta_i. \end{aligned} \tag{12.92}$$

We wish to maximize $\mathcal{Q}_{NG}$ by choosing an appropriate partition of the nodes in the network into communities. This is equivalent to choosing $\mathbf{s}$ to maximize

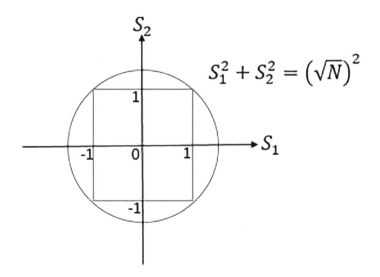

Figure 12.2 Relaxation method for the modularity of two communities. The true optimization occurs at the corners of the hypercube (square), centered at the origin. The relaxation displayed here is optimizing over anywhere on the circumscribed hypersphere (circle) that touches the corners of the hypercube (square).

(12.92). Without any constraints, we would choose $\mathbf{s}$ proportional (i.e., parallel) to $\mathbf{u}_1$. However, each entry of $\mathbf{s}$ is either -1 or $+1$, and so cannot be taken to be parallel to $\mathbf{u}_1$.

Hyperspherical Solution
Relaxation. The N-vector $\mathbf{s} = (s_1, \ldots, s_N)^\tau$, $s_i \in \{-1, +1\}$, $i = 1, 2, \ldots, N$, can be viewed as restricted to the 2^N corners of an N-dimensional hypercube centered at the origin. Because the vector $\mathbf{s}$ has squared length $\mathbf{s}^\tau \mathbf{s} = \sum_i s_i^2 = N$, which restricts each $s_i \in [-\sqrt{N}, +\sqrt{N}]$, one possibility is to maximize $\mathcal{Q}_{NG} = \mathbf{s}^\tau \mathbf{B} \mathbf{s} = \sum_{i,j} B_{ij} s_i s_j$ subject to the hyperspherical constraint $\sum_i s_i^2 = N$. The two-dimensional situation is displayed in **Figure 12.2**, where the four corners of the square touch the circumscribed circle with radius $\sqrt{N}$; this shows that not only do true solutions to the constrained problem lie on the circle, but approximate solutions also lie on the circle. This maximization problem will be referred to as the *hyperspherical model*.

Form the objective function,

$$f(\mathbf{s}, \beta) = \sum_{i,j} B_{ij} s_i s_j - \beta \left(\sum_i s_i^2 - N \right), \tag{12.93}$$

where β is a Lagrange multiplier. Differentiate f wrt s_i, set the result equal to zero, and solve:

$$\frac{\partial f}{\partial s_i} = 2 \sum_j B_{ij} s_j - 2\beta s_i = 0, \tag{12.94}$$

whence

$$\sum_{i,j} B_{ij} s_j = \beta s_i, \quad i = 1, 2, \ldots, N, \tag{12.95}$$

or

$$\mathbf{Bs} = \beta\mathbf{s}. \tag{12.96}$$

So, β is an eigenvalue, with associated eigenvector $\mathbf{s}$, of the modularity matrix $\mathbf{B}$. Thus, the modularity of the network is

$$\mathcal{Q}_{NG} = \mathbf{s}^\tau(\mathbf{Bs}) = \beta\mathbf{s}^\tau\mathbf{s} = N\beta. \tag{12.97}$$

To maximize $\mathcal{Q}_{NG}$, we should choose β to be as large as possible. Thus, we choose $\mathbf{s}$ as the eigenvector $\mathbf{u}_1$ associated with the largest eigenvalue β_1 of $\mathbf{B}$. But $\mathbf{s}$ is constrained by $s_i \in \{-1, +1\}$, $i = 1, 2, \ldots, n$, and so it cannot take values proportional (i.e., parallel) to $\mathbf{u}_1$. We can, however, approximate this solution as close as possible by finding $\mathbf{s}$ to maximize the inner product

$$\mathbf{s}^\tau\mathbf{u}_1 = \sum_i s_i u_{1i}, \tag{12.98}$$

where $\mathbf{u}_1 = (u_{11}, u_{12}, \ldots, u_{1N})^\tau$. As we noted above, the matrix $\mathbf{B}$ has an eigenvalue of 0 with associated eigenvector $\mathbf{1}_N = (1, \ldots, 1)^\tau$, which corresponds to an *undivided network*; that is, one in which all nodes are placed into a single community. Because all the remaining eigenvectors of $\mathbf{B}$ are orthogonal to the eigenvector $\mathbf{1}_N$, they must each contain both positive and negative entries. So, the maximum of (12.98) is attained when each term in the summand is non-negative.

Rounding. The standard solution here is to round off the entries to the specified values of -1 and $+1$. In other words, positive entries are rounded off to $+1$ and negative entries to -1; that is,

$$s_i = \begin{cases} +1, & \text{if } u_{1i} > 0 \\ -1, & \text{if } u_{1i} < 0 \end{cases}, \quad i = 1, 2, \ldots, N. \tag{12.99}$$

To summarize:

> *For the hyperspherical model and two communities, we compute the eigenvector $\mathbf{u}_1$ corresponding to the largest eigenvalue β_1 of the modularity matrix $\mathbf{B}$, and then assign the ith node, v_i, to community 1 if the ith entry of $\mathbf{u}_1$ is positive, and to community 2 otherwise.*

Hyperelliptical Solution

Other solutions may be found by changing the form of the constraint. For example, instead of the hyperspherical constraint, consider the following hyperelliptical constraint (Newman, 2013b),

$$\mathbf{s}^\tau\mathbf{Ds} = L, \tag{12.100}$$

where $\mathbf{D} = \text{diag}\{k_1, \ldots, k_N\}$. The constraint can also be expressed as $\sum_i k_i s_i^2 = L$, which, in two dimensions, can be written as the ellipse

$$\frac{s_1^2}{\left(\sqrt{L/k_1}\right)^2} + \frac{s_2^2}{\left(\sqrt{L/k_2}\right)^2} = 1. \tag{12.101}$$

The situation is displayed in **Figure 12.3**, where the circumscribed ellipsoid with center the origin touches all four corners of the square.

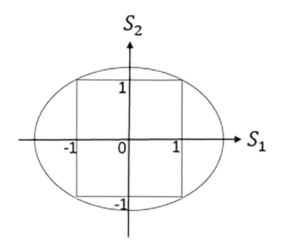

Figure 12.3 Relaxation method for the modularity of two communities. The true optimization occurs at the corners of the hypercube (square), centered at the origin. The relaxation displayed here is optimizing over anywhere on the circumscribed hyperellipse (ellipse) the touches the corners of the hypercube (square).

Form the objective function

$$f(\mathbf{s}, \lambda) = \sum_{i,j} B_{ij} s_i s_j - \lambda \left(\sum_i k_i s_i^2 - L \right), \qquad (12.102)$$

where λ is the Lagrange multiplier. Differentiate wrt s_i, and set the result equal to zero:

$$\frac{\partial f}{\partial s_i} = 2 \sum_j B_{ij} s_j - 2\lambda k_i s_i = 0, \quad i = 1, 2, \ldots, N, \qquad (12.103)$$

whence,

$$\sum_j B_{ij} s_j = \lambda k_i s_i, \quad i = 1, 2, \ldots, N, \qquad (12.104)$$

or

$$\mathbf{Bs} = \lambda \mathbf{Ds}. \qquad (12.105)$$

This is a generalized eigenequation, where λ is an eigenvalue and $\mathbf{s}$ is the associated eigenvector of $\mathbf{B}$ wrt $\mathbf{D}$. Premultiply through by $\mathbf{s}^\tau$ to get

$$\mathcal{Q}_{NG} = \mathbf{s}^\tau \mathbf{Bs} = \lambda(\mathbf{s}^\tau \mathbf{Ds}) = \lambda L. \qquad (12.106)$$

To maximize $\mathcal{Q}_{NG}$, we choose λ to be the top eigenvalue of the generalized eigenequation. However, as we have seen above, the top eigenvalue of $\mathbf{B}$ is 0 and the associated eigenvector is the vectors $\mathbf{1}_N$; so, the solution where $s_i = 1$ for all i corresponds to placing all nodes into the first community and none in the other community!

It turns out that we can do better. Substituting $B_{ij} = Y_{ij} - k_i k_j / L$ into (12.104) yields

$$\sum_j Y_{ij} s_j = \lambda k_i s_i + \frac{k_i k_j}{L} s_j = k_i \left(\lambda s_i + \frac{1}{L} \sum_j k_i s_j \right), \quad i = 1, 2, \ldots, N, \qquad (12.107)$$

or

$$\mathbf{Ys} = \mathbf{D}\left(\lambda\mathbf{s} + \frac{\mathbf{k}^\tau\mathbf{s}}{L}\mathbf{1}_N\right). \tag{12.108}$$

Premultiplying this equation by $\mathbf{1}_N^\tau$ and noting that $\mathbf{1}_N^\tau\mathbf{Y} = \mathbf{1}_N^\tau\mathbf{D} = \mathbf{k}^\tau$ and $\mathbf{k}^\tau\mathbf{1}_N = \sum_i k_i = \sum_{i,j} Y_{ij} = L$, we have that

$$\mathbf{k}^\tau\mathbf{s} = (\mathbf{1}_N^\tau\mathbf{Y})\mathbf{s} = (\mathbf{1}_N^\tau\mathbf{D})\left(\lambda\mathbf{s} + \frac{\mathbf{k}^\tau\mathbf{s}}{L}\mathbf{1}_N\right) = \lambda\mathbf{k}^\tau\mathbf{s} + \mathbf{k}^\tau\mathbf{s}, \tag{12.109}$$

whence $\lambda\mathbf{k}^\tau\mathbf{s} = 0$. Thus, either $\lambda = 0$ or $\mathbf{k}^\tau\mathbf{s} = 0$. But λ is the top eigenvalue of $\mathbf{B}$ wrt $\mathbf{D}$. If $\lambda = 0$, then $\mathbf{s} = \mathbf{1}_N$, and the network will be indivisible. We will assume that this is not the case, that there exists a strictly positive eigenvalue of $\mathbf{Y}$ wrt $\mathbf{D}$. If $\lambda \neq 0$, then $\mathbf{k}^\tau\mathbf{s} = 0$ and, hence,

$$\mathbf{Ys} = \lambda\mathbf{Ds}. \tag{12.110}$$

So, $(\lambda, \mathbf{s})$ forms an eigenvalue–eigenvector pair that solves the generalized eigen-equation of $\mathbf{Y}$ wrt $\mathbf{D}$. This simplifies the previous result (12.105). In principle, then, we should maximize $\mathcal{Q}_{NG}$ by choosing the largest eigenvalue, λ_1, which we assume is positive, of $\mathbf{Y}$ wrt $\mathbf{D}$.

Now, we know that $\mathbf{1}_N$, which has all positive elements, is an eigenvector of $\mathbf{Y}$ wrt $\mathbf{D}$. Therefore, by the *Perron–Frobenius Theorem*,[6] it must have the largest (most-positive) eigenvalue. However, this eigenvector does not satisfy $\mathbf{k}^\tau\mathbf{s} = 0$. So, this does not work.

An approximate solution can be formed by using instead the second-largest eigen-value, λ_2, of $\mathbf{Y}$ wrt $\mathbf{D}$, whose associated eigenvector does satisfy that condition (as do all the other eigenvectors) and is identical to the largest eigenvector of (12.105). As before, we partition the network so that the nodes are divided into two communities, the assignment to community depending upon the sign of the appropriate entry of the eigenvector. In other words,

$$s_i = \begin{cases} +1, & \text{if } v_{2i} > 0 \\ -1, & \text{if } v_{2i} < 0 \end{cases}, \quad i = 1, 2, \ldots, N. \tag{12.111}$$

In practice, this approximation turns out to be quite reasonable.

We can simplify the eigenequation (12.110). Set $\mathbf{w} = \mathbf{D}^{1/2}\mathbf{s}$, so that $w_i = k_i^{-1/2}s_i$, $i = 1, 2, \ldots, N$. Then, it is easy to show that[7]

$$(\mathbf{D}^{-1/2}\mathbf{Y}\mathbf{D}^{-1/2})\mathbf{w} = \lambda\mathbf{w}. \tag{12.112}$$

This is a standard eigenequation in which $(\lambda, \mathbf{w})$ is an eigenvalue–eigenvector pair of the square, symmetric matrix $\mathbf{D}^{-1/2}\mathbf{Y}\mathbf{D}^{-1/2} = \left(Y_{ij}/\sqrt{k_i k_j}\right)$. The eigenvalues of $\mathbf{D}^{-1/2}\mathbf{Y}\mathbf{D}^{-1/2}$ are the same as those of $\mathbf{Y}$ wrt $\mathbf{D}$ and the eigenvectors are related by $\mathbf{s} = \mathbf{D}^{-1/2}\mathbf{w}$.

So, we choose the second-largest eigenvalue, λ_2, and associated eigenvector (the so-called *Fiedler vector*) of the graph Laplacian. The network of nodes is partitioned into

---

[6]The Perron–Frobenius Theorem states that a real square matrix having all positive entries has a unique largest real eigenvalue and that the associated eigenvector can be chosen to have strictly positive entries. This theorem was proved by Oskar Perron (in 1907) and was extended by Georg Frobenius (in 1912).

[7]We will see in Section 13.2.3 that the matrix $\mathbf{L}^{sym} = \mathbf{D}^{-1/2}\mathbf{Y}\mathbf{D}^{-1/2}$ is often referred to as the *normalized graph Laplacian* of the network.

two non-overlapping communities by sorting the entries of the Fiedler vector into decreasing order, and then assigning the ith node to community 1 if the ith vector entry is among the largest (most positive) entries of that vector, and is assigned to community 2 otherwise.

More Than Two Communities

When there are more than two communities, one proposed generalization of the two-community algorithm (Newman, 2006) is first, to divide the nodes into two communities using the above algorithm then, without deleting any edges in the network, to divide each of the two communities into two further communities, and so on, dividing by two at each step of the process until there is reason to stop. This strategy appears to work well for many networks, but there are networks in which it does not yield the best partition.

12.5.5 Spectrum of the Modularity Matrix

For an undirected network, the modularity matrix $\mathbf{B} = (B_{ij})$ is given by

$$\mathbf{B} = \mathbf{Y} - \frac{\mathbf{k}\mathbf{k}^{\tau}}{L}, \tag{12.113}$$

where $\mathbf{Y} = (Y_{ij})$ is the $(N \times N)$ adjacency matrix, $\mathbf{k} = (k_1, \ldots, k_N)^{\tau}$ is an N-vector of node degrees, and $L = 2m$ is the total number of degrees, where $m = \frac{1}{2}\sum_{i=1}^{N} k_i$ is the number of edges. Thus,

$$B_{ij} = Y_{ij} - \frac{k_i k_j}{L}, \quad i, j = 1, 2, \ldots, N. \tag{12.114}$$

Because Y_{ij} has expectation

$$\mathrm{E}\{Y_{ij}\} = \frac{k_i k_j}{L}, \tag{12.115}$$

we see that B_{ij} is the deviation of Y_{ij} from that expectation. In other words,

$$\mathbf{B} = \mathbf{Y} - \mathrm{E}\{\mathbf{Y}\} \tag{12.116}$$

is the centered version of the adjacency matrix $\mathbf{Y}$. Because Y_{ij} is Poisson distributed, its variance is equal to its mean, and so the mean of B_{ij} is zero and its variance is $k_i k_j / L$. As we have seen, the eigenvalues and eigenvectors of $\mathbf{B}$ play a major role in community detection.

The spectral density of $\mathbf{B}$ is given by

$$\rho(z) = \frac{1}{N}\sum_{i=1}^{N} \delta_{z, \lambda_i}, \quad z \in \mathbb{R}, \tag{12.117}$$

where the $\{\lambda_i\}$ are the eigenvalues of $\mathbf{B}$, and $\delta_{a,b}$ is the *Dirac delta function*, equal to 1 if $a = b$, and 0 otherwise. Two related models have been proposed to obtain the spectral density of $\mathbf{B}$:

1. The *configuration model* (see Section 8.2) prespecifies the degree of each of the N nodes using the sequence $k_1, k_2, \ldots, k_N$. This model explores how closely the degree distribution captures the complexities of a real-world network (Bender and Canfield, 1978; Wormald, 1978; Bollobás, 1980; Molloy and Reed, 1995).

The model can be viewed as a fixed set of N nodes, where the ith node has k_i "spokes" (or "stubs") emanating from it, and the spokes need to be connected to the spokes of other nodes to form edges of the network. The total number of spokes has to be even so that no spoke is left unattached. Generating a network based upon the configuration model will always possess the same given degree distribution, but its other properties (e.g., clustering) may not be preserved.

2. An *expected-degree random graph model* (see Section 8.3) is a relaxation of the configuration model in which the number of edges is taken to be a random variable, so that the $k_1, k_2, \ldots, k_N$ are *expected* node degrees (rather than a specified degree sequence) and m is the expected total number of edges (Chung and Lu, 2003).

If the number, N, of nodes is large and the degrees are large, the two models yield the same spectral density (Nadakuditi and Newman, 2013).

First, the *Stieltjes–Perron inversion formula* was used to express the spectral density as the imaginary part of the trace of the expectation (over the ensemble of model networks) of the *resolvent matrix*,

$$\mathbf{R}(z) = (z\mathbf{I}_N - \mathbf{B})^{-1}. \tag{12.118}$$

Note that the matrix $z\mathbf{I}_N - \mathbf{B}$ has off-diagonal entries that are zero-mean random variables. The spectral density is

$$\rho(z) = -\frac{1}{N\pi}\mathcal{I}m[\mathrm{tr}(\mathrm{E}\{\mathbf{R}(z)\})], \tag{12.119}$$

where $\mathcal{I}m[w]$ is the imaginary part of w. The quantity

$$g(z) = \frac{1}{N}\mathrm{tr}[\mathbf{R}(z)] = \frac{1}{N}\mathrm{tr}[(z\mathbf{I}_N - \mathbf{B})^{-1}] \tag{12.120}$$

is the *Stieltjes transform* of $\mathbf{B}$. To evaluate the trace, Nadakuditi and Newman used an approach due to Bai and Silverstein (2010). We give an outline of their arguments.

At each of N steps, one row and corresponding column of $\mathbf{R}(z)$ are removed, and an expression is obtained for $\mathrm{E}\{\mathbf{y}^\tau\mathbf{R}_{-i}(z)\mathbf{y}\}$, where $\mathbf{R}_{-i}(z)$ is the $((N-1)\times(N-1))$-matrix formed from $\mathbf{R}(z)$ without its ith row $\mathbf{y}^\tau$ and ith column $\mathbf{y}$. They then show that

$$\mathrm{E}\{\mathbf{y}^\tau\mathbf{R}_{-i}(z)\mathbf{y}\} = \frac{k_i}{L}\mathrm{tr}[\mathbf{D}_{-i}\mathrm{E}\{\mathbf{R}_{-i}(z)\}], \tag{12.121}$$

where the diagonal matrix $\mathbf{D}_{-i}$ is formed from $\mathbf{D} = \mathrm{diag}\{k_1, \ldots, k_N\}$ by removing its ith row and column. If the network is very large and the node degrees increase with N, then asymptotically (as $N \to \infty$), $\frac{1}{L}\mathrm{tr}[\mathbf{D}_{-i}\mathrm{E}\{\mathbf{R}_{-i}(z)\}]$ will be the same as

$$h(z) = \frac{1}{L}\mathrm{tr}[\mathbf{D}\cdot\mathrm{E}\{\mathbf{R}(z)\}], \tag{12.122}$$

where it is shown that, for large N, $\mathrm{E}\{\mathbf{R}(z)\} = \mathrm{E}\{(z\mathbf{I}_N - \mathbf{B})^{-1}\}$ is a diagonal matrix. Let $c = L/N$ be the average degree (so that $L = Nc$). Now, we use the fact that the trace of a matrix is the sum of its N diagonal elements. If the node degrees $k_1, \ldots, k_N$ are assumed to be iid draws from a given probability distribution $p(k)$, then, for large N, $h(z)$ can be written as

$$h(z) = \frac{1}{c}\int_0^\infty \frac{k\cdot p(k)}{z - k\cdot h(z)}dk. \tag{12.123}$$

Solving this equation for $h(z)$, we have the following result.

Theorem 12.8 *The spectral density function of the modularity matrix* **B** *(i.e., the centered adjacency matrix) is given by*

$$\rho(z) = -\frac{c}{\pi z} \mathcal{I}m[(h(z))^2].$$

(12.124)

Example: Erdős–Renyi Random Graph

As a simple example, consider the standard Poisson random graph (Erdős and Rényi, 1959, 1960). Here, $p(k) = \delta_{k,c}$, where c is the common expected degree. Substituting $p(k)$ into (12.123), noting that $\delta_{k,c} = 1$ if $k = c$, and 0 otherwise, we have

$$h(z) = \frac{1}{z - c \cdot h(z)},$$

(12.125)

which gives the quadratic equation $cx^2 - zx + 1 = 0$, where $x = h(z)$. Solving for x (and taking the negative square root for a positive density) yields

$$h(z) = \frac{z - \sqrt{z^2 - 4c}}{2c}.$$

(12.126)

Substituting $h(z)$ into $\rho(z)$, the spectral density is given by

$$\rho(z) = \frac{1}{2\pi c}\sqrt{4c - z^2},$$

(12.127)

which is *Wigner's semicircle law* for the Erdős–Renyi random graph (see, e.g., Izenman, 2021). See **Exercise 12.3**.

12.5.6 Spectrum of the Adjacency Matrix

This argument for obtaining the spectral density of the modularity matrix has been extended to the problem of community detection via the adjacency matrix **Y** (Zhang, Nadakuditi, and Newman, 2014).

Assume there are K communities in the network. Let

$$\mathbf{k}_i = (k_{1i}, \ldots, k_{Ki})^\tau$$

(12.128)

be a K-vector associated with the ith node v_i, $i = 1, 2, \ldots, N$. In this formulation, k_{ri} denotes the propensity of the ith node to be a member of the rth community. Let

$$\mathbf{K} = (\mathbf{k}_1, \mathbf{k}_2, \ldots, \mathbf{k}_N)$$

(12.129)

be a $(K \times N)$-matrix. Then,

$$\mathrm{E}\{\mathbf{Y}\} = \frac{1}{L}\mathbf{K}^\tau \mathbf{K},$$

(12.130)

so that the number of edges between nodes v_i and v_j has expectation

$$\mathrm{E}\{Y_{ij}\} = \frac{\mathbf{k}_i^\tau \mathbf{k}_j}{L}, \quad i, j = 1, 2, \ldots, N.$$

(12.131)

Because $\mathbf{k}_i$ is a K-vector, it follows that $\mathrm{E}\{\mathbf{Y}\}$ is an $(N \times N)$-matrix with rank K, and so, from the spectral decomposition theorem, we can write

$$E\{\mathbf{Y}\} = \sum_{\ell=1}^{K} \alpha_\ell \mathbf{u}_\ell \mathbf{u}_\ell^\tau, \tag{12.132}$$

where $(\alpha_\ell, \mathbf{u}_\ell)$ is the ℓth eigenvalue–eigenvector pair of the matrix $E\{\mathbf{Y}\}$, $\ell = 1, 2, \ldots, K$, where the eigenvectors, $\{\mathbf{u}_\ell\}$, are normalized. For easy reference, we temporarily write $\mathbf{Y}^{(K)}$ for the adjacency matrix modeled as being composed of K communities. From (12.116), we have that

$$\mathbf{Y}^{(K)} = \mathbf{B} + E\{\mathbf{Y}^{(K)}\} = \mathbf{B} + \sum_{i=1}^{K} \alpha_\ell \mathbf{u}_\ell \mathbf{u}_\ell^\tau. \tag{12.133}$$

A Single Community

We start by assuming that there is only one community, so that $K = 1$. Then

$$\mathbf{Y}^{(1)} = \mathbf{B} + \alpha_1 \mathbf{u}_1 \mathbf{u}_1^\tau. \tag{12.134}$$

Let $(z, \mathbf{v})$ be the eigenvalue–eigenvector pair of this matrix. Then we have the eigenequation

$$(\mathbf{B} + \alpha_1 \mathbf{u}_1 \mathbf{u}_1^\tau)\mathbf{v} = z\mathbf{v}, \tag{12.135}$$

or

$$(z\mathbf{I}_N - \mathbf{B})\mathbf{v} = \alpha_1 (\mathbf{u}_1 \mathbf{u}_1^\tau)\mathbf{v}. \tag{12.136}$$

Premultiplying both sides of (12.136) by $\mathbf{u}_1^\tau (z\mathbf{I}_N - \mathbf{B})^{-1}$, we have that

$$\mathbf{u}_1^\tau (z\mathbf{I}_N - \mathbf{B})^{-1}(z\mathbf{I}_N - \mathbf{B})\mathbf{v} = \mathbf{u}_1^\tau (z\mathbf{I}_N - \mathbf{B})^{-1}\alpha_1 \mathbf{u}_1 \mathbf{u}_1^\tau \mathbf{v}, \tag{12.137}$$

or

$$\mathbf{u}_1^\tau \mathbf{v} = \alpha_1 \mathbf{u}_1^\tau (z\mathbf{I}_N - \mathbf{B})^{-1}\mathbf{u}_1 \mathbf{u}_1^\tau \mathbf{v}. \tag{12.138}$$

This yields the following equation in z:

$$\mathbf{u}_1^\tau (z\mathbf{I}_N - \mathbf{B})^{-1}\mathbf{u}_1 = \frac{1}{\alpha_1}, \tag{12.139}$$

assuming that $\mathbf{u}_1^\tau \mathbf{v} \neq 0$. Next, we express $\mathbf{B}$ in terms of its spectral decomposition as $\mathbf{B} = \sum_{j=1}^{N} \lambda_j \mathbf{w}_j \mathbf{w}_j^\tau$, where the eigenvalues are ordered so that $\lambda_1 \geq \lambda_2 \geq \cdots \geq \lambda_N$. Then (12.139) becomes

$$\begin{aligned}
\frac{1}{\alpha_1} &= \mathbf{u}_1^\tau \left(z\mathbf{I}_N - \sum_j \lambda_j \mathbf{w}_j \mathbf{w}_j^\tau \right)^{-1} \mathbf{u}_1 \\
&= \mathbf{u}_1^\tau \left(\sum_j (z - \lambda_j)\mathbf{w}_j \mathbf{w}_j^\tau \right)^{-1} \mathbf{u}_1 \\
&= \mathbf{u}_1^\tau \left(\sum_j (z - \lambda_j)^{-1}\mathbf{w}_j \mathbf{w}_j^\tau \right) \mathbf{u}_1 \\
&= \sum_{j=1}^{N} \frac{(\mathbf{u}_1^\tau \mathbf{w}_j)^2}{z - \lambda_j}.
\end{aligned} \tag{12.140}$$

To solve (12.140) for z, Zhang, Nadakuditi, and Newman plot the two sides of the equation on the same graph and the points at which they intersect yield the solutions for z. The rhs are curves as functions of z and the lhs is a constant represented by a horizontal line at the value $\frac{1}{\alpha_1}$. The rhs has simple poles at $z = \lambda_j$, $j = 1, 2, \ldots, N$. They then argue that the values of z must fall between consecutive values of the eigenvalues, λ_j, $j = 1, 2, \ldots, N$. In other words,

$$z_1 \geq \lambda_1 \geq z_2 \geq \lambda_2 \geq \cdots \geq z_N \geq \lambda_N. \tag{12.141}$$

Now, as N gets larger, consecutive values of the $\{\lambda_j\}$ get closer to each other, which in turn places tighter bounds on the z_j. As $N \to \infty$, it follows that

$$z_j \to \lambda_j, \quad j = 2, 3, \ldots, N. \tag{12.142}$$

Thus, for large networks, the spectral density of the adjacency matrix, $\mathbf{Y}^{(1)} = \mathbf{B} + \alpha_1 \mathbf{u}_1 \mathbf{u}_1^\tau$, is identical (with one exception) to that of the modularity matrix, $\mathbf{B}$. From (12.141), the exception is that the largest eigenvalue, z_1, of $\mathbf{B}$ has a lower bound but no upper bound, so it may not equal λ_1. In fact, z_1 may lie outside the spectrum values of $\mathbf{B}$.

What can we say about the value of z_1, the largest eigenvalue of the adjacency matrix $\mathbf{Y}^{(1)}$? We have seen from (12.116) that $\mathbf{B}$ is a centered random $(N \times N)$-matrix, with independent entries, each having mean 0. Thus, the eigenvectors, $\{\mathbf{w}_j\}$, of $\mathbf{B}$ are each random N-vectors. Furthermore, $\mathbf{u}_1$, because it is an eigenvector of E$\{\mathbf{Y}\}$, is a fixed N-vector. Taking expectations of (12.140), we need to find E$\{(\mathbf{u}_1^\tau \mathbf{w}_j)^2\}$.

To this end, we note that the probability density of $\mathbf{w}_j$ is invariant under any orthogonal transformation $\mathbf{O}$, so that P$\{\mathbf{w}_j\}$ = P$\{\mathbf{O}\mathbf{w}_j\}$. If $\mathbf{y}_j = \mathbf{O}\mathbf{w}_j$, then $\mathbf{w}_j = \mathbf{O}^\tau \mathbf{y}_j$. Suppose now that $\mathbf{a}$ and $\mathbf{b}$ are two arbitrary fixed N-vectors, both normalized to unity, so that $\mathbf{a}^\tau \mathbf{a} = \mathbf{b}^\tau \mathbf{b}$. Then $\mathbf{a}$ and $\mathbf{b}$ differ by an orthogonal transformation, $\mathbf{b} = \mathbf{O}\mathbf{a}$. It follows that[8]

$$
\begin{aligned}
E\{\mathbf{a}^\tau \mathbf{w}_j\} &= \int_{\mathbf{w}_j} \mathbf{a}^\tau \mathbf{w}_j P\{\mathbf{w}_j\} \\
&= \int_{\mathbf{w}_j} \mathbf{a}^\tau \mathbf{w}_j P\{\mathbf{O}\mathbf{w}_j\} \\
&= \int_{\mathbf{y}_j} \mathbf{a}^\tau (\mathbf{O}^\tau \mathbf{y}_j) P\{\mathbf{y}_j\} \\
&= \int_{\mathbf{y}_j} (\mathbf{O}\mathbf{a})^\tau \mathbf{y}_j P\{\mathbf{y}_j\} \\
&= \int_{\mathbf{y}_j} \mathbf{b}^\tau \mathbf{y}_j P\{\mathbf{y}_j\} \\
&= E\{\mathbf{b}^\tau \mathbf{w}_j\}. \tag{12.143}
\end{aligned}
$$

But if E$\{\mathbf{a}^\tau \mathbf{w}_j\}$ = E$\{\mathbf{b}^\tau \mathbf{w}_j\}$, then $(\mathbf{a} - \mathbf{b})^\tau E\{\mathbf{w}_j\} = 0$ and so E$\{\mathbf{w}_j\}$ = $\mathbf{0}$. Hence, E$\{(\mathbf{a}^\tau \mathbf{w}_j)^2\}$ = var$\{\mathbf{a}^\tau \mathbf{w}_j\}$. Because the distribution of $\mathbf{a}^\tau \mathbf{w}_j$ is the same as the distribution of $\mathbf{1}_N^\tau \mathbf{w}_j$, where $\mathbf{1}_N$ is the unit N-vector, then var$\{\mathbf{a}^\tau \mathbf{w}_j\}$ = var$\{\sum_{i=1}^N w_{ji}\}$ = $\sum_{i=1}^N$ var$\{w_{ji}\}$ = $N \cdot \frac{1}{N^2}$ = $\frac{1}{N}$. From these results, it follows that

---

[8]The author thanks M.E.J. Newman for e-mail correspondence on this issue.

$$E\{\mathbf{u}_1^\tau \mathbf{w}_j\} = 0 \quad \text{and} \quad E\{(\mathbf{u}_1^\tau \mathbf{w}_j)^2\} = \frac{1}{N}. \tag{12.144}$$

Taking expectations of (12.140) over networks, we obtain

$$\begin{aligned}
\frac{1}{\alpha_1} &= \sum_{j=1}^{N} E\left\{\frac{(\mathbf{u}_1^\tau \mathbf{w}_j)^2}{z - \lambda_j}\right\} \\
&= \frac{1}{N} E\left\{\sum_{j=1}^{N} \frac{1}{z - \lambda_j}\right\} \\
&= \frac{1}{N} \mathrm{tr}[E\{(z\mathbf{I}_N - \mathbf{B})^{-1}\}] = E\{g(z)\},
\end{aligned} \tag{12.145}$$

where $g(z)$ is given by (12.118). The second line of (12.145) follows from the first because, even though the numerator, $(\mathbf{u}_1^\tau \mathbf{w}_j)^2$, and the denominator, $z - \lambda_j$, are (possibly correlated) random variables, the eigenvalues vary as $\mathcal{O}(1/N)$, which means they are essentially constant for large N. So, the numerator can be treated independently of the denominator, and its expectation is $E\{(\mathbf{u}_1^\tau \mathbf{w}_j)^2\}$, which equals $1/N$. The general principle here is that if $Y = \mathcal{O}(1/N)$, then as $N \to \infty$, $E\{X/Y\} = E\{X\} \cdot E\{1/Y\}$. The last line assumes that for large networks (i.e., as $N \to \infty$), the matrix $E\{(z\mathbf{I}_N - \mathbf{B})^{-1}\}$ is diagonal, with diagonal entries $E\{1/(z - B_{jj})\}$. Because the eigenvalues of a diagonal matrix are just the diagonal entries, then $B_{jj} = \lambda_j$. Also, expectation is a linear operation and so we can interchange expectation and trace.

We can solve (12.145) for z_1. This yields the spectrum of the adjacency matrix $\mathbf{Y}^{(1)} = \mathbf{B} + \alpha_1 \mathbf{u}_1 \mathbf{u}_1^\tau$ in the case of a single community. Zhang, Nadakuditi, and Newman showed that the spectrum of the adjacency matrix when there is one community, therefore, can be viewed as two parts: (1) a continuous spectral band whose spectral density is the same as that of the spectral density of $\mathbf{B}$; and (2) a single outlying eigenvalue, z_1, whose value can be obtained by solving (12.145).

Two Communities

What does the spectral density of the adjacency matrix look like in the case of two communities? Set $K = 2$ in (12.133) so that

$$\mathbf{Y}^{(2)} = \mathbf{B} + \alpha_1 \mathbf{u}_1 \mathbf{u}_1^\tau + \alpha_2 \mathbf{u}_2 \mathbf{u}_2^\tau, \tag{12.146}$$

where the top two eigenvalue–eigenvector pairs of $E\{\mathbf{Y}^{(2)}\}$ are $(\alpha_1, \mathbf{u}_1)$ and $(\alpha_2, \mathbf{u}_2)$. We apply a similar argument in the case of two communities as we did for one community. We have that

$$\mathbf{Y}^{(2)} = \mathbf{Y}^{(1)} + \alpha_2 \mathbf{u}_2 \mathbf{u}_2^\tau, \tag{12.147}$$

where $\mathbf{Y}^{(1)} = \sum_j z_j \mathbf{v}_j \mathbf{v}_j^\tau$ and $\mathbf{Y}^{(2)} = \sum_j z_j' \mathbf{v}_j' \mathbf{v}_j'^\tau$. The same reasoning as above yields

$$\begin{aligned}
\frac{1}{\alpha_2} &= \sum_{j=1}^{N} \frac{(\mathbf{u}_2^\tau \mathbf{v}_j)^2}{z' - z_j} \\
&= \frac{(\mathbf{u}_2^\tau \mathbf{v}_1)^2}{z' - z_1} + \sum_{j=2}^{N} \frac{(\mathbf{u}_2^\tau \mathbf{v}_j)^2}{z' - z_j}.
\end{aligned} \tag{12.148}$$

Zhang, Nadakuditi, and Newman (2014) then show that the eigenvalues, $\{z'_j\}$, of $\mathbf{Y}^{(2)}$ alternate with the eigenvalues, $\{z_j\}$, of $\mathbf{Y}^{(1)}$. As a result, using similar arguments as for the case of $K = 1$, the continuous part of the spectrum is the same as that of $\mathbf{B}$, except that there is a discrete part that now consists of two outlying eigenvalues, z_1 and z_2.

Repeating this argument in a sequential fashion until there are K outlying eigenvalues (corresponding to the number of communities, $K \ll N$) leads to the same continuous part of the spectrum as above for the bulk of the eigenvalues plus the K outlying eigenvalues. The number of outlying eigenvalues provides information regarding the number of communities that can be extracted from the adjacency matrix $\mathbf{Y}$.

A problem with this method of community detection is that the ease with which the smallest of the K largest eigenvalues can be identified as outlying depends upon the strength of the community structure. If that structure is weak, so that the eigenvalues $\{\alpha_j\}$ will be small, then the smallest of the outlying eigenvalues will decrease towards the bulk spectrum, possibly merging into the bulk spectrum, and thereby disappearing from view. The weaker the structure, the more we could see a number of the smaller eigenvalues repeating this process, which could, therefore, lead to a false perception of the community structure. In fact, when the second-largest eigenvalue gets absorbed into the bulk spectrum, evidence of any community structure no longer exists.

12.6 Optimizing Modularity in Large Networks

In this section, we describe four approximation algorithms, namely, the Newman, CNM, Louvain, and Leiden algorithms. The first three of these approximation algorithms are each greedy algorithms.[9] Each of the last three algorithms, the CNM, the Louvain, and the Leiden, modifies Newman's algorithm to make it more computationally efficient, thereby improving its speed and performance.

12.6.1 Newman's Algorithm

This original and novel algorithm was proposed by Newman (2004b). Newman's partition strategy starts by associating every node with its own community. Then, the pair of these communities that produces the largest increase in Q are merged. The process is repeated until only a single community remains.

Modularity is defined in (12.81) as $Q = \frac{1}{L} \sum_{i,j} (Y_{ij} - L \cdot P_{ij}) \delta_{c_i,c_j}$. If we replace $L \cdot P_{ij}$ by $k_i k_j / L$, and, from (12.79), set $L = 2m$, then we can write Q as follows:

$$Q = \frac{1}{2m} \sum_{i,j=1}^{N} \left(Y_{ij} - \frac{k_i k_j}{2m} \right) \delta_{c_i,c_j}. \tag{12.149}$$

The modularity Q of the partition is the difference between the ratio of the number of edges within each community to the number of edges between communities, and the ratio that would be expected from a completely random partition. If there is no discernible difference between the network and a random network, then $Q = 0$.

---

[9] A *greedy algorithm* refers to a strategy in which at each step the best optimal choice is made that ultimately leads to a global optimum solution, without regard for consequences.

Nonzero values of Q indicate deviations from a random network, with values greater than 0.3 suggesting significant community structure.

Newman re-expressed Q as follows. Let $e_{st} = \frac{1}{2m} \sum_{i,j} Y_{ij} \delta_{c_i, v_s} \delta_{c_j, v_t}$ be the fraction of edges that join nodes in community C_s to nodes in community C_t and let $a_s = \frac{1}{2m} \sum_t k_t \delta_{c_t, v_s}$ be the fraction of ends of edges that are attached to nodes in community C_s. Also, set $\delta_{c_i, c_j} = \delta_{c_i, v_s} \delta_{c_j, v_s}$. Then we can write $Q = \sum_s (e_{ss} - a_s^2)$. Let $\Delta Q_{ij} = e_{ij} - a_i a_j$ denote the change in Q from combining the ith and jth communities. The two communities that are joined are those that provide the largest value of ΔQ_{ij}. This merging process is repeated $N - 1$ times until only a single community that contains all N nodes in the network is left standing. By merging communities in order, a dendrogram can be drawn to show the hierarchical structure of the network. A horizontal cut through the dendrogram can then be chosen at the largest value of Q, which would produce the best network partition.

Newman applied his algorithm to several small networks, including the 34-node Zachary's karate club data (two communities), a 1275-node jazz musician collaboration (two communities), and a 56,276-node network of collaborations between physicists (600 communities), and showed that his algorithm enjoyed significantly faster performances in each case than an algorithm proposed by Girvan and Newman (Girvan and Newman, 2002; Newman and Girvan, 2004).

Comment. Newman shows that his algorithm does very well on small networks. However, his algorithm is a greedy optimizing process and, as such, it suffers from high computational cost and much wasted memory usage caused by unnecessary operations when dealing with a sparse network, where its adjacency matrix has a lot of zero entries. This feature increases computation and, hence, execution time, which becomes a serious problem when dealing with huge networks.

12.6.2 CNM Algorithm

This algorithm, proposed by Clauset, Newman, and Moore (2004), improved Newman's algorithm by fixing the computational problems in the optimization process. The Clauset–Newman–Moore (CNM) algorithm, while retaining Newman's partition strategy, speeds up Newman's algorithm, and allows networks of size at most 500,000 nodes to be analyzed. The CNM algorithm is given in **Algorithm 12.3**.

This process can be visualized as a multigraph: (1) an entire community is represented by a node; (2) groups of edges connect pairs of nodes; and (3) edges between nodes of the same community are represented as self-loops. The multigraph has adjacency matrix $\mathbf{Y}' = (Y'_{ij})$, where $Y'_{ij} = 2m \cdot e_{ij}$. To combine communities C_i and C_j, when at least one of the two communities has nonzero entries, replace the ith and jth rows and columns by their totals (as in step 6 of the algorithm).

Clauset et al. applied the CNM algorithm to a co-purchasing network from Amazon.com, whose largest component consisted of 409,687 items (nodes) and 2,464,630 edges. They found 1684 communities with a mean size of 243 items, and the distribution of community size followed a power law over a certain range.

Comment. The CNM algorithm is a greedy heuristic process, which can yield partitions that do not agree with the optimal solution and where the modularity value is smaller than that found by other methods. In experiments with many real-world

---

Algorithm 12.3 *Clauset–Newman–Moore (CNM) algorithm.*

---

1. Start with an unweighted network $\mathcal{G} = (\mathcal{V}, \mathcal{E})$ of N nodes, where the edge Y_{ij} between the ith node and the jth node either exists ($Y_{ij} = 1$) or does not exist ($Y_{ij} = 0$).
2. Assign each node of the network to a different community. So, we start with N communities.
3. Set $e_{ij} = \frac{1}{2m}$ if nodes v_i and v_j are connected and 0 otherwise, and $a_i = k_i/2m$. For the initial partition, we set $\Delta Q_{ij} = \frac{1}{2m} - k_i k_j/(2m)^2 = e_{ij} - a_i a_j$ if $(v_i, v_j) \in \mathcal{E}$, and 0 otherwise.
4. Form the matrix $\Delta Q = (\Delta Q_{ij})$ that stores the gain in Q caused by the union of two generic communities $\mathcal{C}_i$ and $\mathcal{C}_j$, as long as the two communities are connected by at least one edge.
5. Construct a max-heap H from the largest entry in each row of ΔQ. (A *max-heap* is a complete binary tree in which the value of each internal node is greater than or equal to the values of its child nodes.)
6. Choose the largest entry in H and join the corresponding communities.
7. Update the matrix ΔQ, H, and a_j.
   Suppose we merge community $\mathcal{C}_i$ into community $\mathcal{C}_j$. Then ΔQ_{jk} for each community $\mathcal{C}_k$ that is adjacent to community $\mathcal{C}_j$ is updated as follows:

$$\Delta Q'_{jk} = \begin{cases} \Delta Q_{ik} + \Delta Q_{jk}, & \text{if } \mathcal{C}_k \text{ is connected to } \mathcal{C}_i \text{ and } \mathcal{C}_j \\ \Delta Q_{ik} - 2a_j a_k, & \text{if } \mathcal{C}_k \text{ is connected to } \mathcal{C}_i \text{ but not to } \mathcal{C}_j, \\ \Delta Q_{jk} - 2a_i a_k, & \text{if } \mathcal{C}_k \text{ is connected to } \mathcal{C}_j \text{ but not to } \mathcal{C}_i \end{cases}$$

   $a'_j = a_j + a_i$ (and $a_i = 0$).
8. Repeat steps 6 and 7 $N - 1$ times until only a single community remains.

---

networks, it has been shown that, for larger networks, CNM tends to produce much larger numbers of communities than other methods (Vieira et al., 2014). CNM has also been found to generate "super-communities" containing a large percentage of the nodes, even when the network does not possess any community structure (Wakita and Tsurumi, 2007). In such instances, computational inefficiency can occur when merging unbalanced communities, which leads to very unbalanced dendrograms and a slow-down in the algorithm. Wakita and Tsurumi (2007) introduced modifications of the CNM algorithm to take into account the balance of community pairs, which allows the algorithm to work efficiently with networks of up to 10 million nodes.

12.6.3 Louvain Algorithm

Etienne Lefebvre first developed this algorithm in his Master's thesis in March 2007 at the Université catholique de Louvain (UCL) in Belgium. The algorithm was improved and tested by Blondel et al. (2008), who were each at UCL at that time, and so the algorithm became known after the city, although the authors did not assign it that name in their article.

The Louvain algorithm, which builds upon the CNM algorithm, is given in **Algorithm 12.4**. It is a greedy optimization method that is easy to implement and

---

Algorithm 12.4 *Louvain algorithm*

Phase 1: Modularity optimization

1. Start with a weighted network of N nodes, where the network has weights on its edges. The weight on the edge between the ith and jth nodes is Y_{ij}.
2. To each node of the network, assign a different community. So, we start with N communities.
3. For the ith node, v_i, consider the community, C_i, of v_i. Compute the gain in modularity, written as ΔQ, that would take place if v_i were moved from its community C_i and placed into the community, C_j, of a neighboring node v_j.
4. Compute ΔQ for each of the neighboring nodes of v_i.
5. Place v_i into the community for which ΔQ is a maximum and is positive. Ties can be resolved. If no possibility of a positive gain exists, keep v_i in its original community C_i.
6. Repeat this process for all nodes. Stop when no further improvement is reached and a local maximum of the modularity Q is attained.

Phase 2: Community aggregation

1. Based upon the results of Phase 1, a new weighted network is built by grouping together ("aggregating") the nodes that were located to the same community. Call these groupings of nodes "meta-communities." These new meta-communities become the nodes of the new network. The edge weight between two nodes is the total weight of the edges between nodes in the corresponding meta-communities. Edges between nodes of the same community become self-loops for this community in the new network.
2. Reapply the Phase 1 process to the new meta-communities and iterate.
3. Stop when no further changes occur and a maximum of modularity Q is attained.

---

is very fast, especially for large networks. The Blondel et al. article shows that the algorithm outperforms all other community detection methods (including the CNM algorithm and its modification by Wakita and Tsurumi) for large networks in terms of modularity value and computation time.

The algorithm is divided into "passes," each of which is composed of two "phases." It starts out in the same way as Newman's algorithm and the CNM algorithm by assigning each node in the network to its own community and then merges pairs of communities in a sequential fashion. The first phase is described as *modularity optimization*, in which small communities are created by locally optimizing Q, and the second phase is *community aggregation*, in which a new network is formed whose nodes are the communities created from the first phase. Repeated iterations of these steps take place until a global maximum of Q is obtained.

Features of the algorithm include the following: (1) after a few passes of the algorithm, the number of communities is severely reduced, so that most of the computation time is taken up by the first few passes; (2) the algorithm produces a complete hierarchical community structure of the network based upon the order in which the nodes in the communities are combined; (3) it is an unsupervised

Table 12.5 Computational complexities of spectral method, Newman, CNM, Louvain, and Leiden algorithms. N is the number of nodes and M is the number of edges

Algorithm	Computational complexity
Spectral	$\mathcal{O}(N(N + M))$
Newman	$\mathcal{O}(N + M)$
CNM	$\mathcal{O}(N(\log N)^2)$
Louvain	$\mathcal{O}(N \log N)$
Leiden	Unknown

algorithm in that the final communities may be of unequal size, which parallels real-life experience, and it does not require the number of final communities to be specified in advance; (4) the order in which the nodes are considered does not appear to affect the modularity, but it can influence the computation time; and (5) it is not unusual for a node to be revisited several times within a given pass, which may slow down the algorithm.

The algorithm was applied by Blondel et al. to a number of small and large networks. These include Zachary's karate club network of 34 nodes, a citation network of 9000 scientific papers, a subnetwork of 39 million nodes and 783 million edges, and a network of 118 million nodes and one billion edges. In each case, computation times were much smaller than those of competing algorithms, ranging from 12 minutes to 152 minutes.

A comparison of the computational complexities of the Newman, CNM, and Louvain algorithms is given in **Table 12.5**.

Comment. It turns out that the Louvain algorithm, which is one of the most popular algorithms for detecting communities, has a major flaw that went unnoticed until identified by Traag, Altman, and van Eck (2019). It seems that the Louvain algorithm has a tendency to discover communities that are poorly connected or even internally disconnected. In their experiments, they found that up to 25% of the communities were poorly connected and up to 16% were internally disconnected. The primary problem is that the algorithm keeps moving individual nodes, one at a time, from one community to another, and in doing so, it may move a "bridge" node (a single node that has edges joining two groups of nodes within one community) to a different community, so that it breaks the connectivity of the original community.

12.6.4 Leiden Algorithm

While the Louvain algorithm merges communities, the Leiden algorithm splits communities as well as merging them. This strategy guarantees that communities are internally well-connected, as opposed to communities found by the Louvain algorithm that can be internally disconnected. In fact, the Leiden algorithm was designed specifically to counter the problem of the Louvain algorithm regarding poorly connected communities.

The Leiden algorithm,[10] proposed by Traag, Altman, and van Eck (2019), all of whom are at Leiden University in the Netherlands, is listed in **Algorithm 12.5**. It uses three phases (instead of two as in the Louvain algorithm): in the first phase, a local moving of nodes creates a partition $\mathcal{P}$; in the second phase, a refinement of the partition $\mathcal{P}$, which may include splitting communities into multiple subcommunities that can then be assigned to any of the existing communities, yields a refined partition $\mathcal{P}_{\text{refined}}$; and in the third phase, the network is aggregated based upon the $\mathcal{P}_{\text{refined}}$ partition. The refinement phase does not follow a greedy strategy. Although the Leiden algorithm is much more complicated than the Louvain algorithm, it has demonstrated superior performance in speed and finds high-quality connected communities.

In the first phase of the algorithm, nodes are initially considered as their own communities. Then, each node is moved to a different community, where "different" could, in principle, mean any other community. However, if a node had no link to that community, such a move would always be suboptimal. So, the algorithm would only consider moving the node to the community of a randomly chosen neighboring node.

In the second phase, the algorithm tries to identify communities in $\mathcal{P}$ that should be split up if the communities are not well-connected internally. A refined partition $\mathcal{P}_{\text{refined}}$ starts out as the partition $\mathcal{P}$ from the first phase. Each community in $\mathcal{P}_{\text{refined}}$ is considered to be a subnetwork. Then, a node in a subnetwork is locally moved to the community of a randomly chosen neighboring node for which modularity increases; the larger the increase, the more likely that that community will be chosen. The argument is that the best community for a node to join will be one that contains most of its neighboring nodes. The main action in this phase is that poorly connected communities obtained as $\mathcal{P}$ may be further split into multiple communities in $\mathcal{P}_{\text{refined}}$.

By iterating the Leiden algorithm, communities become "subset optimal," meaning that one cannot improve the quality of the communities by moving nodes from one community to another.

The Leiden algorithm is implemented as the package `leiden` in R; see
`CRAN.R-project.org/package=leiden`.
The full Python and Java implementations are available on GitHub. See
`github.com/vtraag/leidenalg`.
See Traag (2021) for documentation of the Leiden algorithm.

Resolution limit. Graph-partitioning algorithms based upon modularity maximization generally suffer from a *resolution limit* (Fortunato and Barthélemy, 2007; Lancichinetti and Fortunato, 2011). In certain situations, when the algorithm maximizes modularity, it can fail to discover small communities and will force small communities to merge into larger communities. Here, "small" refers to smaller than the resolution limit, which depends upon the number, M, of edges in the network, and on the degree of interconnectivity between the communities. For example, consider the case of two communities, $\mathcal{C}_A$ and $\mathcal{C}_B$, with only one edge between them. If $(k_A k_B)/2M < 1$, then modularity increases by merging $\mathcal{C}_A$ and $\mathcal{C}_B$ to form a single, bigger community, even if $\mathcal{C}_A$ and $\mathcal{C}_B$ are separate and distinct communities. In other situations, the reverse may occur; communities that are larger than the resolution limit may cause the unnecessary splitting of large communities into smaller communities. There are other types of quality functions for which the resolution limit does not occur, such as the so-called constant Potts model (Traag, van Dooren, and Nesterov, 2011).

---

[10] The author thanks Vincent A. Traag for e-mail correspondence on the algorithm.

Algorithm 12.5 *Leiden algorithm*

Phase 1: Local moving of nodes

1. Start with an undirected network, $\mathcal{G} = (\mathcal{V}, \mathcal{E})$, where each node is its own community.
2. Initialize a queue to which all nodes in the network are added in a random order. Remove the first node from the front of the queue and determine whether modularity can be increased by moving this node from its present community to the community of a neighboring node. If the node is moved to the community of a neighboring node, add to the end of the queue all neighbors of the node that do not belong to the node's new community and that are not yet in the queue. After the first iteration in which all nodes have been visited once, only those nodes whose neighborhood has changed can be revisited in subsequent iterations. Continue removing nodes from the front of the queue, possibly moving these nodes to the community of a neighboring node, until the queue is empty. The resulting partition is $\mathcal{P}$.

Phase 2: Refining the partition

1. Consider each community derived in $\mathcal{P}$ as a subnetwork.
2. Partition each subnetwork into subcommunities.
3. In each subcommunity, a node may be moved by selecting a neighboring node at random and checking to see whether modularity increases if the node is moved to that neighbor's subcommunity. Communities in $\mathcal{P}$ that consist of poorly connected sets of nodes may be split into multiple subcommunities. The resulting partition is $\mathcal{P}_{\text{refined}}$.

Phase 3: Network aggregation

1. Consider each community in $\mathcal{P}_{\text{refined}}$ as a node in an "aggregated network" and reapply the Phase 2 refining process within each community. Any subnetworks found are treated as different communities in the next iteration.
2. Repeat until no further improvements are possible.

12.7 Consistency

The notion of *consistency* for classification problems states the following:

Definition 12.16 If we have a sequence of realizations for increasing sample size N, and

$$\tau : [N] = \{1, 2, \ldots, N\} \to [K] = \{1, 2, \ldots, K\} \tag{12.150}$$

is a rule that assigns each item $i \in [N]$ to a class $k \in [K]$, then $\widehat{\tau}$ is a *consistent estimator* of τ if the fraction of misclassified items converges almost surely to zero as $N \to \infty$.

This concept of consistency was applied by Bickel and Chen (2009) to a community-detection criterion Q. Given a network $\mathcal{G} = (\mathcal{V}, \mathcal{E})$, the classification rule $Q : \mathcal{V} \to$

$[K]$ that assigns which of the N nodes belong to which of the K communities is called a *block membership function.*

Let the *block assignment function* $\widehat{\mathcal{Q}} : \mathcal{V} \to [K]$ be an estimator of $\mathcal{Q}$, a process that is often referred to as "block estimation." Then,

Definition 12.17 $\widehat{\mathcal{Q}}$ is said to be a *consistent estimator* of $\mathcal{Q}$ if the fraction of misclassified nodes almost surely goes to zero when the number of nodes, N, goes to infinity.

This definition of consistency was recast as one of "strong" consistency by Zhao, Levina, and Zhu (2012).

Let the community labels be $\{1, 2, \ldots, K\}$ and let $\mathbf{c} = (c_1, c_2, \ldots, c_N)^\tau$ be the true community labels of the N nodes. Let the community labels, which are obtained by maximizing $\widehat{\mathcal{Q}}$, be denoted by $\widehat{\mathbf{c}} = (\widehat{c}_1, \widehat{c}_2, \ldots, \widehat{c}_N)^\tau$. Then, we have the following definition:

Definition 12.18 The criterion $\widehat{\mathcal{Q}}$ is said to be *strongly consistent* for $\mathcal{Q}$ if the community labels $\widehat{\mathbf{c}}$, obtained by maximizing the criterion $\widehat{\mathcal{Q}}$, satisfy

$$\mathrm{P}\{\widehat{\mathbf{c}} = \mathbf{c}\} \to 1, \quad \text{as } N \to \infty. \tag{12.151}$$

Under this definition of strong consistency, Bickel and Chen (2009) show that the Newman–Girvan modularity $\mathcal{Q}_{NG}$ is strongly consistent for $K = 2$, but not for $K > 2$. In a counterexample for three communities, they showed that only when the two sparser communities were merged did $\mathcal{Q}_{NG}$ attain its maximum.

Zhao, Levina, and Zhu (2012) also proposed a definition of "weak" consistency for $\widehat{\mathcal{Q}}$ in the case of stochastic blockmodels:

Definition 12.19 $\widehat{\mathcal{Q}}$ is *weakly consistent* for $\mathcal{Q}$ if, for all $\epsilon > 0$:

$$\mathrm{P}\left\{ N^{-1} \sum_{i=1}^{N} I_{[\widehat{c}_i \neq c_i]} < \epsilon \right\} \to 1, \quad \text{as } N \to \infty. \tag{12.152}$$

Furthermore, to overcome identifiability issues with both definitions of consistency, Zhao, Levina, and Zhu replaced the condition $\widehat{\mathbf{c}} = \mathbf{c}$ by the requirement that $\widehat{\mathbf{c}}$ and $\mathbf{c}$ belong to the same equivalence class of label permutations. They also assumed, within the context of degree-corrected SBM, that the true community labels $\mathbf{c}$ and the degree parameters $\boldsymbol{\theta} = (\theta_1, \ldots, \theta_N)^\tau$ were latent variables (not fixed parameters). Under those assumptions, they then showed that:

- Under the standard SBM, the criterion $\widehat{\mathcal{Q}}_{SBM}$ is strongly consistent when $\lambda_N / \log N \to \infty$ and weakly consistent when $\lambda_N \to \infty$, where λ_N denotes the expected degree of the network.
- Under the degree-corrected SBM, the degree-corrected SBM criterion $\widehat{\mathcal{Q}}_{KN}$ is strongly consistent when $\lambda_N / \log N \to \infty$ and weakly consistent when $\lambda_N \to \infty$, but the criterion $\widehat{\mathcal{Q}}_{SBM}$ is not necessarily consistent under that model.
- If links within communities are assumed to be more likely than links between communities, then Zhao, Levina, and Zhu show that the Newman–Girvan modularity is strongly consistent when $\lambda_N / \log N \to \infty$ and weakly consistent when $\lambda_N \to \infty$.

- The Erdős–Renyi modularity, which assigns nodes with large numbers of degrees to different communities than nodes with small numbers of degrees (not always an appropriate strategy), is strongly consistent under the SBM when $\lambda_N / \log N \to \infty$ and weakly consistent when $\lambda_N \to \infty$, but is not consistent under the degree-corrected SBM.

A concept related to that of consistency has been described by Abbe (2018) in his detailed review of recent developments in community detection. He defines three types of "recovery" for a partition model. Specifically, if there exists an algorithm that takes the network $\mathcal{G}$ as input and the fitted partition as output, then the requirements for the various levels of recovery are:

1. *Exact recovery.* This requires that the entire partition be correctly recovered.
2. *Almost exact recovery.* This allows for a vanishing fraction of misclassified nodes.
3. *Partial recovery.* This allows for a constant fraction of misclassified nodes.

These types of recovery correspond to the types of consistency: exact recovery is equivalent to strong consistency and almost exact recovery is equivalent to weak consistency.

In the case of the graph being an SBM, Abbe describes an approach designed to answer the above recovery problems. He describes a "graph-splitting" algorithm (see **Algorithm 12.6**), which he characterizes as a powerful strategy for achieving exact recovery for an SBM.

Algorithm 12.6 *Graph-splitting algorithm*

---

1. Form two random graphs, $\mathcal{G}_1$ and $\mathcal{G}_2$, on the same set of nodes $\mathcal{V}$ of $\mathcal{G}$, where $\mathcal{G}_1$ is formed by sampling each edge independently with a given probability γ (which works really well if $\gamma = \log \log(N) / \log(N)$), and $\mathcal{G}_2 = \mathcal{G} \backslash \mathcal{G}_1$ consists of the edges of $\mathcal{G}$ that do not appear in $\mathcal{G}_1$. Although $\mathcal{G}_1$ and $\mathcal{G}_2$ are not independent of each other, consider them to be approximately independent.
2. A clustering algorithm is applied to $\mathcal{G}_1$, which obtains a good, but not necessarily exact, clustering. Consider this a preliminary clustering.
3. The results of the preliminary clustering, whose goal is an almost exact recovery, are then applied to $\mathcal{G}_2$.
4. A local algorithm is used on $\mathcal{G}_2$ to "clean up" the first algorithm by reclassifying each node.

Conditions (e.g., node exchangeability and a limit to the growth of node degrees) are described by Abbe under which the solution from graph splitting and the resulting almost exact recovery can be used to obtain exact recovery with computational efficiency. However, finding necessary and sufficient conditions for almost exact recovery to be achieved in the general SBM case remains an open problem.

12.8 Sampling for Network Structure

In this chapter, we discussed various methods for community detection, which is the problem of determining a partition of the network in which nodes that are members of the same community are more densely connected to each other than they are connected to nodes in other communities. However, very little work has appeared on how to deal with the problem of sampling from a network that is composed of a number of communities. In the following, we will assume that the communities represent a disjoint partition of the entire network, rather than allow overlapping communities.

12.8.1 The Sampling Problem

We have to distinguish between two possible scenarios. The first scenario is that the network is completely known and all K communities comprising the network have been identified; either they are known or they have been discovered through the community detection methods of this chapter. The other scenario is that the K communities are completely unknown, as is also the size of each of those communities.

A large part of the sampling problem has to do with identifying what makes a sampled network *representative* of a parent network that has substantial community structure. One view (Maiya and Berger-Wolf, 2010) is that the sampled network should contain nodes from each community (analogous to the practice of stratified sampling in standard populations), or from most of the communities. If a community detection algorithm were to be applied to the parent network and, independently, to the random subnetwork, we would expect communities discovered in the random subnetwork to correspond to communities discovered in the parent network. What this implies is that nodes identified with a community in the random subnetwork should also be connected to the same nodes and identified with the same community in the parent network.

12.8.2 Sampling Strategies

In order to sample from a network that has known or suspected community structure, we have the following scenarios:

- If the K communities are known, and we know which nodes are members of each community, then we have a sampling frame for each community. We then have the following possible sampling strategies:

  - We could sample each community separately and independently. This strategy is analogous to stratified sampling for statistical surveys.
  - We could take a simple random sample of a subset of those K communities and then include all nodes and associated edges in each of those selected communities for the sample. This strategy is analogous to cluster sampling.

- If a community structure in the network is suspected but is otherwise not known, we can apply one of the following sampling strategies:

  - We could, first, carry out a community detection algorithm to define the communities, and then we could sample from each of the identified communities, understanding that some error may be present.
  - We could sample from the network and then carry out community detection on the sampled network. There is no guarantee, however, that nodes, which are

organized in a particular way in the parent network, will be organized in the same way in the sampled network.

It is much more difficult to sample from difficult-to-access populations when the population has several communities. For example, in the RDS procedure, if a participant is asked to recommend another individual to the study, the chance is high that the recommended individual will be a member of the same community as the participant. Individuals who are members of the same community usually have similar features and idiosyncracies, whilst those from differing communities will tend to have different characteristics. The result will be that successive sample individuals will be drawn from the population through a dependent selection process, the majority of participants in the study would likely come from a single community, and the sample composition could quickly become severely unbalanced.

A few articles have appeared that attempt to construct sampling strategies for networks with community structure (see, e.g., Blagus et al., 2015). Essentially, these articles outline the standard network sampling methods that we described in Chapter 9, and then compare their performances with each other when community structure is present in the data. A related issue is to provide statistical inferences from the sampled network to the parent network regarding the community affiliation of nodes that do not appear in the sampled network.

12.8.3 Sampling Communities

There has been very little research on network sampling when the network consists of a number of communities. The following methods represent the state-of-the-art at the present time.

Expansion Sampling

Maiya and Berger-Wolf (2010) proposed a concept of "expansion sampling" in which a sample is constructed with maximal expansion so that it mirrors the known community structure in the parent network. They described two versions of this method.

The first builds on the general idea of snowball sampling. Let S denote the set of sampled nodes and $\mathcal{N}(S)$ the set of neighbors of S. Define the *expansion factor*, $X(S)$, of the sample S of nodes to be

$$X(S) = \frac{|\mathcal{N}(S)|}{|S|}, \tag{12.153}$$

where $|\mathcal{N}(S)|$ is the number of nodes in $\mathcal{N}(S)$, and similarly for $|S|$.

Start by setting S to be the empty set. Add a node at random from $\mathcal{V}$ to S. The impact on the expansion factor if a node v is added to S is that the expansion factor will increase (decrease) if $X(S \cup \{v\})$ is larger (smaller) than $X(S)$. Using the definition (12.152) and some algebra, this condition boils down to whether $|\mathcal{N}(\{v\}) - (\mathcal{N}(S) \cup S)|$ is larger (smaller) than $X(S)$. If v is a member of a community that already exists in S, then the expansion factor will decrease because v will have few new neighbors; on the other hand, if v is a member of a community not included in S, then the expansion factor will increase. Thus, the value of the expansion factor is directly related to the community structure of the network, although the strength of the community structure is important in the notion of expansion. The *snowball expansion sampler* (XSN) proposed by the authors is to select a new node $v \in \mathcal{N}(S)$ to add to S based upon

$$\arg \max_{v \in \mathcal{N}(S)} |\mathcal{N}(\{v\}) - (\mathcal{N}(S) \cup S)|. \tag{12.154}$$

This method always produces sample networks that are connected.

A related sampling procedure also proposed by the authors is *Markov chain Monte Carlo expansion sampling* (XMC) using the Metropolis algorithm. The method starts by constructing a Markov chain in which each state represents a subgraph of a certain size (much smaller than $|\mathcal{V}|$). Next, the user selects a "quality measure" that determines how representative the subgraph sample of nodes is. The quality measure is related to the expansion factor. Then, a random sample of nodes is drawn from the network. We add a single node to it, and accept or reject that perturbed sample based upon the change in the quality measure. Because the Markov chain converges to a stationary distribution, sampling from that Markov chain is equivalent to sampling subgraphs with probability proportional to the scores of the quality measure of the subgraphs. The representative subgraph is then the one with the maximum quality score. Further details may be found in the reference.

Worked examples of XSN and XMC with real-data networks indicate that the expansion sampling approach yields samples that were more representative of community structure in the parent network than samples produced by existing methods.

PageRank Sampling

Salehi, Rabiee, and Rajabi (2012) proposed PageRank sampling (PRS), based upon a modified version of the PageRank criterion. PPR determines the importance of every node to the seed by using local information, and it is used to solve the problem of community detection. The method assumes that an equal number of nodes are drawn from each community.

Suppose it is known that the network contains K disjoint communities. If K is unknown, then approximate it. Start out each community as the empty set $\emptyset$. Let S denote the set of current sample nodes and $\mathcal{N}(S)$ the set of neighboring nodes of S. There are two phases to this algorithm:

1. Select a node $v \in \mathcal{V}$ randomly from the network and add it to S. By approximating the PRS vector, select a new node $v' \in \mathcal{N}(S)$ that has maximum PageRank value. The chosen node, v', has the highest probability amongst all nonvisited nodes to be in the same community as v. Add v' to S and to the appropriate community. This allows the current community to be expanded from a single node to a set of nodes. The process is repeated until the required number of nodes is drawn from the community, and no further node can be added to that particular community.

2. PRS next computes PageRank values for all current sampled nodes and their neighbors, and selects that node that has the lowest PageRank value and the highest number of unknown neighbors. PRS then jumps to that node, which is regarded as the initiator node for a new community.

The PRS procedure turns out to enjoy good properties for networks having strong community structure. In particular, PRS outperforms RDS on all test datasets when community structure is present in the network.

12.9 Further Reading

The graph partitioning methods covered in this chapter are currently under vigorous development. More articles and books on these topics are expected to appear over the next several years in the statistics, statistical physics, computer science, machine learning, and social science literature, which makes keeping track of all the huge amount of research a definite challenge.

Sections or chapters in the following books describe community detection: Newman (2010, Chapter 11), which also describes graph partitioning, graph cuts, and modularity, and Kolaczyk (2009, Section 4.3). An excellent review of recent developments in community detection (or graph clustering), focusing on the SBM and its variations, is given by Abbe (2018). An edited volume on graph partitioning is the book by Bichot and Siarry (2011).

12.10 Exercises

Exercise 12.1 Some authors analyzed Sampson's monk data (see **Table 2.2** for the adjacency matrix) and concluded that there ought to be three blocks (rather than four). Possible reallocations of the members of the Interstitials block are given in footnote 3 in Section 12.3.2. Redo the blockmodel analysis with three blocks and obtain the permuted adjacency matrix, the network graph, the image matrix, and the reduced graph.

Exercise 12.2 Show that the complete-data likelihood of $\mathbf{Y}$ (see (12.20)) can be transformed into (12.22) in which the products are taken over blocks rather than over nodes.

Exercise 12.3 This exercise illustrates *Wigner's semicircle law* (see (12.125)). Fill a single $(N \times N)$-matrix $\mathbf{A} = (A_{ij})$ with iid standard Gaussian deviates, where $A_{ij} \sim \mathcal{N}(0,1)$, $i,j = 1,2,\ldots,N$. Next, compute the random, symmetric matrix $\mathbf{H}_N = \frac{1}{2}(\mathbf{A} + \mathbf{A}^\tau)$, also known as the *Wigner matrix*. Consider the two cases: $N = 1000$ and $N = 25,000$. Compute the N eigenvalues of $\mathbf{H}_N$ for each case. Draw the histogram of those eigenvalues for each case using 100 bins. Plot each histogram separately using the x-axis from -2.5 to 2.5 and the y-axis from 0 to 0.4. These histograms should each resemble a semicircle.

Exercise 12.4 Consider the political blogs network (Adamic and Glance, 2005) consisting of 1494 blogs. The data can be found at the website `www-personal.umich.edu/~mejn/netdata`.

The nodes represent blogs on the topic of the 2004 U.S. Presidential Election. Each blog is labeled as "liberal" (0) or "conservative" (1). There are 759 liberal blogs and 735 conservative blogs. The edges are hyperlinks from one blog to another. Compute the degrees of all nodes and draw the histogram of the degree distribution. Find also the total number of edges.

Exercise 12.5 In Section 12.5.4, the matrix $\mathbf{B} = (B_{ij})$ has entries $B_{ij} = Y_{ij} - (k_i k_j)/L$ for $i,j = 1,2,\ldots,N$. Show that $\sum_i B_{ij} = \sum_j B_{ij} = 0$.

Exercise 12.6 Using the results from Exercise 12.4, compute the modularity matrix **B** (see (12.87)) for the political blogs network data. Compute the resolvent matrix $\mathbf{R}(z)$ (see (12.116)) and the spectral density $\rho(z)$ of the modularity matrix (see (12.117)). Draw the spectral density of the modularity matrix (i.e., plot $\rho(z)$ vs. z).

Exercise 12.7 For Zachary's karate club network data (see Section 2.5.5), where $K = 2$, compute the Newman–Girvan modularity criterion $\mathcal{Q}_{NG}$ (see (12.83)). Comment on your results.

Exercise 12.8 (a) Draw the undirected graph with 8 nodes and 14 edges corresponding to the following adjacency matrix:

$$\mathbf{Y} = \begin{pmatrix} 0 & 1 & 1 & 1 & 0 & 0 & 0 & 1 \\ 1 & 0 & 1 & 1 & 0 & 0 & 0 & 0 \\ 1 & 1 & 0 & 1 & 0 & 0 & 0 & 0 \\ 1 & 1 & 1 & 0 & 1 & 0 & 0 & 0 \\ 0 & 0 & 0 & 1 & 0 & 1 & 1 & 1 \\ 0 & 0 & 0 & 0 & 1 & 0 & 1 & 1 \\ 0 & 0 & 0 & 0 & 1 & 1 & 0 & 1 \\ 1 & 0 & 0 & 0 & 1 & 1 & 1 & 0 \end{pmatrix}.$$

(b) Compute the modularity matrix **B** (see (12.87)). (c) Find the eigenvector of **B** corresponding to the largest eigenvalue. (d) Partition the graph into two communities based upon your results.

Exercise 12.9 Run the CNM, Louvain, and Leiden algorithms on Zachary's karate club network data and report the modularity score and the number of communities.

Exercise 12.10 In the CNM algorithm, show that the modularity criterion (12.149) can be written as $\mathcal{Q} = \sum_s (e_{ss} - a_s^2)$.

CHAPTER THIRTEEN

Graph Partitioning: III. Spectral Clustering

The previous two chapters have focused on the problem of graph partitioning, which has seen enormous interest and research work in recent years. We continue that aspect of network analysis by introducing the notion of spectral clustering. The main tool of this chapter is the graph Laplacian, which can be unnormalized or normalized. Also discussed is a regularized version of the adjacency matrix.

13.1 Introduction

In this chapter, we first introduce unnormalized and normalized graph Laplacians, which are important tools in the spectral clustering of graphs. Then, we discuss various spectral clustering algorithms for graphs. An important issue is whether unnormalized or normalized graph Laplacians are preferable to work with, and this question is discussed and illustrated using the political blogs network. Then, we discuss the regularization of spectral clustering algorithms, in which a regularization parameter is added to the adjacency matrix. Regularized spectral clustering is also illustrated by the political blogs network.

13.2 Graph Laplacians

Spectral clustering is carried out using the eigenvalues and eigenvectors of *graph Laplacian* matrices. Such matrices have advantages over considering only the adjacency matrix.

Suppose $\mathcal{G} = (\mathcal{V}, \mathcal{E})$ is a graph with undirected and unweighted edges. Because $\mathcal{G}$ is undirected, the $(N \times N)$ random adjacency matrix $\mathbf{Y} = (Y_{ij})$ is a symmetric matrix, so that $Y_{ij} = Y_{ji} = 1$ if the edge $(v_i, v_j) \in \mathcal{E}$, and 0 otherwise. As usual, we assume that $Y_{ii} = 0, i = 1, 2, \ldots, N$. Let $\mathbf{D} = (D_{ij})$ denote an $(N \times N)$ diagonal matrix with diagonal entries

$$D_{ii} = \sum_{j=1}^{N} Y_{ij}, \quad i = 1, 2, \ldots, N. \tag{13.1}$$

Thus, D_{ii} is the *degree* of node v_i, which is the number of edges present at that node. If $D_{ii} = 0$ for some i, we say that node v_i is *isolated*. If $\mathbf{J}_N = \mathbf{1}_N \mathbf{1}_N^\tau$ is the $(N \times N)$-matrix all of whose entries are 1 (where $\mathbf{1}_N$ is the N-vector of 1s), then

$$\mathbf{D} = \text{diag}\{\mathbf{YJ}_N\} \tag{13.2}$$

contains information about the degree of every node in the graph $\mathcal{G}$. Note that $\text{tr}(\mathbf{D}) = \sum_i D_{ii} = 2|\mathcal{E}|$.

13.2.1 Unnormalized Graph Laplacians

We have the following definition of an unnormalized graph Laplacian:

Definition 13.1 The *unnormalized graph Laplacian* is

$$\mathbf{L} = \mathbf{D} - \mathbf{Y}. \tag{13.3}$$

If $\mathbf{L} = (L_{ij})$, then

$$L_{ij} = L_{ji} = \begin{cases} D_{ii}, & \text{if } v_i = v_j \\ -1, & \text{if } (v_i, v_j) \in \mathcal{E}, v_i \neq v_j. \\ 0, & \text{otherwise} \end{cases} \tag{13.4}$$

Worked Example 13.1

Consider the four-node graph in **Figure 13.1**. The adjacency matrix $\mathbf{Y}$ and the $\mathbf{D}$ matrix are

$$\mathbf{Y} = \begin{pmatrix} 0 & 1 & 0 & 0 \\ 1 & 0 & 1 & 1 \\ 0 & 1 & 0 & 1 \\ 0 & 1 & 1 & 0 \end{pmatrix}, \quad \mathbf{D} = \begin{pmatrix} 1 & 0 & 0 & 0 \\ 0 & 3 & 0 & 0 \\ 0 & 0 & 2 & 0 \\ 0 & 0 & 0 & 2 \end{pmatrix}, \tag{13.5}$$

and the $\mathbf{L}$ matrix is

$$\mathbf{L} = \mathbf{D} - \mathbf{Y} = \begin{pmatrix} 1 & -1 & 0 & 0 \\ -1 & 3 & -1 & -1 \\ 0 & -1 & 2 & -1 \\ 0 & -1 & -1 & 2 \end{pmatrix}. \tag{13.6}$$

Properties of $\mathbf{L}$. Because each row sum (and each column sum) of $\mathbf{L}$ is zero, the matrix $\mathbf{L}$ is singular (and non-negative definite), and hence its smallest eigenvalue is equal to 0, and its associated eigenvector is $\mathbf{1}_N = (1, 1, \dots, 1)^\tau$. The second smallest

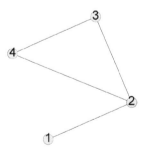

Figure 13.1 Network graph for Worked Example 13.1. There are four nodes and four edges.

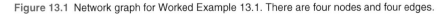

eigenvalue of $\mathbf{L}$ was called by Fiedler (1973) the *algebraic connectivity* of the graph $\mathcal{G}$. The smallest nonzero eigenvalue of $\mathbf{L}$ is called the *spectral gap*. See **Exercise 13.1** for further properties of $\mathbf{L}$.

13.2.2 Example: The Load-Balancing Problem

One of the important problems in the design of distributed and parallel computing is the *load-balancing problem,* in which multiprocessors are managed within a multiprocessor interconnection network. Each processor in the network is assigned a number of elementary tasks (or jobs) that it has to execute. The load-balancing problem entails moving elementary tasks amongst processors so that each processor gets to work on the same number of tasks.

The processors are represented by the nodes of the network, while the connection links between neighboring processors are shown by the edges. The processors within the network communicate by sending or receiving messages through these communication links. Non-adjacent nodes are not permitted to communicate with each other. The problem is to construct algorithms that will achieve such a uniform distribution of tasks amongst the processors.

A spectral clustering approach to the load-balancing problem based upon the eigenvalues and eigenvectors of the Laplacian matrix $\mathbf{L} = \mathbf{D} - \mathbf{Y}$ of the network was proposed by Elsässer, Královič and Monien (2003).

13.2.3 Normalized Graph Laplacians

There are two matrices that are referred to as *normalized graph Laplacians.*

Definition 13.2 The normalized graph Laplacian has two related forms:

$$\widetilde{\mathbf{L}}^{rw} = \mathbf{D}^{-1}\mathbf{L} = \mathbf{I}_N - \mathbf{D}^{-1}\mathbf{Y}, \tag{13.7}$$

$$\widetilde{\mathbf{L}}^{sym} = \mathbf{D}^{-1/2}\mathbf{L}\mathbf{D}^{-1/2} = \mathbf{I}_N - \mathbf{D}^{-1/2}\mathbf{Y}\mathbf{D}^{-1/2}, \tag{13.8}$$

where $\mathbf{I}_N$ is the $(N \times N)$ identity matrix.

Both $\widetilde{\mathbf{L}}^{rw}$ and $\widetilde{\mathbf{L}}^{sym}$ are non-negative-definite. The matrix $\widetilde{\mathbf{L}}^{rw}$ is a matrix related to a random walk (see below). The entries in the matrix $\widetilde{\mathbf{L}}^{rw} = (\widetilde{L}_{ij}^{rw})$ are given by the following:

$$\widetilde{L}_{ij}^{rw} = \begin{cases} 1, & \text{if } v_i = v_j, \, D_{ii} \neq 0 \\ -\frac{1}{D_{ii}}, & \text{if } (v_i, v_j) \in \mathcal{E}, \, v_i \neq v_j. \\ 0, & \text{otherwise} \end{cases} \tag{13.9}$$

The more useful of the two matrices is $\widetilde{\mathbf{L}}^{sym} = (\widetilde{L}_{ij}^{sym})$ because it is symmetric and its entries are given by the following:

$$\widetilde{L}_{ij}^{sym} = \widetilde{L}_{ji}^{sym} = \begin{cases} 1, & \text{if } v_i = v_j, \, D_{ii} > 0, \text{ all } i \\ -\frac{1}{\sqrt{D_{ii}D_{jj}}}, & \text{if } (v_i, v_j) \in \mathcal{E}, \, v_i \neq v_j \\ 0, & \text{otherwise} \end{cases}. \tag{13.10}$$

Worked Example 13.1 (continued)

From the above example, the two normalized graph Laplacians are

$$
\widetilde{\mathbf{L}}^{rw} = \begin{pmatrix} 1 & -1 & 0 & 0 \\ -\frac{1}{3} & 1 & -\frac{1}{3} & -\frac{1}{3} \\ 0 & -\frac{1}{2} & 1 & -\frac{1}{2} \\ 0 & -\frac{1}{2} & -\frac{1}{2} & 1 \end{pmatrix}, \quad
\widetilde{\mathbf{L}}^{sym} = \begin{pmatrix} 1 & -\frac{1}{\sqrt{3}} & 0 & 0 \\ -\frac{1}{\sqrt{3}} & 1 & -\frac{1}{\sqrt{6}} & -\frac{1}{\sqrt{6}} \\ 0 & -\frac{1}{\sqrt{6}} & 1 & -\frac{1}{2} \\ 0 & -\frac{1}{\sqrt{6}} & -\frac{1}{2} & 1 \end{pmatrix}.
\tag{13.11}
$$

The matrix $\widetilde{\mathbf{L}}^{rw}$ is not symmetric. However, because $\widetilde{\mathbf{L}}^{sym}$ is symmetric, its eigenvalues are real and non-negative. The eigenvectors of both $\widetilde{\mathbf{L}}^{sym}$ and $\mathbf{I}_n - \widetilde{\mathbf{L}}^{sym}$ are identical,[1] and so the difference has no impact on spectral clustering. See **Exercise 13.3** for properties of $\widetilde{\mathbf{L}}^{rw}$ and $\widetilde{\mathbf{L}}^{sym}$ and relations between them.

As a result, statisticians tend to use

$$
\mathbf{L}^{rw} = \mathbf{D}^{-1}\mathbf{Y},
\tag{13.12}
$$

$$
\mathbf{L}^{sym} = \mathbf{D}^{-1/2}\mathbf{Y}\mathbf{D}^{-1/2},
\tag{13.13}
$$

in place of (13.7) and (13.8), respectively.

Worked Example 13.1 (continued)

By subtracting each of the matrices in (13.11) from the identity matrix, we have that

$$
\mathbf{L}^{rw} = \begin{pmatrix} 0 & 1 & 0 & 0 \\ \frac{1}{3} & 0 & \frac{1}{3} & \frac{1}{3} \\ 0 & \frac{1}{2} & 0 & \frac{1}{2} \\ 0 & \frac{1}{2} & \frac{1}{2} & 0 \end{pmatrix}, \quad
\mathbf{L}^{sym} = \begin{pmatrix} 0 & \frac{1}{\sqrt{3}} & 0 & 0 \\ \frac{1}{\sqrt{3}} & 0 & \frac{1}{\sqrt{6}} & \frac{1}{\sqrt{6}} \\ 0 & \frac{1}{\sqrt{6}} & 0 & \frac{1}{2} \\ 0 & \frac{1}{\sqrt{6}} & \frac{1}{2} & 0 \end{pmatrix}.
\tag{13.14}
$$

The matrix $\mathbf{L}^{rw}$ is often called the *random-walk normalized Laplacian* (and denoted by $\mathbf{P}$) because it is the transition matrix of a Markov chain that has the same eigenvalues as $\mathbf{L}^{sym}$.

All eigenvalues of $\mathbf{L}^{sym}$ lie in the interval $[-1, 1]$. This can be shown as follows. Let $\mathbf{M}$ be a matrix whose eigenvalue λ and associated eigenvector $\mathbf{y}$ satisfy the equation $\mathbf{M}\mathbf{y} = \lambda\mathbf{y}$. Premultiplying by $\mathbf{y}^\tau$, we get $\mathbf{y}^\tau\mathbf{M}\mathbf{y} = \lambda\mathbf{y}^\tau\mathbf{y}$, whence, $\lambda = \mathbf{y}^\tau\mathbf{M}\mathbf{y}/\mathbf{y}^\tau\mathbf{y}$.[2] So, the eigenvalues of $\mathbf{L}^{sym}$ will lie in $[-1, 1]$ iff

$$
-1 \leq \frac{\mathbf{z}^\tau\mathbf{L}^{sym}\mathbf{z}}{\mathbf{z}^\tau\mathbf{z}} \leq 1, \quad \text{for all } \mathbf{z} \in \mathbb{R}^N, \mathbf{z} \neq \mathbf{0},
\tag{13.15}
$$

where $\mathbf{z} = \mathbf{D}^{1/2}\mathbf{x}$ and $\mathbf{x}$ is the solution to the generalized eigenproblem $\mathbf{A}\mathbf{x} = \lambda\mathbf{D}\mathbf{x}$. Substituting for $\mathbf{z}$ in (13.15), this will be true iff

$$
-1 \leq \frac{\mathbf{x}^\tau\mathbf{Y}\mathbf{x}}{\mathbf{x}^\tau\mathbf{D}\mathbf{x}} \leq 1, \quad \text{for all } \mathbf{x} \in \mathbb{R}^N, \mathbf{x} \neq \mathbf{0}.
\tag{13.16}
$$

---

[1] If the eigenvalues of $\mathbf{L}^{sym} = \mathbf{I}_N - \widetilde{\mathbf{L}}^{sym}$ are $\{\lambda_i\}$, then the eigenvalues of $\widetilde{\mathbf{L}}^{sym}$ are $\{1 - \lambda_i\}$. Because 0 is always an eigenvalue of $\widetilde{\mathbf{L}}^{sym}$, then 1 is always an eigenvalue of $\mathbf{L}^{sym} = \mathbf{I}_N - \widetilde{\mathbf{L}}^{sym}$.

[2] The ratio $\mathbf{y}^\tau\mathbf{M}\mathbf{y}/\mathbf{y}^\tau\mathbf{y}$ is usually referred to as the *Raleigh quotient* of a vector $\mathbf{y}$ wrt a matrix $\mathbf{M}$ and is defined for any vector $\mathbf{y}$ (not just eigenvectors) and matrix $\mathbf{M}$. A generalized version of this ratio is given by $\mathbf{y}^\tau\mathbf{M}\mathbf{y}/\mathbf{y}^\tau\mathbf{N}\mathbf{y}$ for a conformable matrix $\mathbf{N}$.

Because $\mathbf{D}$ is positive-definite, (13.16) will be true iff

$$-\mathbf{x}^\tau \mathbf{D}\mathbf{x} \le \mathbf{x}^\tau \mathbf{Y}\mathbf{x} \le \mathbf{x}^\tau \mathbf{D}\mathbf{x}, \quad \text{for all } \mathbf{x} \in \mathbb{R}^N, \ \mathbf{x} \ne \mathbf{0}, \tag{13.17}$$

or iff

$$\mathbf{x}^\tau (\mathbf{D} + \mathbf{Y})\mathbf{x} \ge 0 \ \text{ and } \ \mathbf{x}^\tau (\mathbf{D} - \mathbf{Y})\mathbf{x} \ge 0, \quad \text{for all } \mathbf{x} \in \mathbb{R}^N, \ \mathbf{x} \ne \mathbf{0}. \tag{13.18}$$

But these last conditions follow because, by expanding each expression, substituting for D_{ii} and D_{jj}, and completing the square, we see that

$$\mathbf{x}^\tau (\mathbf{D} + \mathbf{Y})\mathbf{x} = \frac{1}{2} \sum_{i=1}^{N} \sum_{j=1}^{N} Y_{ij}(x_i + x_j)^2 \ge 0, \tag{13.19}$$

$$\mathbf{x}^\tau (\mathbf{D} - \mathbf{Y})\mathbf{x} = \frac{1}{2} \sum_{i=1}^{N} \sum_{j=1}^{N} Y_{ij}(x_i - x_j)^2 \ge 0. \tag{13.20}$$

See **Exercise 13.4**. Thus, (13.15) follows, and all eigenvalues of $\mathbf{L}^{sym}$ lie in $[-1, 1]$. It follows from this result that all eigenvalues of $\mathbf{L}$ lie in $[0, 2]$.

The normalized graph Laplacian $\mathbf{L}^{sym}$ has been used to define algorithms for nonlinear manifold learning methods such as Laplacian eigenmaps and diffusion maps (see, e.g., Izenman, 2012a,b).

13.3 Spectral Clustering of Networks

Spectral clustering of networks is an outgrowth of the interplay between *spectral graph theory*, which looks at how the eigenvalues of a matrix representation of a graph relate to the structure of that graph, and the partitioning, cutting, and clustering of graphs into communities. Spectral clustering is a nonparametric algorithm that has been described as "a convex relaxation of the Normalized Cut optimization problem" (Rohe, Chatterjee, and Yu, 2011).[3]

There is a huge literature on spectral graph theory (see, e.g., Chung, 1997), having either an algebraic flavor or, more recently, a geometric flavor (emphasizing the relationship between spectral graph theory and differential geometry). The spectral approach to graphs, once the province of those in physics and chemistry, has now caught the attention of those in the mathematics, probability, statistics, and computer science disciplines, not to mention those in biology, engineering, and the social sciences, and has helped push the development of graph theory towards new directions.

The essential features of spectral clustering (although not under that name) were introduced in articles by Donath and Hoffman (1973), Fiedler (1973), and Anderson and Morley (1985),[4]. Donath and Hoffman considered the *graph partitioning problem* in which the N nodes of an undirected graph $\mathcal{G} = (\mathcal{V}, \mathcal{E})$ are partitioned (in some optimal manner) into a given number, K, of non-empty, disjoint subsets, $\mathcal{V}_1, \ldots, \mathcal{V}_K$, where $\mathcal{V}_i \cap \mathcal{V}_j = \emptyset$ and $\cup_{k=1}^{K} \mathcal{V}_k = \mathcal{V}$. They focused on the case of $K = 2$ subsets and used the eigenvalues and eigenvectors of $\mathbf{D} - \mathbf{Y}$ to find graph partitions. Anderson and Morley had previously called $\mathbf{D} - \mathbf{Y}$ the *Laplacian matrix* of a graph, because

---

[3]See Section 11.2.5 for a discussion of the normalized-cut method of graph partitioning.

[4]This article was widely circulated in October 1971 as a technical report and published without change in 1985.

it arises as a discrete analogue of the Laplacian–Beltrami operator of Riemannian manifolds. At the same time, Fiedler realized that the eigenvector corresponding to the second smallest eigenvalue (the smallest is zero) of the Laplacian matrix $\mathbf{D} - \mathbf{Y}$ could be associated with a graph's connectivity.

Much of the early research in this area consisted of eigenanalyses together with linear and convex programming techniques, which proved to be impractical for large problems, and the development of upper and lower bounds on the second smallest and largest eigenvalues of the Laplacian matrix $\mathbf{D} - \mathbf{Y}$. Spectral clustering of graphs and networks has proved to be successful because it is easy to implement, is reasonably fast and efficient, especially when applied to large, sparse datasets, and does not make many assumptions for deriving properties of the network clustering algorithms. However, many theoretical issues are not understood well.

We will need the following `kmeans` algorithm for the spectral clustering algorithms.

13.3.1 The `kmeans` Algorithm

Given a set of points $\mathbf{x}_1, \ldots, \mathbf{x}_N \in \mathbb{R}^p$, the `kmeans` algorithm (MacQueen, 1967) seeks to assign each point to one and only one of K non-overlapping clusters by minimizing the error sum of squares,

$$\text{ESS} = \sum_{k=1}^{K} \sum_{\mathbf{x}_i \in C_k} \|\mathbf{x}_i - \mathbf{m}_k\|_2^2, \tag{13.21}$$

of each point from its current cluster centroid, where, for $\mathbf{y} \in \mathbb{R}^p$, $\|\mathbf{y}\|_2^2 = \mathbf{y}^\tau \mathbf{y}$ is the squared Euclidean distance, $\mathbf{m}_k$ is the kth cluster centroid, and C_k is the cluster containing $\mathbf{x}_i$. The algorithm works in an iterative fashion in which each point $\mathbf{x}_i$ is reassigned to its nearest cluster centroid so that at each iteration, ESS is reduced in magnitude, and the cluster centroids are updated after each reassignment until convergence (see, e.g., Izenman, 2013, Section 12.4). It is known that the `kmeans` algorithm has to be restarted many times using different random initial values in order to attain a solution close to the global, rather than a local, minimum of ESS; furthermore, it tends to fail when clusters do not form convex-shaped regions.

13.3.2 Spectral Clustering Algorithms

Algorithm 13.1 gives the algorithm for *unnormalized spectral clustering* based upon the unnormalized graph Laplacian.

Algorithms 13.2a and **13.2b** show the two algorithms for *normalized spectral clustering* corresponding to the two versions of the normalized graph Laplacian.

A variant of normalized spectral clustering using $\mathbf{L}^{sym}$ (**Algorithms 13.2b**) that has been proposed is the following substitution for step 2:

2′ Compute the *K largest (ordered by absolute value)* eigenvalues and associated orthogonal eigenvectors, $\mathbf{u}_1, \ldots, \mathbf{u}_K \in \mathbb{R}^N$, of $\mathbf{L}^{sym}$. The absolute eigenvalues of $\mathbf{L}^{sym}$ are all in $[0, 1]$.

See, for example, Rohe, Chatterjee, and Yu (2011) and Amini et al. (2013).

Algorithm 13.1 *Unnormalized spectral clustering using* **L**

---

1. Input: N nodes, K groups, adjacency matrix **Y**.
2. Compute the K *smallest* eigenvalues and their associated orthogonal eigenvectors, $\mathbf{u}_1, \dots, \mathbf{u}_K \in \mathbb{R}^K$, of **L**.
3. Form the $(N \times K)$-matrix $\mathbf{U} = (\mathbf{u}_1, \dots, \mathbf{u}_K)$.
4. Cluster the rows of **U** as N points in $\mathbb{R}^K$ using the kmeans algorithm.
5. Output: Clusters $B_1, B_2, \dots, B_K$. Assign node v_i to cluster B_a iff the ith row of the matrix **U** is assigned to cluster B_a.

Algorithm 13.2a *Normalized spectral clustering using* $\mathbf{L}^{rw}$ *(Shi and Malik, 2000)*

---

1. Input: N nodes, K groups, adjacency matrix **Y**.
2. Compute the K *largest* eigenvalues and associated generalized eigenvectors, $\mathbf{u}_1, \dots, \mathbf{u}_K \in \mathbb{R}^N$, of the generalized eigenproblem $\mathbf{L}^{rw}\mathbf{u} = \lambda \mathbf{D}\mathbf{u}$.
3. Form the $(N \times K)$-matrix $\mathbf{U} = (\mathbf{u}_1, \dots, \mathbf{u}_K)$.
4. Cluster the rows of **U** as N points in $\mathbb{R}^K$ using the kmeans algorithm.
5. Output: Clusters $B_1, B_2, \dots, B_K$. Assign node v_i to cluster B_a iff the ith row of the matrix **U** is assigned to cluster B_a.

13.3.3 Normalized or Unnormalized Laplacian?

There have been a number of articles that use the normalized Laplacian (e.g., Shi and Malik, 2000; Ng, Jordan, and Weiss, 2002) and other articles that use the unnormalized version (e.g., Barnard, Pothen, and Simon, 1995; Gauttery and Miller, 1998). An important question of interest is whether we should prefer unnormalized or normalized spectral clustering, and if the normalized is preferred, which of its two versions is more appropriate?

A first step in this direction is to check the degree distribution of the graph, namely, the diagonal entries of the matrix **D**. If most nodes have approximately the same

---

Algorithm 13.2b *Normalized spectral clustering using* $\mathbf{L}^{sym}$ *(Ng, Jordan, and Weiss, 2002)*

---

1. Input: N nodes, K groups, adjacency matrix **Y**.
2. Compute the K *largest* eigenvalues and associated orthogonal eigenvectors, $\mathbf{u}_1, \dots, \mathbf{u}_K \in \mathbb{R}^N$, of $\mathbf{L}^{sym}$.
3. Form the $(N \times K)$-matrix $\mathbf{U} = (\mathbf{u}_1, \dots, \mathbf{u}_K)$, $\mathbf{u}_j = (u_{1j}, \dots, u_{Nj})^{\tau}$.
4. Normalize the rows to unit norm to form the matrix $\mathbf{T} = (t_{ij})$, $t_{ij} = u_{ij}/(\sum_k u_{ik}^2)^{1/2}$.
5. Cluster the rows of **T** as N points in $\mathbb{R}^K$ using the kmeans algorithm.
6. Output: Clusters $B_1, B_2, \dots, B_K$. Assign node v_i to cluster B_a iff the ith row of the matrix **T** is assigned to cluster B_a.

---

degree, then the three Laplacians will look alike, and will prove to be indistinguishable with regards to clustering.

If, on the other hand, the diagonal entries of $\mathbf{D}$ are wildly different, then the Laplacians will be very different from each other. In this case, there are theoretical and experimental reasons (including applications to real network data) to prefer a normalized graph Laplacian (and, hence, normalized spectral clustering) to the unnormalized version for use in identifying blocks in stochastic blockmodels (see Section 12.4). In the normalized case, von Luxburg (2007) recommends using the eigenvectors of $\mathbf{L}^{rw}$ rather than those of $\mathbf{L}^{sym}$. Although such results were obtained for the two-block network, it is believed that they may hold more generally.

The normalized spectral clustering algorithms were also preferred when the statistical aspects of the normalized and unnormalized versions were compared in a more general context than graph partitioning. von Luxburg, Belkin, and Bousquet (2008) derived conditions needed to show consistency and rates of convergence of the eigenvectors of the unnormalized and normalized spectral clustering algorithms as the number of nodes increases:

- Under certain mild conditions, the first t eigenvectors of the *normalized* Laplacian converge (a.s.) to eigenfunctions of some limit operator. The conditions are that the eigenvalues, $\{\lambda_i\}$, of the limit operator have to be simple (i.e., each has multiplicity 1) and satisfy $\lambda_i \neq 1$, all i.
- For the *unnormalized* Laplacian, convergence is a more complicated matter: convergence only holds if the eigenvalues of the limit operator satisfy some specific conditions, which can fail to hold in theory and in many practical examples. If the convergence conditions are violated, the eigenvectors would provide no help in graph partitioning and would mislead if used; in such instances, unnormalized spectral clustering would completely fail.

For both cases, the conditions boil down to showing that the eigenvalues of the limit operator are isolated in the spectrum and satisfy an additional restriction as to their values or range of values.

13.3.4 Consistency of Spectral Partitioning

For networks generated by a stochastic blockmodel, Rohe, Chatterjee, and Yu (2011) considered the scenario in which the number of blocks is allowed to increase with the number of nodes. This condition was introduced so that the block sizes do not get too big. They showed that block estimation (of the unknown SBM parameters) based upon spectral partitioning of the normalized graph Laplacian yields an upper bound on the number of misclassified nodes. Because the partition of the N nodes into K blocks is unique only up to permutation of the K block labels, the authors provide a definition of "misclassified" nodes. Unfortunately, as the authors note, the bound cannot be used to show that the proportion of misclassified nodes goes to zero (almost surely) as the number of nodes tends to infinity, and so their theoretical results are "not sharp enough to prove consistency." Their simulations, however, show some evidence of consistency.

This scenario, which is analogous to high-dimensional clustering, was extended by Sussman et al. (2012), who used a least-squares error criterion (rather than using the kmeans clustering algorithm) for clustering the nodes into K blocks. They proved that an SBM estimator based upon spectral partitioning of the *adjacency matrix* yields consistent block estimation.

Both Rohe et al. and Sussman et al. assume that K and rank$(\mathbf{P})$ are known, where $\mathbf{P} = (p_{ij}) \in [0,1]^{K \times K}$ is the $(K \times K)$ connection probability matrix (see Section 12.4.1). When K and rank$(\mathbf{P})$ are unknown, and only an upper bound is known for rank$(\mathbf{P})$, Fishkind et al. (2012) show that block estimation is again consistent. So, it does not matter whether rank$(\mathbf{P})$ is known or overestimated, the consistency results still hold. The authors also provide a consistent estimator for K and an estimator $\widehat{\mathcal{Q}}$ for the block assignment function $\mathcal{Q}$ (see Section 12.6) that minimizes the number of misclassified nodes.

Empirical evidence indicates that these spectral partitioning algorithms are computationally fast and efficient. Fishkind et al. found no discernible difference when carrying out spectral partitioning either on the normalized Laplacian or on the adjacency matrix. Sussman et al., however, showed that the adjacency matrix provided improvements in performance over the normalized Laplacian in some situations. More research work needs to be done on this issue.

13.3.5 Example: Political Blogs

The political blogs network, compiled by Adamic and Glance (2005), has been studied to assess the role of normalization in spectral clustering. The nodes are posts of blogs whose authors wrote about U.S. politics over the two months prior to the 2004 U.S. Presidential Election, and each node has been labeled as either "liberal" (0) or "conservative" (1). The edges are hyperlinks from one blog to another, and the actual content of the blogs is ignored. The network originally contained 1494 blogs, 759 liberal and 735 conservative.

Most researchers prefer to work with a subset of these data by removing nodes with no edges (degree-0 nodes), one edge (degree-1 nodes), and two blogs that were linked together with no connection to the largest component.[5] Attention is focused on the largest connected component, which has 1222 nodes (586 liberal blogs and 636 conservative blogs) and 19,024 directed edges; see **Figure 13.2**. Only nodes with degree at least one are displayed. The network graph shows two dominant clusters that appear to be evenly balanced and reasonably separated from each other.

The node-degree distribution (see **Figure 13.3**) is heavily skewed to the right, suggesting a power-law degree distribution with exponential cutoff, with average node degree 27.36, median 13, and maximum 351.

Misclassification rates have been computed using spectral clustering for the normalized and unnormalized Laplacians for the full and subset data. Sarkar and Bickel (2013) showed that for the full data (1494 nodes), the unnormalized Laplacian had an error rate of 40% and the normalized Laplacian 50%. For the subset data (1222 nodes), the unnormalized Laplacian yielded an error rate of 37%, whilst the normalized Laplacian yielded an error rate of 4%. According to Cai, Ackerman, and Freer (2015), who studied the robustness properties of community detection procedures, these results show that nodes that have very few (or no) connections to

---

[5]The 2004 U.S. political blogs data as compiled by Adamic and Glance (2005) can be found at the website www-personal.umich.edu/~mejn/netdata.

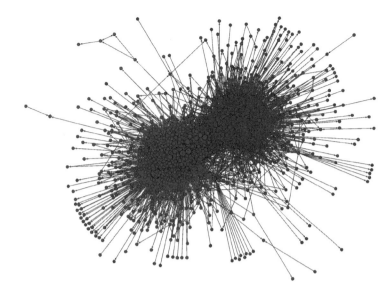

Figure 13.2 The political blog network prior to the 2004 U.S. Presidential Election. There are 1222 blogs consisting of 586 liberal-leaning blogs (blue nodes) and 636 conservative-leaning blogs (red nodes). Only nodes with degree at least one are displayed; singleton nodes (nodes with degree 0) have been removed from the network graph.

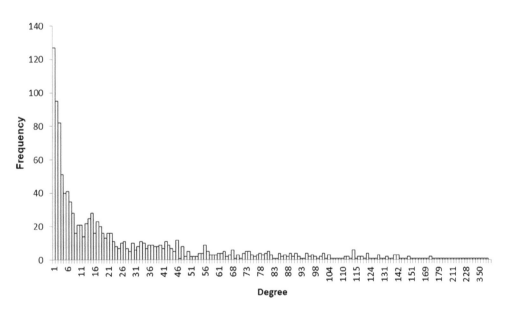

Figure 13.3 The degree distribution of the political blog network.

other nodes can significantly affect the performance of spectral clustering, and hence may be viewed as "outliers."

13.4 Regularized Spectral Clustering

Spectral clustering tends to perform very poorly for networks with severe node-degree heterogeneity (e.g., the presence of "hubs," nodes with ultra-high degrees) or for sparse networks with many isolated components. Spectral clustering uses the eigenvectors corresponding to the largest eigenvalues of the graph Laplacian to separate the network into clusters or blocks of nodes. The presence of low-degree nodes (e.g., isolated nodes) or ultra-high-degree nodes (e.g., hubs) stifles this ability to cluster the nodes successfully. Consequently, there has been interest in determining the impact of regularizing the algorithm for spectral clustering. Regularization is achieved by incorporating a regularization parameter σ into the graph Laplacian $\mathbf{L}^{sym}$.

As before, $\mathbf{Y}$ is the $(N \times N)$ adjacency matrix. Recall that the normalized symmetric graph Laplacian is given by

$$\mathbf{L}^{sym} = \mathbf{D}^{-1/2}\mathbf{Y}\mathbf{D}^{-1/2}, \tag{13.22}$$

where $\mathbf{D} = \mathrm{diag}(D_{ii})$, D_{ii} is the degree of node v_i. Let $\mathbf{1}_N$ denote the N-vector all of whose elements are 1s, and let

$$\mathbf{K} = \frac{1}{N}\mathbf{J}_N, \tag{13.23}$$

where $\mathbf{J}_N = \mathbf{1}_N\mathbf{1}_N^{\tau}$ is a matrix of 1s and $\mathbf{A}^{\tau}$ represents the transpose of the matrix $\mathbf{A}$. Then every element of $\mathbf{K}$ is $1/N$. Define $\mathbf{Y}_{\sigma}$ as

$$\mathbf{Y}_{\sigma} = \mathbf{Y} + \sigma\mathbf{K}, \tag{13.24}$$

where $\sigma > 0$ is a regularization parameter. This adjustment to the adjacency matrix $\mathbf{Y}$ adds a constant matrix of small values (i.e., σ/N) to every 1 and every 0 in $\mathbf{Y}$, which then connects – through artificially constructed "weak" links – all nodes to each other and every isolated component to the main community.

> **Definition 13.3** The regularized version of the normalized symmetric graph Laplacian (13.22) is given by
>
> $$\mathbf{L}_{\sigma}^{sym} = \mathbf{D}_{\sigma}^{-1/2}\mathbf{Y}_{\sigma}\mathbf{D}_{\sigma}^{-1/2}, \tag{13.25}$$
>
> where $\mathbf{D}_{\sigma} = \mathrm{diag}(D_{ii,\sigma})$ is the diagonal matrix whose diagonal elements are the "degrees" of the matrix $\mathbf{Y}_{\sigma}$, and $D_{ii,\sigma} = D_{ii} + \sigma$, $i = 1, 2, \ldots, N$.

The regularized spectral clustering (RSC(σ)) algorithm is given as **Algorithm 13.3**. Clearly, if $\sigma = 0$, the RSC algorithm reverts to **Algorithm 13.2b**.

13.4.1 RSC Under the Stochastic Blockmodel

The effect of regularizing a spectral clustering algorithm has been studied under the assumption that the network is randomly generated from either the stochastic blockmodel (SBM) or the degree-corrected stochastic blockmodel (DC-SBM), where the number, K, of well-defined communities is known.

Algorithm 13.3 *Regularized spectral clustering (RSC(σ)) using* $\mathbf{L}_\sigma^{sym}$ (Amini et al., 2013; Qin and Rohe, 2013; Joseph and Yu, 2016)

---

1. Input: N nodes, K groups, adjacency matrix $\mathbf{Y}$, regularization parameter σ.
2. Compute the regularized graph Laplacian $\mathbf{L}_\sigma^{sym}$.
3. Compute the K *largest* eigenvalues and associated orthogonal eigenvectors, $\mathbf{u}_{1,\sigma}, \ldots, \mathbf{u}_{K,\sigma} \in \mathbb{R}^N$, of $\mathbf{L}_\sigma^{sym}$.
4. Form the $(N \times K)$-matrix of eigenvectors $\mathbf{U}_\sigma = (\mathbf{u}_{1,\sigma}, \ldots, \mathbf{u}_{K,\sigma})$, where $\mathbf{u}_{j,\sigma} = (u_{1j,\sigma}, \ldots, u_{Nj,\sigma})^\tau$.
5. Normalize the rows to unit norm to form the $(N \times K)$-matrix $\mathbf{T}_\sigma = (t_{ij,\sigma})$, where $t_{ij,\sigma} = u_{ij,\sigma}/(\sum_k u_{ik,\sigma}^2)^{1/2}$.
6. Consider the rows of $\mathbf{T}_\sigma$ as N points in $\mathbb{R}^K$, and assign them into K clusters using the kmeans algorithm.
7. Output: Clusters $B_1, B_2, \ldots, B_K$. Assign node v_i to cluster B_a iff the ith row of the matrix $\mathbf{T}_\sigma$ is assigned to cluster B_a.

We start by defining $\mathcal{Y} = \mathrm{E}\{\mathbf{Y}\}$. Next, we set up population versions:

$\mathcal{D} = \mathrm{diag}(\mathcal{D}_{ii})$, where $\mathcal{D}_{ii} = \sum_j \mathcal{Y}_{ij}$ is the degree of the ith node v_i,

$\mathcal{D}_\sigma = \mathcal{D} + \sigma \mathbf{I}_N$,

$\mathcal{L}^{sym} = \mathcal{D}^{-1/2} \mathcal{Y} \mathcal{D}^{-1/2}$,

$\mathcal{L}_\sigma^{sym} = \mathcal{D}_\sigma^{-1/2} \mathcal{Y}_\sigma \mathcal{D}_\sigma^{-1/2}$, where $\mathcal{Y}_\sigma = \mathcal{Y} + \sigma \mathbf{J}_N$,

$\mathcal{L}_{deg,\sigma}^{sym} = \mathcal{D}_\sigma^{-1/2} \mathcal{Y} \mathcal{D}_\sigma^{-1/2}$,

of the network matrices, $\mathbf{D}$, $\mathbf{D}_\sigma$, $\mathbf{L}^{sym}$, $\mathbf{L}_\sigma^{sym}$, and $\mathbf{L}_{deg,\sigma}^{sym}$, respectively, for the regularized spectral clustering (RSC) model.

Qin and Rohe (2013) showed that under a K-block DC-SBM, and with a condition on the minimum expected degree, these population versions, with a suitable choice of σ, should determine a perfect reconstruction of the true partition of the network. The condition involving the minimum expected degree, however, was regarded by Joseph and Yu (2016) as "highly restrictive" for the clustering process, and was replaced instead by a condition on the maximum expected degree. For a two-block SBM, Joseph and Yu show that recovering the population clusters requires that the maximum degree of the network should be "large" in the sense that it should grow faster than $\log(N)$ as N increases.

For an SBM with K blocks, Joseph and Yu define the "clustering error" $\widehat{f}$ of RSC and show that it is dependent upon the square of the ratio

$$\frac{\| \mathbf{L}_\sigma^{sym} - \mathcal{L}_\sigma^{sym} \|}{\mu_{K,\sigma}}, \tag{13.26}$$

where the norm is the spectral norm[6] and $\mu_{K,\sigma}$ is the Kth largest eigenvalue of $\mathcal{L}_\sigma^{sym}$. The smaller the ratio (13.26), the more accurately the top K eigenvectors can be estimated (Davis and Kahan, 1970). As we noted above, the *eigengap* is the smallest

---

[6]The spectral norm of a square matrix $\mathbf{A}$ is defined as $\sqrt{\lambda_1(\mathbf{A}\mathbf{A}^\tau)}$, where $\lambda_1(\mathbf{A}\mathbf{A}^\tau)$ is the largest eigenvalue of $\mathbf{A}\mathbf{A}^\tau$.

nonzero eigenvalue of $\mathcal{L}_\sigma^{sym}$. In this case, because the graph Laplacian $\mathcal{L}_\sigma^{sym}$ has rank K, it follows that $\mu_{K+1,\sigma} = \cdots = \mu_{N,\sigma} = 0$, and so $\mu_{K,\sigma}$ is the eigengap, which should be quite large.

An optimal value of σ can be obtained by minimizing (13.26) wrt σ. However, $\mathcal{L}_\sigma^{sym}$ and $\mu_{K,\sigma}$ are unknown, and so this is not a feasible approach. Instead, large-σ results are used by Joseph and Yu to derive a high-probability bound on the numerator of (13.26), which they show depends upon $1/\sigma$ for large σ; similarly, they show that the denominator also behaves like $1/\sigma$ for large σ. The ratio (13.26), therefore, converges for large values of σ. They also derived a bound on the limiting value.

These results provide motivation for studying a large-σ version of RSC(σ). Joseph and Yu show that (13.26) is bounded above by a quantity they call $\delta_{\sigma,N}$. Let $\delta_N = \lim_{\sigma\to\infty} \delta_{\sigma,N}$. For a K-block SBM, under the following conditions,

(1) $\delta_N \to 0$ as $N \to \infty$,

(2) $\{\sigma_N\}$ is a sequence of regularity parameters satisfying a certain condition,

(3) cluster sizes are of the same order of magnitude,

(4) the block probability matrix $\mathbf{P}$ has a certain structure,

they show that the sequence $\{\text{RSC}(\sigma_N)\}$ yields consistent cluster estimates. Even though these results are for large σ, the authors report that there does not appear to be much change in algorithmic performance for σ larger than a certain value.

13.4.2 Choice of Regularization Parameter

Empirical evidence (Amini et al., 2013) suggests that large values of σ in the RSC algorithm tend to yield more correct clusters than do the unregularized spectral clustering algorithms, which generally perform poorly. Other empirical work (Joseph and Yu, 2016) indicates that intermediate values of σ may produce even better clustering performance.

One way of selecting σ is to use the Newman–Girvan modularity. See Section 12.5. For a specified value of σ taken from a grid of such values, the Newman–Girvan modularity is computed for the clusters obtained from the RSC algorithm. Then, the σ that maximizes the maximum modularity value is the chosen value.

An alternative regularization of spectral clustering was proposed by Chaudhuri, Chung, and Tsiatas (2012), in which they substitute $\mathbf{Y}$ for $\mathbf{Y}_\sigma$, so that (13.25) is replaced by

$$\mathbf{L}_{deg,\sigma}^{sym} = \mathbf{D}_\sigma^{-1/2}\mathbf{Y}\mathbf{D}_\sigma^{-1/2}, \tag{13.27}$$

where *deg* is used to show that the Laplacian contains a modified version of the degree matrix $\mathbf{D}$. Good empirical results can be obtained from this type of regularization if σ is chosen to be proportional to the average node degree (i.e., $\sigma \propto d/N$, where $d = \sum_i D_{ii}$ is the sum of the node degrees); too small a value of σ, it is argued, would lead to "insufficient regularization" to satisfy certain conditions, and too large a value would "wash out significant eigenvalues" (Qin and Rohe, 2013).

13.4.3 Example: Political Blogs (continued)

Spectral clustering works if the top eigenvectors of the graph Laplacian are able to distinguish between the clusters. For example, consider the political blogs dataset described in Section 13.3.5. The RSC(0) algorithm applied to the political blogs data assigned 1144 of the 1222 nodes into the same cluster, clearly missing the whole point of the exercise. The RSC(σ) algorithm, with $\sigma \in [1, 30]$, only misclassified about 80 of the 1222 nodes (an error rate of 6.5%), and the results were insensitive to the value of σ (Qin, 2015).

Furthermore, in networks with a few weak "outlier" nodes, unregularized spectral clustering may not work well. Accordingly, it was found that for the network of political blogs, the first eigenvector of the normalized symmetric graph Laplacian in both unregularized (13.22) and regularized (13.25) cases was not able to discriminate at all between the two types of blogs (Joseph and Yu, 2016). In the regularized case, however, the second eigenvector discriminates well between the clusters. In the unregularized case, the second eigenvector still does not discriminate, but the third does. See **Exercises 13.10** and **13.11**. Thus, in the presence of outlier nodes, looking at the first two eigenvectors of the unregularized graph Laplacian can yield poor (if not bad) clustering performance.

13.5 Further Reading

Section 13.2. An excellent monograph on spectral graph theory is Chung (1997). For a survey of applications of spectral graph theory in computer science, see Arsić et al. (2012).

Section 13.2.2. For a discussion of the load-balancing problem, see van Driessche and Roose (1995), Elsässer, Královič and Monien (2003), and Cvetković and Davidović (2009).

Section 13.3. Chapters or sections on spectral clustering in books include Kolaczyk (2009, Section 4.3.3) and Newman (2010, Chapter 11). There is a good tutorial on spectral clustering by von Luxburg (2007). For a review of spectral clustering, see Spielman and Teng (1996).

Sections 13.3.5 and 13.4.3. For analyses of the political blog network, see Amini et al. (2013) and Sarkar and Bickel (2013).

Section 13.4 For a discussion of regularized spectral clustering, see Amini et al. (2013), Qin and Rohe (2013), and Joseph and Yu (2016).

13.6 Exercises

Exercise 13.1 Let $\mathbf{Y} = (Y_{ij})$ denote the adjacency matrix of the graph $\mathcal{G}$, and let $\mathbf{D} = (D_{ij})$ be a diagonal matrix where $D_{ii} = \sum_j Y_{ij}$. Show that the unnormalized graph Laplacian matrix $\mathbf{L} = \mathbf{D} - \mathbf{Y}$ satisfies the following properties:

(a) For every vector $\mathbf{x} \in \mathbb{R}^N$, $\mathbf{x}^\tau \mathbf{L} \mathbf{x} = \frac{1}{2} \sum_{i,j=1}^N Y_{ij}(x_i - x_j)^2$.
(b) $\mathbf{L}$ is symmetric and non-negative-definite.
(c) The smallest eigenvalue of $\mathbf{L}$ is 0 with associated eigenvector $\mathbf{1}_N$, a vector of 1s.

(d) $\mathbf{L}$ has N non-negative, real-valued eigenvalues $0 = \lambda_1 \leq \lambda_2 \leq \cdots \leq \lambda_N$.

Exercise 13.2 Show the following properties of the normalized graph Laplacian matrices $\mathbf{L}_1$ and $\mathbf{L}_2$:

(a) λ is an eigenvalue of $\mathbf{L}_1$ with associated eigenvector $\mathbf{u}$ iff λ is an eigenvalue of $\mathbf{L}_2$ with associated eigenvector $\mathbf{w} = \mathbf{D}^{1/2}\mathbf{u}$.

(b) λ is an eigenvalue of $\mathbf{L}_1$ with associated eigenvector $\mathbf{u}$ iff λ and $\mathbf{u}$ solve the generalized eigenvalue problem $\mathbf{Lu} = \lambda\mathbf{Du}$.

(c) 0 is an eigenvalue of $\mathbf{L}_1$ with associated eigenvector $\mathbf{1}_N$.

(d) 0 is an eigenvalue of $\mathbf{L}_2$ with associated eigenvector $\mathbf{D}^{1/2}\mathbf{1}_N$.

(e) $\mathbf{L}_1$ and $\mathbf{L}_2$ are both non-negative-definite and have non-negative, real-valued eigenvalues $0 = \lambda_1 \leq \lambda_2 \leq \cdots \leq \lambda_N$.

Exercise 13.3 From the worked example in Section 13.2.1, find the eigenvalues and eigenvectors of the matrices $\mathbf{L}^{rw}$ and $\mathbf{L}^{sym}$.

Exercise 13.4 From the worked example in Section 13.2.2, find the eigenvalues and eigenvectors of $\widetilde{\mathbf{L}}^{rw}$ and $\widetilde{\mathbf{L}}^{sym}$.

Exercise 13.5 In the case of a fully connected, undirected graph of N nodes (i.e., every node is connected to every other node), write out the form of the Laplacian matrix $\mathbf{L}$ and show that its smallest eigenvalue is 0 with the remainding $N - 1$ eigenvalues each equal to N.

Exercise 13.6 In the case of a star-shaped, undirected graph (i.e., the N nodes are arranged such that a center node is connected to each of the $N - 1$ surrounding nodes, but no other edges between nodes), write out the form of the Laplacian matrix $\mathbf{L}$ and show that its smallest eigenvalue is 0, the next $N - 2$ (i.e., the second smallest to the second largest) eigenvalues are each equal to 1, and the largest eigenvalue is equal to N.

Exercise 13.7 Let $\mathbf{B} = (B_{ij})$, where $B_{ij} = Y_{ij} - (k_i k_j)/L$. Show that $\sum_i B_{ij} = \sum_j B_{ij} = 0$.

Exercise 13.8 Maximize $\mathbf{s}^\tau \mathbf{Bs}$ subject to the constraints $\mathbf{s}^\tau \mathbf{s} = N$ and $\mathbf{1}^\tau \mathbf{s} = N_1 - N_2$, where N_i is the number of nodes that belong in the ith community, $i = 1, 2$.

Exercise 13.9 Show (13.19) and (13.20).

Exercise 13.10 Consider the political blogs network data. Compute the normalized symmetric graph Laplacian of those data and find its eigenvalues and eigenvectors. Plot the first eigenvector against the second eigenvector. Show that neither the first nor the second eigenvectors can discriminate between the two types of blogs. Plot each of the first three eigenvectors against the node indices, and show again that the first and second eigenvectors cannot discriminate between the two types of blogs, but that third eigenvector can discriminate.

Exercise 13.11 For the political blogs network data, compute the regularized version, $RSC(\sigma)$, of the normalized symmetric graph Laplacian. Take (a) $\sigma = 26.5$, (b) $\sigma =$

N. For each case, plot the first eigenvector against the second eigenvector. Also, plot each eigenvector against its node indices. Show that the first eigenvector does not discriminate between the two kinds of blog, but that the second eigenvector can discriminate between them.

Exercise 13.12 Using the adjacency matrix of the graph $\mathcal{G}$ in Exercise 12.8:

(a) Compute the graph Laplacian.
(b) Find the second smallest eigenvalue (Fiedler's *algebraic connectivity* of $\mathcal{G}$) of the graph Laplacian $\mathbf{L}$ and its associated eigenvector.
(c) Partition the graph into two equal-sized communities using methods of spectral clustering.

Exercise 13.13 Let $\mathcal{G}$ be a graph. Show that there is an eigenvalue of the unnormalized graph Laplacian $\mathbf{L}$ that is zero.

Exercise 13.14 Let $\mathcal{G}$ be a graph. Show that the number of connected components of $\mathcal{G}$ is the same as the number of orthonormal eigenvectors associated with the eigenvalue zero of $\mathbf{L}$.

Exercise 13.15 Let $\mathcal{G}$ be an undirected graph having N nodes. Suppose there are k connected components in $\mathcal{G}$. Show that $\text{rank}\{\mathbf{L}\} = N - k$.

CHAPTER FOURTEEN

Graph Partitioning: IV. Overlapping Communities

So far, we have assumed that if there are several communities in a network, then those communities are distinct and non-overlapping. In this chapter, we discuss situations in which communities overlap with each other. We describe a number of algorithms for modeling overlapping communities, such as mixed-membership SBMs, link-based clustering, overlapping SBMs, the community-affiliation graph model, and the latent cluster random-effects model.

14.1 Introduction

We have described several tools for partitioning a network or graph, using various "cutting" tools such as minimum cuts, ratio cuts, normalized cuts, and multiway cuts, where each of those tools was designed to divide a graph into non-overlapping communities. We have also described stochastic blockodels and their variations, and the modularity approach, each of which again assumes disjoint communities. In many diverse circumstances, however, overlapping communities actually do occur and should be viewed as worth studying. This implies that nodes can each be members of more than just a single community. In this chapter, we discuss the problems involved in modeling overlapping communities.

14.1.1 Examples

Some prominent examples of networks with overlapping communities are:

- In a *social network*, people tend to belong to many different kinds of groups (e.g., family, school, clubs, friends, professional, sports, charity organizations, Facebook, LinkedIn), not just to one; they include *transaction networks* involving "senders," who post comments, or "recipients," who reply to comments, on a wide range of topics on social news websites (see, e.g., Shafiei and Chipman, 2010).
- In a *document network*, language is heterogeneous and documents (nodes) may involve multiple themes or topics (e.g., news items, business, health, fashion, sports), where each theme is a collection of words from a vocabulary and the links (edges) between documents are hyperlinks or citations (Blei, Ng, and Jordan, 2003). This example is closely related to probabilistic latent semantic analysis (PLSA) and latent Dirichlet allocation (LDA).

- In a *molecular-biological network* of protein–protein interactions, many proteins take part in multiple functions or processes at different times or biological conditions (Airoldi et al., 2014).
- *Ecological food webs*, such as those in the Florida Bay, heavily overlap while forming a dominant core, due to the closed nature of the ecosystem (Ulanowicz, Bondavilli, and Egnotovich, 1998).
- *Biochemical pathways in metabolic networks* play various roles in metabolic cycles (Palla et al., 2005; Lancichinetti, Fortunato, and Kertész, 2009).
- In studies of the *structural connectivity networks of the human brain* (the so-called "connectomes"), the evidence suggests that nodes (i.e, regions of interest) are each affiliated with multiple groups or clusters (Moyer et al., 2015).
- In a huge *network of scientific articles* from the arXiv preprint server, where articles cite or are cited by other articles, hundreds of overlapping communities have been found amongst the citations (Gopalan and Blei, 2013).

The transition from non-overlapping communities to (possibly) overlapping communities is technically difficult. Although there has been much research on community detection when communities overlap, the vast majority of research has focused on constructing algorithms, whilst little work has been carried out on the development of theory.

14.2 Mixed-Membership SBMs

Assume we have a directed graph $\mathcal{G} = (\mathcal{V}, \mathcal{E})$, where $\mathcal{V} = \{v_i\}$ is the set of N nodes and $\mathcal{E} = \{(v_i, v_j) \in \mathcal{V} \times \mathcal{V}\}$, where $\mathcal{V} \times \mathcal{V}$ is the set of all possible edges, whether they exist or not. Let K be the number of communities (or latent groups) that are to be identified.

14.2.1 The Generative Process

The *mixed-membership stochastic blockmodel* (Airoldi et al., 2008) is an hierarchical Bayesian model that proposes that $\mathcal{G}$ is generated by the following sequential procedure:

1. For each node, draw a mixed-membership K-vector.

Let π_{ik} denote the probability that the ith node, v_i, belongs to the kth community, and let

$$\boldsymbol{\pi}_i = (\pi_{i1}, \pi_{i2}, \ldots, \pi_{iK})^{\tau} \tag{14.1}$$

be the K-vector of mixture probabilities for v_i, $i = 1, 2, \ldots, N$, where

$$\sum_{k=1}^{K} \pi_{ik} = 1. \tag{14.2}$$

For an SBM with non-overlapping communities, one entry in $\boldsymbol{\pi}_i$ would be 1 and all other entries would be 0. In the present formulation, we allow a more general assignment of probabilities so that each node can be a member of all K communities. Assume that $\boldsymbol{\pi}_i$ is randomly drawn from a Dirichlet distribution,

$$\boldsymbol{\pi}_i \sim \mathcal{D}irichlet(\boldsymbol{\alpha}), \quad i = 1, 2, \ldots, N, \tag{14.3}$$

with probability density given by

$$p(\boldsymbol{\pi}_i|\boldsymbol{\alpha}) = \frac{\Gamma\left(\sum_{k=1}^{K} \alpha_k\right)}{\prod_{k=1}^{K} \Gamma(\alpha_k)} \prod_{k=1}^{K} \pi_{ik}^{\alpha_k - 1}, \tag{14.4}$$

where $\boldsymbol{\alpha} = (\alpha_1, \alpha_2, \ldots, \alpha_K)^\tau$ is a K-vector of hyperparameters and $\Gamma(a)$ is the gamma function. The Dirichlet distribution, which is the multivariate version of the Beta distribution, can be expressed as a member of the exponential family and is conjugate to the multinomial distribution. The vector $\boldsymbol{\alpha}$ is referred to as the *mixed-membership vector*.

2. For each pair of nodes, draw an indicator vector for the initiator's community membership.
 The K-vector

$$\mathbf{Z}_{i \to j} \sim \mathcal{M}(1, \boldsymbol{\pi}_i) \tag{14.5}$$

is an indicator vector that is randomly drawn from the multinomial distribution, where $\boldsymbol{\pi}_i$ indicates the community membership of v_i (the *initiator*) to v_j according to the directed edge $v_i \to v_j$. The indicator vector $\mathbf{Z}_{i \to j}$ has a 1 in the position corresponding to the indicated community and zeroes in all other positions.

3. For each pair of nodes, draw an indicator vector for the receiver's community membership.
 Similarly, the K-vector

$$\mathbf{Z}_{i \leftarrow j} \sim \mathcal{M}(1, \boldsymbol{\pi}_j) \tag{14.6}$$

is an indicator vector that is randomly drawn from the multinomial distribution, where $\boldsymbol{\pi}_j$ indicates the community membership of v_j to v_i (the *receiver*) according to the directed edge $v_i \leftarrow v_j$. The indicator vector $\mathbf{Z}_{i \leftarrow j}$ has a 1 in the position corresponding to the indicated community and zeroes in all other positions.
 The multinomial distribution in (14.5) has probability density

$$p(\mathbf{z}|\boldsymbol{\pi}_i) = \frac{N!}{\prod_{k=1}^{K} z_{ik}!} \prod_{k=1}^{K} \pi_{ik}^{z_{ik}}, \tag{14.7}$$

and similarly for $p(\mathbf{z}|\boldsymbol{\pi}_j)$ in (14.6).

4. For each pair of nodes, sample their interaction.
 Let $\mathbf{Y} = (Y_{ij})$, denote the $(N \times N)$ adjacency matrix of $\mathcal{G}$, where $Y_{ij} \in \{0, 1\}$ gives the status of an edge (i.e., present or absent) between the pair of nodes v_i and v_j, $i, j = 1, 2, \ldots, N$. The Bernoulli probability of an edge from v_i to v_j is given by

$$P\{Y_{ij} = 1\} = 1 - P\{Y_{ij} = 0\} = p_{ij}, \tag{14.8}$$

which we can write as

$$Y_{ij} \sim \text{Bernoulli}(p_{ij}), \tag{14.9}$$

where the edge that links v_i to v_j exists with probability

$$p_{ij} = \mathbf{Z}_{i \to j}^\tau \mathbf{B} \mathbf{Z}_{i \leftarrow j}. \tag{14.10}$$

The probability p_{ij} is also referred to as a *compatibility function* over the communities (Fu, Song, and Xing, 2009). The matrix $\mathbf{B} = (B_{jk})$ is a $(K \times K)$-matrix of block-interaction probabilities, where B_{jk} represents the probability that a member of the jth community is linked to a member of the kth community. The matrix $\mathbf{B}$, therefore, plays the same role that the connection probabilities played in the $\mathbf{P}$-matrix for an SBM. If we assume that members of the same community are more likely to interact with each other, we take $\mathbf{B}$ to be a diagonal matrix. On the other hand, if we assume a more complex set of interactions between members of different communities, then $\mathbf{B}$ can have a much richer structure.

The indicator variables, $\mathbf{Z}_{i \rightarrow j}$ and $\mathbf{Z}_{i \leftarrow j}$, belong to the two sets, $\mathbf{Z}_{\rightarrow} = \{\mathbf{Z}_{i \rightarrow j}, i, j \in \mathcal{V}\}$ and $\mathbf{Z}_{\leftarrow} = \{\mathbf{Z}_{i \leftarrow j}, i, j \in \mathcal{V}\}$, respectively.

Note also that if the edges are undirected, we do not need to distinguish between "$\rightarrow$" and "$\leftarrow$."

14.2.2 Posterior Inference

The parameters are, therefore, a K-vector $\boldsymbol{\alpha}$ and a $(K \times K)$-matrix $\mathbf{B}$. The joint probability distribution of $\mathbf{Y}$ and the latent variables $\{\boldsymbol{\pi}_i\}$, $\mathbf{Z}_{\rightarrow}$, $\mathbf{Z}_{\leftarrow}$, given $\boldsymbol{\alpha}$ and $\mathbf{B}$, is given by

$$p(\mathbf{y}, \{\boldsymbol{\pi}_i\}, \mathbf{z}_{i \rightarrow j}, \mathbf{z}_{i \leftarrow j} | \boldsymbol{\alpha}, \mathbf{B}) =$$

$$\prod_{i \neq j} p(y_{ij} | \mathbf{z}_{i \rightarrow j}, \mathbf{z}_{i \leftarrow j}, \mathbf{B}) p(\mathbf{z}_{i \rightarrow j} | \boldsymbol{\pi}_i) p(\mathbf{z}_{i \leftarrow j} | \boldsymbol{\pi}_j) \prod_{i=1}^{N} p(\boldsymbol{\pi}_i | \boldsymbol{\alpha}). \qquad (14.11)$$

The goal, then, is to estimate the posterior distribution of the community labels so that the probability of the observed data $\mathbf{Y}$ is maximized. However, this is not feasible because the normalization constant, which is the marginal probability of the data,

$$p(\mathbf{y} | \Theta) = \int_{\boldsymbol{\pi}} \sum_{\mathbf{z}_{i \rightarrow j}, \mathbf{z}_{i \leftarrow j}} p(\mathbf{y}, \{\boldsymbol{\pi}_i\}, \mathbf{z}_{i \rightarrow j}, \mathbf{z}_{i \leftarrow j} | \Theta) \, d\boldsymbol{\pi}, \quad \Theta = (\boldsymbol{\alpha}, \mathbf{B}), \qquad (14.12)$$

is computationally intractable, involving a multidimensional integral and summations.

Several approximation methods (e.g., mean-field variational methods, expectation propagation, and MCMC) have been proposed in the statistical and machine-learning literature to overcome the difficulty of computing the normalization constant in various complex problems. However, some of them (e.g., MCMC) do not scale well to large problems and are much too slow to be practical in such situations.

For this particular problem, Airoldi et al. (2008) propose a variational EM approach. Here, $\mathbf{Y}$ is the adjacency matrix representing the data, $\mathbf{X} = (\boldsymbol{\pi}, \mathbf{Z})$ are the latent variables, where $\mathbf{Z} = (\mathbf{Z}_{\rightarrow}, \mathbf{Z}_{\leftarrow})$, and $\Theta = (\boldsymbol{\alpha}, \mathbf{B})$ are the parameters to be estimated. In their variational approach, there are two EM steps: a variational-expectation step (vE-step) and a maximization step (M-step).

vE-step. Set up a simple approximate variational distribution, $q_\Delta(\mathbf{x})$, say, over the latent variables $\mathbf{X}$, that depends upon a set of free variational parameters Δ, say. Then, fit those free parameters so that the variational distribution, $q_\Delta(\mathbf{x})$, closely approximates the true posterior distribution in terms of Kullback–Leibler divergence.

M-step. Because the fitted variational distribution is simpler than the true posterior, the next part of the EM process – maximization – can be carried out. Jensen's inequality is used to provide a lower bound on the log-likelihood, $\log p(\mathbf{y}|\Theta)$, as follows:

$$
\begin{aligned}
\log p(\mathbf{y}|\Theta) &= \log \int_{\mathcal{X}} p(\mathbf{y}, \mathbf{x}|\Theta) d\mathbf{x} \\
&= \log \int_{\mathcal{X}} q(\mathbf{x}) \frac{p(\mathbf{y}, \mathbf{x}|\Theta)}{q(\mathbf{x})} d\mathbf{x} \\
&\geq \log \int_{\mathcal{X}} q(\mathbf{x}) \log \left(\frac{p(\mathbf{y}, \mathbf{x}|\Theta)}{q(\mathbf{x})} \right) d\mathbf{x} \\
&= E_q \{ \log p(\mathbf{Y}, \mathbf{X}|\Theta) - \log q(\mathbf{X}) \},
\end{aligned}
\tag{14.13}
$$

for any distribution q. The lower bound (14.13), which is denoted by $\mathcal{L}(q, \Theta)$, is used as if it were the log-likelihood. We approximate the posterior distribution by the variational distribution,

$$
q(\mathbf{x}) = p(\mathbf{x}|\mathbf{y}, \Theta), \tag{14.14}
$$

which is chosen to be tractable for purposes of optimization.

The algorithm maximizes the lower bound $\mathcal{L}(q, \Theta)$ over (q, Θ), iterating between fitting q and estimating Θ until convergence. Using $q = q_\Delta$, we estimate Δ and Θ as follows:

$$
\Delta^* = \arg \max_\Delta \mathcal{L}(q_\Delta, \Theta), \tag{14.15}
$$

$$
\Theta^* = \arg \max_\Theta \mathcal{L}(q_{\Delta^*}, \Theta), \tag{14.16}
$$

and we iterate between these two maximizations until convergence.

This vEM process yields the empirical Bayes estimates of the parameters Θ. The M-step is equivalent to finding the MLE of $\Theta = (\boldsymbol{\alpha}, \mathbf{B})$ using expected sufficient statistics, where the expectation is taken under the variational distribution q.

Comment. In a critical study, Yang and Leskovec (2014) show through examples of social, information, food web, and biological networks (with a wide range of sizes and edge densities and community overlap) that the mixed-membership SBM (and also the SBM and degree-corrected SBM) fails to identify communities with dense overlaps. For example, in a Facebook friendship network of a single user's friends, consisting of 183 nodes and 2873 edges, the user organized his friends into $K = 4$ communities: high-school friends, workplace colleagues, and two communities of university friends with whom the user plays basketball and squash. The MMSB model was applied to this friendship network and completely failed to detect the communities. In a small simulated example with two overlapping communities, Yang and Leskovec show that all three models (mixed-membership SBM, regular SBM, and degree-corrected SBM) identify three blocks, where the overlap is confused as a separate block.

14.3 Alternative Algorithms

There have been several published studies in which different algorithms are derived to detect and identify overlapping communities in undirected or directed networks.

For overviews and recent theoretical developments of such methods, see Xie, Kelley, and Szymanski (2013) and Abbe (2018).

One of the issues in comparing algorithms for community detection is the ambiguities in the various definitions of what exactly constitutes a community, and this question is made even more complicated if communities are allowed to overlap with each other. In some applications, nodes that belong to multiple communities may be relatively few in number, a feature that is often present in real-life social networks. In other types of networks, the extent of overlapping communities may be quite prominent. With this in mind, how should we define how significant (in a statistical sense) the overlap is? Furthermore, how should one define a measure of an algorithm's accuracy in identifying overlapping communities? Little attention has been paid so far to these important questions.

In this section, we outline some of these algorithms that propose alternative ways of attacking this problem. On the one hand, we describe how, by relaxing some of the model constraints, the variational EM method can be adapted to apply to overlapping communities, while on the other hand, we describe a completely different way of looking at the problem by clustering edges instead of nodes.

14.3.1 Link-Based Clustering

Ball, Karrer, and Newman (2011) suggest that instead of focusing on clustering the nodes of a network (what is referred to as "node communities"), when overlapping is suspected, it should be more appropriate to cluster edges of the network (leading to "link communities"). Their underlying premise is that edges can be of different types, each type of which reflects membership in a specific community. Thus, a node is a member of multiple communities if it has multiple types of edges.

To differentiate the different types of edges, Ball et al. color the nodes in such a way that the colors correspond to the K communities. Let θ_{ib} be the probability that the node v_i has an edge of color b (blue, say). Then $\theta_{ib}\theta_{jb}$ is the expected number of edges of color b that connect nodes v_i and v_j. The exact number of such edges follows the Poisson distribution with mean $\theta_{ib}\theta_{jb}$.[1] Thus, the expected total number of edges of all colors that join nodes v_i and v_j is $\sum_b \theta_{ib}\theta_{jb}$, and half that number for self-edges. The exact number of such edges is the sum of independent Poisson variables, which is again a Poisson variable (see **Exercise 14.3**).

Putting all this together, we have that the graph $\mathcal{G}$ with adjacency matrix $\mathbf{Y} = (Y_{ij})$, where $Y_{ij} = 1$ if an edge is present between nodes v_i and v_j, $Y_{ii} = 2$ for a self-edge, and 0 otherwise, has probability

$$P\{\mathcal{G}|\boldsymbol{\theta}\} = \tag{14.17}$$

$$\prod_{i<j} \frac{(\sum_b \theta_{ib}\theta_{jb})^{Y_{ij}}}{Y_{ij}!} \exp\left\{-\sum_b \theta_{ib}\theta_{jb}\right\} \times \tag{14.18}$$

$$\times \prod_i \frac{(\frac{1}{2}\sum_b \theta_{ib}\theta_{ib})^{Y_{ii}/2}}{Y_{ii}!/2} \exp\left\{-\frac{1}{2}\sum_b \theta_{ib}\theta_{ib}\right\}. \tag{14.19}$$

---

[1]Although this formulation allows multi-edges (more than one edge between two nodes) and self-edges (edges that connect to the same node at both ends), it is argued that θ_{ib} will be small and so error due to including multi-edges will also be small, which simplifies the model.

This is the likelihood function. Taking logs and ignoring all terms in the log-likelihood that do not involve the θs, we have that

$$\log P\{\mathcal{G}|\boldsymbol{\theta}\} = \sum_{ij} Y_{ij} \log \left\{ \sum_b \theta_{ib}\theta_{jb} \right\} - \sum_{ij}\sum_b \theta_{ib}\theta_{jb}. \tag{14.20}$$

Rather than maximizing the log-likelihood wrt θ_{ib}, which results in a set of difficult equations to solve, Ball et al. apply Jensen's inequality[2] to obtain equations that are simpler to solve. From (14.20), this yields

$$\log P\{\mathcal{G}|\boldsymbol{\theta}\} \geq \sum_{ij} Y_{ij} \left\{ \sum_b q_{ij}(b) \log \left(\frac{\theta_{ib}\theta_{jb}}{q_{ij}(b)} \right) \right\} - \sum_{ij}\sum_b \theta_{ib}\theta_{jb}$$

$$= \sum_{ijb} \left\{ Y_{ij}q_{ij}(b) \log \left(\frac{\theta_{ib}\theta_{jb}}{q_{ij}(b)} \right) - \theta_{ib}\theta_{jb} \right\}, \tag{14.21}$$

where

$$q_{ij}(b) = \frac{\theta_{ib}\theta_{jb}}{\sum_b \theta_{ib}\theta_{jb}}. \tag{14.22}$$

Optimizing (14.21), the result is

$$\theta_{ib} = \frac{\sum_j Y_{ij}q_{ij}(b)}{\sqrt{\sum_{ij} Y_{ij}q_{ij}(b)}}. \tag{14.23}$$

See Ball et al. for details.

Equations (14.22) and (14.23) are solved simultaneously by iterating between them, starting with random initial values. Ball et al. also suggest ideas for speeding up the computations. Because this is essentially a version of the EM algorithm, the usual warnings about the iterations getting trapped in a local maximum apply, so that restarting the algorithm using different sets of starting values is recommended.

As a measure of how accurately the algorithm detects overlapping communities, Ball et al. used the following Jaccard index.

Definition 14.1 If S is the set of nodes in the true overlap and V is the set of nodes that the algorithm identifies as being in the overlap, then the *Jaccard index* is defined as the ratio $\mathcal{J} = |S \cap V|/|S \cup V|$.

This measure rewards accurate identification of the overlap, and penalizes false positives and false negatives.

Ball et al. use the algorithm to study a number of synthetic and real-world networks (Zachary's karate club; interactions between fictional characters in *Les Misérables* by Victor Hugo; and a transportation network involving flights between U.S. airports). Each of these networks had good reasons for incorporating overlapping communities. They also show that this algorithm correctly places nodes into their appropriate communities when the communities are non-overlapping. Based upon these and other examples, Ball et al. report that this algorithm works well for moderate-size networks,

---

[2]Jensen's inequality can be stated as $\log(\sum_k a_k) \geq \sum_k q_k \log(a_k/q_k)$, where the $\{a_k\}$ are any set of positive numbers and $\{q_k\}$ are any probabilities whose sum is 1.

in the range of tens of thousands of nodes (and possibly millions of nodes). The algorithm, however, does not provide a way for choosing the value of K.

14.3.2 Overlapping SBM

How could Section 12.4.1 be adapted to incorporate overlapping communities? From (12.9), we have that the ith node is associated with a latent K-vector $\mathbf{Z}_i \sim \mathcal{M}(1, \boldsymbol{\pi})$, $i = 1, 2, \ldots, N$, where $\boldsymbol{\pi} = (\pi_1, \ldots, \pi_K)^\tau$ and $Z_{ki} \in \{0, 1\}$. Unlike (12.8), however, because nodes can be members of more than a single community, each $\mathbf{Z}_i$ can have multiple elements equal to one, and so we cannot assume that $Z_{+i} = \sum_{k=1}^{K} Z_{ki} = 1$. Furthermore, unlike (12.11), we also cannot assume that $\sum_{k=1}^{K} \pi_k = 1$.

Latouche, Birmelé, and Ambroise (2011) describe an overlapping stochastic block-model (OSBM) for a directed network (without self-loops) with overlapping communities. Assume that

$$\mathbf{Z}_i \sim q(\mathbf{Z}_i) = \prod_{k=1}^{K} \text{Bernoulli}(Z_{ki}; \pi_k) = \prod_{k=1}^{K} \pi_k^{Z_{ki}} (1 - \pi_k)^{1 - Z_{ki}}. \tag{14.24}$$

Furthermore, assume that the edge Y_{ij} has probability

$$Y_{ij} | \mathbf{Z}_i, \mathbf{Z}_j \sim \text{Bernoulli}(Y_{ij}; g(a_{ij})) = e^{Y_{ij} a_{ij}} g(-a_{ij}), \tag{14.25}$$

where $g(x) = (1 + e^{-x})^{-1}$ and

$$a_{ij} = \mathbf{Z}_i^\tau \mathbf{W} \mathbf{Z}_j + \mathbf{Z}_i^\tau \mathbf{u} + \mathbf{v}^\tau \mathbf{Z}_j + w^*, \tag{14.26}$$

where $\mathbf{W}$ is a $(K \times K)$-matrix and $\mathbf{u}$ and $\mathbf{v}$ are K-vectors, but otherwise have no special structure. The first term on the rhs of (14.26) shows interactions between nodes v_i and v_j, the second term represents the tendency for v_i to link to other nodes, the third term represents the tendency for v_j to receive links from other nodes, and the last term represents a bias term. To simplify (14.26), set

$$a_{ij} = \widetilde{\mathbf{Z}}_i^\tau \widetilde{\mathbf{W}} \widetilde{\mathbf{Z}}_j, \tag{14.27}$$

where $\widetilde{\mathbf{Z}}_i = (\mathbf{Z}_i, 1)^\tau$ is a $(K + 1)$-vector and

$$\widetilde{\mathbf{W}} = \begin{pmatrix} \mathbf{W} & \mathbf{u} \\ \mathbf{v}^\tau & w^* \end{pmatrix} \tag{14.28}$$

is a $(K + 1) \times (K + 1)$ matrix.

In the spirit of (12.29), the log-likelihood of the observed $\mathbf{Y}$ can be written as

$$\mathcal{J}(q_{\mathbf{Y}}) = \log p(\mathbf{Y} | \boldsymbol{\pi}, \widetilde{\mathbf{W}}) - KL(q_{\mathbf{Y}}(\cdot) \parallel p(\cdot | \mathbf{Y}, \boldsymbol{\pi}, \widetilde{\mathbf{W}})). \tag{14.29}$$

One way to proceed is to find a variational approximation to $q_{\mathbf{Y}}$ that will minimize the KL term in (14.29). Following (12.32), choose $q_{\mathbf{Y}}(\mathbf{Z})$ to be the closest distribution to the posterior distribution $p(\mathbf{Z} | \mathbf{Y}, \boldsymbol{\pi}, \widetilde{\mathbf{W}})$ among all distributions that can be expressed as a product,

$$q_{\mathbf{Y}}(\mathbf{Z}) = \prod_{i=1}^{N} q(\mathbf{Z}_i), \tag{14.30}$$

where $q(\mathbf{Z}_i)$ is given by

$$\mathbf{Z}_i \sim q(\mathbf{Z}_i) = \prod_{k=1}^{K} \mathcal{B}(Z_{ki}; \eta_{ki}) = \prod_{k=1}^{K} \eta_{ki}^{Z_{ki}} (1 - \eta_{ki})^{1-Z_{ki}}, \qquad (14.31)$$

where η_{ki} is a variational parameter that corresponds to the posterior probability that node v_i is a member of community k.

Unfortunately, the lower bound to the log-likelihood is intractible. So, form another set of variational parameters $\boldsymbol{\xi} = (\xi_{ij})$ that yield a further lower bound on the first lower bound, thereby providing a second level of approximation, which is now tractible. Then, use a variational EM algorithm to compute the posterior probabilities of the latent variables, $\{\mathbf{Z}_i\}$, and the model parameters, π and $\widehat{\mathbf{W}}$, given the edges. The resulting OSBM algorithm (see Algorithm 1 in Latouche et al.) computes maximum *à posteriori* (or MAP) estimates of the class membership vectors $\{\mathbf{Z}_i\}$. A feature of the resulting OSBM algorithm is the ability to identify "outlier" nodes, which are nodes that are assigned the value $\mathbf{Z}_i = \mathbf{0}$ by the algorithm so that they are not assigned to any of the K communities.

Simulations and real-data examples. Several experiments were carried out by Latouche et al. to compare OSBM with other algorithms (including non-overlapping SBM and MMSB) on simulated networks and real-life networks. Because SBM and MMSB cannot deal with outliers, an extra class was added to those models (i.e., run them with $K + 1$ classes). For certain types of simulated networks, MMSB was found to be an improvement over SBM, which cannot deal with overlapping classes. In those cases, both MMSB and OSBM return accurate estimates of the true clustering matrix $\mathbf{Z}$. For more complicated types of simulated networks, OSBM performed well, while although MMSB found the correct classes, it misclassified some of the overlapping classes.

Latouche et al. applied their methods for overlapping SBM (OSBM) to the French political blogosphere data (196 nodes, 2864 edges) that were recorded over a single day in October 2006 and contained network data on political issues. OSBM used $K = 4$ clusters, corresponding to the four main political parties: Gauche (French democrat, 36% of blogs), Divers Centre (moderate party, 23%), Droit (French republican, 21%), and Liberal (supporters of economic-liberalism, 8%), plus a number of blogs that overlapped certain clusters. There were two other minor parties, Ecologiste (green, 8%) and Analysts (4%). The blogs that overlapped clusters were not surprising because of relational ties between the parties in question. The MMSB results were similar to those of OSBM, with slightly different overlapping blogs. SBM, of course, could not identify overlapping blogs and placed all blogs identified by OSBM as overlapping into the same cluster.

Latouche et al. also studied the *Saccharomyces cerevisiae* (yeast) transcriptional regulatory network (197 nodes, 303 edges), which is a sparse network, and which was arranged into $K = 6$ clusters of operons corresponding to well-known biological functions. There were two overlapping clusters, but 112 operons with low degrees were not assigned to any cluster. Although MMSB and SBM found the same six clusters as OSBM, MMSB did not identify any overlap.

14.3.3 Community-Affiliation Graph Model

Yang and Leskovec (2012, 2014) propose a different algorithm for detecting overlapping communities that they call AGM (community-affiliation graph model). Their AGM algorithm visualizes communities as overlapping "tiles," where tile density corresponds to edge density. The authors claim that the AGM algorithm can decompose a network into communities that may be overlapping, non-overlapping, or hierarchically organized.

They observe that, for many different types of networks (e.g., biological, social, collaboration), nodes that lie in the overlapping parts of the network tend to be more densely connected than nodes in the non-overlapping parts of the network. They also find that protein communities have small amounts of overlap, food-web communities have large amounts of overlap (due to the closed Florida Bay ecosystem), and Web communities overlap in a moderate fashion.

For an unlabeled, undirected network, the AGM uses a combination of MLE with convex optimization and the Metropolis–Hastings algorithm over the space of community-affiliation graphs. The goal is to find that community-affiliation graph that yields the largest likelihood for the observed network. For a network of about a few thousand nodes, the authors report that the AGM algorithm completes the fitting process in about an hour.

In empirical studies, the authors show that the AGM algorithm does a much better job (50% better on average, using four different criteria) of community detection of overlapping social, collaboration, and biological networks than all other existing methods. In the example described above of a single Facebook friendship network, the AGM automatically identified four communities, whose node composition almost perfectly aligns (94%) with the composition of the communities defined by the user.

14.4 Latent Cluster Random-Effects Model

The latent space model of Section 10.7 has been extended to include model-based clustering of the latent-space positions (Hoff, 2005; Handcock, Raftery, and Tantrum, 2007; Krivitsky et al., 2009). These methods were proposed to detect community structure amongst the nodes.

The *latent cluster random-effects model* is the first such model that incorporates four features that are common to social networks.

Transitivity. If node A links to node B, and node B links to node C, then node A is more likely to link to node C.

Homophily. Nodes with similar characteristics are more likely to link to each other.

Community structure. Nodes cluster into groups or communities, where links are more dense within the communities than between them.

Degree heterogeneity. Some nodes tend to send and/or receive more links than other nodes.

Let $\{\mathbf{Z}_i\}$ denote N latent K-vectors. Clustering these vectors is carried out by using a multivariate Gaussian mixture model,

$$\mathbf{Z}_i \sim \sum_{k=1}^{K} \alpha_k \mathcal{N}_K(\boldsymbol{\mu}_k, \sigma_k^2 \mathbf{I}_K), \quad i = 1, 2, \ldots, N, \tag{14.32}$$

where α_k is the probability that a node belongs to the kth group. Hence, $\alpha_k \geq 0$, $k = 1, 2, \ldots, K$, and $\sum_{k=1}^{K} \alpha_k = 1$. Thus, each node is modeled as being randomly generated from one of these K groups, where the kth group is characterized by a mean (μ_k) and variance (σ_k^2) specific to that group.

As before, η_{ij} is the logit of the probability $P\{Y_{ij} = 1 | z_i, z_j, x_{ij}, \theta\}$ and is modeled as

$$\eta_{ij} = \sum_{\ell=1}^{L} \theta_\ell x_{\ell,ij} - d_{ij} + h_{ij}, \qquad (14.33)$$

where L is the number of covariates, $x_{ij} = (x_{1,ij}, \ldots, x_{L,ij})^\tau$. The pairwise distances between latent positions are $d_{ij} = \| z_i - z_j \|$, and h_{ij} depends upon the type of network. If the network is undirected, then $h_{ij} = \delta_i + \delta_j$, where δ_i captures the latent "sociality" of node v_i (the propensity of node v_i to form links with other nodes); if the network is directed, then $h_{ij} = \delta_i + \gamma_j$, where δ_i represents the propensity of node v_i to send links to other nodes and γ_j represents the propensity of node v_j to receive links from other nodes. The terms δ_i and δ_j in the undirected model and δ_i and γ_j in the directed model are each random effects having univariate Gaussian distributions with mean zero and variance specific to the effect,

$$\delta_i \sim \mathcal{N}(0, \sigma_\delta^2), \quad \gamma_i \sim \mathcal{N}(0, \sigma_\gamma^2), \quad i = 1, 2, \ldots, N. \qquad (14.34)$$

The likelihood includes the latent positions, $\{z_i\}$, only through their pairwise distances, and so it is invariant to reflections, rotations, and translations of the latent positions, which, in turn, introduce identifiability problems.

Simultaneous estimation of the parameters, $\{\alpha_g\}$, $\{\mu_g\}$, $\{\sigma_g^2\}$, $\{\theta_\ell\}$, σ_δ^2, σ_γ^2, in the Gaussian mixture model is carried out through a Bayesian approach, where each parameter has a prior distribution characterized by prespecified hyperparameters. Computation is based upon MCMC methods and Metropolis–Hastings update algorithms, and initialization uses multidimensional scaling. Details, including a discussion of how to overcome identifiability issues, can be found in Krivitsky et al. (2009).

There is empirical evidence, however, that this estimation method is computationally slow and expensive, that MCMC does not scale well, and that a variational Bayesian approach may be preferable (Salter-Townshend and Murphy, 2012).

14.5 Further Reading

An edited volume of essays for the mixed-membership SBM is Airoldi et al. (2014).

An overview and ranking of 14 algorithms published primarily in the physics and computer science literature for identifying overlapping communities in networks is given by Xie, Kelley, and Szymanski (2013), but which ignores all the work on mixed-membership SBMs and barely mentions OSBM.

Books on Gaussian mixture models include Everitt and Hand (1984), Titterington, Smith, and Makov (1985), and McLachlan and Basford (1988).

14.6 Exercises

Exercise 14.1 Suppose X is a random vector with probability density $p(x)$. (a) If a and b are two different fixed vectors such that $a^\tau a = b^\tau b$, show that $b = Ua$, where

U is an orthogonal matrix. (b) Show that $p(\mathbf{x})$ is invariant under any orthogonal transformation $\mathbf{Y} = \mathbf{UX}$. (c) If $\mathbf{Y} = \mathbf{UX}$ and $\mathbf{b} = \mathbf{Ua}$, show that (i) $E\{\mathbf{a}^\tau\mathbf{X}\} = E\{\mathbf{b}^\tau\mathbf{X}\}$ and hence that $E\{\mathbf{X}\} = \mathbf{0}$, and (ii) $\mathrm{var}\{\mathbf{a}^\tau\mathbf{X}\} = \mathrm{var}\{\mathbf{b}^\tau\mathbf{X}\}$.

Exercise 14.2 Let $\mathbf{w}_j$ be the jth eigenvector, with $\mathbf{w}_j^\tau\mathbf{w}_j = 1$, of the zero-mean matrix $\mathbf{B}$, and let $\mathbf{u}_1$ be a fixed vector. (a) Show that the distribution of $\mathbf{x}_j^\tau\mathbf{u}_1$ is the same as the distribution of $\mathbf{x}_j^\tau\mathbf{1}$, where $\mathbf{1} = (1, 1, \ldots, 1)^\tau$. (b) Use this result and the central limit theorem to show that $E\{\mathbf{u}_1^\tau\mathbf{w}_j\} = 0$ and $E\{(\mathbf{u}_1^\tau\mathbf{w}_j)^2\} = 1/N$.

Exercise 14.3 Show that the sum of m independent Poisson variables is a Poisson variable.

Exercise 14.4 The gamma function $\Gamma(x)$ is defined by $\Gamma(x) = \int_0^\infty y^{x-1}e^{-y}dy$, $x = 0, 1, 2, \ldots$. Show that it satisfies the relation $\Gamma(x+1) = x\Gamma(x)$. Show that $\Gamma(x+1) = x!$, where $0! = 1$.

Exercise 14.5 Use the OSBM software R-package (or any other appropriate software) to fit overlapping communities to Sampson's monk network data. Take $K = 3$ and $K = 4$. Which value of K seems to be most appropriate for these data? Is there any evidence for overlapping membership in the communities?

Exercise 14.6 Use the OSBM software R-package (or any other appropriate software) to fit overlapping communities to Zachary's karate club data (see Section 2.5.4). Take $K = 2$. Is there any evidence for overlapping membership in the two communities?

Exercise 14.7 Use the OSBM software R-package (or any other appropriate software) to fit overlapping communities to the Enron e-mail network (see Section 2.5.2). Which value of K do you think is most appropriate for these data?

CHAPTER FIFTEEN

Examining Network Properties

As we have seen, networks, such as the Internet and World Wide Web, social networks (e.g., Facebook and LinkedIn), biological networks (e.g., gene regulatory networks, PPI networks, networks of the brain), transportation networks, and ecological networks are becoming larger and larger in today's interconnected world. Some of these networks are truly huge and difficult, if not impossible, to analyze completely and efficiently. In this chapter, we discuss some of the issues involving comparing networks for similarity or differences, including choice of similarity measures, exchangeable random structures of networks, and property testing in networks.

15.1 Introduction

The articles by László Lovász and his colleagues (Christian Borgs, Jennifer Chayes, Vera Sós, and Katalin Vesztergombi, and others)[1] and his recent book (Lovász, 2012) have had a major impact on how theorists envision large graphs or networks, how sequences of networks can be modeled and approximated, and how to think about the limits of such networks. Lovász writes in the preface to his book that the development of these topics resulted from "unexpected connections," on the one hand, between a number of different questions on graphs raised at a meeting of the Theory Group and, on the other hand, between different areas in mathematics that would be needed to investigate these questions further.

Many of the interesting questions relate to the comparison of networks. How can we determine the difference or similarity of two networks? Are there measures of distance between two networks that we can use to compare networks? Can we say anything about the growth of a network that is observed at many different time points?

First, we will need the following definitions of dense and sparse networks:

Definition 15.1 A *dense network* is one in which the number of edges is at least proportional to $|\mathcal{V}|^2$, while a *sparse network* is one in which the number of edges is of order $o(|\mathcal{V}|^2)$, where $|\mathcal{V}|$ is the number of nodes in the network.

---

[1] All are or were members of (or visitors to) the Theory Group of Microsoft Research in Redmond, Washington. Some were statistical physicists and others were graph theorists.

In this and the next chapter, we discuss ways of answering some of the following important questions.

How Similar are Networks?

We first describe different structural similarity measures for finite networks, such as the *edit distance* and the *cut distance*, the latter of which is based upon the Frieze–Kannan *cut norm* of a matrix.

What is the Role of Exchangeability?

Next, we outline types of exchangeability for random structures, such as exchangeable random sequences and exchangeable arrays, including the Aldous–Hoover theorem. As formalized by de Finetti, the notion of exchangeability involves the invariance of probabilities under permutation of their random arguments. Exchangeability of random quantities is generally used when an assumption of independence is too strong to make.

How Well do Network Properties Translate to a Network Sample?

We consider properties of mappings between two networks, $\mathcal{F}$ and $\mathcal{G}$, say. We first describe homomorphisms and isomorphisms, where both types of mappings preserve edge adjacency. The difference between these mappings from $\mathcal{F}$ to $\mathcal{G}$ is that isomorphisms relate to $\mathcal{F}$ and $\mathcal{G}$ being of the same size, while homomorphisms deal with mappings where $\mathcal{F}$ could be much smaller than $\mathcal{G}$. The interesting problem of *graph coloring*, including the *4-color problem* of whether a map composed of contiguous regions can be colored with four different colors, is described as an application of homomorphisms.

We next discuss the important and popular topic of *property testing* in very large networks, where we draw a random subnetwork from the parent network and use it to test whether a particular network property holds for the parent network in question. Sampling for property testing is carried out by selecting a random subset of nodes and then filling in the edges using the appropriate edges induced from the parent network.[2]

15.2 Similarity Measures for Large Networks

An important area of interest relates to comparing two networks. How can we show that two networks are "similar" to each other, or "far away" from each other? Networks can be quite similar in some respects, but different in others. In particular, we would like to know how similar a network generated from an algorithm is to a real-world network. Among the many aspects of network comparisons that are of interest, we mention questions such as: How can we compare two networks of different sizes? What if we do not know the sizes of the networks because they are so huge?

One type of comparison was carried out by Faust (2006), who used dyad and triad distributions to compare similarities between a variety of social networks, including those of humans. In that study, standard multivariate techniques were used to provide a low-dimensional representation of these similarities. A more graph-theoretic and network-oriented approach was initiated by van Wijk, Stam, and Daffertshofer (2010), who compared brain networks of different sizes using network properties and ERGMs.

---

[2]It is worth rereading Section 9.3 on node sampling.

15.2.1 The Edit Distance

To pursue questions about the closeness or similarity of networks, we need to define the concept of *distance* between two networks.

First, we consider two networks, $\mathcal{F}$ and $\mathcal{G}$, that have a common set of nodes $\mathcal{V} = \{v_i\}$, and so they have the same size. Clearly, if these two networks were identical (i.e., the edge $(v_i, v_j) \in \mathcal{E}(\mathcal{F})$ iff $(v_i, v_j) \in \mathcal{E}(\mathcal{G})$ for all $1 \leq i < j \leq n$), then any distance between them should be zero.

Definition 15.2 The *symmetric difference* between sets A and B is defined as

$$A \triangle B = (A \backslash B) \cup (B \backslash A) = (A \cup B) \backslash (A \cap B), \tag{15.1}$$

which is the set of elements that are in either of the sets A and B, but not in their intersection.

In the following, $\mathcal{F} = (\mathcal{V}(\mathcal{F}), \mathcal{E}(\mathcal{F}))$ and $\mathcal{G} = (\mathcal{V}(\mathcal{G}), \mathcal{E}(\mathcal{G}))$ are two unweighted networks.

Definition 15.3 Let $\mathcal{F}$ and $\mathcal{G}$ be two unweighted networks defined on the same set of n nodes; i.e., $\mathcal{V}(\mathcal{F}) = \mathcal{V}(\mathcal{G})$. Consider all those edges that are in $\mathcal{E}(\mathcal{F})$ but not in $\mathcal{E}(\mathcal{G})$, and vice versa. This set of edges is given by $\mathcal{E}(\mathcal{F}) \triangle \mathcal{E}(\mathcal{G})$. The *edit distance* (or *Hamming distance*) between $\mathcal{F}$ and $\mathcal{G}$ is $|\mathcal{E}(\mathcal{F}) \triangle \mathcal{E}(\mathcal{G})|$, and its normalized version is given by

$$d_{\text{edit}}(\mathcal{F}, \mathcal{G}) = \frac{1}{\binom{n}{2}} |\mathcal{E}(\mathcal{F}) \triangle \mathcal{E}(\mathcal{G})|. \tag{15.2}$$

This distance can be viewed as the minimum number of edge changes (additions or deletions) that need to be made to turn $\mathcal{F}$ into $\mathcal{G}$, or vice versa. The edit distance is often defined either without the normalization term, or, if the network is very large and dense so that $\binom{n}{2} \sim n^2$, by taking the normalization to be $1/n^2$.

15.2.2 The Cut Norm of a Matrix

The cut norm of a matrix, which was introduced and studied by Frieze and Kannan (1999), has been used to provide efficient approximation algorithms for various combinatorial problems and was used in a central role by Borgs et al. (2008) for the problem of determining the similarities of dense networks.

Definition 15.4 (Frieze and Kannan, 1999) Let $\mathbf{A} = (A_{ij})$ be a real matrix, with the set of rows denoted by R and the set of columns denoted by C. The (Frieze–Kannan) *cut norm* of $\mathbf{A}$ is defined as

$$\| \mathbf{A} \|_{\square} = \frac{1}{|S| \cdot |T|} \max_{S \subset R, T \subset C} \left| \sum_{i \in S, j \in T} A_{ij} \right|. \tag{15.3}$$

If $\mathbf{A}$ is a symmetric $(n \times n)$-matrix (so that $R = C = [n]$, where $[n]$ denotes $\{1, 2, \ldots, n\}$), then $|R| = |C| = n$, the maximum is taken over subsets $S, T \subset [n]$, and the normalization term reduces to $1/n^2$.

The cut norm partitions $\mathbf{A}$ into two parts that have the largest cut between them. It looks at the maximum sum of the entries of $\mathbf{A}$ over all submatrices of $\mathbf{A}$, normalized by the size of the submatrix.

Computing the cut norm of a matrix (15.3) turns out to be an $\mathcal{NP}$-hard problem. Frieze and Kannan devised an efficient algorithm to approximate the cut norm. Variations of this algorithm have since appeared. For example, Alon and Noar (2004) showed that an approximation to the cut norm can be computed in polynomial time by combining semidefinite programming with a rounding technique based upon the matrix version of Grothendieck's inequality (Grothendieck, 1953). They also described applications of the approximation algorithm to graph theory and to solve a certain bi-clustering problem in computational biology.

15.2.3 The Cut Distance

A more useful distance function (rather than the edit distance) between two networks is the cut distance, which is adapted from the cut norm of the adjacency matrix (15.3). We set out two possible scenarios for the cut distance when comparing networks.

1. *Two Networks Having the Same Set of Nodes.*
   We first need the following definition of cut distance:

   Definition 15.5 Let $\mathcal{F}$ and $\mathcal{G}$ be two unweighted finite networks defined on the same set of n nodes (i.e., $\mathcal{V}(\mathcal{F}) = \mathcal{V}(\mathcal{G}) = \mathcal{V}$ and $|\mathcal{V}| = n$). Let $S, T \subseteq \mathcal{V}$ be two subsets of nodes. Let $e_\mathcal{F}(S, T)$ denote the number of edges in $\mathcal{F}$ with one end-node in S and the other in T (edges in $S \cap T$ are counted twice), with a similar definition for $e_\mathcal{G}(S, T)$. The *cut distance* between $\mathcal{F}$ and $\mathcal{G}$ is

$$d_\square(\mathcal{F}, \mathcal{G}) = \frac{1}{n^2} \max_{S, T \subseteq \mathcal{V}} |e_\mathcal{F}(S, T) - e_\mathcal{G}(S, T)| = \| \mathbf{Y}_\mathcal{F} - \mathbf{Y}_\mathcal{G} \|_\square , \qquad (15.4)$$

   where $\mathbf{Y}_\mathcal{F}$ and $\mathbf{Y}_\mathcal{G}$ are the adjacency matrices of $\mathcal{F}$ and $\mathcal{G}$, respectively.

We see that the cut distance (15.4) is obtained by taking the difference of the two adjacency matrices, and then computing the Frieze–Kannan cut norm on that difference. In an unlabeled network (or a network in which labels are not important), we can minimize the cut distance over all permutations of the nodes. This yields a generalized version of the cut distance.

Note that the normalization in (15.4) does not depend upon the sizes of S and T. As a result, the cut distance is biased away from small subsets of nodes and towards larger subsets of nodes.

2. *Two Networks Having Different Numbers of Nodes.*
   Now, we deal with the distance between two arbitrary networks. This is a difficult issue, but an important idea allows us to define an appropriate distance.

   Suppose the network $\mathcal{F}$ has m nodes and the network $\mathcal{G}$ has n nodes. We "blow up" each network to their least common multiple as follows: replace each of the m nodes in $\mathcal{F}$ by n new nodes (an *n-fold blow-up*, which we denote by $\mathcal{F}[n]$) and replace each of the n nodes in $\mathcal{G}$ by m new nodes (an *m-fold blow-up*, denoted by $\mathcal{G}[m]$). So, now, both $\mathcal{F}$ and $\mathcal{G}$ have mn nodes, or, more generally, an integer multiple, knm, $k \in \mathbb{N}$, of those numbers of nodes. For each blown-up network, a pair of the new nodes is connected by an edge iff their original nodes were connected.

The next step is to compare the two blown-up versions, $\mathcal{F}[n]$ and $\mathcal{G}[m]$, of the networks, now with the same number of nodes. If the original networks, $\mathcal{F}$ and $\mathcal{G}$, have the same number of nodes, then their cut distance can be defined by

$$\widehat{\delta}_\square(\mathcal{F},\mathcal{G}) = \min_{\widetilde{\mathcal{F}},\widetilde{\mathcal{G}}} d_\square(\widetilde{\mathcal{F}},\widetilde{\mathcal{G}}), \tag{15.5}$$

where the minimum is taken over all "overlays," $\widetilde{\mathcal{F}}$ and $\widetilde{\mathcal{G}}$, of $\mathcal{F}$ and $\mathcal{G}$, respectively. The hat over the δ indicates that computation of this expression is difficult (Lovász, 2012, p. 129) because of the min–max operation in combining (15.4) and (15.5). Comparing now $\mathcal{F}[n]$ and $\mathcal{G}[m]$, the cut distance becomes

$$\delta_\square(\mathcal{F},\mathcal{G}) = \lim_{n,m\to\infty} \widehat{\delta}_\square(\mathcal{F}[n],\mathcal{G}[m]). \tag{15.6}$$

This definition can be extended to the comparison of weighted graphs.

15.3 Exchangeable Random Structures

Random structures include random variables, graphs, networks, and functions. We will be interested in such structures that enjoy an exchangeability property, in which certain elements of the structure can be permuted without disturbing the underlying distribution of the structure. Exchangeability is a fundamental assumption used in most models of Bayesian inference. Following Aldous (2010) and Orbanz and Roy (2015), we call them *exchangeable random structures*. We describe exchangeable random variables, exchangeable arrays, and then edge exchangeability.

15.3.1 Exchangeable Sequences

Exchangeability has a long history in the foundations of probability and statistics. The honor of introducing the idea of "exchangeability" goes to William Ernest Johnson in his 1924 book *Logic, Part III: The Logical Foundations of Science*. Additional early references on exchangeability include Haag (1924) and Khintchine (1932). However, some of the biggest advances to the field were made by Bruno de Finetti during the 1930s.[3] Amongst a number of topics on which he published, de Finetti studied the exchangeability of sequences of random variables, and his 1937 article (de Finetti, 1937) made him a leader in this field.

> **Definition 15.6** A *finite* sequence, $Y_1, Y_2, \ldots, Y_n$, of random variables is said to be *finitely exchangeable* (or *n-exchangeable* to show the number of random variables) under probability P if its joint distribution is invariant under permutation of the subscripts. In other words, for every $n \in \mathbb{N}$ and for every permutation σ defined on the set $[n] = \{1, 2, \ldots, n\}$,

---

[3]Bruno De Finetti (1906–1985) was born in Innsbruck, Austria, and studied applied mathematics, later working as an actuary and statistician in Rome and Trieste, Italy. Over the years, he developed a lifelong interest in probability, which led to his subjective view of probability (and which influenced the Bayesian approach to statistics). His reputation improved substantially when, in the 1950s, L.J. Savage, Dennis Lindley, and others discovered his work and introduced it to a wider international audience.

$$P(Y_1 = y_1, \ldots, Y_n = y_n) = P(Y_1 = y_{\sigma(1)}, \ldots, Y_n = y_{\sigma(n)}), \qquad (15.7)$$

where $y_{\sigma(1)}, \ldots, y_{\sigma(n)}$ is a permutation of $y_1, \ldots, y_n$. We can express this condition as

$$(Y_1, \ldots, Y_n) \stackrel{d}{=} (Y_{\sigma(1)}, \ldots, Y_{\sigma(n)}), \qquad (15.8)$$

where $\stackrel{d}{=}$ means that the sequences have identical distributions.

Definition 15.7 An *infinite* sequence, $Y_1, Y_2, \ldots$, of random variables is said to be *infinitely exchangeable* under probability P,

$$(Y_1, Y_2, \ldots) \stackrel{d}{=} (Y_{\sigma(1)}, Y_{\sigma(2)}, \ldots), \qquad (15.9)$$

if the joint distribution of every finite subsequence is finitely exchangeable (in the sense of Definition 15.6) with the remaining terms in the sequence being held fixed.

Exchangeability is not always compatible with certain types of data, especially time-series data, where the ordering of the observations is fundamental to understanding the underlying stochastic process. However, in certain situations, the increments of the process can be viewed as exchangeable (see, e.g., Orbanz and Roy, 2015). Furthermore, an infinite exchangeable process is strictly stationary.

Any iid sequence $Y_1, Y_2, \ldots$ is exchangeable; however, the converse does not necessarily hold. For example, a sequence of random variables can be exchangeable but not independent. A simple example is provided by Heath and Sudderth (1976). Imagine you have an urn with m balls, r of which have "1" written on them, and the rest, $m - r$, have "0" written on them. Draw balls, one at a time, without replacement, from the urn. Let X_k denote the number on the kth ball. The sequence $X_1, X_2, \ldots, X_m$ is exchangeable, but not independent. So, exchangeability is a weaker notion than iid. What does hold is that even if exchangeable random variables turn out to be correlated, they will have identical distributions.

Exchangeability is also quirky in how it can or cannot be extended. For example, embedding a finite exchangeable sequence into a larger finite exchangeable sequence may not preserve exchangeability of the new larger sequence, and similarly for embedding it in an infinite exchangeable sequence (Bernardo and Smith, 1994, Section 4.2.2).

15.3.2 De Finetti's Theorem

De Finetti proved the following result:

Theorem 15.1 (de Finetti, 1937) *An infinite binary (0–1) sequence of random variables is exchangeable for each $n \in \mathbb{N}$ iff it can be expressed as a probability mixture of sequences of iid random variables.*

The theorem, however, does not hold for a finite sequence. A more general result is that *a sequence of exchangeable (not necessarily binary) random variables is a mixture of sequences of iid random variables.*

Many of the proofs of this theorem are complicated. The proof we present here is much simpler and is based upon Heath and Sudderth (1976).

Proof Let $X_1, X_2, \ldots, X_N$ denote a finite sequence of N exchangeable binary random variables, where each X_i takes the values 0 or 1. Let

$$E_r = \left\{ \sum_{i=1}^{N} X_i = r \right\} \tag{15.10}$$

denote the event that exactly r of the Xs are all 1s and the rest, $N - r$, are all 0s, and let $q_r = P\{E_r\}, r = 0, 1, 2, \ldots, N$. Consider the N Xs as balls in an urn, with r of them labeled with 1s and the rest labeled with 0s. Draw n balls without replacement from the urn, $n \leq N$. Let $p_{k,n}$ denote the probability of getting k 1s, $k \leq n$. There are $\binom{n}{k}$ ways of arranging k 1s and $n - k$ 0s. Exchangeability then implies that

$$p_{k,n} = \binom{n}{k} P\{X_1 = 1, \ldots, X_k = 1, X_{k+1} = 0, \ldots, X_n = 0\}. \tag{15.11}$$

Conditioning this probability on the event E_r yields

$$p_{k,n} = \binom{n}{k} \sum_{r=0}^{n} P\{X_1 = 1, \ldots, X_k = 1, X_{k+1} = 0, \ldots, X_n = 0 \mid E_r\} \cdot P\{E_r\}. \tag{15.12}$$

Now

$$P\{X_1 = 1, \ldots, X_k = 1, X_{k+1} = 0, \ldots, X_n = 0 \mid E_r\} =$$

$$\left(\frac{r}{N} \frac{r-1}{N-1} \cdots \frac{r-(k-1)}{N-(k-1)} \right) \cdot \left(\frac{N-r}{N-k} \frac{N-r-1}{N-k-1} \cdots \frac{N-r-(n-k-1)}{N-(n-1)} \right)$$

$$= \frac{(r)_k \cdot (N-r)_{n-k}}{(N)_n}, \tag{15.13}$$

where $(a)_b = \prod_{j=0}^{b-1}(a - j)$. Subsituting (15.13) into (15.12) yields

$$p_{k,n} = \binom{n}{k} \sum_{r=0}^{N} \frac{(r)_k \cdot (N-r)_{n-k}}{(N)_n} \cdot q_r, \quad 0 \leq k \leq n \leq N, \tag{15.14}$$

where $q_r = P\{E_r\}$, which can be written as

$$p_{k,n} = \binom{n}{k} \int_0^1 \frac{(\theta N)_k \cdot ((1-\theta)N)_{n-k}}{(N)_n} F_N(d\theta), \tag{15.15}$$

where F_N is a probability distribution concentrated at the frequencies $\{\frac{r}{N}, 0 \leq r \leq N\}$, with jumps of q_r at $\theta = r/N$.

Now, we let $N \to \infty$. Helly's *selection theorem*[4] provides a subsequence $\{F_{N_i}\}$ that converges in distribution to a limit F, say. So, as $N \to \infty$, the integrand in (15.15) converges uniformly in θ to $\theta^k(1 - \theta)^{n-k}$ and hence, for all n and $0 \leq k \leq n$:

---

[4]*Helly's selection theorem: Every sequence $\{F_N\}$ of probability distributions on $[0, 1]$ contains a subsequence $F_{N_1}, F_{N_2}, \ldots$ that converges uniformly to a nondecreasing, right-continuous function F at continuity points of F.* See, e.g., Billingsley (1979, p. 289).

$$p_{k,n} = \binom{n}{k} \int_0^1 \theta^k (1-\theta)^{n-k} F(d\theta). \tag{15.16}$$

The result follows. □

15.3.3 Exchangeable Arrays

One of the major generalizations of the definition of an infinite exchangeable sequence is to an infinite, exchangeable, two-dimensional array. Interest in exchangeable arrays derives from the possibility that if the nodes of a network were labeled in some arbitrary fashion, it could affect the way in which network models are formed and analyzed. Assuming that the nodes are exchangeable is a way of removing any concern regarding arbitrary labeling of the nodes.

We have the following definition (Diaconis and Janson, 2007).

Definition 15.8 Let $\{Y_{ij}, 1 \leq i, j \leq \infty\}$ be an infinite two-dimensional array of binary random variables. These random variables are said to be *jointly exchangeable* if

$$P\{Y_{ij} = y_{ij}, 1 \leq i, j \leq n\} = P\{Y_{ij} = y_{\sigma(i)\sigma(j)}, 1 \leq i, j \leq n\}, \tag{15.17}$$

for every positive integer $n \in \mathbb{N}$, all permutations σ, and all $y_{ij} \in \{0, 1\}$. Any random array whose entries are jointly exchangeable is called an *exchangeable random array*. If the exchangeable random array is the adjacency matrix of a graph, it is referred to as an *exchangeable graph model (ExGM)*.

In other words, the array $\{Y_{ij}\}$ is exchangeable if the array $\{Y_{\sigma(i)\sigma(j)}\}$ has the same distribution as $\{Y_{ij}\}$ for every permutation σ of $[n] = \{1, 2, \ldots, n\}$, $n \in \mathbb{N}$, where the remaining nodes are kept fixed at their current positions. In short,

$$(Y_{ij}) \stackrel{d}{=} (Y_{\sigma(i)\sigma(j)}), \tag{15.18}$$

for every permutation σ of $[n]$. This implies that relabeling the nodes has no effect on the distribution of the random network. This definition is known as *joint exchangeability* because the same permutation is applied simultaneously to both rows and columns of the array (Aldous, 1985; Kallenberg, 2005). In fact, exchangeability is the same as what many call *permutation invariance*.

We will see that exchangeable random arrays play a major role in defining a nonparametric model of a graph limit.

For an infinite adjacency matrix (derived from a network having an infinite number of nodes), this definition implies that exchangeability is defined in terms of the permutation invariance of the distribution of every finite submatrix (corresponding to a finite subset of the nodes) of that adjacency matrix.

Definition 15.9 If arbitrary permutations, σ and τ, are applied to rows and columns, respectively, which would be appropriate in the case of bipartite graphs, then the condition

$$(Y_{ij}) \stackrel{d}{=} (Y_{\sigma(i)\tau(j)}) \tag{15.19}$$

—— **381** ——

is called *separate exchangeability* or *row and column exchangeability* (see, e.g., Aldous, 1981).

15.3.4 The Aldous–Hoover Theorem

The Aldous–Hoover theorem, which is a generalization of de Finetti's results, enables us to characterize an exchangeable array.

Theorem 15.2 (Hoover, 1979; Aldous, 1981) *An infinite random two-dimensional array $\{Y_{ij}\}$ is said to be jointly exchangeable iff there exists a random measurable function $f : [0,1]^3 \rightarrow \{0,1\}$ such that*

$$(Y_{ij}) \stackrel{d}{=} (f(U_i, U_j, U_{ij})), \tag{15.20}$$

where the random sequence $\{U_i\}$ and the random matrix $\{U_{ij}\}$ are all iid from the uniform distribution, $U[0,1]$, on the interval $[0,1]$, and $U_{ij} = U_{ji}$ for $i < j \in \mathbb{N}$.

In the same way that de Finetti's theorem fails for finite sequences, the Aldous–Hoover theorem also fails for finite arrays.

For a network adjacency matrix $\mathbf{Y} = (Y_{ij})$, the $\{U_i\}$ are uniform random variables in $[0,1]$ independently assigned to the nodes and the $\{U_{ij}\}$ are uniform random variables in $[0,1]$ independently assigned to the edges. We will see in Chapter 16 that if we define the function f in the Aldous–Hoover theorem as a specific indicator function (see (16.5)), then that f will define a graphon.

An unfortunate statistical consequence of the Aldous–Hoover theorem is that networks represented by a node-exchangeable random array are either empty or dense with probability one (Caron and Fox, 2017; Crane, 2018, Section 6.5.1). Thus, exchangeability of nodes (a Bayesian modeling requirement) turns out to be incompatible with the notion of sparse networks (a typical real-world situation).

This dilemma has led to a body of work that asks the following question: What can one achieve by sacrificing exchangeability in order to model real-world networks? Indeed, there is a growing collection of research articles that seeks to extend results on large dense networks to sparse networks. Furthermore, the nodes of most real-life networks tend to be labeled, where nodes refer to specific people, places, genes, etc., so that exchangeability of nodes cannot necessarily be viewed as realistic.

15.3.5 Other Types of Exchangeability

There are several other types of exchangeability that deserve attention, namely, edge exchangeability, relative exchangeability, and relational exchangeability. We outline here their basic ideas. See Crane (2018, Chapters 8–10) for further details.

Edge Exchangeability

Instead of defining exchangeability in terms of the nodes of a network, some work has been carried out by Harry Crane on modeling the exchangeability of the *edges* of a network (Crane and Dempsey, 2016; Crane, 2018, Chapter 9). Such an approach is motivated by the realization that network models that are built around node-exchangeability are not able to portray certain properties (e.g., large-sample behavior of sparseness and power-law degree distribution) often found in real-data networks.

Edge exchangeability, which is most appropriate when the real interest is in interaction, cooperation, collaboration, or connection between nodes, offers the possibility of modeling sparse networks and/or networks that may have power-law structure. Examples include citation networks (see Section 2.3.2), e-mail networks (see Section 2.5.1), scientific collaboration networks (see Section 2.5.2), and telephone-call networks (see Section 3.5.1). In each of these examples, nodes are important only because of their relationship to other nodes. For these types of networks, the edges (e.g., phone calls, e-mails) rather than the nodes (e.g., those who call or receive phone calls, or those who send e-mails or receive them) define the relationships.

Edge exchangeability deals with those networks in which permuting the edges would be more appropriate than permuting the nodes. Crane makes the point that nodes have "names" and "identities." Given the network structure, although the node names may consist of arbitrary and unimportant labels, their identities could be meaningful and important. Even if the node names were permuted, their identities would remain the same, and information about the edges would not change. Furthermore, if we assume that the edges are exchangeable, then the probability of a permuted set of edges would be the same as the probability of the original set of edges.

Suppose we sample edges randomly from the set of edges $\mathcal{E}$ and suppose that these edges are exchangeable. Let $X_1, X_2, \ldots$ denote an exchangeable sequence of edges in $\mathcal{E}$ so that every finite subsequence $X_1, X_2, \ldots, X_n$ is finitely exchangeable; that is,

$$\mathbf{X} \stackrel{d}{=} \mathbf{X}^\sigma, \tag{15.21}$$

where $\mathbf{X} = (X_1, \ldots, X_n)$ and $\mathbf{X}^\sigma = (X_{\sigma(1)}, \ldots, X_{\sigma(n)})$, and σ is a permutation of $\{1, 2, \ldots, n\}$. In terms of probabilities,

$$P\{X_i = (v_i, v_i'), i = 1, 2, \ldots, n\} = P\{X_i = (v_{\sigma(i)}, v_{\sigma(i)}'), i = 1, 2, \ldots, n\}, \tag{15.22}$$

where $v_i, v_i' \in \mathcal{V}$ and $(v_i, v_i') \in \mathcal{E}$, $i = 1, 2, \ldots, n$. In other words, the edges of a randomly labeled graph are exchangeable iff the distribution of those edges is invariant under a σ-permutation.

Relative Exchangeability

The previous descriptions of exchangeability have assumed that the nodes in the sample or population are homogeneous. As we have seen, this is often not the case. In most real-world networks, the nodes tend to be heterogeneous: they may form disjoint communities, where the nodes within a community interact more with other nodes in the same community than they do with nodes in different communities. It is, therefore, important to identify these communities through techniques such as stochastic blockmodels by partitioning the nodes into reasonably homogeneous and non-overlapping clusters.

How can we modify our definition of exchangeability to take into account similarities and differences in groups of nodes? One possible modification was proposed by Crane and Dempsey (2016). Let $\mathbf{Y} = (Y_{ij})$ be the adjacency matrix corresponding to a network, where $Y_{ij} = 1$ or 0 depending upon whether or not an edge is present between nodes v_i and v_j, and let $X = (X_i)$, where $X : \mathcal{V} \to \{1, 2, \ldots, K\}$, be a (nonrandom) classifier that assigns each node in $\mathcal{V}$ to one of the K classes. Then $\mathbf{Y}$ is said to be *relatively exchangeable with respect to X* if

$$\mathbf{Y}^\sigma \stackrel{d}{=} \mathbf{Y} \text{ for all permutations } \sigma : \mathbb{N} \to \mathbb{N} \text{ such that } X^\sigma = X, \qquad (15.23)$$

where $\mathbf{Y}^\sigma = (Y_{\sigma(i),\sigma(j)})$ is a row and column permutation of $\mathbf{Y}$ and $X^\sigma = (X_{\sigma(i)})$ is the same permutation of the elements of X. For example, if the nodes represent college students and links between pairs of nodes represent friendships, it is realistic to expect such friendships to be governed to a large extent by college year: students in year A are more likely to be friends with other students in year A than with students in year B. Thus, the SBM would consider students in year A to be exchangeable with other students in year A, and students in year B to be exchangeable with other students in year B. But students in year A would not be exchangeable with students in year B. In this example, X is college year. So, in general, exchangeability would be viewed as being conditional upon community membership.

When community structure is unknown and has to be discovered, we are in the area of community detection. Then, conditional on a particular (unknown) community $X \in \{1, 2, \ldots, K\}$, the network adjacency matrix $\mathbf{Y}$ is relatively exchangeable with respect to X. Instead of X being nonstochastic as above, Snijders and Nowicki (1997) constructed a Bayesian formulation of the problem in which X is taken to be a random parameter having an exchangeable prior on the K communities. Then, $\mathbf{Y}$ given X is relatively exchangeable. Statistical inference for X given $\mathbf{Y} = \mathbf{y}$ can be found from the resulting posterior distribution.

The mathematical theory of this version of exchangeability is complicated and beyond the scope of this book. We refer the interested reader to Crane (2018, Chapter 8) for technical details and extensions of the SBM to the degree-corrected SBM, to the concept of relative exchangeability of a network with respect to another network, and to the latent space models, where covariate information is available instead of X.

Relational Exchangeability

This version of exchangeability extends the ideas of edge exchangeability to samples of edges drawn from networks that show relational structure. The relational sampling process samples the relations between nodes, rather than first sampling the nodes and then dealing with the relationships between those nodes. Edge exchangeability deals with interactions (edges) between pairwise nodes of random graphs, whilst *hyperedge exchangeability* deals with multiway interactions (hyperedges) in the case of random hypergraphs. Together, edge and hyperedge exchangeability (and a number of other types of exchangeability) can be regarded as examples of *relational exchangeability*.

Consider, for example, actors who appear in movies. Suppose we sample, without replacement, n movies, which are represented by the set $\{X_1, \ldots, X_n\}$, from the Internet Movie Database. The movies $X_1, \ldots, X_n$, and the actors who appear in them, can be viewed as a network. The jth actor starring in the ith movie is denoted by $X_i(j)$, $j = 1, 2, \ldots, k_i$, $i = 1, 2, \ldots, n$, where k_i denotes the number of actors in the ith movie. Then, the k_i different actors in the ith movie X_i are listed as $X_i(1), X_i(2), \ldots, X_i(k_i)$, $i = 1, 2, \ldots, n$, which represent a subset of nodes in the network.

In this example. an actor could appear in several different movies, or even several times in the same movie. In fact, famous actors who star in many movies tend to appear in more of the sampled movies than lesser-known actors who appear in fewer movies. It is also quite possible, due to the randomness of the sampling process, that a famous actor would appear only once in the sample. The nodes (actors in a movie), therefore, are linked by edges that record the order in which the movies were sampled. The movies (i.e., the $\{X_i\}$) constitute the *hyperedges* of the network, where the ith movie has edges linking an ordered set of pairs of the k_i nodes (rather than the $\binom{k_i}{2}$ possible pairwise edges). For example, if $k_i = 2$, the movie is a line, if $k_i = 3$, it is a triangle, and if $k_i = 4$, it is a rectangle, and so on.

An *edge-labeled hypergraph* is said to be *hyperedge exchangeable* if its distribution is invariant under relabeling of its hyperedges. In other words,

$$\mathbf{X} \stackrel{d}{=} \mathbf{X}^\sigma \text{ for all permutations } \sigma : \{1, 2, \ldots, n\} \to \{1, 2, \ldots, n\}, \tag{15.24}$$

where $\mathbf{X} = (X_1, X_2, \ldots, X_n)$ and $\mathbf{X}^\sigma = (X_{\sigma(1)}, X_{\sigma(2)}, \ldots, X_{\sigma(n)})$.

We refer the interested reader to the theory, extensions, and further examples of relational exchangeability in Crane (2018, Chapter 10).

15.4 Homomorphisms and Isomorphisms

The ideas of homomorphisms and isomorphisms are very important in mathematics. We first give informal descriptions of these two concepts. A *homomorphism* is a transformation of a set $\mathcal{A}$ into another set $\mathcal{B}$ that preserves in $\mathcal{B}$ the structure of relationships between entities in $\mathcal{A}$. Note that matrix exponential is not a homomorphism (e.g., $e^{\mathbf{A}+\mathbf{B}}$ is not typically equal to $e^{\mathbf{A}}e^{\mathbf{B}}$). When two entities within the same mathematical system can be placed in a 1–1 correspondence so that all structural features are preserved, then the two entities are said to be *isomorphic*. In particular, a homomorphism can be an isomorphism, but the converse is not necessarily true. Note that the sets of real numbers and of complex numbers are not isomorphic, nor are the sets of reals and rationals.

15.4.1 Homomorphism Densities

In graph theory, a homomorphism is referred to as a *graph homomorphism*, which generalizes concepts such as graph coloring. We have the following definition of a homomorphism:

Definition 15.10 Let $\mathcal{F} = (\mathcal{V}(\mathcal{F}), \mathcal{E}(\mathcal{F}))$ and $\mathcal{G} = (\mathcal{V}(\mathcal{G}), \mathcal{E}(\mathcal{G}))$ be two finite networks. Suppose $\mathcal{F}$ is a small network while $\mathcal{G}$ is a large, dense network. Let $\phi : \mathcal{V}(\mathcal{F}) \to \mathcal{V}(\mathcal{G})$. Then ϕ is a *homomorphism* if it preserves edge adjacency. In other words, for every edge $(u, v) \in \mathcal{E}(\mathcal{F})$, $(\phi(u), \phi(v))$ is an edge in $\mathcal{E}(\mathcal{G})$.

Note that because homomorphisms preserve edges, it follows that the image of a homomorphic map of a connected graph is also connected.

Definition 15.11 The set of all homomorphisms from $\mathcal{F}$ to $\mathcal{G}$ is denoted by $\mathrm{Hom}(\mathcal{F}, \mathcal{G})$ and the cardinality of that set is denoted by $\mathrm{hom}(\mathcal{F}, \mathcal{G})$. The *homomorphism density* from $\mathcal{F}$ to $\mathcal{G}$ is defined as the normalized number of homomorphisms between $\mathcal{F}$ and $\mathcal{G}$,

$$t(\mathcal{F},\mathcal{G}) = \frac{\hom(\mathcal{F},\mathcal{G})}{|\mathcal{V}(\mathcal{G})|^{|\mathcal{V}(\mathcal{F})|}}, \qquad (15.25)$$

where $|\mathcal{V}|$ is the cardinality of $\mathcal{V}$.

The denominator of $t(\mathcal{F},\mathcal{G})$ is the total number of mappings from $\mathcal{V}(\mathcal{F})$ to $\mathcal{V}(\mathcal{G})$. So, $t(\mathcal{F},\mathcal{G})$, which lies in $[0,1]$, is the probability that a random map of $\mathcal{V}(\mathcal{F})$ into $\mathcal{V}(\mathcal{G})$ is a homomorphism. In other words, if k is the number of nodes in $\mathcal{F}$ (i.e., $k = |\mathcal{V}(\mathcal{F})|$), then $t(\mathcal{F},\mathcal{G})$ is the probability that $\mathcal{F}$ is a subnetwork of $\mathcal{G}$ if $\mathcal{G}$ were restricted to the same k nodes as $\mathcal{F}$ (Diaconis and Janson, 2007).

15.4.2 Isomorphisms

We need the following two definitions:

Definition 15.12 Let $\mathcal{F}$ and $\mathcal{G}$ be two networks having the same number of nodes. A *node* (or *vertex*) *bijection*, $f : \mathcal{V}(\mathcal{F}) \rightarrow \mathcal{V}(\mathcal{G})$, is a function between the nodes of those networks such that each node of $\mathcal{F}$ is paired with exactly one node of $\mathcal{G}$, and vice versa. There are no unpaired nodes in either $\mathcal{F}$ or $\mathcal{G}$.

A more formal definition is given by the following:

Definition 15.13 A function $f : \mathcal{V}(\mathcal{F}) \rightarrow \mathcal{V}(\mathcal{G})$ is a *bijection* iff it is one-to-one (*injective function*) and onto (*surjective function*).

Injective means for all $u, v \in \mathcal{V}(\mathcal{F})$, if $f(u) = f(v)$, then $u = v$, and surjective means for all $v \in \mathcal{V}(\mathcal{G})$, there exists an $u \in \mathcal{V}(\mathcal{F})$ such that $f(u) = v$.

As an example of a bijection, consider a class of students and a room with chairs such that the number of students equals the number of chairs. There are no empty chairs or students who have to stand.

Definition 15.14 Consider two networks $\mathcal{F}$ and $\mathcal{G}$ that have the same number of nodes and the same number of edges. An *isomorphism* between $\mathcal{F}$ and $\mathcal{G}$ is a bijection between their node sets, $\psi : \mathcal{V}(\mathcal{F}) \rightarrow \mathcal{V}(\mathcal{G})$, such that adjacent nodes in $\mathcal{F}$ are mapped to adjacent nodes in $\mathcal{G}$; that is, $(u,v) \in \mathcal{E}(\mathcal{F})$ iff $(\psi(u), \psi(v)) \in \mathcal{E}(\mathcal{G})$. Also, non-adjacent nodes in $\mathcal{F}$ are mapped by ψ to non-adjacent nodes in $\mathcal{G}$. If an isomorphism exists between the two networks, then the networks are said to be *isomorphic* and their relationship is written as $\mathcal{F} \cong \mathcal{G}$.

Note that if we are dealing with infinite networks, it is not obvious how to define an isomorphism.

If two networks $\mathcal{G}$ and $\mathcal{G}'$ are related by $\hom(\mathcal{F},\mathcal{G}) = \hom(\mathcal{F},\mathcal{G}')$, for every finite network $\mathcal{F}$, then, under an additional condition, $\mathcal{G}$ and $\mathcal{G}'$ are isomorphic (Lovász, 1967).

15.4.3 Graph Coloring

A *proper graph coloring* is a coloring of the nodes of a graph $\mathcal{G}$ in such a way that no two adjacent nodes can have the same color. This is an $\mathcal{NP}$-complete problem.

If there are r colors used to color the nodes of a graph, then the coloring can be viewed as a homomorphism from $\mathcal{G}$ into a complete graph, usually denoted by K_r, with r nodes. So, every node in $\mathcal{G}$ that has color c is mapped to that node of K_r that corresponds to the color c, and every edge in $\mathcal{G}$ is mapped to an adjacent pair of nodes in K_r. Note that K_1 is a graph with one node and no edges. The smallest number of colors needed for a proper coloring of $\mathcal{G}$ is the *chromatic number*, $\chi(\mathcal{G})$, of $\mathcal{G}$; that is, $\chi(\mathcal{G})$ is the smallest r so that there is a homomorphism from $\mathcal{G}$ onto K_r.

The *graph coloring problem* is to determine whether or not $\mathcal{G}$ is colorable for a given value of r and, if so, how many proper colorings of $\mathcal{G}$ can be created with r colors. This is equivalent to asking how many distinct homomorphisms there are from $\mathcal{G}$ to K_r. The number of colorings of $\mathcal{G}$ with r colors is, therefore, $\mathrm{hom}(\mathcal{G}, K_r)$, which is known to be a polynomial in r, and is called the *chromatic polynomial*.

The 4-Color Map Problem

A famous problem in graph theory is the *4-color map problem*, introduced by DeMorgan in 1852. Can a planar map (i.e., a map that can be drawn in the plane) composed of contiguous regions (e.g., countries) be colored by four colors so that no two adjacent regions sharing a common boundary are colored with the same color? Think of the nodes as regions and each edge represents a pair of regions that share a partial or complete boundary. This proved to be a very difficult problem to solve. After several false proofs by eminent mathematicians, Kenneth Appel and Wolfgang Haken of the University of Illinois showed in 1976 (see Appel and Haken, 1977) that four colors did indeed suffice to color such a map. This was the first important mathematical result that used a computer for its proof. The result was "proved" by combining theory that first reduced the problem to a more manageable size (essentially to a series of about 1500 special cases that needed to be checked individually), which was then completed using a specialized computer program for testing all remaining configurations.[5] Several improvements to the proof have since been made, including reducing the number of configurations to be checked to 633 and improving the efficiency of the computer algorithm.

Applications

There are many different applications of the graph coloring problem, other than just assigning colors to nodes of a graph. One of them is the problem of assigning different communication channels to access points in a wireless Local Area Network, where channels with similar frequencies interfere with each other (Aggarwal et al., 2006). Using ideas from graph coloring, the authors assigned far-apart channels to nearby access points where each node represented an access point and two potentially interfering access points are connected by an edge.

Another such application deals with scheduling students taking exams, where students need to be prevented from being assigned to take two exams at the same time. Thus, exams form the nodes (i.e., day, time, room) and the edges record which students are taking two different exams. The problem is to make sure that exams are assigned to the nodes in such a way that no adjacent nodes have exam assignments at the same time. This corresponds to a proper coloring of $\mathcal{G}$, where the colors correspond

---

[5]The Mathematics Department at the University of Illinois then stamped their outgoing mail with the message "Four colors suffice."

to different exam times, and the homomorphism from $\mathcal{G}$ to K_r represents an exam scheduling solution.

15.5 Property Testing in Networks

Property testing is currently an area of great interest in computer science and statistical physics. The topic was introduced by Blum, Luby, and Rubinfeld (1993) to check whether the linearity of a function could be tested by a constant number of queries of the function. This was followed by work of Rubinfeld and Sudan (1996), who formalized the idea of property testing for the specific problem of testing whether a function was a low-degree polynomial. Property testing, more generally, was extended by Goldreich, Goldwasser, and Ron (1998), who also applied their ideas to the testing of combinatorial graph properties. Many other extensions of the theory have taken place since then. For graphs, "queries" involve looking at a randomly selected number $k \ll |\mathcal{V}|$ of distinct nodes in $\mathcal{V}$ and making a decision on whether $\mathcal{G}$ has some specified property based upon how the nodes connect to each other in $\mathcal{E}$.

Property testing was designed to develop "super-fast" algorithms (i.e., sublinear-time algorithms) for making "approximate" decisions (rather than exact decisions). The concept has been applied to testing whether a function, a graph, or bit-strings of fixed length possess properties of interest. Properties of functions that have been studied include monotonicity and linearity, and properties of bit-strings, such as palindromes, are of interest in testing membership in a regular language $\mathcal{L}$.

15.5.1 Network Properties

For our purposes, we are interested in testing whether a huge, dense network $\mathcal{G} = (\mathcal{V}, \mathcal{E})$ has some specific property P, or is too far from having that property (Goldreich, Goldwasser, and Ron, 1998). A dense network is considered to have property P if it is relatively close to another network that has property P. By "far," we mean that the network in question is far away from any other network that has property P. An important example of a property P is the following:

Definition 15.15 A property P of network $\mathcal{G}$ is called *hereditary* if every sample subnetwork of $\mathcal{G}$ also has property P.

When we talk about a network property P, we are really dealing with a collection of networks, which is closed under isomorphism (i.e., relabeling of nodes). For example, all bipartite networks (i.e., where the nodes can be divided into two parts, A and B, say, and all edges connect an A-node to a B-node; there are no $A - A$ or $B - B$ edges) or all connected networks are collections of such networks.

Examples of queries about network properties include:

Is $\mathcal{G}$ connected? Does $\mathcal{G}$ contain a triangle? Is $\mathcal{G}$ cycle-free? Is $\mathcal{G}$ bipartite? Can the nodes of $\mathcal{G}$ be colored with r colors? Does $\mathcal{G}$ contain a clique of size k (i.e., a complete subnetwork of k nodes)?

Some of these questions relate to dense networks (where $|\mathcal{E}| \sim |\mathcal{V}|^2$), whilst others relate to sparse networks, such as those having bounded degree.

———— **388** ————

Relationship to Hypothesis Testing

The problem of assessing whether a network belongs to a certain class of networks or is far away from that class raises issues connected to the statistical testing of competing hypotheses (see, e.g., Lehmann, 1986). In this case, the *null* and *alternative hypotheses* are given by the following:

$\mathcal{H}_0$: The network $\mathcal{G}$ belongs to a specified class of networks.
$\mathcal{H}_1$: The network $\mathcal{G}$ is far from that class.

There are various ways of defining these hypotheses and each such formulation has appeared in the network literature.

15.5.2 Network Sampling

If a network is huge, examining all of it for a specified property would be infeasible. Could we, however, make some intelligent statements regarding whether $\mathcal{G}$ has property P by using only a "small part" of $\mathcal{G}$, such as a random sample of nodes and their induced subnetwork? This is equivalent to knowing the homomorphism densities $t(\mathcal{F},\mathcal{G})$ for small networks $\mathcal{F}$ (Borgs et al., 2006).

Property testing was created to develop randomized algorithms that are able to differentiate between the two competing hypotheses, using only a randomly sampled fragment of $\mathcal{G}$. Any answer to these questions would, by necessity, be approximate and would, therefore, have to employ some probabilistic measure of uncertainty. Regardless of our answer to the question, we would like some assurance that we are making a correct decision, say with probability greater than 3/4.

The process starts by drawing a random sample of k nodes from $\mathcal{G}$. Based upon the resulting induced subnetwork, the algorithm makes a decision YES or NO as to whether or not $\mathcal{G}$ has property P. If $\mathcal{G}$ actually has property P, we would like the algorithm to output YES with high probability, whereas if $\mathcal{G}$ does not have property P, the algorithm should say NO with high probability. Clearly, this is unrealistic to expect. Suppose we perturb $\mathcal{G}$ by changing a tiny fraction of its edges, thereby creating $\mathcal{G}'$. Suppose $\mathcal{G}$ has property P, while $\mathcal{G}'$ does not. A random sample of nodes drawn independently from $\mathcal{G}$ and $\mathcal{G}'$ will yield approximately the same distribution, and so the algorithm will make the same inference for each network (Borgs et al., 2006).

This scenario can be avoided by controlling the number of edge changes. Let $\mathcal{A}$ denote a randomized algorithm, and let $\epsilon > 0$ be a *tolerance* (or *proximity*) *parameter* that is fixed in advance. Think of ϵ as a small number such as 0.1 or 0.01. The notion of a network being ϵ-*far from having property* P is determined by modifying (either by adding or deleting) at least[6] $\epsilon |\mathcal{V}|^2$ edges of the sample subnetwork. In other words, if the sample subnetwork is represented by its adjacency matrix, we modify at least $\epsilon |\mathcal{V}|^2$ entries in the adjacency matrix, by changing 0s to 1s or 1s to 0s.

15.5.3 Testing Network Properties

We start by assuming that $\mathcal{G}$ is a large, dense network. Let $\delta > 0$ denote a measure of confidence. The algorithm $\mathcal{A}$ (also called a *property tester*) is allowed

---

[6]Actually, because of the symmetry of the adjacency matrix, it is $\epsilon \binom{|\mathcal{V}|}{2}$.

to make incorrect decisions (i.e., false positives and false negatives) with probability at most δ:[7]

- If the sample subnetwork has property P, then, with probability at least $1 - \delta$, the algorithm will decide YES, that $\mathcal{G}$ has property P.
- If it takes at least $\epsilon|\mathcal{V}|^2$ edge changes (either adding or deleting edges) for the sample subnetwork to have property P, then, with probability at least $1 - \delta$, the algorithm will decide NO, that $\mathcal{G}$ does not have property P.

Between these two extremes, the algorithm would decide between YES and NO in some arbitrary manner. Thus, whatever the outcome, we know that we can change at most $\epsilon|\mathcal{V}|^2$ edges to get the correct answer.

If the randomized algorithm $\mathcal{A}$ is able to decide, from a sample-induced subnetwork of $\mathcal{G}$, whether $\mathcal{G}$ satisfies property P or not, then we say that network property P is *testable* (Borgs et al., 2006; Lovász and Szegedy, 2010). For example, every hereditary network property is testable (Alon and Shapira, 2008).

One-sided error. If $\mathcal{A}$ always decides YES with probability 1, we say that P is *testable with one-sided error*. That is, given a random subnetwork $\mathcal{G}'$ of $\mathcal{G}$ and parameters ϵ and δ:

$$\text{If } \mathcal{G}' \text{ has property P, then P}\{\mathcal{A} \text{ decides YES}\} = 1. \tag{15.26}$$

$$\text{If } \mathcal{G}' \text{ is } \epsilon-\text{far from P, then P}\{\mathcal{A} \text{ decides NO}\} \geq 1 - \eta. \tag{15.27}$$

An example is querying whether $\mathcal{G}$ is bipartite: if $\mathcal{G}'$ is bipartite, then $\mathcal{A}$ will always decide YES.

Two-sided error. If $\mathcal{A}$ allows the possibility of an erroneous decision in either direction, we say that P is *testable with two-sided error*. That is:

$$\text{If } \mathcal{G}' \text{ has property P, then P}\{\mathcal{A} \text{ decides YES}\} \geq 1 - \eta. \tag{15.28}$$

$$\text{If } \mathcal{G}' \text{ is } \epsilon-\text{far from P, then P}\{\mathcal{A} \text{ decides NO}\} \geq 1 - \eta. \tag{15.29}$$

Sparse Networks

Suppose, instead, that $\mathcal{G}$ is a sparse network with bounded degree d (i.e., every node in $\mathcal{G}$ has degree at most $d \geq 2$). Then, the sample subnetwork is said to be ϵ-*far from having property* P if we need to modify at least $\epsilon d|\mathcal{V}|$ of the edges of the sample subnetwork to obtain a network having property P. If the sample subnetwork has property P, then decide YES that $\mathcal{G}$ has property P, whilst if the sample subnetwork is ϵ-far from property P, then decide NO.

15.5.4 Relationship to Learning Theory

What is the relationship between property testing and learning theory? Property testing is designed to be used prior to carrying out a learning algorithm (Goldreich, Goldwasser, and Ron, 1998). The goal of learning is to discover a good approximation

---

[7]In the property-testing literature, $1 - \delta$ is typically assigned a value of $2/3$, while δ is assigned $1/3$, but neither values alter the concept of testability.

to some target object (which could be a function or a graph) that is a member of some class of interest. On the other hand, the point of a testing algorithm is to determine whether the target object is actually in that class or is far away from it. The learning algorithm should perform well if the target object is in the desired class. The testing algorithm, however, has to perform well even if the target object is far away from the desired class.

15.6 Further Reading

The primary reference for the topics in this chapter is Lovász (2012), where cut norm and cut distance (Chapter 8) and graph homomorphisms and isomorphisms (Chapter 5) are discussed.

Section 15.2. The monograph by Janson (2013) includes discussion on cut norm and cut distance. See also the series of papers by Borgs et al. (2006, 2008, 2012). A brief mention of cut distance is given by Kolaczyk (2017, p. 15).

Section 15.3. The literature on exchangeability has exploded since the work of de Finetti (1937). Detailed studies of exchangeability can be found in Aldous (1985) and Kallenberg (2005). Excellent discussions on the use of exchangeable random variables for nonparametric Bayesian inference are given in Bernardo (1997, Chapter 4), Press (1989, Section 2.9.2), and Berger (1985, Section 3.5.7). See also Crane (2018, Chapters 6 and 7) for an excellent description of the different types of exchangeability and their advantages and disadvantages. A very good survey of the uses of exchangeability was given by Kingman (1978) in the 1977 Wald Memorial Lectures. The relationship of exchangeability to causality is discussed by Pearl (2000, Section 6.1.3).

Section 15.4.3. There are several excellent books on the history, development, and solution of the 4-color problem, including Biggs, Lloyd, and Wilson (1998), Katz (1998), Fritsch and Fritsch (2000), and Wilson (2003).

Section 15.5. A good monograph on property testing is Goldreich (2017). Another survey is by Ron (2010). Property testing is described in Lovász (2012, Section 15.3). See also Ron (2008) for a connection of property testing to machine learning. A useful blog on property testing, which lists and describes the latest papers on the topic, can be found at `ptreview.sublinear.info/`.

15.7 Exercises

Exercise 15.1 Let $\mathbf{A} = \begin{pmatrix} 1 & 1 \\ 0 & 1 \end{pmatrix}$ and let $\mathbf{B} = \begin{pmatrix} 1 & 0 \\ 1 & 1 \end{pmatrix}$. Compute $\mathbf{AB}$. Let $f(\mathbf{M}) = \mathbf{M}^2$, for any square matrix $\mathbf{M}$. Compute $f(\mathbf{A})f(\mathbf{B})$. Is $f(\mathbf{AB}) = f(\mathbf{A})f(\mathbf{B})$? Is f a homomorphism?

Exercise 15.2 Let $\mathbf{A} = \begin{pmatrix} a & b \\ 0 & 1 \end{pmatrix}$ and let $\mathbf{B} = \begin{pmatrix} c & d \\ 0 & 1 \end{pmatrix}$, where $a, c > 0$ and $b, d \in \mathbb{R}$. Let $f(\mathbf{M}) = e^{\mathbf{M}}$, for any square matrix $\mathbf{M}$. Is $f(\mathbf{A} + \mathbf{B}) = f(\mathbf{A})f(\mathbf{B})$? Is f a homomorphism?

Exercise 15.3 Draw a graph that has chromatic number four.

Exercise 15.4 Show that the symmetric difference $A \triangle B = (A \backslash B) \cup (B \backslash A)$ of two matrices A and B can be written as $A \triangle B = (A \cup B) \backslash (A \cap B)$.

Exercise 15.5 Let A, B, and C be three sets. Show that

(a) $(A \triangle B) \triangle (B \triangle C) = A \triangle C$.
(b) $A \cap (B \triangle C) = (A \cap B) \triangle (A \cap C)$.

Exercise 15.6 Suppose the two graphs $\mathcal{F}$ and $\mathcal{G}$ are isomorphic. Show that their complements, $\mathcal{F}^c$ and $\mathcal{G}^c$, are also isomorphic.

Exercise 15.7 Let $\mathbf{A}$ be a symmetric matrix. Consider the cut norm given by (15.3). What would be the effect on the cut norm if we restricted the two sets S and T to be the same (i.e., $S = T$)? What would be the effect if S and T were required to be disjoint (i.e., $S \cap T = \emptyset$)?

Exercise 15.8 Although an iid sequence of random variables is exchangeable, the converse does not necessarily hold. With that in mind, give some conditions under which a sequence of exchangeable random variables can be assumed to be approximately independent.

CHAPTER SIXTEEN

Graphons as Limits of Networks

In this chapter, we discuss various issues that arise when networks increase in size. What does it mean for a network to increase in size and how would we visualize that process? Can a sequence of networks, increasing in size, converge to a limit, and what would such a limit look like? We discuss the transformation of an adjacency matrix to a pixel picture and what it means for a sequence of pixel pictures to increase in size. If a limit exists, the resulting function is called a limit graphon, but it is not itself a network. Estimation of a graphon is also discussed and methods described include an approximation by SBM and a network histogram.

16.1 Introduction

As we have seen, a network $\mathcal{G} = (\mathcal{V}, \mathcal{E})$ can be described formally through its $(N \times N)$ adjacency matrix $\mathbf{Y} = (Y_{ij})$, where Y_{ij} is 1 if an edge exists between the pair of nodes v_i and v_j, and is 0 if no such edge exists. We now assume that we are dealing with a sample of n nodes (and corresponding edges) from the parent network $\mathcal{G}$. We will refer to this sample network as $\mathcal{G}_n$.

Consider now the following transformation of the adjacency matrix. We convert the adjacency matrix to a so-called *pixel picture*, in which we replace each 1 by a black square and each 0 by a white square. These types of arrays go back to Diaconis and Freedman (1981), who simulated exchangeable arrays in the context of the Julesz[1] conjecture in visual perception. This conjecture states the following.

Julesz Conjecture: Two "random patterns" (matrices with random 1 or 0 entries, converted into dots and no-dots, respectively) with the same first- and second-order statistics cannot be visually distinguished.

Diaconis and Freedman gave several counterexamples of the conjecture. This black-square/white-square representation of a network, which resembles a step function, is called an *empirical graphon*. The word "graphon" is abbreviated from the term *graph function*.

---

[1]Béla Julesz (1928–2003) was a Hungarian-born American scientist who specialized in visual neuroscience and experimental psychology. He was a member of Bell Laboratories and led the Sensory and Perceptual Processes Department until 1982 and then the Visual Perception Research Department until 1989.

Next, suppose we rescale the pixel-picture version of the adjacency matrix so that it occupies the unit square $[0, 1] \times [0, 1]$. Partition $(0, 1]$ into n intervals $(0, 1] = \cup_{i=1}^{n} I_{in}$, where $I_{in} = \left(\frac{i-1}{n}, \frac{i}{n}\right]$, $i = 1, 2, \ldots, n$, so that each cell has side-length $\frac{1}{n}$. The step function is now defined on the unit square. We can ask the following question:

What would this pixel picture look like if the network increased in size? As n increases, the cell dimensions will be reduced in size and, for very large n, the cells become tiny pixel-dots within the unit square. As $n \to \infty$, the pixel-dots blend into one another, producing a smooth and continuous pattern of light areas and dark areas within the unit square.[2]

This scenario introduces a number of issues. Answering the following questions is the subject of this chapter.

How does an empirical graphon (or pixel picture) increase in size? If the network $\mathcal{G}_n$ has n nodes, we form $\mathcal{G}_{n+1}$ by adding an additional node (i.e., a new row and column for the adjacency matrix and, hence, for the pixel picture). Its associated edges can then be added in a number of ways. For example, we add a new node to $\mathcal{G}_n$ and then either

(a) Randomly choose an existing node and copy its edge pattern onto the new node, or

(b) Connect the new node to each of the other existing nodes wpr $p_n = 1/n$.
The only published strategy of how to add edges when a new node is added to $\mathcal{G}_n$ in this context is given by Borgs et al. (2016, Example 8):

(c) Connect every pair of previously non-adjacent nodes wpr $p_n = 1/n$. The resulting graph is called the *uniform attachment graph*.

If any of these transformations from $\mathcal{G}_n$ to $\mathcal{G}_{n+1}$ is carried out, then, as n gets larger, we have an increasing sequence of empirical graphons. The problem, however, for any of these methods is whether the resulting sequence of empirical graphons $\{\mathcal{G}_n\}$ converges (to a graphon).

What does it mean for a sequence of networks to converge? If the networks in the sequence get bigger and bigger because the number of nodes, n, increases to infinity, then we can think of each of the networks in the sequence as a subgraph sample (selected by randomly sampling a finite number of nodes and returning its induced subgraph) from an infinite network with an infinite number of nodes. In this way, we are taking the convergent sequence of discrete adjacency matrices of a growing network and representing the limit of that sequence as a smooth and continuous display, while retaining the major properties of the network.

If a sequence of networks converges, to what does it converge? If we let the number of cells in the rescaled adjacency matrix grow to infinity (i.e., $n \to \infty$) while fixing its physical size, we arrive at an infinite two-dimensional array $\{Y_{ij}\}, 1 \leq i, j \leq \infty$, of binary random variables. As $n \to \infty$, the cells of the matrix get smaller and smaller until, in the limit, their size becomes infinitesimal. Assuming this limit exists, the result is a function W defined on the unit square $[0, 1] \times [0, 1]$. Note that the limiting function W, which is called a *limit graphon* of converging networks, will not itself be a network.

---

[2]Rather than deal with certain technical (i.e., measurability) aspects of this development, we refer the interested reader to the discussion in Lovász (2012, Remark 1.4).

This is analogous to the situation in which a sequence of rational numbers converges to an irrational number.

What is a graphon? Graphons are limiting objects of sequences of dense networks. They are symmetric measurable functions from the unit square $[0, 1] \times [0, 1]$ into the unit interval $[0, 1]$, so their values are probabilities. They are a special case of *kernels*, which are symmetric measurable functions from the product space $\Omega \times \Omega$, where Ω is an arbitrary probability space, to the real line $\mathbb{R}$.

As we described above, graphons are derived from the adjacency matrix of a network, which is rescaled and transformed into a *pixel picture*. We define the *cut norm* for a matrix and the *cut distance* between two graphons. The graphon space endowed with the cut distance turns out to be a compact metric space. Our discussion includes Szemerédi's regularity lemma for partitioning a network into disjoint parts and an algorithm on how to generate networks from graphons.

How does one estimate a graphon? The final collection of topics in this chapter involves graphon approximation through histograms and stochastic blockmodels.

The mathematical development of graph limits and graphons has been exciting and thought-provoking for the network science community, but, so far, it has not been more than a mathematical theory. The literature on this topic has had no application to a real-data scenario.

One might argue that networks can be observed over a period of time and that it is of interest to discuss their growth. In most real-world network applications, however, the number of nodes and the structure of relationships between the nodes do not remain constant over time. Nodes get added to a network while others leave, only to return, possibly later, and edges are observed to do the same. This situation can be seen in the growth of the Internet, the Web, and social graphs such as Facebook, where, over time, new friends can appear and new profiles are created, and old friends can disappear or are unfriended. The same lack of constancy in network structure can also be observed in all the other types of networks discussed in this book.

To the author's knowledge, not a single real-data application of graphons (as the limit of a sequence of graphs) has appeared, so far, in the literature. There has also been little or no discussion in the network science literature of the applied side of the theory of graphons. The real question then is: Are graphons just artificial mathematical constructs that have no practical application whatsoever to real-data networks? Or can one show that they are more than that?

With those questions and concerns in mind, we describe in this chapter the mathematical tools used to develop the theory of graphons.

16.2 Kernels and Graphons

Suppose $\mathbf{Y} = (Y_{ij})$ is the $(N \times N)$ adjacency matrix associated with a very large network $\mathcal{G} = (\mathcal{V}, \mathcal{E})$, where $Y_{ij} = 1$ if there is an edge between nodes v_i and v_j and is 0 otherwise, and where $N = |\mathcal{V}|$ is the number of nodes and $M = |\mathcal{E}|$ is the number of edges. For undirected networks, $Y_{ij} = Y_{ji}$, so that the adjacency matrix is symmetric. We assume, for the moment, that M and N are finite and that the network $\mathcal{G}$ is *simple*, without self-loops (i.e., $Y_{ii} = 0$, for all i) and multiple edges.

In the following, $\mathcal{G}_n$ denotes a sample network of n nodes (and corresponding edges) drawn from $\mathcal{G}$.

Definition 16.1 Let $\{\mathcal{G}_n\}$ be a sequence of networks, where $n = |\mathcal{V}(\mathcal{G}_n)|$ is increasing. Let $t(\mathcal{F}, \mathcal{G}_n)$ denote the probability that $\mathcal{F}$ is an embedding into $\mathcal{G}_n$. We say that the sequence $\{\mathcal{G}_n\}$ *converges* (as $n \to \infty$) if $t(\mathcal{F}, \mathcal{G}_n)$ converges for every finite network $\mathcal{F}$.

We have the following theorem:

Theorem 16.1 (Lovász and Szegedy, 2007) *If there exists a function W such that* $\lim_{n \to \infty} t(\mathcal{F}, \mathcal{G}_n) = t(\mathcal{F}, W)$ *for every finite network* $\mathcal{F}$, *then W is the limit of the convergent sequence* $\{\mathcal{G}_n\}$, *and* $W = \lim_{n \to \infty} \mathcal{G}_n$ *is called a graphon.*

In general, computing $t(\mathcal{F}, \mathcal{G}_n)$ for every finite $\mathcal{F}$ is very complicated.

16.2.1 Kernels

Before we formally define a graphon, we first define a more general kernel function.

Definition 16.2 Let Ω represent a probability space (usually together with a sigma-field $\mathcal{A}$ and a probability measure P). A *kernel function* on Ω is a symmetric measurable function, $W : \Omega \times \Omega \to \mathbb{R}_+ = [0, \infty)$. By *symmetric function*, we mean that $W(x, y) = W(y, x)$ for $x, y \in \Omega$.

Recent work shows that kernels do not have to be non-negative and bounded (see, e.g., Janson, 2016; Borgs et al., 2018). Some kernels are defined having range $\mathbb{R}$ rather than $[0, \infty)$. If the range of W is taken to be $[0, 1]$ instead of $[0, \infty)$, then W is called a *standard kernel* or *graphon*. If $\Omega = [0, 1]$, then graphons are sometimes called *simple graphons*.

Definition 16.3 We denote by $\mathcal{W}$ the space of symmetric, bounded, measurable kernels $W : [0, 1]^2 \to \mathbb{R}$. The space $\mathcal{W}$ is a linear space and comes equipped with a pseudonorm named the *Frieze–Kannan cut norm*, defined as

$$\| W \|_{\square} = \sup_{S, T} \left| \int_{S \times T} W(x, y) dx dy \right|, \tag{16.1}$$

where the supremum is taken over all pairs of measurable subsets S and T. The sup is known to exist because $S \times T$ and $W(x, y)$ are bounded.

Several different versions of the cut norm have been defined (see, e.g., Janson, 2013). The *degree function* is defined as $d_W(x) = \int_0^1 W(x, y) dy$.

The notion of a measure-preserving transformation is given by the following definition:

Definition 16.4 A transformation $\phi : [0, 1] \to [0, 1]$ is said to be *measure-preserving* (wrt a measure μ) if, for any measurable subset $A \subset [0, 1]$, the measures of A and $\phi^{-1}(A)$ are identical; i.e., $\mu(A) = \mu(\phi^{-1}(A))$, where $\phi^{-1}(A) = \{x \in [0, 1] : \phi(x) \in A\}$.

Note that the inverse of an invertible measure-preserving transformation $\phi : [0,1] \rightarrow [0,1]$ is also measure-preserving.

We have the following definition of isomorphism between kernels (Lovász, 2012, p. 121):

Definition 16.5 Two kernels $W_1, W_2 \in \mathcal{W}$ are said to be *isomorphic up to a nullset* if there is an invertible measure-preserving transformation $\phi : [0,1] \rightarrow [0,1]$ such that $W_1(\phi(x), \phi(y)) = W_2(x, y)$, almost everywhere.

Hence, it follows that an isomorphism up to a nullset is an equivalence relation.

Definition 16.6 Two kernels $W_1, W_2 \in \mathcal{W}$ are said to be *weakly isomorphic* if $t(\mathcal{F}, W_1) = t(\mathcal{F}, W_2)$ for every simple network $\mathcal{F}$, where t is defined in Definition 16.1.

16.2.2 Graphons

We can now formally define a graphon. We will use the following terminology:

Definition 16.7 A *graphon* W is a symmetric measurable function

$$W : [0,1]^2 \rightarrow [0,1], \tag{16.2}$$

where $W(x, y) = W(y, x)$ for $x, y \in [0,1]$ and $[0,1]^2 = [0,1] \times [0,1]$.

The name *graphon* appears first in Borgs et al., 2008), but the general concept (not the name) can be found in the earlier articles of Frieze and Kannan (1999) and Lovász and Szegedy (2006).

Definition 16.8 Let $\mathcal{W}_0 \subset \mathcal{W}$ denote the *space of all graphons*; that is, $\mathcal{W}_0$ is the space of all symmetric, measurable functions:

$$\mathcal{W}_0 = \{W \in \mathcal{W} : W \in [0,1]\}. \tag{16.3}$$

The Frieze–Kannan cut norm for $W \in \mathcal{W}_0$ is

$$\| W \|_\square = \sup_{S, T \subset [0,1]} \left| \int_{S \times T} W(x, y) dx dy \right|. \tag{16.4}$$

Recent work by Borgs et al. (2018) has extended the definition of a graphon from being defined on a probability space to allowing the underlying measure space to be a σ-finite measure space[3] in which the space can have infinite total measure. In this way, graphons defined on σ-finite spaces can be considered as limits of sequences of sparse graphs, just as graphons defined on probability spaces are considered as limits of dense graphs.

If the network is undirected and $\mathbf{Y} = (Y_{ij})$ is a symmetric, binary, adjacency matrix, then the Aldous–Hoover theorem (Theorem 15.2) ensures the existence of a graphon. Recall that $\{U_i\}$ represent uniform random variables in $[0,1]$ independently

---

[3] A σ-finite measure space $(\mathcal{X}, \mathcal{F}, \mu)$ is a space $\mathcal{X}$ with σ-field $\mathcal{F}$ and measure μ such that any set $A \in \mathcal{F}$ can be written as a countable union of sets $B_i \in \mathcal{F}$ with measure $\mu(B_i) < \infty$.

Table 16.1   Examples of continuous graphons $W(x, y)$, $x, y \in [0, 1]$

$W(x, y)$	Reference
$1 - \max(x, y)$	Lovász (2012, p. 17)
$1 - e^{-2xy}$	Caron and Fox (2017)
$e^{-(x^a + y^a)}$, $a = 0.7$	Chan and Airoldi (2014)
$I_{[y \geq x + \frac{1}{2} \text{ or } x \geq y + \frac{1}{2}]}$	Lovász (2012, p. 17)
$\frac{1}{2}[(1 - x) + (1 - y)]$	Cai, Ackerman, and Freer (2015)
$c(\log x)(\log y)$	Borgs et al. (2006)
$[1 + \exp\{-c(x^2 + y^2)\}]^{-1}$	Chatterjee (2015)

assigned to the nodes, while $\{U_{ij}\}$ represent uniform random variables independently assigned to the edges of the graph. From (15.13), the graphon can be expressed through the indicator function as

$$Y_{ij} = f(U_i, U_j, U_{ij}) = I_{[U_{ij} < W(U_i, U_j)]}, \tag{16.5}$$

which equals 1 if $U_{ij} < W(U_i, U_j)$ and 0 otherwise.

Note that the graph-theory literature has not settled on a consistent terminology, with authors and articles often using different names for the same thing or the same name for different things. For example, it is not unusual to see the terms *kernel* and *graphon* used interchangeably.

Examples of continuous graphons are given in **Table 16.1**. Other examples are given in Chan and Airoldi (2014). Two special examples are the following:

- For the Erdős–Rényi random graph $\mathcal{G}(n, p)$, the graphon is constant; if $p = \frac{1}{2}$, then the constant is $W \equiv \frac{1}{2}$.
- For the stochastic blockmodel, the graphon is piecewise constant with a finite number of pieces.

Because $W(x, y) \in [0, 1]$, it can be viewed either as a probability or as a weight associated with the edge, (x, y), joining nodes x and y.

16.2.3   Identifiability of Graphons

Graphons are not, in general, unique. For example, because $U \sim U(0, 1)$ and $1 - U$ are equal in distribution, $W(1 - x, 1 - y)$ will yield the same graph characteristics as $W(x, y)$. Thus, there is an identifiability problem for graphons, which makes for theoretical difficulties. If W and W' define the same exchangeable graph model, a measure-preserving transformation ϕ' such that $W(x, y) = W'(\phi'(x), \phi'(y))$ may not exist, so that the converse of Definition 16.4 generally will not hold (Diaconis, 2008).

To state conditions under which a converse to Definition 16.4 holds, we combine the cut distance between two graphs introduced in (15.4) and the cut norm (16.1) to define the distance between two graphons (Frieze and Kannan, 1999).

Definition 16.9 The *cut metric* (or *cut distance*) between two graphons, $W_1, W_2 \in \mathcal{W}_0$, is

$$\delta_\square(W_1, W_2) = \inf_{\psi \in \mathcal{M}} d_\square(W_1, W_2^\psi), \qquad (16.6)$$

where

$$d_\square(W_1, W_2^\psi) = \| W_1 - W_2^\psi \|_\square, \qquad (16.7)$$

and where the infimum is taken over the set $\mathcal{M}$ of all measure-preserving bijections, $\psi : [0,1] \to [0,1]$, and $W^\psi(x, y) = W(\psi(x), \psi(y))$. Two graphons, W_1 and W_2, are said to be in the same *equivalence class* if $\delta_\square(W_1, W_2) = 0$.

The infimum in (16.6) is known to exist. The cut metric $\delta_\square$ satisfies the triangle inequality,[4]

$$d_\square(W_1, W_3) \le d_\square(W_1, W_2) + d_\square(W_2, W_3), \qquad (16.8)$$

and if we identify graphons that have a cut metric of zero, the space $(\mathcal{W}_0, \delta_\square)$ becomes a metric space (Borgs et al., 2012).

There is a slight technical glitch to this result. Suppose W is the limit graphon to a convergent sequence of graphs and suppose W' (different from W) is such that the cut metric $d_\square(W, W') = 0$. Then W' can also be viewed as the limit graphon of that sequence. In other words, two different graphons, W and W', can have $d_\square(W, W') = 0$. It follows that $d_\square$ is really a pseudometric (rather than a metric). However, we continue to call this pseudometric the cut metric.

We have the following important fact:

Theorem 16.2 (Lovász and Szegedy, 2007) *The metric space $(\mathcal{W}_0, \delta_\square)$ is compact.*

Now, a metric space is compact iff it is complete and totally bounded (see, e.g., Simmons, 1963, p. 125). So, $\mathcal{W}_0$ is the completion of the space of finite graphs with the cut metric $\delta_\square$.

We have the following result (Diaconis and Janson, 2007, Theorem 7.1), due to Kallenberg (2005) and Borgs et al. (2008), which characterizes when $\delta_\square(W, W') = 0$. See also Borgs et al. (2010).

Theorem 16.3 *The cut metric between two graphons, W_1 and W_2, is zero [that is, $\delta_\square(W_1, W_2) = 0$] iff there exist a pair of measure-preserving transformations $\psi, \psi' : [0,1] \to [0,1]$ such that $W_1(\psi(u), \psi(v)) = W_2(\psi'(u), \psi'(v))$ a.s. on $[0,1]^2$.*

16.3 Szemerédi's Regularity Lemma

In 1975, Endre Szemerédi, in proving an old (dating back to 1936) conjecture of Erdös and Turán on arithmetic progressions (now called Szemerédi's theorem), proved an auxiliary lemma on bipartite graphs. A year later, Szemerédi generalized this result to simple graphs (undirected graphs without self-loops and multiple edges). This last result became known as *Szemerédi's regularity lemma*. This lemma provides a way of approximating the structure of a large, dense graph by a union

---

[4]For a proof see, e.g., Janson (2013, Lemma 6.5).

of a small number of random bipartite graphs, and this approximation is referred to as a "regularity" graph. The lemma has been widely used in many different areas of discrete mathematics, including combinatorics, extremal graph theory, and complexity theory. Although the regularity lemma is an existence result, it has been converted into a reasonably efficient computer algorithm for partitioning graphs.

Let $\mathcal{G} = (\mathcal{V}, \mathcal{E})$ be a simple, undirected network. We first need the following three definitions:

Definition 16.10 Let $A, B \subset \mathcal{V}$ be disjoint subsets of nodes in $\mathcal{V}$. Define *edge density*, $d(A, B)$, as the ratio between the number of edges, $e(A, B)$, with one node in A and the other node in B, and the maximum number of edges it could have; that is,

$$d(A, B) = \frac{e(A, B)}{|A||B|}, \tag{16.9}$$

where $|A|$ is the number of nodes in set A.

Edge density is the ratio of the number of edges between A and B divided by the product of the total number of nodes in A and in B, and so takes values in $[0, 1]$. For example, suppose set A has 4 nodes and set B has 6 nodes, and there are 15 edges between A and B. Then, the edge density is $d(A, B) = 15/(4 \cdot 6) = 5/8$. If A and B are part of a complete bipartite graph, the edge density is 1, while if no edges link A and B, the edge density is 0.

Definition 16.11 Fix $\epsilon > 0$. Let $A, B \subset \mathcal{V}$ be finite, non-empty, disjoint subsets of nodes in $\mathcal{V}$. We say that A and B are ϵ-*regular* if, for all $A' \subset A$ with size $|A'| \geq \epsilon |A|$ and $B' \subset B$ with size $|B'| \geq \epsilon |B|$, we have

$$|d(A, B) - d(A', B')| \leq \epsilon. \tag{16.10}$$

The definition of ϵ-regularity shows that the edges of a network are distributed almost randomly across A and B with density $d(A, B)$.

Definition 16.12 Let $\mathcal{G} = (\mathcal{V}, \mathcal{E})$ be a network and let $\mathcal{P} = \{\mathcal{V}_0, \mathcal{V}_1, \ldots, \mathcal{V}_k\}$ be a collection of pairwise-disjoint subsets of $\mathcal{V}$. Then $\mathcal{V}$ is said to admit an ϵ-*regular partition* if it satisfies the following three conditions:

1. At most ϵk^2 pairs $\{\mathcal{V}_i, \mathcal{V}_j\}$ are not ϵ-regular, $1 \leq i < j \leq k$,
2. $|\mathcal{V}_1| = \cdots = |\mathcal{V}_k|$,
3. $|\mathcal{V}_0| \leq \epsilon |\mathcal{V}|$,

where $\mathcal{V}_0$ is an *exceptional set* that is allowed to have size different from the others.

This turns out to be a very strong condition for a partition of a graph $\mathcal{G}$. Now, we are ready to state Szemerédi's[5] regularity lemma:

---

[5]Endre Szemerédi was awarded the Abel Prize in Mathematics in 2012 for his work on arithmetic progressions based upon his regularity lemma.

Lemma 16.1 (Szemerédi, 1978) *For every $\epsilon > 0$, every network $\mathcal{G} = (\mathcal{V}, \mathcal{E})$, and positive integer t, there exist two integers $S = S(\epsilon, t)$ and $T = T(\epsilon, t)$ such that for every $n \geq S$, every n-node network $\mathcal{G}$ admits an ϵ-regular partition $\mathcal{V}_0 \cup \mathcal{V}_1 \cup \cdots \cup \mathcal{V}_k$, for some $k \in [t, T]$. The partition is denoted by $\mathcal{P} = \{\mathcal{V}_0, \mathcal{V}_1, \ldots, \mathcal{V}_k\}$.*

A simpler statement of this important tool is the following: For every large network $\mathcal{G} = (\mathcal{V}, \mathcal{E})$, the set of nodes $\mathcal{V}$ can be partitioned into at most k subsets of nodes, $\mathcal{V} = \mathcal{V}_0 \cup \mathcal{V}_1 \cup \cdots \cup \mathcal{V}_k$, of exactly equal size (plus an exceptional subset, $\mathcal{V}_0$, whose size is small), so that the edges between different subsets behave almost randomly. The point of including the exceptional set $\mathcal{V}_0$ is so that all the other $\mathcal{V}_i, i = 1, 2, \ldots, k$, have exactly the same size. Furthermore, the $|\mathcal{V}_0|$ bound in Definition 16.12 is needed so that there is a negligible number of edges inside each of the $\{\mathcal{V}_i\}$.

The regularity lemma turns out to be equivalent to the compactness of the space of graphons in an appropriate metric. It also implies that any network can be approximated arbitrarily closely by a stochastic blockmodel that has a very large number of blocks.

16.4 Sampling of Graphons

16.4.1 Generating Networks from Graphons

To generate an n-node, exchangeable, random network from a graphon W, we carry out the following sampling process:

Definition 16.13 Let $W \in \mathcal{W}_0$ be a graphon. Fix n.

1. Draw n unobserved (latent) random variables $U_1, U_2, \ldots, U_n$ iid from the uniform distribution $U[0, 1]$, where U_i is a random label assigned to node v_i.
2. With probability $W(U_i, U_j)$, join the pair of nodes, v_i, v_j, by the edge $(v_i, v_j), i \neq j$. Thus, $W(U_i, U_j) = \mathrm{P}\{Y_{ij} = 1\}$.

Then, the resulting network $\mathcal{G}(n, W)$ is called a *random network induced by W*.

We can view the entries of the adjacency matrix $\mathbf{Y}$ as a realization of $\binom{n}{2}$ independent Bernoulli trials, where Y_{ij} has success probability p_{ij}. In other words,

$$Y_{ij} | p_{ij} \overset{ind}{\sim} \text{Bernoulli}(p_{ij}), \quad i \leq i < j \leq n, \tag{16.11}$$

where $Y_{ij} = Y_{ji}, Y_{ii} = 0$, and

$$p_{ij} = \mathrm{P}\{Y_{ij} = 1 | U_i, U_j\} = W(U_i, U_j). \tag{16.12}$$

So, (16.11) can be re-expressed as

$$Y_{ij} | U_i, U_j \sim \text{Bernoulli}(W(U_i, U_j)), \quad i \leq i < j \leq n. \tag{16.13}$$

16.4.2 Bipartite Graphons

An important class of graphons are bipartite graphons.

Definition 16.14 A graphon $W \in \mathcal{W}_0$ is said to be *bipartite* if the set of nodes $\mathcal{V}$ can be partitioned into two disjoint subsets of nodes, $\mathcal{V}_1$ and $\mathcal{V}_2$, as $\mathcal{V} = \mathcal{V}_1 \cup \mathcal{V}_2$ such that $W(U_1, U_2') = 0$ for almost all $(U_1, U_2') \in \mathcal{V}_1 \times \mathcal{V}_2$.

Bipartite graphons are graphs in which each edge connects a node in $\mathcal{V}_1$ only to a node in $\mathcal{V}_2$. There are no edges from a node in $\mathcal{V}_1$ to a node in $\mathcal{V}_1$ or from a node in $\mathcal{V}_2$ to a node in $\mathcal{V}_2$. Furthermore, there is no reason for the number of nodes in $\mathcal{V}_1$ to be the same as the number of nodes in $\mathcal{V}_2$. If there are m nodes in $\mathcal{V}_1$ and n nodes in $\mathcal{V}_2$, then the adjacency matrix $\mathbf{Y} = (Y_{ij})$ has m rows and n columns, where $Y_{ij} = 1$ if an edge connects the node $v_i \in \mathcal{V}_1$ to the node $v_j' \in \mathcal{V}_2$, and $Y_{ij} = 0$ otherwise.

To generate a bipartite graphon, we carry out the following sampling process:

Definition 16.15 Let $W \in \mathcal{W}_0$ be a graphon. Fix m and n.

1. Draw m unobserved (latent) random variables $U_1, U_2, \ldots, U_m$ iid from $U[0, 1]$, where U_i is a random label assigned to node $v_i \in \mathcal{V}_1$.
2. Draw n unobserved (latent) random variables $U_1', U_2', \ldots, U_n'$ iid from $U[0, 1]$, where U_j' is a random label assigned to node $v_j' \in \mathcal{V}_2$.
3. With probability $W(U_i, U_j')$, join the pair of nodes $v_i \in \mathcal{V}_1$ and $v_j' \in \mathcal{V}_2$ by the edge (v_i, v_j'), $i = 1, 2, \ldots, m$, $j = 1, 2, \ldots, n$. Thus, $W(U_i, U_j') = \mathrm{P}\{Y_{ij} = 1\}$.

The resulting network, $\mathcal{G}(m, n, W)$, is called a *random bipartite network induced by W*.

Bipartite graphons are discussed by Choi (2017), who gives an example of product–customer interactions, where products and customers may be derived as samples from some population.

16.4.3 Scaled Graphons

If we allow the success probability p_{ij} to depend upon n (Bickel and Chen, 2009), then we can define a *scaled graphon*, $\rho_n W(x, y)$, where $\rho_n \in (0, 1]$ is a scale parameter. The probability of a network edge is, therefore,

$$p_{ij} = \rho_n W(U_i, U_j), \quad i \leq i < j \leq n, \tag{16.14}$$

where $U_1, \ldots, U_n \overset{iid}{\sim} U[0, 1]$ and $\int \int W(x, y) \, dx \, dy = 1$. The parameter ρ_n can be interpreted as the expected proportion of nonzero edges,

$$\rho_n = \mathrm{E} \left\{ \frac{1}{\binom{n}{2}} \sum_{i<j} Y_{ij} \right\}. \tag{16.15}$$

It can also be interpreted as a sparse graphon model that has had its edges independently deleted wpr $1 - \rho_n$ and retained wpr ρ_n. The sequence $\{\rho_n\}$ is assumed to be fixed and monotone non-increasing. A dense network corresponds to $\rho_n = 1$, while sparse networks correspond to $\rho_n \to 0$ as $n \to \infty$. Note that

$$\mathrm{P}\{Y_{ij} = 1\} = \mathrm{E}\{Y_{ij}\} = \rho_n \int_0^1 \int_0^1 W(x, y) \, dx \, dy = \rho_n. \tag{16.16}$$

So, we can estimate ρ_n by the sample version of (16.15), namely,

$$\widehat{\rho}_n = \frac{1}{\binom{n}{2}} \sum_{i<j} Y_{ij}. \tag{16.17}$$

16.4.4 Comparing Graphon Estimates

Before we describe estimation methods for a graphon, it will be helpful first to understand how simulations have been used to compare various graphon estimation methods. Simulations of a graphon estimator use the following steps:

1. Choose a specific graphon W. (Examples of continuous graphons are given in **Table 16.1**. In the case of a piecewise-constant function, the graphon will take the form of a $(K \times K)$-matrix.)
2. Generate a sample network from that graphon using Definition 16.13.
3. Apply the proposed graphon estimator $\widehat{W}$ to the sample network, and estimate the graphon W (see Section 16.5).
4. Assess the closeness of the estimated graphon $\widehat{W}$ to the true underlying graphon W using an appropriate error criterion. Popular choices of error criterion are *mean-squared error*, which is defined as

$$MSE(\widehat{W}) = \frac{1}{n^2} \sum_{i=1}^{n} \sum_{j=1}^{n} \left(\widehat{W}(U_i, U_j) - W(U_j, U_j) \right)^2, \tag{16.18}$$

and *mean absolute error*,

$$MAE(\widehat{W}) = \frac{1}{n^2} \sum_{i=1}^{n} \sum_{j=1}^{n} \left| \widehat{W}(U_i, U_j) - W(U_i, U_j) \right|, \tag{16.19}$$

where U_i and U_j are independently sampled from $U[0, 1]$.

The MSE and MAE have been used as criteria for evaluating nonparametric function estimators at a single point $x \in \mathbb{R}$ (see, e.g., Izenman, 1991). In graphon estimation, however, $\widehat{W}$ is a function of the latent random variables $\{U_i\}$, and so we take the expected value of the MSE and MAE, namely, $E\{MSE(\widehat{W})\}$ and $E\{MAE(\widehat{W})\}$, as the error criteria, recognizing that the $\{U_i\}$ (and hence also $\widehat{W}$) are random variables.

Comparisons of graphon estimators can be made based upon this last step and a visual comparison of $\widehat{W}$ to W.

16.5 Graphon Estimation

Estimating graphons has become a popular research topic in network theory. A number of articles have provided a theoretical and algorithmic basis for graphon estimation.

There have been two main structures used to estimate a graphon's latent value $W(U_i, U_j)$, where the $\{U_i\}$ are independently sampled from $U[0, 1]$:

- The *stochastic blockmodel* (Airoldi, Costa, and Chan, 2013; Wolfe and Olhede, 2014; Cai, Ackerman, and Freer, 2015; Chatterjee, 2015; Latouche and Robin, 2015).
- The *network histogram* (Chan and Airoldi, 2014; Olhede and Wolfe, 2014).

Methods proposed for developing either of these two techniques have primarily been nonparametric, including the use of nonparametric Bayesian techniques (Lloyd et al., 2012; Latouche and Robin, 2015; Orbanz and Roy, 2015; Caron and Fox, 2017). Most of these methods and algorithms tend to be computationally intensive; for example, requiring knowledge of the entire network structure (e.g., Bickel, Chen, and Levina, 2011) or using an MCMC sampling algorithm for Bayesian statistical inference (e.g., Lloyd et al., 2012). However, recent efforts have tried to make such algorithms more efficient.

16.5.1 Stochastic Blockmodel Approximation

Suppose the nodes in a network are not constrained to possess any special ordering. Then, the most popular estimation method of a continuous graphon W is to approximate W by a piecewise-constant function with a finite number of classes – the stochastic blockmodel (SBM) – which is an example of an exchangeable graph model. In the simplest case, if the graphon is taken to be a constant, then the resulting graph is the Erdős–Rényi graph. Approximating an arbitrary graphon by an SBM can be explained by Szemerédi's regularity lemma, which basically says that any large network can be viewed, for some K, approximately as a blockmodel.

Recall that the essential properties of an SBM are that (1) each network node belongs to one and only one of K classes or communities, and (2) the probability that an edge exists between any two nodes is dependent only upon the community to which the nodes belong; this is what we have referred to in Chapter 11 as "community detection." Here, we are not claiming that the graphon structure was generated by an SBM or that we are doing community detection; the SBM is just proposed as an approximation to the true underlying structure so that nodes can be grouped together according to the patterns of their interactions. One way to think about this is as a problem in dimensionality reduction: we are replacing an $(n \times n)$ adjacency matrix by a much smaller $(K \times K)$ block matrix, where K is the number of blocks, and the connection pattern of nodes only depends upon their membership in a particular block.

When a network becomes very large, describing its structure by a blockmodel may not really be appropriate, especially if we require the number of blocks, K, to be fixed. If we allow $K = K(n)$ to increase as the size, n, of the network increases (Choi, Wolfe, and Airoldi, 2012), the blockmodel approximation to the graphon will improve because we would be explaining a greater amount of the underlying structure; however, the complexity of the model and estimation error will also increase, and the model would be prone to overfitting.

We first describe a method proposed by Airoldi, Costa, and Chan (2013) in which multiple sample graphs can be used to approximate a continuous graphon; they call it a *stochastic blockmodel approximation (SBA)* algorithm. The SBA algorithm proceeds in two stages:

1. Cluster the unknown node labels $U_1, U_2, \ldots, U_n$ into non-overlapping blocks $\widehat{B}_1, \widehat{B}_2, \ldots, \widehat{B}_K$ by using an empirical estimate of the distance between two graphon cross-sections (or "slices") as a clustering criterion.
2. Determine the bin width for an empirical bivariate histogram and use it to provide an approximation to the graphon W by a piecewise-constant function $\widehat{W}$.

The SBA algorithm requires that W be piecewise Lipschitz.

Definition 16.16 A function W is said to be *piecewise Lipschitz* if it satisfies the following conditions. (1) There exists a sequence of K disjoint subintervals $I_k = [\alpha_{k-1}, \alpha_k]$ of the unit interval, where $0 = \alpha_0 < \alpha_1 < \cdots < \alpha_K = 1$. (2) There exists a constant $L > 0$, such that, for any $(x, y), (x', y') \in I_i \times I_j$, $|W(x, y) - W(x', y')| \leq L(|x - x'| + |y - y'|)$.

In other words, if we partition the unit square into K^2 disjoint blocks, $I_{ij} = I_j \times I_j$, $i, j = 1, 2, \ldots, K$, then W is a piecewise Lipschitz function over the unit square if it is Lipschitz on every such block.

Consider two labels U_i and U_j. Suppose, for generality, the graphon W is not symmetric: $W(x, y) \neq W(y, x)$. Then any graph generated by W will not have symmetric (undirected) edges and can, therefore, be considered as a directed graph.

Define a horizontal cross-section ("row slice") of W as $|W(U_i, \cdot) - W(U_j, \cdot)|$ and a vertical cross-section ("column slice") of W as $|W(\cdot, U_i) - W(\cdot, U_j)|$. For a smooth graphon, if U_i and U_j are close (i.e., $|U_i - U_j| \approx 0$), then the neighboring row and column slices will each be small and hence "similar" to each other. Define the distance between the two graphon slices as the average of the total squared row slices and total squared column slices of W,

$$d_{ij} = \frac{1}{2} \left(\int_0^1 [W(U_i, y) - W(U_j, y)]^2 dy + \int_0^1 [W(x, U_i) - W(x, U_j)]^2 dx \right). \quad (16.20)$$

Then, use d_{ij} to cluster nodes into blocks. If the slices are similar, then d_{ij} will be small. To simplify d_{ij}, we define

$$r_{ij} = \int_0^1 W(U_i, y) W(U_j, y) dy, \quad (16.21)$$

$$c_{ij} = \int_0^1 W(x, U_i) W(x, U_j) dx. \quad (16.22)$$

Expanding the expression for d_{ij}, it can be written as

$$d_{ij} = \frac{1}{2} \{(r_{ii} - r_{ij} - r_{ji} + r_{jj}) + (c_{ii} - c_{ij} - c_{ji} + c_{jj})\}. \quad (16.23)$$

Because c_{ij}, r_{ij}, and hence d_{ij}, are unknown, they each have to be estimated.

This estimation problem is hard to resolve with only a single graph. Airoldi et al. instead assumed that a sequence of $2T$ independently observed graphs, $\mathcal{G}_1, \mathcal{G}_2, \ldots, \mathcal{G}_{2T}$, $T \geq 1$, is available, each defined on the same set of nodes. (So, at least two graphs are required for the estimation process.)

This estimation method appears to have application primarily for simulation purposes. Multiple graphs on the same set of nodes are easily obtained in simulation mode: just repeat steps 1 and 2 in Section 16.4.4 as many times as necessary, thereby generating multiple independent sample graphs from a given graphon. It is harder, however, to imagine situations (and the authors did not give any examples) when multiple *independent* graphs would be available from a real network.

Let $\mathbf{Y}^{(t)} = (Y_{ij}^{(t)})$ denote the adjacency matrix of the tth graph, $\mathcal{G}^{(t)}, t = 1, 2, \ldots, 2T$. They also assume that the graphs are directed (so that $\mathbf{Y}^{(t)}$ is not symmetric); however, undirected graphs with symmetric $\mathbf{Y}^{(t)}$, $t = 1, 2, \ldots, 2T$, can be treated as a special case covered by the development of the theory.

First, $W(U_i, U_k)$ is approximated by Y_{ik} and $W(U_j, U_k)$ is approximated by Y_{jk}, and so on. Then, r_{ij} and c_{ij} are estimated by

$$\widehat{r}_{ij,k} = \frac{1}{T^2} \left(\sum_{t_1 \in [1,T]} Y_{ik}^{(t_1)} \right) \left(\sum_{t_2 \in (T,2T]} Y_{jk}^{(t_2)} \right), \tag{16.24}$$

$$\widehat{c}_{ij,k} = \frac{1}{T^2} \left(\sum_{t_1 \in [1,T]} Y_{ki}^{(t_1)} \right) \left(\sum_{t_2 \in (T,2T]} Y_{kj}^{(t_2)} \right), \tag{16.25}$$

respectively, where k is used as a dummy variable akin to x and y in (16.21) and (16.22). Those estimates can now be used to provide an unbiased estimate,

$$\widehat{d}_{ij} = \frac{1}{2} \left[\frac{1}{|S|} \sum_{k \in S} \{ (\widehat{r}_{ii,k} - \widehat{r}_{ij,k} - \widehat{r}_{ji,k} + \widehat{r}_{jj,k}) + (\widehat{c}_{ii,k} - \widehat{c}_{ij,k} - \widehat{c}_{ji,k} + \widehat{c}_{jj,k}) \} \right] \tag{16.26}$$

of d_{ij}, where $S = \{v_1, v_2, \ldots, v_n\} \setminus \{v_i, v_j\}$.

Next, the unknown labels, $U_1, U_2, \ldots, U_n$, are clustered into K blocks, $\widehat{B}_1$, $\widehat{B}_2$, $\ldots, \widehat{B}_K$, using the greedy algorithm in **Algorithm 16.1**, where a key quantity is a precision parameter $\Delta > 0$. Note that a large Δ leads to a small K (where the blocks are large), while a small Δ gives a large K (where the blocks are small). In essence, then, the value of Δ determines the number, K, of blocks. So, $K = K(\Delta)$. For a reliable estimate of the value of a block, an appropriate number of nodes are needed. Too few nodes (i.e., large K) will undermine the estimation process. Hence, some form of objective choice of the value of Δ is needed.

To estimate Δ, Airoldi et al. proposed a cross-validation procedure for choosing the optimal bin width for a bivariate histogram. See **Algorithm 16.2**.

When the K blocks, $\widehat{B}_1, \widehat{B}_2, \ldots, \widehat{B}_K$, have been determined, an estimate, $\widehat{W}(U_i, U_j)$, of the graphon, $W(U_i, U_j)$, can be computed from the empirical frequency of edges that exist between blocks $\widehat{B}_i$ and $\widehat{B}_j$:

$$\widehat{W}(U_i, U_j) = \frac{1}{|\widehat{B}_i||\widehat{B}_j|} \sum_{i_x \in \widehat{B}_i} \sum_{j_y \in \widehat{B}_j} \left(\frac{1}{2T} \sum_{t=1}^{2T} Y_{i_x, j_y}^{(t)} \right). \tag{16.27}$$

Airoldi et al. proved that, under certain conditions, the estimated graphon, $\widehat{W}$, is a consistent estimator of W. Specifically, they showed that $\widehat{W}$ converges to W, in the sense that $E\{MSE(\widehat{W})\} \to 0$ and $E\{MAE(\widehat{W})\} \to 0$, as $n \to \infty$. Further details may be found in the article.

16.5.2 Network Histogram

A different approach was taken by Olhede and Wolfe (2014), who constructed a bivariate histogram approximation (which they called a *network histogram*) for the graphon W, using only a single network dataset and, hence, a single $(n \times n)$ adjacency matrix $\mathbf{Y} = (Y_{ij})$. The graphon is assumed to be Hölder-continuous and bounded away from 0.

Algorithm 16.1 *Clustering n nodes into K blocks*

---

1. Set $\Omega = \{v_1, v_2, \ldots, v_n\}$ and let $\Delta > 0$ be a precision parameter.
2. Choose a random node v_{i_p} from Ω and designate it as the "pivot" node. Set $v_{i_p} \in \widehat{B}_1$.
3. Let $v_{i_a} \in \Omega \setminus \{v_{i_p}\}$ be a node in Ω that is different from the pivot node.
4. Compute the distance, $\widehat{d}_{v_{i_p}, v_{i_a}}$, of v_{i_a} to the pivot node.
5. If $\widehat{d}_{v_{i_p}, v_{i_a}} \leq \Delta^2$, assign node v_{i_a} to the same block $\widehat{B}_1$ as the pivot node.
6. Keep repeating this process by checking all nodes in Ω until the block $\widehat{B}_1 = \{v_{i_p}, v_{i_{a_1}}, v_{i_{a_2}}, \ldots\}$ is completed.
7. Now, consider $\Omega \setminus \widehat{B}_1$.
8. Repeat the process, updating $\Omega \leftarrow \Omega \setminus \widehat{B}_1$.
9. Continue to update using $\Omega \leftarrow \Omega \setminus \widehat{B}_k$, for $k = 1, 2, \ldots, K$, until $\Omega = \emptyset$.

Algorithm 16.2 *Cross-validation estimation of* Δ

---

1. Set up a grid of possible values of Δ.
2. Compute the cross-validation risk

$$\widehat{J}(\Delta) = \frac{2}{h(n-1)} - \frac{n+1}{h(n-1)} \sum_{j=1}^{K} \widehat{p}_j^2 \tag{16.28}$$

for each value in the grid, where $\widehat{p}_j = |\widehat{B}_j|/n$ and bandwidth $h = 1/K$. Note that h does not depend upon n as would usually occur in nonparametric function estimation.
3. Find the value of Δ that minimizes $\widehat{J}(\Delta)$.
4. Determine its associated blocks $\widehat{B}_1, \widehat{B}_2, \ldots, \widehat{B}_K$.

They constructed their network histogram using a single uniform "community size" h defined by the user, $2 \leq h \leq n$. The value of h implies a specific number of bins K. The n nodes are evenly divided up according to $n = hK + r$. There are $K - 1$ bins, each containing h nodes, and one bin of $h + r$ nodes. The nodes are grouped together according to an n-vector of labels, $\mathbf{Z} = (Z_1, Z_2, \ldots, Z_n)^\tau$, where $Z_i \in \{1, 2, \ldots, K\}$, which identifies which nodes fall into which bins. The vector $\mathbf{Z}$, therefore, contains h components having values from 1 to $K - 1$ and $h + r$ components equal to K. The assignment of labels to nodes is defined only up to a permutation of bins, where the actual assignment may depend upon covariates recorded during the data-collection phase. For a given n, let $\mathcal{Z}_K \subseteq \{1, 2, \ldots, K\}^n$ consist of all possible bin-assignment vectors $\mathbf{Z}$ with $n = hK + r$. The problem is to learn $\mathbf{Z}$ from $\mathbf{Y}$.

This prescription may sound a lot like the stochastic blockmodel approach to community detection of Chapter 12. There are similarities, but there are also differences between the two approaches. The use of a single h-value that assigns the same number, h, of nodes to each bin (except for the bin assigned $h + r$ nodes) differs from the stochastic blockmodel, where block sizes are arbitrary and depend upon the observed network structure, and the number of blocks is fixed. In the Olhed–Wolfe approach,

---

407

---

the nodes are clustered together into groups (i.e., bins) of the same size, where the nodes in each group have relationship patterns that are "similar" to the other nodes in the group, and the number of bins K is allowed to increase with n.

The Bernoulli likelihood function is given by

$$\mathcal{L} = \prod_{i<j} Y^{*}_{Z_i Z_j}{}^{Y_{ij}} (1 - Y^{*}_{Z_i Z_j})^{1-Y_{ij}}, \tag{16.29}$$

where

$$Y^{*}_{ab} = \frac{\sum_{i<j} Y_{ij} I_{[\hat{Z}_i=a]} I_{[\hat{Z}_j=b]}}{\sum_{i<j} I_{[\hat{Z}_i=a]} I_{[\hat{Z}_j=b]}}, \quad 1 \le a,b \le K, \tag{16.30}$$

is the histogram bin-height for the abth block, which is defined as the proportion of edges in a histogram bin corresponding to the ab block of Bernoulli trials. Because $\mathbf{Y}$ is symmetric, $Y^{*}_{ab} = Y^{*}_{ba}$. The maximum likelihood estimator, $\hat{\mathbf{Z}} = (\hat{Z}_1, \dots, \hat{Z}_n)^{\tau}$, of the vector $\mathbf{Z}$ is obtained by maximizing the log-likelihood function

$$\hat{\mathbf{Z}} = \arg \max_{\mathbf{Z} \in \mathcal{Z}_K} \sum_{i<j} \{ Y_{ij} \log Y^{*}_{Z_i Z_j} + (1 - Y_{ij}) \log(1 - Y^{*}_{Z_i Z_j}) \}. \tag{16.31}$$

In practice, the ML procedure (16.31) turns out to be $\mathcal{NP}$-hard and, hence, is computationally infeasible, and so an alternative search algorithm was used to find a local maximum of the log-likelihood function.

Putting it all together, the graphon W is estimated by

$$\widehat{W}_h(x,y) = \hat{\rho}_n^+ Y^{*}_{\min(\lceil nx/h \rceil, k), \min(\lceil ny/h \rceil, k)}, \tag{16.32}$$

where $\hat{\rho}_n^+$ is the generalized inverse of $\hat{\rho}_n$, and $\lceil x \rceil$ is the smallest integer that is greater than or equal to x.

Assuming that $h = h_n$ grows slower than n, and assuming certain smoothness conditions on the graphon W (including the existence of an upper bound M on the gradient of W), Olhed and Wolfe were able to derive an upper bound for the MISE of $\widehat{W}_h$, and find the h^* (depending upon M^2) that minimized that upper bound. By rearranging the rows and columns of $\mathbf{Y}$ so that the components of the vector $\mathbf{d} = (d_1, d_2, \dots, d_n)^{\tau}$ of node degrees, where $d_i = \sum_{j=1}^{n} Y_{ij}$, are ordered from smallest to largest, and by regressing the components of $\mathbf{d}$ on their sequence number, they then derived a nonparametric estimate, $\widehat{M^2}$, of M^2. Substituting the estimate $\widehat{M^2}$ into the expression for h^*, the $\hat{h}^*$ that provides a consistent estimator of W was obtained. Improvements to the convergence rate of the ML estimator of W have since been derived by Klopp, Tsybakov, and Verzelen (2017).

In a different approach to the same problem, Chan and Airoldi (2014) sought to provide a graphon estimation algorithm that satisfied the dual demands of being both computationally efficient and consistent. The data again consisted of a single adjacency matrix $\mathbf{Y} = (Y_{ij})$. They proposed a *sorting-and-smoothing algorithm (SAS)* for estimating a continuous graphon W with absolutely continuous degree distribution. The SAS algorithm produces a bivariate histogram and proceeds in two stages:

1. *Sorting.* Compute the empirical degree of the ith row of the adjacency matrix by

$$d_i = \sum_{j=1}^{n} Y_{ij}, \quad i = 1, 2, \ldots, n. \tag{16.33}$$

Let $\widehat{\sigma}$ be a permutation that sorts the empirical degrees into monotonically increasing order, so that $d_{\widehat{\sigma}(1)} < d_{\widehat{\sigma}(2)} < \cdots < d_{\widehat{\sigma}(n)}$. Do the same for the columns. Denote the sorted adjacency matrix by $\widehat{\mathbf{Y}} = (\widehat{Y}_{ij}) = (Y_{\widehat{\sigma}(i)\widehat{\sigma}(j)})$.

2. *Smoothing.* The next step is to find a smooth surface that is the best fit to $\widehat{\mathbf{Y}}$. Like the Olhed–Wolfe method, their version of a network histogram uses a uniform block size $h > 0$. First, compute the bivariate histogram $\widehat{\mathbf{H}} = (\widehat{H}_{ij})$, where

$$\widehat{H}_{ij} = \frac{1}{h^2} \sum_{i_1=1}^{h} \sum_{j_1=1}^{h} \widehat{Y}_{ih+i_1, jh+j_1}, \quad i, j = 1, 2, \ldots, K. \tag{16.34}$$

Because this method is a simplified version of the SBA algorithm, all function values in the same block are the same, and so the effective degrees of freedom are $K \times K$, where K is the number of blocks. Then, assuming that graphons are sparse in the gradients, smooth the sorted adjacency matrix $\widehat{\mathbf{Y}}$ by minimizing the total variation, defined as

$$\| \mathbf{W} \|_{TV} = \sum_{i=1}^{K} \sum_{j=1}^{K} \left\{ (D_x W)_{ij}^2 + (D_y W)_{ij}^2 \right\}^{1/2}, \tag{16.35}$$

where $D_x W$ and $D_y W$ are the horizontal and vertical finite differences of W, respectively, subject to a penalty function that controls how close the total variation solution is to the histogram $\widehat{\mathbf{H}}$. This yields an estimated $(K \times K)$ graphon $\widehat{\mathbf{W}}_{tv}$.

The overall complexity of the SAS algorithm is $\mathcal{O}(n \log(n) + K^2 \log(K^2))$ multiplications plus $\mathcal{O}(n^2)$ additions.

Finally, to have the estimated graphon match up correctly to the true graphon $\mathbf{W}$, which has size $(n \times n)$, we compute the $(n \times n)$ Kronecker product, $\widehat{\mathbf{W}}_{est} = \widehat{\mathbf{W}}_{tv} \otimes \mathbf{J}_h$, where $\mathbf{J}_h = \mathbf{1}_h \mathbf{1}_h^{\tau}$, and $\mathbf{1}_h$ is an h-vector of 1s. Under certain conditions on $\mathbf{W}$, Chan and Airoldi showed that the MSE of the SAS estimator of $\mathbf{W}$ converged to 0 as $n \to \infty$ and $K/n \to 0$, and hence is a consistent estimator of $\mathbf{W}$.

16.6 Further Reading

An excellent survey of graphons, cut norm, cut distance, and related topics is given in the monograph by Svante Janson (2013). The book by Lovász (2012) is the classic reference for the material in this chapter. An excellent review of Szemerédi's regularity lemma is given by Komlós et al. (2002).

16.7 Exercises

Exercise 16.1 Show that the inverse of an invertible measure-preserving transformation $\phi : [0, 1] \to [0, 1]$ is also measure preserving.

Exercise 16.2 Define the edge density of a simple, undirected graph with N nodes and M edges as the ratio between the number of edges it has and the maximum number it could have. Find a working expression for the edge density involving node degrees.

Exercise 16.3 Construct a sequence of graphs as follows. Initialize the graph $\mathcal{G}_1$ with a single node. Introduce a new node and then toss a fair coin. If the coin lands heads, link the new node to the other node; if the coin lands tails, do nothing. This yields the graph $\mathcal{G}_2$. Continue to do this, adding a new node each time and joining up the new node to all previous nodes depending upon the result of the coin toss. So, $\{\mathcal{G}_n\}$ forms a quasi-random sequence of graphs with $|\mathcal{G}_n| = n$.

(a) Show that the sequence of graphs $\mathcal{G}_1, \mathcal{G}_2, \ldots$ converges almost surely to the graphon $W(x, y) = 1$ if $x + y \leq 1$, and $W(x, y) = 0$ otherwise.
(b) Draw a picture of this graphon.
(c) Does this result hold if the coin is not fair and instead turns up heads wpr p and tails wpr $q = 1 - p$ $(p \neq \frac{1}{2})$?

Exercise 16.4 Construct a sequence of graphs as follows. Initialize the graph $\mathcal{G}_1$ with a single node. Introduce a new node and, with probability 1, place an edge between the two nodes. This yields the graph $\mathcal{G}_2$. Add a third node and place an edge, with probability $1/2$, between each pair of previously unattached nodes. This yields $\mathcal{G}_3$. Continue the process so that at the $(n + 1)$st step, add a node to $\mathcal{G}_n$, and add edges, each with probability $1/n$, to all previously unattached pairs of nodes. Does the sequence of empirical graphons converge, and, if so, to what does it converge?

CHAPTER SEVENTEEN

Dynamic Networks

In this chapter, we recognize that the configurations of almost all networks vary with time. We define dynamic networks, which can be observed in discrete or continuous time. Discrete-time dynamic networks can be visualized as a sequence of snapshots of the network taken at different points in time. Continuous-time dynamic networks are more complicated, both visually and theoretically, and assume that edges can appear and disappear continuously through time. We discuss the idea of dynamic community discovery in which community detection strategies are applied to dynamic networks. Information can then be obtained on the types of structural changes that occur to the network over time. We illustrate these ideas with longitudinal social networks, with their emphasis on monitoring networks for change, and dynamic biological networks, with special attention to finding and counting motifs and modeling the spread of epidemics.

17.1 Introduction

In most of this book, we have focused on networks that exist at a given point in time. These "frozen in time" networks are referred to as *static networks*. In Chapter 15, we went further by considering ways of comparing two large networks through various similarity measures, such as edit distance and cut distance, and we discussed the concepts of homomorphisms and isomorphisms between two different graphs. We now expand upon this idea of comparing different networks by considering the possibility that the configuration of a network can vary over time. After all, most real-world networks evolve over time. Think of Facebook or LinkedIn networks, or even an e-mail network, which form, grow, and maybe dissolve, while infectious diseases appear, spread themselves around, and, in many cases, disappear. The structure of real-world networks does not stay the same over time, but they keep changing, and it is of interest to identify how and when they change.

A network can change from one time period to the next in several ways. The most obvious changes are that nodes can be added or removed and edges can be added or removed. In certain models, if a node were added to an existing network, an edge would also be added. For example, this can happen in citation networks (Section 2.3.2), where nodes represent articles and edges represent cited articles. We have also seen this happen in the preferential attachment model (Section 8.4). There are many types of networks in which only nodes may be added; these are called *node-dynamic* networks.

We define a network that varies over time as follows:

Definition 17.1 A *dynamic network* is a collection of undirected graphs that take the form

$$\mathcal{G}_t = (\mathcal{V}_t, \mathcal{E}_t), \ t \in \mathcal{T}, \tag{17.1}$$

where $\mathcal{V}_t$ and $\mathcal{E}_t$ constitute the set of nodes and set of edges, respectively, that form the network $\mathcal{G}_t$ at time $t \in \mathcal{T}$, where time can be continuous or discrete.

Dynamic networks are also known as *evolutionary networks* or *time-varying networks* in the literature of various disciplines. The relatively new field of *dynamic network analysis* has become a very popular research topic in network science and data science. Interpreting the behavior of time-varying networks has also been described as *dynamic network tomography*, a term borrowed from magnetic resonance imaging (Xing et al., 2010). Dynamic networks can be described by certain characteristics such as the dependence within and between networks in the sequence and the types of changes to the network that can occur from one time point to the next. Dynamic networks can be deterministic or stochastic.

In some applications, the entire network $\mathcal{G}_t$ will be observed at each point in time, whilst in other situations, either the sequence of nodes $\mathcal{V}_t$ or the sequence of edges $\mathcal{E}_t$ are observed through time, or some combination of those two scenarios. In many instances, an edge between two nodes may not be binary (0/1) valued, but may instead be represented by the probability that there is an edge.

There is still much that is unknown about dynamic networks, primarily because real-world dynamic networks have not become readily available yet, and those that are available tend to be small, not well organized, and not easy to visualize. Articles that studied the problems of constructing and evaluating algorithms that produce visual layouts for dynamic networks date back to the 1990s, and are described in the surveys by Vehlow et al. (2015) and Beck et al. (2017).

Notation. How does one represent nodes and edges over time? One way is to follow Rossetti and Cazabet (2018), who use the following notation: A node of a dynamic network graph is represented as (v, t_a, t_b), $t_a, t_b \in \mathcal{T}$, where t_a is the start and t_b is the end time point of the existence of node v. An edge of the graph is represented as (v, v', t_a, t_b), where t_a and t_b are the start and end, respectively, of the existence of the edge (v, v'), v and v' being a pair of nodes. For both node and edge, $t_a \leq t_b$.

17.2 Networks with a Time Component

There are different types of networks that depend upon time, including many of those described in earlier chapters of this book. Examples of dynamic networks include the explosive growth of both the Internet and the World Wide Web, disruption of the North American power grid due to weather-related events, the flow of traffic in airline networks that describe the connections between airports, expansions of financial networks, ever-changing patterns of social networks, the spread of infectious diseases and epidemics, interactions of species in ecological networks and food webs, and changes in loyalties and coalitions of terrorist networks.

Networks with a time component can be divided into three categories: edge-weighted, discrete, and continuous, which we now describe.

17.2.1 Edge-Weighted Static Networks

A different way of representing the time feature in an undirected network is to start with a single network $\mathcal{G} = (\mathcal{V}, \mathcal{E})$ and associate each edge e with a parameter called a *time label* λ_e, where $\lambda_e : \mathcal{E} \to \mathbb{R}$, which specifies the time at which its two endpoints, v and v', communicate with each other (Kempe, Kleinberg, and Kumar, 2002).

Sometimes, the time label is a member of an interval I_e, where we know that communication took place sometime during that interval, but we do not know exactly when. If we set I_e to be a constant, we know, a priori, the precise time that the communication happened. An example of such a network is a telephone network, where records will include information on the phone call, such as origination and termination telephone numbers, the date and time of the call, and the duration of the call (Hill et al., 2006).

This structure, which has been called an *edge-weighted static network*, is, however, not a dynamic network (Skarding, Gabrys, and Musial, 2021).

17.2.2 Discrete-Time Dynamic Networks

The majority of time-varying networks that we deal with are described as *discrete-time dynamic networks*, in which the data consist of a sequence of "snapshots" of the network taken at different times. Examples include: Sampson's monk data (see Section 2.5.4), which were collected as snapshots of the relationships between the novices taken at three discrete times with a view to studying how those relationships evolved over time, and the e-mails by Enron employees (see Section 2.5.2), transmitted every month for a year (i.e., 12 time points) before and during the collapse of the company.

Definition 17.2 We can represent discrete *snapshots* as an ordered sequence of networks, which we can write as $\{\mathcal{G}_1, \mathcal{G}_2, \ldots, \mathcal{G}_T\}$, where T is the number of snapshots and $\mathcal{G}_t = (\mathcal{V}_t, \mathcal{E}_t)$ represents the network configuration at time t, $t = 1, 2, \ldots, T$.

The discrete sequence of snapshots then allows the network at each snapshot to be analyzed as a static network, perhaps followed by some form of video animation that connects them together. Combining the static network analyses enables one to approximate the process of a continuous dynamic network, which is the most complex type of dynamic network, but also the most informative. Clearly, the greater the number of snapshots, the better the approximation. Furthermore, the time interval between snapshots should be kept small, but not too small, which might prevent any network structure from being discovered.

In many published studies, a time-dependent discrete sequence of networks is represented as

$$\{\mathcal{G}_1, \mathcal{G}_2, \ldots, \mathcal{G}_T\}, \quad \mathcal{G}_t = (\mathcal{V}, \mathcal{E}_t). \tag{17.2}$$

Notice that, unlike Definition 17.2, it is assumed in (17.2) that the set of nodes remains the same for all time points (i.e., $\mathcal{V}_1 = \mathcal{V}_2 = \cdots = \mathcal{V}_T = \mathcal{V}$), with the number of nodes $N = |\mathcal{V}|$ fixed, while the set of edges can change at each time point. For such a network model, let $\mathcal{G} = (\mathcal{V}, \mathcal{E})$ denote a given network and set $\mathcal{G}_0 = \mathcal{G}$ and $\mathcal{E}_0 = \mathcal{E}$. Consider addition/deletion operations on the edges of $\mathcal{G}$ that alter its topology.

Let $e_{\pi_t} \in \mathcal{E}$ denote an edge to be added or deleted at time t, depending upon its status at time $t - 1$. At time point t, there have already been $t - 1$ edge changes, either additions or deletions; these can be denoted by $e_{\pi_1}, e_{\pi_2}, \ldots, e_{\pi_{t-1}}$, for all $e_{\pi_i} \in \mathcal{E}$. An edge addition at time t takes the form $\mathcal{E}_t = \mathcal{E}_{t-1} \cup \{e_{\pi_t}\}$, assuming that $e_{\pi_t} \notin \mathcal{E}_{t-1}$; otherwise, an edge deletion at time t takes the form $\mathcal{E}_t = \mathcal{E}_{t-1} \setminus \{e_{\pi_t}\}$.

Comment. Although discrete-time dynamic networks are quite popular, we note that there are certain complications that arise in working with such networks, especially when dealing with large networks. These complications include the loss of information from using snapshots of a changing network over time that approximate a continuously evolving network, and the computational requirements of having to use enormous amounts of computational memory and processor time.

17.2.3 Continuous-Time Dynamic Networks

Improvements in data collection, advances in computational facilities, and the extension of network techniques to many different disciplines have provided opportunities of working instead with *continuous-time dynamic networks*, which are considered as the most complicated of the types of dynamic networks, yet have the most potential. Although studies of continuous-time dynamic networks go back to Holland and Leinhardt (1977), the theoretical and statistical development of the area is still in its infancy.

There are several ways of defining a continuous-time dynamic network. One such definition is the following (Nguyen et al., 2018):

Definition 17.3 A *continuous-time dynamic network* can be represented as

$$\mathcal{G} = (\mathcal{V}, \mathcal{E}_T, \mathcal{T}), \tag{17.3}$$

where $\mathcal{V}$ is a set of nodes, $\mathcal{E}_T \subseteq \mathcal{V} \times \mathcal{V} \times \mathbb{R}^+$ is the set of edges between pairs of nodes in $\mathcal{V}$, and $\mathcal{T} : \mathcal{E} \to \mathbb{R}^+$ is a function that maps each edge to a corresponding time point.

Visualizing a continuous-time dynamic network can take different forms (Skarding, Gabrys, and Musial, 2021). In a *graph stream representation*, edges (viewed as "events") appear continuously and at a fast rate over time, but tend to last for a while. (Nodes that appear or disappear over time can also be regarded as events.) In an *event-based representation* of a continuous-time dynamic network, the time at which the event appeared and the duration of the event are its main characteristics.

An event can be viewed as a behavior (e.g., edge deletion, edge addition, node deletion, node addition) carried out at a certain point in time. Note that when deleting a node, all edges linked to that node will also be deleted. In some cases, such as e-mail and text-messaging networks, the event may be instantaneous or of short duration (lasting a second or millisecond), or perhaps the event is not considered to be important, in which case, no event duration is given. In other situations, it is important to define what constitutes an event: for example, in a graph stream representation, edge appearance and edge disappearance are often considered to be two different and separate events. In summary, the type of event determines the type of dynamic network.

17.3 Dynamic Community Discovery

When building static network models, probabilistic rules are used to determine how edges of a network are formed, and specific features found in real-world networks, such as node-degree distribution or community structure, are included in the model. Incorporating time into a network model introduces a new dimension into the model, especially if we are interested in tracking changes in network structure over time.

How should we deal with changes in a network that occur over time? What sorts of structural changes should we expect to see over time? In Table 17.1, we list the terminology of the lifecycle of a dynamic community (Rossetti and Cazabet, 2018) that explains in a simple way the various possible structural changes of a network over time, which they refer to as *dynamic community discovery*.

In some of these scenarios (e.g., Merge, Split, and Resurgence), there are decisions that have to be made about how to deal with dynamic community life (e.g., which communities persist and which communities disappear). Furthermore, different community detection algorithms may produce different possible partitions of an existing network, depending upon whether the partition yields disjoint or overlapping communities, or whether the algorithm has an instability problem that produces inconsistent solutions.

Although dynamic community discovery is a relatively new field of study, there is every possibility that it will expand in the future. There is still much work to be accomplished because it is important to understand how communities function and change over time. With this in mind, we present the following brief description of dynamic community discovery.

17.3.1 Modeling Dynamic Network Structure

Consider a discrete-time sequence of networks of the form (17.2), where the nodes remain the same for all time points. The goal is to develop a statistical model that tracks how nodes behave differently over time whilst carrying out different roles or functions. Research on this problem has centered around providing dynamic versions of many of the static community detection methodologies discussed in this book, including modularity maximization (e.g., Louvain algorithm), degree-corrected SBM, and spectral clustering. For a survey, see Dall'Amico, Couillet, and Tremblay (2020).

Table 17.1 Lifecycle of a dynamic community. Adapted from Rossetti (2010)

Term	Definition
Birth	a new community appears containing a number of nodes
Death	an existing community disappears
Growth	an existing community expands by adding nodes
Contraction	some nodes are dropped from a community
Merge	two or more communities merge together to form a single community
Split	an existing community breaks into two or more communities
Continue	no change to an existing community
Resurgence	a community vanishes for some time, then returns

Dynamic Mixed-Membership SBM

The most interesting of these methods was developed in a series of articles by Fu, Song, and Xing (2009), Xing et al. (2010), and Ho, Song, and Xing (2011). They introduced a time factor into the mixed-membership SBM for a static network (Airoldi et al., 2008), and addressed the question of how to think about dynamic networks with community structure.

In the first two of these articles, Fu et al. and Xing et al. presented an algorithm, which they called dMMSB, for a dynamic (discrete-time) extension of the static *mixed-membership SBM* (see Section 14.2). At each time point, this model allows each of the N nodes in the network to belong to each of $K > 1$ communities. What this approach does is to allow individual nodes to play multiple roles, where a role is associated with a given community. The question is to determine how likely it is that a node is a member of each community and, hence, plays the role associated with that community. Over time, we should not be surprised to see some nodes changing their community membership and their role in favor of a different community and role. Note that the roles in the model are not independent of each other.

Recall that the static mixed-membership SBM uses a Bayesian approach with a Dirichlet prior because it is conjugate to the multinomial distribution. However, when this approach is applied to the vector of proportions of the mixed-membership SBM, it leads to an intractable computational problem.

The dMMSB model closely follows the general outline of the mixed-measurement SBM, but with certain modifications due to the time component. They also used a Bayesian approach, but now the ith mixed-membership K-vector $\boldsymbol{\pi}_{i,t} = (\pi_{i1,t}, \ldots, \pi_{iK,t})^{\tau}$, $i = 1, 2, \ldots, N$, the $(K \times K)$-matrix $\mathbf{B}_t = (B_{ij,t})$, and the prior distributions each vary over time.

The K-vector $\boldsymbol{\pi}_{i,t}$ is a vector of mixture probabilities for the ith node at time t. For example, if there are $K = 3$ communities, then the ith mixed-membership vector at time t could be $\boldsymbol{\pi}_{i,t} = (0.1, 0.6, 0.3)^{\tau}$, meaning that, on average, at time t, the ith node has a 10% chance of being a member of community 1, a 60% chance of being a member of community 2, and a 30% chance of being a member of community 3.

Instead of a Dirichelet prior as in (14.3), they used a logistic-normal prior. The dynamic networks are generated using the following sequential process:

1. For the ith node, $v_{i,t}$, at time point t, draw N mixed-membership K-vectors from the logistic-normal distribution.

$$\boldsymbol{\pi}_{i,t} \sim \mathcal{LN}_K(\boldsymbol{\mu}_t, \Sigma), \quad i = 1, 2, \ldots, N. \tag{17.4}$$

The logistic-normal distribution is obtained in two steps as follows:

1. For the ith node, draw a K-vector $\boldsymbol{\gamma}_{i,t} = (\gamma_{i1,t}, \ldots, \gamma_{iK,t})^{\tau}$ according to

$$\boldsymbol{\gamma}_{i,t} \sim \mathcal{N}_K(\boldsymbol{\mu}_t, \Sigma), \quad i = 1, 2, \ldots, N. \tag{17.5}$$

2. Take a logistic transformation of each element of $\boldsymbol{\gamma}_{i,t}$ to form $\pi_{ik,t}$:

$$\pi_{ik,t} = \exp\{\gamma_{ik,t} - C(\boldsymbol{\gamma}_{i,t})\}, \quad k = 1, 2, \ldots, K, \ i = 1, 2, \ldots, N, \tag{17.6}$$

where the normalizing constant is

$$C(\boldsymbol{\gamma}_{i,t}) = \log\left\{\sum_{k=1}^{K} \exp\{\gamma_{ik,t}\}\right\}. \tag{17.7}$$

From (17.6) and (17.7), we see that

$$\pi_{ik,t} = \frac{e^{\gamma_{ik,t}}}{\sum_{\ell=1}^{K} e^{\gamma_{i\ell,t}}}, \quad k = 1, 2, \ldots, K, \ i = 1, 2, \ldots, N. \tag{17.8}$$

2. Draw a sample of means from the dMMSB priors at time t.
Based upon the state-space model, it is assumed that the mean vector $\boldsymbol{\mu}_t$ in (17.4) varies over time as a linear normal model, where

$$\boldsymbol{\mu}_t = \mathbf{A}\boldsymbol{\mu}_{t-1} + \mathbf{w}_t, \quad \mathbf{w}_t \sim \mathcal{N}_K(\mathbf{0}, \Phi), \tag{17.9}$$

and $\mathbf{A}$ is a transition matrix (in the simplest case, $\mathbf{A}$ can be set to the identity matrix $\mathbf{I}_K$ and Φ can be set to $\sigma \mathbf{I}_K$). In other words:

$$\boldsymbol{\mu}_t \sim \mathcal{N}_K(\mathbf{A}\boldsymbol{\mu}_{t-1}, \Phi), \quad t > 1, \tag{17.10}$$

and for $t = 1$, we have that $\boldsymbol{\mu}_1 \sim \mathcal{N}_K(\boldsymbol{\theta}, \Phi)$, where $\boldsymbol{\theta} = \boldsymbol{\mu}_0$.

3. Compute probabilities using a logistic transformation.
Similarly, it is assumed that each entry of $\mathbf{B}_t$ is distributed as a logistic-normal prior

$$B_{ij,t} \sim \mathcal{LN}(\eta_{ij,t}, \Psi), \quad i, j = 1, 2, \ldots, N, \tag{17.11}$$

where the state-space model assumes that

$$\eta_{ij,t} = b\eta_{ij,t-1} + \xi_{ij,t}, \quad \xi_{ij,t} \sim \mathcal{N}(0, \psi), \tag{17.12}$$

and b is a scalar. In other words,

$$\eta_{ij,t} \sim \mathcal{N}(b\eta_{ij,t-1}, \psi), \quad t > 1, \tag{17.13}$$

and for $t = 1$, we have that $\eta_{ij,1} \sim \mathcal{N}(\beta, \psi)$, where $\beta = \eta_{ij,0}$.
The entries of $\mathbf{B}_t$ are, therefore, given by a logistic transformation of $\eta_{ij,t}$:

$$B_{ij,t} = \frac{e^{\eta_{ij,t}}}{e^{\eta_{ij,t}} + 1}, \quad i, j = 1, 2, \ldots, N. \tag{17.14}$$

The covariances, Σ and Ψ, are assumed to be independent over time.
The $(K \times K)$-matrix $\mathbf{B}_t$ relates the different latent roles to each other. If $\mathbf{B}_t$ is diagonal, then nodes are more likely to interact only with other nodes that are members of the same community (i.e., with nodes playing the same role), whilst if $\mathbf{B}_t$ is not diagonal, nodes are likely to interact with nodes from different communities (i.e., with nodes playing different roles). Hence, $\mathbf{B}_t$ has been referred to as a *role-compatibility matrix*.
Extending (14.5) and (14.6) to show dependence upon time yields indicator vectors for each initiator $\mathbf{Z}_{i \to j,t}$ and receiver $\mathbf{Z}_{i \leftarrow j,t}$:

4. For each pair of nodes, $v_{i,t}$ and $v_{j,t}$, at time t, draw a sample of the membership indicator vectors for the initiator and receiver nodes.

$$\mathbf{Z}_{i \to j,t} \sim \mathcal{M}(1, \boldsymbol{\pi}_{i,t}), \quad \mathbf{Z}_{i \leftarrow j,t} \sim \mathcal{M}(1, \boldsymbol{\pi}_{j,t}), \quad i, j = 1, 2, \ldots, N, \tag{17.15}$$

respectively, where $\mathcal{M}(1, \boldsymbol{\pi})$ represents a single observation taken from the multinomial distribution with probability vector $\boldsymbol{\pi}$.

5. For each pair of nodes, $v_{i,t}$ and $v_{j,t}$, at time t, draw a sample of their interaction.

$$Y_{ij,t} \sim \text{Bernoulli}(p_{ij,t}), \quad i, j = 1, 2, \ldots, N, \tag{17.16}$$

where $Y_{ij,t}$ is the element in the ith row and jth column of the adjacency matrix $\mathbf{Y}_t$ at the tth time point (i.e., $Y_{ij,t} = 1$ if there is an edge between nodes $v_{i,t}$ and $v_{j,t}$, and $Y_{ij,t} = 0$ otherwise), and because $\mathbf{Z}_{i \to j,t}$ and $\mathbf{Z}_{i \leftarrow j,t}$ are indicator vectors, $p_{ij,t} = \mathbf{Z}_{i \to j,t}^{\tau} \mathbf{B}_t \mathbf{Z}_{i \leftarrow j,t}$ chooses a single element of $\mathbf{B}_t$.

6. Statistical inference.

For the static MMSB, where $\boldsymbol{\pi}_i$ is drawn from a Dirichlet prior, exact statistical inference is not feasible. Marginalization of the posterior distribution over a huge state space for the model parameters (i.e., $\boldsymbol{\alpha}$, $\mathbf{B}$) and latent variables (i.e., $\mathbf{Z}_\to$, $\mathbf{Z}_\leftarrow$, $\{\boldsymbol{\pi}_i\}$) using the EM algorithm turns out to be computationally intractable. As described in Section 14.2, the variational EM algorithm was developed to carry out an approximate inference procedure.

For the time-varying version dMMSB, the model parameters are $\boldsymbol{\mu}_t$, Σ, and $\mathbf{B}_t$, and the latent variables are $\mathbf{Z}_{\to,t} = \{\mathbf{Z}_{i \to j,t}\}$, $\mathbf{Z}_{\leftarrow,t} = \{\mathbf{Z}_{i \leftarrow j,t}\}$, and $\{\boldsymbol{\pi}_{i,t}\}$. The parameters are to be estimated at time t from the data in the set of edges $\mathcal{E}_t$. Using a logistic-normal prior for $\boldsymbol{\pi}_{i,t}$ makes integration for estimating the model parameters even more complicated. So, an extra approximation step was added to the variational EM procedure. A Laplace approximation to (17.7) was used to approximate the joint posterior distribution of the mixed-membership vector of roles.

The variational inference approach is based upon the mean-field method, which approximates the joint posterior distribution by a factored approximate distribution. The variational parameters are estimated by minimizing the Kullback–Leibler divergence between the true posterior distributions and the approximate posterior over arbitrary choices of the factored distributions. For details, see Xing et al. This approximation, referred to as a *Laplace variational EM algorithm*, provides a method for inferring the latent variables and for estimating the model parameters.

Xing et al. also showed, using simulated network data, that the dMMSB model has advantages over the static MMSB model with a Dirichlet prior that ignores the time component linking the networks and when different roles are correlated. Furthermore, if the matrix $\mathbf{B}_t$ is not taken to be diagonal (i.e., roles are assumed to be correlated), they show that the logistic-normal prior provides a better fit than using a Dirichlet prior.

Extending the dMMSB Model by Mixtures of Distributions

Because the logistic-normal prior is unimodal, it cannot provide a good fit to complex, multimodal data. So, this approach was extended by Ho, Song, and Xing (2011) to a multimodal prior distribution consisting of a *mixture* of discrete-time logistic-normal distributions. However, they note that there are integrability problems with this mixture prior, the likelihood tends to have multiple local maxima, and there are computational difficulties due to the fact that the state space is exponentially large in the number of nodes N and time points T. They overcome these problems by developing an approximation using a variational EM procedure based upon the generalized mean field algorithm, similar to that proposed by Xing et al.

17.3.2 Example: Simpson's Monk Data

We next present the results of the algorithm applied by Xing et al. to Simpson's monk network and compare those results to the static results from Section 2.5.4.

Xing et al. (2010, Figure 7) assessed that the 18 novice monks could be grouped into the three communities of Young Turks (YT), Outcasts (O), and Loyal Opposition (LO). In **Table 17.2**, we display the memberships of the communities formed by the "liking" relationships at each of three time points, T_1, T_2, and T_3. Memberships of these communities were determined by Xing et al., who applied to the data a dynamic discrete-time version of the mixed-membership SBM with three communities (see Section 14.2). For each novice at each time point, a probability was calculated for each of the three roles of YT, O, and LO. **Table 17.2** displays the community that has the maximum probability for each novice and time point. For comparison, we display in the last column the static classifications for four roles, including the Interstitials (I), at time point T_3, which were obtained from Section 2.5.4.

We see a striking difference between the static snapshot and the dynamic classifications for many of the novices. Novices 1–3, 5, 7–10, 12–14 each had different static and dynamic classifications. We also note that most of the novices were consistent in their roles over time; only novices 4, 9, and 14 showed abrupt changes in their roles from time point T_1 to time point T_2. The above static classifications agreed with the static classifications displayed in the three corners of a triangle in Xing et al. (2010, Figure 5), except for the interstitial role, which they did not consider. The three

Table 17.2 Static and dynamic classifications of Simpson's monk network. Data were collected at three time points, T_1, T_2, and T_3. At each of these time points, the communities of Young Turks (YT), Outcasts (O), and Loyal Opposition (LO) were formed. The memberships of these communities, which are taken from Xing et al. (2010, Figure 7), are determined by the highest probability roles for each monk based upon a dynamic mixed-membership stochastic blockmodel. The static classifications, including Interstitials (I), are for time point T_3 and were taken from Section 2.5.4

No.	Name	T_1	T_2	T_3	Static
1	John	LO	LO	LO	YT
2	Gregory	LO	LO	LO	YT
3	Basil	YT	YT	YT	O
4	Peter	O	LO	LO	LO
5	Bonaventure	YT	YT	YT	LO
6	Berthold	LO	LO	LO	LO
7	Mark	O	O	O	YT
8	Victor	YT	YT	YT	I
9	Ambrose	YT	O	O	LO
10	Ramuald	YT	YT	YT	I
11	Louis	LO	LO	LO	LO
12	Winfrid	LO	LO	LO	YT
13	Armund	YT	YT	YT	I
14	Hugh	O	LO	LO	YT
15	Boniface	YT	YT	YT	YT
16	Albert	YT	YT	YT	YT
17	Elias	O	O	O	O
18	Simplicius	O	O	O	O

interstitials are Victor (8), Ramuald (10), and Armund (13). In a static analysis, Xing et al. placed all three of them in the YT group. If we now substitute YT for each of the three I classifications, we have agreement by 10/18 of the novices.

17.4 Longitudinal Social Networks

Much of social science research has been on studying social interactions (e.g., written communications, Facebook friendships, Twitter or LinkedIn professional connections, scientific collaborations, or physical contacts) in a wide variety of fields. The study of such social interactions had much to do with the growth in popularity of the cross-disciplinary science of *social network analysis* or SNA. The resulting applications of SNA in the "big data" era has encouraged the development of specialized software and algorithms.

Many real-world social networks, however, are not static. We see this in Facebook, Twitter, and LinkedIn, whose memberships grow and change frequently over time. So, the next stage in the evolution of SNA proceeded with the study of time-varying social networks with a view to understanding trends, changes, and evolution of those networks. These include research on such diverse areas as the collaborations of employees in a large organization, the spread of viruses in computer networks, polling and survey work, and the activities of many small, interconnected terrorist groups. Such networks are generally referred to as *longitudinal social networks*.

Various disciplines have since taken longitudinal social networks in a number of different directions, including an emphasis on extracting information on the structure of complex and dynamic social networks through computational and statistical approaches. These types of studies have expanded greatly due to the huge volume of data collected using electronic communications. In particular, the area of social networks has seen many books and articles on the subject published since the late 1950s. However, it took several decades for the time component to be taken into account in the modeling of longitudinal social networks. Articles by Holland and Leinhardt (1977) and Wasserman (1980) introduced continuous-time Markov chains for modeling longitudinal social networks, which was followed by other Markovian methods for analyzing such data. See, for example, the book on dynamic network theory relevant for social networks by Westerby (2012).

Recent research efforts in social networks have focused on statistical work on estimation, inference, and Bayesian methods, so that static network methods can be extended to time-varying networks (see Bartlett, Kosmidis, and Silva, 2021, for references). This also includes the detection of changes in network community structure over time. For example, the static SBM methodology was extended to longitudinal social networks (Xu and Hero III, 2014). There has also been a recognition of the need for good ideas on dealing with measurement error and goodness-of-fit statistics.

We now outline recent work on network monitoring and change detection, which has become an important research field in social network analysis.

17.4.1 Network Monitoring and Change Detection

Much of the research on longitudinal social networks has centered on developing statistical methods for monitoring network change and network evolution.

Types of dynamic network behaviors. Adapted from McCulloh and Carley (2011)

Type of network change	Definition
Network stability	network relationships do not change
Endogenous change	the network changes due to the evolution of goals and motives
Exogenous change	a source outside the network causes a change in the network relationships
Initiated change	network change is initiated by a sequence of outside changes

In particular, interest has been shown in detecting an abrupt change in a social network, such as organizational behavior, over time. Organizations tend to change their structure, their composition, and their focus gradually (or maybe quickly) over time to reflect changes in personnel or direction, and often it is of interest to identify when such changes occurred.

With this in mind, McCulloh and Carley (2011) identified four types of dynamic network behaviors, which they called *network stability*, *endogenous change*, *exogenous change*, and *initiated change*. Descriptions of these types are given in **Table 17.3**. If a network is stable, changes tend to occur only through observation error, which translates, for example, into minor variability in daily work schedules. Endogenous change can occur in a network, for example, when individuals interact with each other and share beliefs and experiences that cause relationships to evolve. Exogenous change can occur when an external event, such as layoffs due to a financial downturn in the economy, causes a change in the relationships between individuals in a company network. Initiated change can occur when an exogenous change, such as certain bad actions of an individual, causes a sequence of endogenous changes, such as how the relationships of other members of the network are affected.

Statistical Process Control

Statistical measures of change detection in longitudinal social networks have been borrowed from *statistical process control*, which uses control charts to detect changes in industrial production. A control chart plots a summary statistical measure, such as a mean, against time. If the plotted points stay between some designated upper and lower control limits, then it is said that the process is "in control." If a plotted point falls outside of either control limit, then the process is declared to be "out of control," implying that a change probably occurred at that point in time; the manufacturing process is then stopped and adjustments are made to the process so that a reset can take place.

In the case of a sequence of networks, there are certain features of such measures that are deemed to be important. These include the need for scale invariance across different time points, which would have to take into account different network sizes, and so any measures adopted should be standardized.

McCulloh and Carley (2008, 2011) studied the ability of different types of statistical measures (e.g., average closeness, average betweenness) to monitor network summary statistics and detect network changes. Of special interest were the *cumulative sum* (CUSUM) and the *exponentially weighted moving average* (EWMA).

Suppose a process operates in such a way that the mean of the measurements is μ_0 and the standard deviation is σ when the process is "in control," and suppose also that μ_0 and σ are known constants. Interest focuses on detecting changes in the mean.

CUSUM. The CUSUM chart procedure uses the statistic

$$C_t = \sum_{j=1}^{t} (Z_j - k), \quad t = 1, 2, \ldots, T, \tag{17.17}$$

where Z_j is a standardized version of the mean, and k is taken to be 0.5. It is assumed that $Z_j \sim \mathcal{N}(0, 1)$, $j = 1, 2, \ldots, t$. Two statistics are used to detect a change over time:

- $C_t^+ = \max\{0, C_{t-1}^+ + (Z_t - \mu_0) - k\}$, to be compared to an upper control limit h^+,
- $C_t^- = \min\{0, C_{t-1}^- + (Z_t - \mu_0) + k\}$, to be compared to a lower control limit h^-,

$t = 1, 2, \ldots, T$. The initial values are $C_0^+ = c$ and $C_0^- = -c$. In each case, if either $C_t^+ > h^+$ or $C_t^- < -h^-$, this is evidence that the mean has changed over time; in the first case, there is evidence that the mean has increased over time, and in the second case, that it has decreased over time. The main problem with this CUSUM methodology is that the normality assumption is violated for network analysis, which tends to introduce bias into the ability to detect change.

EWMA. Let

$$w_t = \lambda Z_t + (1 - \lambda) w_{t-1}, \quad t > 1, \tag{17.18}$$

denote the EWMA statistic associated with the network at time $t > 1$, where $\lambda \in (0, 1)$ is a weight and when $t = 1$, then $w_0 = \mu_0$. Typically, $\lambda \in [0.1, 0.3]$ performs best in detecting small changes in a process mean. Then, after observing T time points, the statistic w_T is plotted against the interval

$$\mu_0 \pm L\sigma \left(\frac{\lambda}{2 - \lambda} \left(1 - (1 - \lambda)^{2T} \right) \right)^{1/2}, \tag{17.19}$$

where L is a scaling constant.

See Hawkins (2014) for a comparison of CUSUM and EWMA control charts. Adjustments and improvements to these procedures are given by Miller, Beard, and Bliss (2011, 2013) and Azarnoush et al. (2016), who incorporated covariate information and logistic regression into their models. An excellent survey of social network monitoring is given by Woodall et al. (2017).

17.5 Graph Distances for Comparing Networks

A different approach to the problem of monitoring a sequence of networks through time and detecting changes in those networks is described by Donnat and Holmes (2018). Their method for dealing with time-varying networks does not consider statistical process control techniques. Instead, they describe several types of distances between different graphs, such as the same network recorded at two successive time points. Amongst the graph distances that they study are the *Hamming distance*, the *Jaccard distance*, and *spectral distances* (e.g., ℓ_p distances on eigenvalues, *spanning tree dissimilarities*, and the *IM distance*).

In the following definitions, $\mathcal{G} = (\mathcal{V}, \mathcal{E})$ and $\mathcal{G}' = (\mathcal{V}, \mathcal{E}')$ denote two different network graphs, with $N = |\mathcal{V}|$ the number of nodes. In the context of this chapter, think of $\mathcal{G}$ as $\mathcal{G}_t$ and $\mathcal{G}'$ as $\mathcal{G}_{t+1}$. We also set $\mathbf{Y} = (Y_{ij})$ to be the adjacency matrix corresponding to $\mathcal{G}$ and $\mathbf{Y}' = (Y'_{ij})$ the adjacency matrix for $\mathcal{G}'$.

17.5.1 Hamming Distance

This distance is a measure of how much $\mathcal{G}$ and $\mathcal{G}'$ differ from one another. It counts how many edge additions or deletions it takes to convert $\mathcal{G}$ into $\mathcal{G}'$. The normalized Hamming distance between these two networks is given by the following:

$$d_H(\mathcal{G},\mathcal{G}') = \sum_{i,j} \frac{|Y_{ij} - Y'_{ij}|}{N(N-1)} = \frac{1}{N(N-1)} \parallel \mathbf{Y} - \mathbf{Y}' \parallel_1 , \qquad (17.20)$$

where $\parallel \cdot \parallel_1$ is the ℓ_1 norm. The d_H distance lies between 0 and 1.

17.5.2 Jaccard Distance

This distance is a more appropriate measure of distance between networks. It can be viewed as the proportion of edges that have either been added or deleted relative to the total number of edges in either network. It is defined by the following:

$$d_J(\mathcal{G},\mathcal{G}') = \frac{|\mathcal{G} \cup \mathcal{G}'| - |\mathcal{G} \cap \mathcal{G}'|}{|\mathcal{G} \cup \mathcal{G}'|} = \frac{\sum_{i,j} |Y_{ij} - Y'_{ij}|}{\sum_{i,j} \max(Y_{ij}, Y'_{ij})} = \frac{\parallel \mathbf{Y} - \mathbf{Y}' \parallel_1}{\parallel \mathbf{Y} + \mathbf{Y}' \parallel_*}, \qquad (17.21)$$

where $\parallel \cdot \parallel_*$ is the nuclear norm. The Jaccard distance compensates for network density by normalizing by a quantity that contains the average sparsity of the two networks.

In an empirical study of the Hamming and Jaccard distances using real data, Donnat and Holmes found that the Hamming distance has substantial flaws. The flaws in the Hamming distance include an inability to recognize similarities in two networks, an inability to distinguish the difference in importance between additions and deletions, even if they have different effects on network structure, and an inability to identify structural differences between two closely related networks. The Jaccard distance also treats all edges in the same way, and is not sensitive to the level of sparsity in a network. While the Hamming distance is sensitive to the density of a network and is strongly affected by network sparsity, the Jaccard distance, in certain applications, is more appropriate for comparing sparse networks.

17.5.3 Spectral Distances

These distances are based upon the eigenvalues of either the adjacency matrix $\mathbf{Y}$, or the unnormalized graph Laplacian $\mathbf{L} = \mathbf{D} - \mathbf{Y}$, or its normalized version, $\widehat{\mathbf{L}}^{sym} = \mathbf{I}_N - \mathbf{D}^{-1/2}\mathbf{Y}\mathbf{D}^{-1/2}$, which lies between 0 and 2, where $\mathbf{D} = \text{diag}\{\mathbf{Y}\mathbf{J}_N\}$ contains information about the degree of every node in $\mathcal{G}$ and $\mathbf{J}_N = \mathbf{1}_N\mathbf{1}_N^\top$. See Chapter 13. Spectral distances are not influenced by the identity of each node and, hence, are permutation invariant. They can be used to compare different networks as long as N is common to the networks. Let $\{\lambda_i\}$ and $\{\lambda'_i\}$ denote the eigenvalues of the (unnormalized or normalized) graph Laplacian of $\mathcal{G}$ and $\mathcal{G}'$, respectively. These eigenvalues are all real and non-negative, and $\lambda_0 = \lambda'_0 = 0$.

- The simplest version of spectral distance between $\mathcal{G}$ and $\mathcal{G}'$ is the ℓ_p distance given by

$$d(\mathcal{G},\mathcal{G}') = \left\{ \sum_{i=0}^{N-1} |\lambda_i - \lambda_i'|^p \right\}^{1/p}. \tag{17.22}$$

Various transformations of the eigenvalues are considered by the authors using $p = 2$.

- Another choice of spectral distance deals with the use of a *spanning tree* that distinguishes between a network and a transformation of that network. A spanning tree can be viewed as a measure of the connectivity of a network and of the effect an edge deletion has on that connectivity. The number of spanning trees for a connected network can be computed as follows. Consider the ordered eigenvalues, $0 = \lambda_0 < \lambda_1 \leq \cdots \leq \lambda_{N-1}$, of the (unnormalized) graph Laplacian $\mathbf{L} = \mathbf{D} - \mathbf{Y}$. The matrix-tree theorem (Kelmans, 1976, 1997) says that the number of spanning trees for the network $\mathcal{G}$ is given by

$$\mathcal{T}_\mathcal{G} = \frac{1}{N} \prod_{i=1}^{N-1} \lambda_i. \tag{17.23}$$

Then, the *spanning tree dissimilarity* between the two graphs $\mathcal{G}$ and $\mathcal{G}'$ is

$$d_{ST}(\mathcal{G},\mathcal{G}') = |\log(\mathcal{T}_\mathcal{G}) - \log(\mathcal{T}_{\mathcal{G}'})|. \tag{17.24}$$

There is empirical evidence that, when comparing two weakly connected networks, the ST distance (17.24) can identify certain structural information depending upon the values of the eigenvalues.

- The *IM distance* is a spectral measure that was proposed by Ipsen and Mikhailov (2002) for reconstructing graphs from their Laplacian spectra. It is defined as the L_2 measure of the difference of the spectral densities of the two networks:

$$d_\gamma(\mathcal{G},\mathcal{G}') = \left\{ \int_0^\infty [\rho_\mathcal{G}(\omega,\gamma) - \rho_{\mathcal{G}'}(\omega,\gamma)]^2 d\omega \right\}^{1/2}, \tag{17.25}$$

which lies between 0 and 1, where the spectral density of a network graph is defined as

$$\rho(\omega,\gamma) = K \sum_{i=1}^{N-1} \frac{\gamma}{\gamma^2 + (\omega - \omega_i)^2}, \quad \omega \in [0,\infty]. \tag{17.26}$$

In (17.26), $\omega_i^2 = \lambda_i$, $i = 1,2,\ldots,N-1$, with $\omega_0 = \lambda_0 = 0$, γ is a scale parameter to be determined (although Ipsen and Mikhailov use $\gamma = 0.08$), and K is a normalization constant defined to make $\int_0^\infty \rho(\omega,\gamma)d\omega = 1$. In empirical studies, Jurman, Visintainer, and Furlanello (2011) found that the IM distance appears to be almost independent of the number of nodes.

- None of the above spectral distances is able to identify specific changes from one network to another. So, Jurman et al. (2015) proposed the *HIM distance*, a linear combination of the normalized Hamming distance (17.20) and the normalized IM distance (17.25), which overcomes the flaws of both types of distances. The HIM distance lies between 0 and 1. Unfortunately, the HIM distance does not appear to be computationally feasible when used with large networks.

17.6 Dynamic Biological Networks

A biological network can be represented by a graph $\mathcal{G} = (\mathcal{V}, \mathcal{E})$, where $\mathcal{V}$ is a set of nodes that identify with molecular components (e.g., genes, proteins, metabolites) that perform various cellular functions, and $\mathcal{E}$ is a set of (directed or undirected) edges that identify with regulatory relationships between those molecular components (often called the *topology* of the network). Although there are at least 100,000 cellular components represented by nodes in the network known as the *human interactome*, accurate maps of the interactome remain incomplete and noisy (Barabási, Gulbahce, and Loscalzo, 2011). Often, the edge structure of the network may be uncertain, and a biological goal would consist of trying to characterize the edge structure appropriate to the context through biochemical data.

As we saw in Section 2.6, there are different types of biological networks, such as a *gene regulatory network* (with directed edges) or a *protein–protein interaction network* (with undirected edges). These individual networks are not independent of each other, but actually form a "network of networks" that play various roles in determining how a cell behaves (Barabási and Oltvai, 2004).

However, to understand the trajectory of diseases, a static graph of a biological network may be useful, but is severely limited because it cannot show the major changes that can occur in the temporal, context, or condition of the topology of that network. Although biological processes are typically time-varying, much of what has been established on network models has been of the static variety. This is where understanding the time-dependency of biological networks becomes important. In fact, the biological networks observed at each time point are not iid samples from some probability distribution.

The dynamic behavior of biological networks is especially important for detecting cellular defects connected with serious diseases, such as cancer. Studies have shown, for example, that cancer (and certain other diseases) rarely results from an abnormality in a single component of a network; more likely, it responds to an external perturbation of that system (Barabási, Gulbahce, and Loscalzo, 2011). Furthermore, depending upon the state that a network is in at a given time, such information can determine which drug therapies would be most effective and what types of laboratory intervention studies should be planned (Green et al., 2018).

Unlike the case of longitudinal social networks, a sequence of discrete snapshots of a time-varying biological network will not be available. Current technology has not yet provided us with the ability to allow experiments to be carried out with the aim of providing time-dependent networks over a sequence of time points.

Comment. Unfortunately, very little attention has been given to modeling random network models of dynamic biological networks. This leaves open a whole range of possible research areas for such networks.

17.6.1 Finding and Counting Dynamic Network Motifs

It is well known that complex networks contain various subnetworks of a small number (usually consisting of three or four nodes) of highly linked groups of nodes. If a small subnetwork appears frequently within a given biological network, it can be viewed as a basic building block of that network. These subnetworks identify patterns of connections that locally characterize the network. These subnetworks are referred to as *motifs*, with each network characterized by its own set of motifs (Milo et al., 2002).

Static Network Motifs

The application of the term "motif" to static networks was introduced by Uri Alon and coworkers while studying gene regulatory networks and their structures. They defined a network motif as follows:

> **Definition 17.1** A *network motif* is a pattern of interconnections between a small group of nodes that recurs in many different parts of a network much more often than those found in randomized networks.

This suggests that such motifs may carry out special functions in information processing across the network. In directed networks, triangle motifs are also referred to as *feed-forward loops*.

This concept was applied by Shen-Orr et al. (2002) to the *Escherichia coli* transcriptional regulation network, a directed network, which was compared to randomized networks that were generated to have the same characteristics as the *E. coli* network. The nodes in the network represented operons and the edges represented direct transcriptional interactions. In the *E. coli* network, three types of motifs were found, each one of which has a specific function in determining gene expression. From this, it was possible for Shen-Orr et al. to visualize the entire transcriptional network of *E. coli*. Network motifs were then found in other biological networks, ranging from bacteria to humans.

Milo et al. (2002) studied network motifs from a number of types of real-world biological networks. These included transcriptional gene regulation networks (nodes are genes, edges are directed from a gene that encodes for a transcription factor protein to a gene transcriptionally regulated by that transcription factor) and ecosystem food webs (nodes represent groups of species, edges are directed from a predator node to its prey node). In particular, they analyzed the yeast *Saccharomyces cerevisiae* network, a bacterium (*E. coli*) network, and the nematode *Caenorhabditis elegans* network (nodes are neurons and edges are synaptic connections between the neurons). In each example, there were three-node motifs and four-node motifs, and they appeared numerous times in each network. A similar result occurred for the food webs, where also three-node and four-node motifs were found.

Such network motifs, which have been characterized as *basic computational elements*, are often studied with a view to being used for *reverse engineering* (Green et al., 2018). This idea looks at an existing item, where little is known about the structure of the item, breaks it down into its component elements, and then attempts to build a similar item. This approach has been found to be useful for understanding certain aspects of biological behavior. Furthermore, certain types of motifs can join together to form large motif clusters (Mukherjee et al., 2018). Hence, it is important to discover their biological functions, their local properties, and how they interact with each other. However, finding network motifs is a computationally hard problem, especially because the number of nodes increases as the network evolves, as does also the number of motifs. Furthermore, the role of network motifs is still not completely understood.

One of the important goals when analyzing biological networks is to try to count the number of motifs that have a specified pattern, such as triangle, square, pentagon (Mukherjee et al., 2018). There are three ways of counting motifs in a given network, which are listed from least restrictive to most restrictive:

1. Allow overlap between the subnetworks so that they can share nodes or edges.
2. Count only disjoint embeddings of each motif, so that no two embeddings of the same motif share an edge.
3. Count only those motifs where no two embeddings of the same motif share a node.

In the *E. coli* transcription-regulatory network, for example, most motifs overlap, so that individual motifs are not easily identified (Barabási and Oltvai, 2004). The numbers of motifs found by these three methods are assigned the counts $F1$, $F2$, and $F3$, respectively. However, the majority of motif-counting algorithms compute only the $F1$ total.

Dynamic Network Motifs

In a dynamic biological network, the topology of the network changes over time. Certain biological measures may be time-dependent so that an important node at one time point (e.g., a node having the highest degree) may lose its importance at another time point. Furthermore, the motif-counting problem becomes more and more difficult. So far, there has not been much research on dynamic networks, and what there has been, has not attempted to solve the motif-counting problem.

Even though it is important to update the motif count at each time point so that it corresponds with that obtained from a current snapshot of the network, this strategy quickly becomes very difficult and expensive. To overcome this difficulty, various algorithms have been proposed for counting motifs in static and dynamic networks; see Mukherjee et al. (2018) for a description of such algorithms.

There are many nontrivial difficulties (while retaining some element of practicality) when introducing a dynamic feature into a network analysis. Different studies tend to model dynamic network motifs in different ways. One model's definition of a motif may not be valid for another model because of the limitations placed on motifs in the latter model. Also, different disciplines may not be aware of each other's work when forming models of dynamic network motifs, which means that a unified approach to building models does not currently exist.

Temporal Network Motifs

So, how should one incorporate time into the definition of network motifs when studying dynamic biological networks? Researchers have proceeded to work around this difficult problem by replacing the dynamic network by a *temporal network* to describe a biological network with a time component. Although a temporal network is not a dynamic network, a great deal of attention has been focused on temporal networks, and thus we state its definition as follows:

Definition 17.2 Let $\mathcal{G} = (\mathcal{V}, \mathcal{E})$ be a *temporal network*, where $\mathcal{V}$ is a set of nodes and $\mathcal{E}$ is a series of directed *timestamped edges*, in which the ith edge $e_i = (v_i, v_i') \in \mathcal{E}$ is a static projection of an "event" (v_i, v_i', t_i), and the time interval $[t_a, t_b]$ includes all temporal events (v_i, v_i', t_i) in $\mathcal{G}$ such that $v_i, v_i' \in \mathcal{V}$ and $t_a \leq t_i \leq t_b$, for all $i = 1, 2, \ldots, N$. The timestamps $\{t_i\}$ are unique so that events can be time-ordered.

Thus, a temporal network is similar to the edge-weighted static network defined in Section 17.2.1.

There are several models that have been used to define *temporal network motifs*. One such model (Kovanen et al., 2011) assumes each event possesses two distinguishing features: first, the time difference between each pair of consecutive events is smaller than a threshold parameter value Δ_C, and second, for each node in the motif, its adjacent events in the motif are consecutive (i.e., the node does not participate in any other event occurring between its events in the motif). Both of these features have been relaxed in other articles. See, for example, Liu, Guarrasi, and Sanyüce (2015) for references.

17.6.2 Epidemics and Disease Transmission Models

Models of epidemics have been of high interest in various disciplines for a long time, especially since the early 1980s. In fact, early attempts at modeling the spread of infectious diseases derived from attempts to understand certain diseases, such as smallpox and malaria. As a result, the topic of epidemic modeling has now become a popular scientific field of research. The whole concept of epidemics covers such diverse fields as computer viruses, sexually transmitted diseases, and contagion in financial markets.

Which factors influence the spread of a disease? How is a disease transmitted in a population? These are two of the important questions regarding epidemic modeling. Two of the major factors are the relation between disease transmission and the structure of the population, and how the immune system reacts to exposure of the disease. A disease can be transmitted by direct contact of a susceptible individual by an infected individual (infected by, for example, the COVID-19 pandemic, the Ebola outbreak of 2014, the H1N1 influenza pandemic of 2009, the SARS epidemic of 2002–2003, HIV, or gonorrhea) or through food or environmental contamination. Models for disease transmission were surveyed by Kiss et al. (2012).

This topic has a huge literature and we are only able to describe briefly the main points. We have divided the topic up into continuous-time deterministic and stochastic models, discrete-time models, and network models.

Continuous-Time Deterministic Models

First, we describe continuous-time deterministic models. Let N denote the number of individuals in the population, which is assumed to be closed; that is, no individuals can enter or leave the population. At time $t \geq 0$, each individual is assigned a disease state, which is characterized using a compartmental model. The simplest scenario is that there are only two possible states for an individual, *susceptible* (those who are capable of being infected with the disease) or *infected* (those who have the disease and are able to infect others). At time $t = 0$, the population consists of a few infected individuals, but most of the population are susceptibles.

The basic "homogeneous" epidemic model is where every individual is assumed to be potentially linked to everyone else in the population (referred to as *uniform mixing*), so that the set of possible contacts is the entire population. The basic epidemic model also assumes that every individual has the same number of contacts as everyone else in the population. We denote by S_t the set of individuals at time t who are susceptible to a disease and I_t by the set of individuals who arrive and are already infected with the disease and are contagious.

The edges SS, SI, and II represent transitions from one state to another (or to the same type of state). Each of these three types of transitions can be activated or deleted at random with certain probabilities. If an SI transition is deleted, this will slow down or even stop transmission of the disease. Over time, and under certain conditions, the SI transitions will decrease in number, splitting the population into two loosely connected groups, a community of susceptibles and a community of infected individuals.

If an infected individual recovers, one type of model assumes that that individual becomes susceptible again, and the dynamic is viewed as SIS for susceptible–infected–susceptible. The SIS model is appropriate for bacterial infections – meningitis, plague, and sexually transmitted diseases (such as chlamydia or gonorrhoea) – and to parasites such as lice and all types of diseases where repeat infections are common. In another type of model, a community R is created for those individuals who recover and receive lifelong immunity from the disease, which leads to a SIR model for susceptible–infected–recovered individuals.

The SIR model (appropriate, for example, to measles, mumps, rubella) was introduced by Kermack and McKendrick (1927) and has since become a popular modeling technique in epidemiology. In fact, it has been used as mathematical model in the case of COVID-19. The model assumes that a susceptible individual becomes infected and recovers, and then is removed permanently from the network, possibly through death or immunization. This is in contrast to the SIS model in which the individual, after being infected and recovered, is still available to be infected again, which can continue *ad infinitem.*

Let $|S_t|$, $|I_t|$, and $|R_t|$ denote the number of susceptible, infected, and recovered individuals in $\mathcal{G}_t$ at time t, respectively, and let $s_t = |S_t|/N \in [0, 1]$, $i_t = |I_t|/N \in [0, 1]$, and $r_t = |R_t|/N \in [0, 1]$ denote the proportions in each of those states. For the SIS model, $s_t + i_t = 1$ (i.e., all individuals are either susceptible or infected), and for the SIR model, $s_t + i_t + r_t = 1$ (i.e., all individuals are susceptible, infected, or recovered).

Epidemic models incorporate certain parameters that help describe the epidemics as either the SIR or the SIS models. These parameters are rates of disease transmission. Infected individuals who transmit the disease do so at rate β (called the *infection rate*), while individuals who recover and stop being infectious do so at rate γ (the *recovery rate*).

The SIS and SIR models are given by the following schematic view:

$$SIS\ model: \quad S_t \xrightarrow{\beta i_t s_t} I_t \xrightarrow{\gamma i_t} S_t.$$

$$SIR\ model: \quad S_t \xrightarrow{\beta i_t s_t} I_t \xrightarrow{\gamma i_t} R_t.$$

The SIS model. In a continuous-time basic epidemic model, the nonlinear SIS differential equations are

$$\frac{ds_t}{dt} = -\beta i_t s_t + \gamma i_t, \tag{17.27}$$

$$\frac{di_t}{dt} = \beta i_t s_t - \gamma i_t = \beta i_t (1 - i_t) - \gamma i_t = (\beta - \gamma - \beta i_t) i_t. \tag{17.28}$$

Note that these two equations satisfy $\frac{ds_t}{dt} + \frac{di_t}{dt} = 0$, and so we only need (17.28). The solution to (17.28) is (Newman, 2010, Section 17.4)

$$i_t = \left(1 - \frac{\gamma}{\beta}\right) \frac{Ce^{(\beta - \gamma)t}}{1 + Ce^{(\beta - \gamma)t}}, \tag{17.29}$$

where the integration constant is

$$C = \frac{\beta i_0}{\beta - \gamma - \beta i_0}. \tag{17.30}$$

The SIR model. The nonlinear SIR differential equations are (see Kermack and McKendrick, 1927)

$$\frac{ds_t}{dt} = -\beta i_t s_t = -\beta i_t (1 - i_t - r_t), \tag{17.31}$$

$$\frac{di_t}{dt} = \beta i_t s_t - \gamma i_t = \beta i_t (1 - i_t - r_t) - \gamma i_t, \tag{17.32}$$

$$\frac{dr_t}{dt} = \gamma i_t. \tag{17.33}$$

These three equations are linearly dependent because $\frac{ds_t}{dt} + \frac{di_t}{dt} + \frac{dr_t}{dt} = 0$. Substituting (17.33) into (17.31) yields $\frac{ds_t}{dt} = -\beta i_t s_t = -\frac{\beta}{\gamma} \frac{dr_t}{dt} s_t$, whence,

$$\frac{1}{s_t} \frac{ds_t}{dt} = -\frac{\beta}{\gamma} \frac{dr_t}{dt}. \tag{17.34}$$

Integrating both sides of (17.34) wrt t yields $s_t = e^{-\frac{\beta}{\gamma}(r_t + C)} = e^{-\frac{\beta r_t}{\gamma}} e^{-\frac{\beta}{\gamma} C}$, where C is the integration constant. Thus, at $t = 0$, assuming that $r_0 = 0$, we have that $s_0 = e^{-\frac{\beta}{\gamma} C}$, whence,

$$s_t = s_0 e^{-\beta r_t / \gamma}. \tag{17.35}$$

We can solve for i_t by substituting $i_t = 1 - s_t - r_t$ into (17.33) and then substituting (17.35) into the result. This yields $\frac{dr_t}{dt} = \gamma(1 - r_t - s_0 e^{-\beta r_t / \gamma})$. Integrating the result wrt t should yield the solution for r, but a closed-form solution cannot be obtained. A numerical evaluation can be computed instead (Newman, 2010, Section 17.3).

The basic reproduction number. Both the SIS and SIR models use a timescale of $1/\gamma$. It is usual to rescale these equations so that the average infection duration becomes 1. For the SIS model, the equations become

$$\frac{ds_t}{dt} = -R_0 i_t s_t + i_t = -R_0 i_t (1 - i_t) + i_t, \tag{17.36}$$

$$\frac{di_t}{dt} = R_0 i_t s_t - i_t = R_0 i_t (1 - i_t) - i_t, \tag{17.37}$$

while for the SIR model,

$$\frac{ds_t}{dt} = -R_0 i_t s_t = R_0 i_t (1 - i_t - r_t), \tag{17.38}$$

$$\frac{di_t}{dt} = R_0 i_t s_t - i_t = R_0 i_t (1 - i_t - r_t) - i_t, \tag{17.39}$$

$$\frac{dr_t}{dt} = i_t, \tag{17.40}$$

where

$$R_0 = \beta / \gamma \tag{17.41}$$

is termed the *basic reproduction number*. The basic reproduction number can be defined as the expected number of infections caused by a single infected individual who is introduced into a fully susceptible population. We have the following result:

If $R_0 = c > 1$, then each infected individual would infect, on average, c other individuals and the epidemic would spread exponentially, while if $R_0 = c < 1$, then each infected individual would infect, on average, c other individuals and the epidemic would then die out exponentially. The *epidemic threshold* occurs when $R_0 = 1$.

An explanation of this result is as follows. If $R_0 = \beta / \gamma < 1$, then $\beta < \gamma$, which says that the infection rate is smaller than the recovery rate. In other words, susceptible individuals become infected at a slower rate than infected individuals recover, so that the epidemic dies out. On the other hand, if $R_0 = c > 1$, then an infected individual would infect c susceptible individuals, who, in turn, would each infect c further susceptibles, and so on. Thus, at each time point, the number of infected would increase in an exponential fashion, $c \to c^2 \to c^3 \to \cdots$.

If we assume further that deaths occur with probability ω, then $R_0 = \beta / (\gamma + \omega)$.

Early published estimates of R_0 in the SARS epidemic of 2002–2003 ranged from 2.2 to 3.6, but later estimates revised that down to 1.2–1.6 (Craig et al., 2002). Estimates of R_0 for COVID-19 in Western Europe put it at 2.2–2.6 (but not the same in different countries), which is significantly lower than the estimate of 3.32 in China (Locatelli, Trachsel, and Rousson, 2021). Depending upon the variables considered, the time period, the country, and the model used, higher and lower values of R_0 have been published.

Such models have been shown to produce good agreement between complex simulations and ordinary differential equation (ODE) models, which are more rigorous for studying the full range of behaviors of disease transmissions.

Continuous-Time Stochastic Models

The deterministic approach to dynamic modeling of epidemics has some serious deficiencies. One of the main problems of the deterministic view is that it assumes that everyone in (some large and closed) population can have contact with everyone else in the population. However, real life is not like that. The deterministic approach ignores the fact that the social behavior of individuals can be very different due to factors such as numbers of family and work contacts, and differences in geographical location, which would influence the behavior of the model and alter the direction of the spread of the epidemic. Other problems with the deterministic approach include the fact that it has no ability to describe any random noise in the epidemic process, and that it is not appropriate for application to small populations because, in essence, the spread of an infectious disease is a stochastic process.

There are issues in the modeling process of determining when a major outbreak of a disease will occur, and if it does occur, how long should one expect it to remain in the population. Work on providing stochastic models for epidemics has been available for some time, going back to the Galton–Watson process (Watson and Galton, 1874; Galton, 1889). Transmission of a disease can be viewed as a stochastic event that depends upon several factors, including how frequently individuals socialize with other individuals in the population. One reason to study the stochastic model is that when N, the size of the population, is large, the stochastic model shows the limit of the deterministic model, which is based upon the assumption of large N.

The stochastic SIR model can be described as follows. Using the same notation as above, an individual remains infected for some random period of time, after which the individual recovers and becomes immune to the disease. The infectious period is an iid random variable with distribution F_{inf} with mean $1/\gamma$. One choice of F_{inf} is the exponential distribution with parameter γ. Then, the model is Markovian and the Markov process $\{S_t, I_t, R_t\}$ has jump intensities closely related to the equations (17.31)–(17.33) of the continuous-time deterministic model. Unfortunately, an exact closed-form expression for the dynamics of the epidemic based upon this distributional assumption (or, in fact, any alternative distributional assumption) cannot be obtained.

Discrete-Time Epidemic Model

The SIR epidemic model was specialized to discrete time by Lowell Reed and Wade Hampton Frost (Frost, 1976). This model was introduced by Reed and Frost in a series of unpublished lectures in 1928, but an article on the topic was not published until the 1950s. The so-called *Reed–Frost model* is a special case of a *chain-binomial model*. The chain-binomial model is a dynamic model that was introduced by En'ko (1889), but not published until 1989. Other contributors include Greenwood (1931) and Frost (1976).

The assumptions of the Reed–Frost model include the following:

1. The population of size N is homogeneous and isolated, and there is no contact with any individual outside the population.
2. An infection spreads from an infected individual to a susceptible individual with probability p.
3. Exposure to two or more infected individuals at the same time are independent exposures.
4. The infectious period is specified to last one unit of time.

There are variations on these assumptions. For example, extending assumption 2, von Bahr and Martin-Löf (1980) allow each infected individual in the population to have a different probability of infecting susceptibles.

We describe briefly the stochastic version of the discrete-time Reed–Frost model, which is represented by $\{S_t, I_t, R_t; t = 0, 1, 2, \ldots, T\}$. (There is also a continuous-time version of the Reed–Frost model.) At time $t = 0$, an individual is either susceptible, infected, or recovered. Now, a susceptible individual at time t remains susceptible at the next time point $t+1$ as long as that individual is able to avoid infection produced by all infectives. Infections are assumed to last a short while. Let $|I_t|$ denote the number of infectives at time t. Then an individual who is susceptible at time $t + 1$ will remain susceptible if $|I_t|$ Bernoulli experiments with probability of success $p \in (0, 1)$ are failures. So, given $|S_t|$ and $|I_t|$, the number of susceptibles and the number of infected individuals at time $t + 1$ are distributed as binomial random variables:

$$|S_{t+1}| \sim \text{Bin}(|S_t|, q^{|I_t|}), \tag{17.42}$$

$$|I_{t+1}| \sim \text{Bin}(|S_t|, 1 - q^{|I_t|}), \tag{17.43}$$

where $q = 1 - p$ and $p = |I_t|/(N - 1)$. Hence, the chain-binomial expressions are given by the following expected values:

$$\text{E}\{|S_{t+1}|\} = |S_t| q^{|I_t|}, \tag{17.44}$$

$$\text{E}\{|I_{t+1}|\} = |S_t|(1 - q^{|I_t|}) = |S_t| - \text{E}\{|S_{t+1}|\}. \tag{17.45}$$

The Reed–Frost model includes the following updates for susceptibles and recovered individuals at time $t + 1$: $|S_{t+1}| = |S_t| - |I_{t+1}|$ and $|R_{t+1}| = |R_t| + |I_t| = \sum_{r=0}^{t} |I_r|$, where $N = |S_t| + |I_t| + |R_t|$, for all t. The epidemic stops at time T when $|I_T| = 0$ (i.e., when no new infections occur). At that point, the only individuals who remain in the population are the susceptibles and those who have recovered. Of interest is the path of the epidemic, which is given by the numbers of infectives at each time point, $|I_1|, |I_2|, \ldots, |I_T|$. The total number of individuals who have been infected during the course of the epidemic is called the *final size*.

There has been some controversy regarding the formulation of the Reed–Frost epidemic model. The main issue is some researchers consider that the assumptions of the model may be unreasonable. Jacquez (1987), for example, constructed a more general model by making alternative assumptions, and others have suggested assumptions that lead to slightly different results (see, e.g., Lefevre and Picard, 1989). Furthermore, the Reed–Frost model is not recommended for large populations; it is primarily applicable to small populations, where $|S_0|$, the number of susceptibles at the start of the epidemic, is a small value.

17.6.3 Network Approach to Epidemic Modeling

Recent studies have provided a network approach to the study of infectious diseases, which has become a rapidly developing field of study. These studies are generally of two types. The first type centers on how a network approach can help to understand the dynamics by which an epidemic spreads amongst a population. The second type builds models to simulate how an epidemic outbreak can occur in a complex network. We shall discuss only the first type of study.

Let $\mathcal{G}_t = (\mathcal{V}, \mathcal{E}_t)$ denote an undirected network observed at time t, where $\mathcal{V}$ is the set of nodes (representing individuals or households), where the nodes are the same for all times t, and $\mathcal{E}_t$ is the set of edges (representing contacts between individuals that may spread a disease) at time t. Allowing $\mathcal{V}$ to be partitioned into households led Goldstein et al. (1980) to redefine R_0. In some variations of the model, the edges may be weighted, where the weights depend upon the severity of the infection. We assume that the network has connected components so that all nodes in a component can be reachable from all other nodes in that same component. If the component is large, then an infected individual in that component will generate a major outbreak, while if the component is small, the outbreak will be minor.

Let $N = |\mathcal{V}|$ denote the number of nodes in $\mathcal{V}$. The adjacency matrix at time t is denoted by $\mathbf{Y}_t = (Y_{ij,t})$, where $Y_{ij,t} = 1$ if there exists an edge between nodes v_i and v_j at time t, and 0 otherwise. The *degree* of a node in the network expresses the issue of who has contact with whom and also how many work or social contacts (i.e., other nodes in the network) each individual (node) has. The *degree distribution* tells us the number of contacts for each individual, which is a fundamental concept for understanding how a disease spreads. An important feature of such networks is the concept of a "hub," which is a node that has a large number of links to other nodes in the network (see Section 3.7).

Suppose we start out with a large population (i.e., N is large), where the number of infected individuals, $|I_t|$, is small and the number of susceptible individuals, $|S_t|$, is large. The probability that a susceptible individual is infected by an infected individual is proportional to the number of infected neighbors of the susceptible. Clearly, the more infected neighbors one has, the higher the chance of becoming infected. If the epidemic reaches a stage in which the number of infected individuals is increasing to a large number, then we have a major outbreak of the disease; if the number of infected individuals does not increase very much above the initial number of cases, then we have a minor outbreak.

Research on the spread of disease typically assumes that network models use the most accurate information on the social structure of a given population. However, if a network model and parameter values that are assumed for a large population are inaccurate regarding the social structure and contact interactions of individuals in that population, this could result in predictions of infections being severely skewed. If that happens, healthcare systems and services could have very different expectations of future growth trends in the epidemic than the one that actually occurs. We have seen this happen, for example, with predictions of the course of COVID-19.

Probability Generating Functions

Let p_k be the probability that a randomly selected individual in the contact network has k contacts (i.e., has degree k), with $\sum_{k=0}^{\infty} p_k = 1$. This means that a single infected individual who has k contacts transmits the infection to each of those contacts, where transmission works along every edge to whom the individual is connected. The extent of the social contacts for each individual is reflected by the *degree distribution*, $\{p_k\}$, of the network. Knowing the degree distribution allows us to calculate many different features of the network.

Much of the theory of the early stages of a large outbreak of an epidemic has been approximated by a *Galton–Watson branching process*, which was described in Section 4.3.5. It involves the application of stochastic branching processes to networks. In this particular context, branching processes assume the following:

1. The susceptible population is large, while the number of infecteds is small relative to population size.
2. Infected individuals behave independently of each other.
3. Infected individuals transmit the infection with the same probability, and they recover with the same probability.

Using the theory of branching processes for modeling epidemics, the idea of an individual producing "offspring" translates into an infected individual infecting susceptibles.

An extremely useful tool for determining probabilities is the *probability generating function* (PGF). See Section 8.2.6. The PGF for a network degree distribution (see, e.g., Miller, 2018) is a polynomial whose coefficients are the probabilities

$$G_0(x) = E\{x^k\} = \sum_{k=0}^{\infty} x^k p_k, \quad x \in [0,1], \tag{17.46}$$

where the probability distribution, $\{p_k\}$, is normalized so that $G_0(1) = 1$. Probability distributions, $\{p_k\}$, often used in real-world networks include the Poisson and negative binomial distributions. The variable x in the generating function is considered to be a dummy variable or a place-holder. For example, if the probability distribution is Poisson, where $p_k = \lambda^k e^{-\lambda}/k!$, then $G_0(x) = e^{\lambda(x-1)}$. See (8.28). Thus, the average degree of a node is $G_0'(1) = dG_0(x)/dx|_{x=1} = \lambda$. To find an expression for p_k, we differentiate the sum in (17.46) term-by-term, so that

$$p_k = \frac{G_0^{(k)}(0)}{k!}, \quad k = 0, 1, 2, \ldots, \tag{17.47}$$

where $G^{(k)}(x)$ is the kth derivative of $G(x)$ wrt x. Note that the probability that an outbreak goes extinct within k time points is $G_0^{(k)}(0)$. We are often interested in

$$G_0'(1) = \sum_k k p_k = E\{k\} = \mu, \tag{17.48}$$

where μ is the expected number of infections caused by an infected individual early in the outbreak (when the fraction infected is still small), otherwise known as R_0. Furthermore, $G_0(0) = 0$, and $G_0''(z) > 0$.

More generally, suppose p_{abm} denotes the probability that a node in our network has a incoming edges, b outgoing edges, and m undirected edges. Then

$$G_0(x_1, x_2, x_3) = \sum_{a=0}^{\infty} \sum_{b=0}^{\infty} \sum_{m=0}^{\infty} x_1^a x_2^b x_3^m p_{abm} \tag{17.49}$$

is the probability generating function for the degree distribution of the network.

If the degree distribution of the network is known, it can be applied to obtain the following expression for R_0 (Meyers, 2007):

$$R_0 = T\left(\frac{E\{k^2\} - E\{k\}}{E\{k\}}\right) = T \cdot \frac{f_0''(1)}{f_0'(1)}, \tag{17.50}$$

where $T = \beta/(\beta + \gamma)$ is the probability of an infection being transmitted to a susceptible (Miller, Beard, and Bliss, 2011). Note the resemblance of (17.50) to the expressions in Section 8.2. Li, Blakeley, and Smith? (2011)[1] noted that if two networks have the same expected degree $E\{k\}$, then the network that will be most vulnerable to the spread of a disease will be the network with the largest degree variance, $\text{var}\{k\} = E\{k^2\} - (E\{k\})^2$. Of course, if the degree distribution is unknown, then it has to be determined from the network degree distribution.

Recent interest in network models (including lattice models) has been to explain how diseases spread within a heterogeneous population over a period of time. Transmission of a disease in such a population can only take place between individuals

---

[1]The question mark is part of the name.

who share an edge in a network. Examples include individuals living within the same household or who live close to each other, those who attend the same school or college, or those who work in the same organization. Such models might include deterministic or stochastic models that account for different subpopulations.

So far, most research work on network models for epidemics uses computational algorithms that are applied to simulated data with sequences of different values assigned to the parameters. Simulations of epidemic spread on networks have been examined assuming different types of network structures, including Poisson, Erdős–Rényi (Neal, 2003), scale-free, Watts–Strogatz small-world networks, configuration networks (Britton, Juher, and Saldaña, 2016; Yao and Durrett, 2020), and the preferential attachment model (Rao and Durvasula, 2013). These different network structures provide different dynamics of the spread of epidemics and disease transmission within a population.

Unfortunately, theoretical results on the application of dynamic networks for modeling the spread of an infectious disease have not yet reached the level of such studies of static networks. This is now regarded as a major challenge in epidemology (Enright and Kao, 2018).

17.6.4 The Effect of Time on the Spread of Disease

Time can have a major influence on how an epidemic spreads through a population. Despite the importance of time in the spread of disease to epidemiologists and those in related fields, however, there has not been much theoretical research work on how disease spreads through a population (or a network) over time.

Individuals differ in many things related to the time progression of a disease. These include aging, their responses to vaccinations or other types of medical treatment, changes in their local environment, and the length of time they remain infectious. Individuals also modify their behavioral patterns by a mixture of temporary and permanent changes, such as wearing face-masks, socially distancing, getting vaccinated, handwashing, quarantining, and avoiding contact with infected individuals or anyone perceived as possibly being infected. This behavior translates into the removal (with some probability) of edges in a network that link susceptible individuals with infected individuals and "rewires" them to other susceptible individuals chosen at random. Several models have been proposed that attempt to take into consideration each of the preventative behaviors mentioned above.

Whereas the models discussed above assume closed populations, today we see that epidemics have become international, especially with the ease of travel from city to city and country to country. This factor has had a great impact on the spread of diseases around the world, which results in the considerable variability of measures of the spread of an epidemic.

Basic Reproduction Number

There is evidence that time introduces trends in certain quantities that are important in network models. Probably the most important example of this phenomenon is how the basic reproduction number, R_0, varies through time (Holme and Masuda, 2015). It is one of the most widely used types of measurement in epidemiological research. R_0, which is pronounced "R-naught," is defined as the expected number of other individuals that an infected individual will infect (i.e., secondary infections) if he or she joins a completely susceptible population. Its evaluation depends upon the quality of the data (which may be poor) that are collected to estimate the numerous epidemiological parameters in the model. These parameters include the duration of contagiousness after an individual becomes infected, the likelihood of infection of a susceptible by an infected individual, and the rate at which contact occurs between individuals. Other factors known to influence the value of R_0 include differences in population density (e.g., rural, urban, or suburban regions), types of social and professional organizations, and seasonal variation.

A simple transformation of R_0 yields the percentage of the population that should be vaccinated if the epidemic is to be stopped. The transformation is $1 - (1/R_0)$. For example, if $R_0 = 2$, it would take 50% of the population to be vaccinated to stop an outbreak. In Wuhan, where COVID-19 began (see Section 5.2.2), R_0 was 2.5. Therefore, the proportion that needed to be vaccinated to stop COVID-19 was $1 - 1/2.5 = 0.6$, or 60% of the population. If $R_0 = 3$, it would take 67% vaccinated to stop the epidemic.

Complicating issues. There are, however, problems with R_0, even though it is a convenient measure and is easy to use. Many different definitions of R_0 have been proposed, as well as many different methods for estimating it, each of which depends upon what is assumed about the population structure. First, these differences often yield different values of R_0 for the same epidemic, as well as different measures of uncertainty regarding those estimates. Second, because models for R_0 may be complicated, thereby containing many parameters whose values may be unknown or impossible to measure directly, most values assigned to them may just be educated guesses (Delamater, 2019). Third, values of R_0 for the COVID-19 pandemic, for example, have been found to vary considerably from model to model, and are often misinterpreted and misapplied. In fact, Li, Blakeley, and Smith? (2011), who give a detailed survey of methods for determining R_0, go further and argue that the concept of R_0 is "deeply flawed," that it does not, as claimed, actually measure the number of secondary infections, that it is never calculated consistently, and that many diseases can persist even when $R_0 < 1$ and die out even when $R_0 > 1$, which, they claim, tends to limit its usefulness.

Effective Reproduction Number

If we take time into consideration, especially in situations in which there are a large number of infected individuals, a modified version of R_0 has become popular as a summary statistic of the state of the epidemic at time t since the start of the epidemic. It is referred to as the *effective reproduction number* and is denoted by R_t.[2] The value of R_t is used to trace changes in the spread of a disease over time as the number of susceptibles in a population is reduced, and also to determine the effectiveness of interventions.

There are many ways that have been proposed for estimating R_t, each based upon a different theoretical model. One way of computing R_t at time t is the following product:

$$R_t = R_0 \cdot s_t, \qquad (17.51)$$

---

[2]This R_t is not to be confused with the number of individuals who had been recovered at time t.

where $s_t = |S_t|/N$ is the proportion of the population at time t that is susceptible. It always satisfies $R_t \leq R_0$, and equality holds if the population consists only of susceptibles. During the first few generations of a disease, $s_0 \approx 1$, and so $R_t \approx R_0$. If the proportion s_t decreases, then the fraction of the population that is immune increases, and R_t decreases until its value is less than 1. This would trigger "herd immunity" in the population, which would, in turn, reduce the number of infected to zero. We should also expect that certain parameters, such as the infection rate β, would not be constant, but would depend upon time just because of individual behavior and other social interactions. R_t is often used to assess the effectiveness of a program of vaccination or certain restrictive measures such as a lockdown.

Alternative methods for estimating R_t include maximum-likelihood estimation (White and Pagano, 2008) and a sequential Bayesian method (Bettencourt and Ribeiro, 2008), each of which assumes the degree distribution to have a Poisson or negative-binomial distribution. Estimates depend upon data obtained from the number of infected cases (to compute the mean generation time and initial growth rates in the infected population), the number of deaths that can be attributed to the infection (to compute the time from onset of symptoms to death), and a survey of the population for signs of infection.

Complicating issues. There are several reasons for being cautious about using R_t, mostly related to the introduction of bias into the estimation process. First, the data may not be publicly available during epidemics or during the emergence of an infectious disease, and hence the absence of such data would lead to a biased estimate of R_t. (We note, however, that in the earliest part of the COVID-19 pandemic, more epidemiological and genomic data were recorded than for any previous infectious disease.) Second, delays in data collection can lead to a time lag in reporting the current state of infections, which in turn would lead to estimates of an earlier time point than the current time. Third, when estimating R_t, because we expect the data to be noisy, a careful decision should be made regarding the length of the time window for estimation. The shorter the time window, the greater the level of uncertainty in estimating R_t, while, on the other hand, a lower level of uncertainty can be attained by using a longer time window, but not too long. Fourth, depending upon which factors to include in an estimate, the amount of uncertainty in the estimate will also vary. Fifth, another important issue is that most individuals who have an infection like COVID-19 do not know they have it because they show no clinical symptoms of the disease. This makes detection more difficult and makes it easier to spread the infection. Sixth, another issue is that, because of its relationship with R_0, R_t inherits the same problems that have been attributed to R_0. With all this in mind, there is no reason to expect the resulting estimates from these different estimation methods to agree with each other. See, for example, Biggerstaff et al. (2014) for a list of different R_t (and R_0) values calculated from each of the worldwide influenza pandemics of 1918, 1957, 1968, and 2009.

17.7 Further Reading

Most publications on dynamic networks appear in the physics, computer science, and machine learning literatures. So far as is known, there is no survey article in the statistical literature on models for dynamic networks. A review article on SBMs that includes some discussion of modeling dynamic networks is that of Lee and Wilkinson (2019). However, there are chapters or sections on dynamic networks in the books by Kolaczyk (2009, Section 8.6; 2017, Section 5.2), Newman (2010, Chapter 18), and Barrat, Barthélemy, and Vespignani (2013). There is also a special section of the June 2010 issue of *The Annals of Applied Statistics* that deals with network modeling, including dynamic networks. A survey of dynamic networks and its application to neural networks is given by Skarding, Gabrys, and Musial (2021).

Many other articles on longitudinal social networks can be found in the journal *Social Networks*. We also mention the edited volume on the evolution of social networks by Doreian and Stokman (1997).

Excellent discussions of network models for epidemics can be found in Barrat, Barthélemy, and Vespignani (2013, Chapter 9) and Newman (2010, Chapter 17). A special section on inference for infectious disease dynamics can be found in the February 2018 issue of *Statistical Science*, with a focus on Bayesian methods. However, it does not mention network modeling of epidemics. A useful survey of epidemic threshold models on dynamic networks has been given by Leitch, Alexander, and Sengupta (2019).

17.8 Exercises

Exercise 17.1 Draw the network graphs of the three Sampson monk datasets. Label all nodes in each graph. Can you say something about the progression of relationships between the novices?

Exercise 17.2 Use (17.6) and (17.7) to prove (17.8).

Exercise 17.3 Start with (17.33) for the SIR model and follow the directions following that equation. Obtaining an integral equation for t. Use numerical methods to compute a curve for each of the fractions of the population that fall into the susceptibles, the infected, and the recovered groups. Draw graphs of the time evolution of the three groups of individuals. What do you conclude from these graphs?

Exercise 17.4 Prove that (17.29) and (17.30) is the solution to (17.28).

Exercise 17.5 Consider the probability generating function (17.45) for the Poisson distribution, where $p_k = \lambda^k e^{-\lambda}/k!$. Show that $f_0(x) = e^{\lambda(x-1)}$. Show also that the average degree of a node is $f_0'(1) = \lambda$.

Exercise 17.6 Set up a network with $N = 10$ nodes and $M = 20$ edges. Real-world networks are typically dynamic in that edges can disappear (with a given probability) and new edges can appear (with a different probability), and individuals (nodes) can leave or enter the network (again, with different probabilities). Write a computer program to carry out such movements in your network. Run the process over a sequence of time points and then evaluate what changes in the network occurred. In particular, find the changes in the degree distribution and the average degree.

References

Abbe, E. (2018). Community detection and stochastic block models: recent developments, *Journal of Machine Learning Research*, **18** (Special Issue), 1–86.

Abbe, E., Bandeira, A.S., and Hall, G. (2016). Exact recovery in the stochastic block model, *IEEE Transactions on Information Theory*, **62**, 471–487.

Abello, J., Pardalos, P., and Resende, M.G.C. (1999). On maximum clique problems in very large graphs. In: *DIMACS Workshop on External Memory Algorithms and Visualization, DIMACS Series on Discrete Mathematics and Theoretical Computer Science* (J. Abello and J.S. Vitter, eds.), **50**, 119–130. Boston, MA: American Mathematical Society.

Abelson, H., Allen, D., Coore, D., Hanson, C., Homsy, G., Knight, T.F., et al. (2000). Amorphous computing, *Communications of the ACM*, **43**, 74–82.

Abelson, H., Beal, J., and Sussman, G.J. (2007). *Amorphous Computing*, Computer Science and Artificial Intelligence Laboratory, Massachusetts Institute of Technology, Technical Report No. MIT-CSAIL-TR-2007-030.

Adamic, L.A. and Glance, N. (2005). The political blogosphere and the 2004 U.S. election: divided they blog. In: *Proceedings of the WWW-2005 Workshop on the Weblogging Ecosystem*, 10 May, Chiba, Japan.

Adamic, L.A., Lukose, R.M., Puniyani, A.R., and Huberman, B.A. (2001). Search in power-law networks, *Physical Review E*, **64**, pp. 046135.

Adams, J. (2012). The rise of research networks, *Nature*, **490**, 335–336.

Adams, J.D., Black, G.C., Clemmons, J.R., and Stephan, P.E. (2004). Scientific teams and institutional collaborations: evidence from U.S. universities, 1981–1999, NBER Working Paper No. 10640, Cambridge, MA: National Bureau of Economic Research.

Aggarwal, G., Feder, T., Motwani, R., and Zhu, A. (2006). Channel assignment in wireless networks and classification of minimum graph homomorphism, unpublished Stanford University ECCC Computer Science Department Technical Report 06.

Agresti, A. (1990). *Categorical Data Analysis*, Hoboken, NJ: Wiley.

Ahmed, A. and Xing, E.P. (2009). Recovering time-varying networks of dependencies in social and biological studies, *Proceedings of the National Academy of Sciences*, **106**, 11878–11883.

Ahmed, N.K., Neville, J., and Kompella, R. (2012). Network sampling for relational classification, In: *Sixth International AAAI Conference on Weblogs and Social Media*.

Ahmed, N.K., Neville, J., and Kompella, R. (2014). Network sampling: from static to streaming graphs, *ACM Transactions in Knowledge Discovery From Data*, **8**, 1–56.

Aiello, W., Chung, F., and Lu, L. (2000). A random graph model for massive graphs, *Proceedings of the 32nd Annual ACM Symposium on Theory of Computing*, pp. 171–180.

Aigner, M. and Ziegler, G.M. (2010). *Proofs from THE BOOK*, New York: Springer.

Airoldi, E.M., Blei, D.M., Erosheva, E.A., and Fienberg, S.E. (2014). *Handbook of Mixed Membership Models and Their Applications*, New York: Chapman & Hall/CRC.

Airoldi, E.M., Blei, D.M., Fienberg, S.E., and Xing, E.P. (2008). Mixed-membership stochastic blockmodels, *Journal of Machine Learning Research*, **9**, 1981–2014.

Airoldi, E.M., Choi, D.S., and Wolfe, P.J. (2011). Confidence sets for network structure, *Statistical Analysis and Data Mining*, **4**, 461–469.

Airoldi, E.M., Costa, T.B., and Chan, S.H. (2013). Stochastic blockmodel approximation of a graphon: theory and consistent estimation, *Advances in Neural Information Processing Systems (NIPS)*, **26**, 692–700.

Aizenman, M. and Barsky, D.J. (1987). Sharpness of the phase transition in percolation models, *Communications in Mathematical Physics*, **108**, 489–526.

Aizenman, M. and Newman, C.M. (1984). Tree graph inequalities and critical behavior in percolation models, *Journal of Statistical Physics*, **36**, 107–143.

Aizenman, M. and Newman, C.M. (1986). Discontinuity of the percolation density in one-dimensional $1/|x − y|^2$ percolation models, *Communications in Mathematical Physics*, **107**, 611–647.

Aizenman, M., Kesten, H., and Newman, C.M. (1987). Uniqueness of the infinite cluster and continuity of connectivity functions for short and long range percolation, *Communications in Mathematical Physics*, **111**, 505–531.

Albert, R. and Barabási, A.-L. (2000). Topology of evolving networks: local events and universality, *Physical Review Letters*, **85**, 5234–5237.

Albert, R. and Barabási, A.-L. (2002). Statistical mechanics of complex networks, *Review of Modern Physics*, **74**, 47–97.

Albert, R., Jeong, H., and Barabási, A.-L. (1999). Diameter of the World-Wide Web, *Nature*, **401**, 130.

Albert, R., Jeong, H., and Barabási, A.-L. (2000). Error and attack tolerance of complex networks, *Nature*, **406**, 378–382.

Aldous, D.J. (1981). Representations for partially exchangeable arrays of random variables, *Journal of Multivariate Analysis*, **11**, 581–598.

Aldous, D.J. (1985). Exchangeability and related topics, *Lecture Notes in Mathematics*, **1117**, 1–198.

Aldous, D.J. (2010). Exchangeability and continuum limits of discrete random structures. In: *Proceedings of the International Congress of Mathematicians*, Hyderabad, India, **1**, 141–153.

Allen, F. and Babus, A. (2008). Networks in finance. In: *Network-Based Strategies and Competencies* (P. Kleindorfer and J. Wind, eds.), Chapter 21, Philadelphia, PA: Wharton School Publishing.

Allman, E., Matias, C., and Rhodes, J. (2009). Identifiability of parameters in latent structure models with many observed variables, *The Annals of Statistics*, **37**, 3099–3132.

Allman, E., Matias, C., and Rhodes, J. (2011). Parameter identifiability in a class of random graph mixture models, *Journal of Statistical Planning and Inference*, **141**, 1719–1736.

Alon, N. and Noar, A. (2004). Approximating the cut-norm via Grothendieck's inequality, *Proceedings of the 36th Annual ACM Symposium on Theory of Computing*.

Alon, N. and Shapira, A. (2008). Every monotone graph property is testable, *SIAM Journal on Computing*, **38**, 505–522.

Amaral, L., Scala, A., Barthélemy, M., and Stanley, H.E. (2000). Classes of behavior of small-world networks, *Proceedings of the National Academy of Sciences*, **97**, 11149–11152.

Amari, A.-I. (1982). Differential geometry of curved exponential families – curvatures and information loss, *The Annals of Statistics*, **10**, 357–385.

Amini, A.A. and Levina, E. (2018). On semidefinite relaxations for the block model, *The Annals of Statistics*, **46**, 148–179.

Amini, A.A., Chen, A., Bickel, P.J., and Levina, E. (2013). Pseudo-likelihood methods for community detection in large sparse networks, *Annals of Statistics*, **41**, 2097–2122.

Amini, A.A. and Levina, E. (2018). On semidefinite relaxations for the block model, *The Annals of Statistics*, **46**, 148–179.

Anderlucci, L., Montanari, A., and Viroli, C. (2019). The importance of being clustered: uncluttering the trends of statistics from 1970 to 2015, *Statistical Science*, **34**, 280–300.

Anderson, T.W. (1984). *An Introduction to Multivariate Statistical Analysis*, 2nd ed., New York: Wiley.

Anderson, W.N. and Morley, T.D. (1985). Eigenvalues of the Laplacian of a graph, *Linear and Multilinear Algebra*, **18**, 141–145.

Andrieu, C., De Freitas, N., Doucet, A., and Jordan, M.I. (2003). An introduction to MCMC for machine learning, *Machine Learning*, **50**, 5–43.

Antonopoulos, A.M. (2015). *Mastering Bitcoin*. Sebastopol, CA: O'Reilly Media, Inc.

Appel, K. and Haken, W. (1977). Every map is four-colorable I. Discharging, II. Reducibility. *Illinois Journal of Mathematics*, **21**, 429–490, 491–567.

Aqui, K.A. (2014). *Internal Revenue Service Notice 2014-21*.

Arabie, P., Boorman, S.A., and Levitt, P.R. (1978). Constructing blockmodels: how and why. *Journal of Mathematical Psychology*, **17**, 21–63.

Arsić, B., Cvetković, D., Simić, S.K., and Skarić, M. (2012). Graph spectral techniques in computer sciences, *Applicable Analysis and Discrete Mathematics*, **6**, 1–30.

Arunkumar, B.R. and Komala, R. (2015). Applications of bivariate graphs in diverse fields including cloud computing, *International Journal of Modern Engineering Research*, **5**, 1–7.

Azarnoush, B., Paynabar, K., Bekki, J., and Runger, G.C. (2016). Monitoring temporal homogeneity in attribute network streams, *Journal of Quality Technology*, **28**, 28–43.

Babu, M.M. and Teichmann, S.A. (2003). Evolution of transcription factors and the gene regulatory network in *Escherichia coli*, *Nucleic Acids Research*, **31**, 1234–1244.

Bach, F.R. and Jordan, M.I. (2006). Learning spectral clustering, with application to speech separation, *Journal of Machine Learning Research*, **7**, 1963–2001.

Bader, J.S. (2006). The *Drosophila* protein interaction network may be neither power-law nor scale-free. In: *Power Laws, Scale-Free Networks, and Genome Biology* (E.V. Koonin, Y.I. Wolf, and G.P. Karev, eds.), Chapter 5, pp. 53–64, New York: Springer Science+Business Media.

Bagrow, J.P. (2012). Communities and bottlenecks: trees and treelike networks have high modularity, *Physical Review E*, **85**, 066118, 1–9.

Bai, Z. and Silverstein, J.W. (2010). *Spectral Analysis of Large Dimensional Random Matrices*, New York: Springer.

Balberg, I. (1987). Recent developments in continuum percolation, *Philosophical Magazine, B*, **56**, 991–1003.

Ball, B., Karrer, B., and Newman, M.E.J. (2011). An efficient and principled method for detecting communities in networks, *Physical Review E*, **84**, 036103.

Ball, B. and Newman, M.E.J. (2013). Friendship networks and social status, *Network Science*, **1**, 16–30.

Ball, F., Sirl, D., and Trapman, P. (2009). Threshold behaviour and final outcome of an epidemic on a random network with household structure, *Advances in Applied Probability*, **41**, 765–796.

Ball, F., Sirl, D., and Trapman, P. (2010). Analysis of a stochastic SIR epidemic on a random network with household structure, *Mathematical Biosciences*, **224**, 53–73.

Barabási, A.-L. (2002). *Linked: The New Science of Networks*, London: Perseus Books Group.

Barabási, A.-L. (2016). *Network Science*, Cambridge: Cambridge University Press.

Barabási, A.-L. (2018). Love is all you need: Clauset's fruitless search for scale-free networks, unpublished blog (www.barabasilab.com/post/love-is-all-you-need).

Barabási, A.-L. and Albert, R. (1999). Emergence of scaling in random networks, *Science*, **286**, 509–512.

Barabási, A.-L., Albert, R., and Jeong, H. (2000). Scale-free characteristics of random networks: the topology of the World Wide Web, *Physica A: Statistical Mechanics and Its Applications*, **281**, 69–77.

Barabási, A.-L. and Bonabeau, E. (2003). Scale-free networks, *Scientific American*, **288**, 60–69.

Barabási, A.-L., Dezso, Z., Ravasz, E., Yook, S.-H., and Oltvai, Z. (2003). Seventh Granada Lectures: scale-free and hierarchical structures in complex networks, *Proceedings of the American Institute of Physics Conference on the Modeling of Complex Systems*, Melville, NY, 1–16.

Barabási, A.-L., Gulbahce, N., and Loscalzo, J. (2011). Network medicine: a network-based approach to human disease, *Nature Reviews: Genetics*, **12**, 56–68.

Barabási, A.-L., Jeong, H., Néda, Z., Ravasz, E., Schubert, A., and Vicsek, T. (2002). Evolution of the social network of scientific collaborations, *Physica A: Statistical Mechanics and its Applications*, **311**, 590–614.

Barabási, A.-L. and Oltvai, Z.N. (2004). Network biology: understanding the cell's functional organization, *Nature Reviews: Genetics*, **5**, 101–113.

Barnard, S.T., Pothen, A., and Simon, H.D. (1995). A spectral algorithm for envelope reduction of sparse matrices, *Numerical Linear Algebra With Applications*, **2**, 317–334.

Barndorff-Nielsen, O. (1978). *Information and Exponential Families: In Statistical Theory*, New York: Wiley (reissued 2014).

Barrat, A., Barthélemy, M., and Vespignani, A. (2013). *Dynamical Processes on Complex Networks*, Cambridge: Cambridge University Press.

Barrat, A. and Weigt, M. (2000). On the properties of small-world network models, *The European Physical Journal B*, **13**, 547–560.

Bartlett, T.E., Kosmidis, I., and Silva, R. (2021). Two-way sparsity for time-varying networks with applications in genomics, *The Annals of Applied Statistics*, **15**, 856–879.

Bashan, A., Berezin, Y., Buldyrev, S.V., and Havlin, S. (2013). The extreme vulnerability of interdependent spatially embedded networks, *Nature Physics*, **9**, 667–672.

Bassett, D.S. and Lynall, M.-E. (2013). Network methods to characterize brain structure and function. In *Cognitive Neurosciences: The Biology of the Mind*, 5th ed. (M. Gazzaniga, R.B. Ivry, and G.R. Mangun, eds.), New York: WW Norton & Co., pp. 1–27.

Bebek, G., Berenbrink, P., Cooper, C., Friedetzky, T., Nadeau, J.H., and Sahinalp, S.C. (2007). Not all scale free networks are born equal: the role of the seed graph in PPI network emulation, *PLoS Computational Biology*, **3**, 1373–1384.

Becchetti, L. and Castillo, C. (2006). The distribution of PageRank follows a power-law only for particular values of the damping factor, *Proceedings of the 15th International Conference on World Wide Web*, pp. 941–942.

Beck, F., Burch, M., Diehl, S., and Weiskopf, D. (2017). A taxonomy and survey of dynamic graph visualization, *Computer Graphics Forum*, **36**, 133–159.

Bender, E.A. and Canfield, E.R. (1978). The asymptotic number of labeled graphs with given degree sequences, *Journal of Combinatorial Theory, Series A*, **24**, 296–307.

Benjamini, I., Lyons, R., Peres, Y., and Schramm, O. (1999). Critical percolation on any nonamenable group has no infinite clusters, *The Annals of Probability*, **27**, 1347–1356.

Benjamini, I. and Schramm, O. (1996). Percolation beyond $\mathbb{Z}^d$, many questions and a few answers, *Electronic Communications in Probability*, **1**, 71–82.

Ben-Naim, E. and Krapivsky, P.L. (2005). Polymerization with freezing, arXiv manuscript (arxiv.org/pdf/cond-mat/0507313.pdf).

Berg, B.A. (2004). *Markov Chain Monte Carlo Simulations and Their Statistical Analysis*, Singapore: World Scientific.

Berger, J.O, (1985). *Statistical Decision Theory and Bayesian Analysis*, 2nd ed., New York: Springer.

Berman, A. and Plemmons, R.J. (1979). *Nonnegative Matrices in the Mathematical Sciences*, New York: Academic Press.

Bernardo, J.M. and Smith, A.F.M. (1994). *Bayesian Theory*, New York: Wiley.

Besag, J. (1974). Spatial interaction and the statistical analysis of lattice systems, *Journal of the Royal Statistical Society, Series B*, **36**, 192–236.

Besag, J. (1975). Statistical analysis of non-lattice data, *Journal of the Royal Statistical Society, Series D (The Statistician)*, **24**, 179–195.

Besag, J., Green, P., Higdon, D., and Mengersen, K. (1995). Bayesian computation and stochastic systems, *Statistical Science*, **10**, 3–41.

Bettencourt, L.M.A. and Ribeiro, R.M. (2008). Real-time Bayesian estimation of the epidemic potential of emerging infectious diseases, *PLoS ONE*, **3**, e2185-10.1371/journal.pone.0002185.

Bhan, A., Galas, D.J., and Dewey, T.G. (2002). A duplication growth model of gene expression networks, *Bioinformatics*, **18**, 1486–1493.

Bianco, K.M. (2008). The subprime lending crisis: causes and effects of the mortgage meltdown, CCH, unpublished manuscript (`business.cch.com/images/banner/subprime.pdf`).

Bichot, C.-E. and Durand, N. (2006). Airspace block organization with metaheuristics and partitioning packages. In: *Proceedings of the 2nd International Conference on Research in Air Transportation (ICRAT)*, Belgrade, Serbia, pp. 103–110.

Bichot, C.-E. and Siarry, P. (2011). *Graph Partitioning*, New York: Wiley.

Bickel, P.J. and Chen, A. (2009). A nonparametric view of network models and Newman–Girvan and other modularities, *Proceedings of the National Academy of Sciences of the USA*, **106**, 21068–21073.

Bickel, P.J., Chen, A., and Levina, E. (2011). The method of moments and degree distributions for network models, *The Annals of Statistics*, **39**, 2280–2301.

Bickel, P.J., Choi, D.S., Chang, X., and Zhang, H. (2013). Asymptotic normality of maximum likelihood and variational approximation for stochastic blockmodels, *Annals of Statistics*, **41**, 1922–1943.

Biemar, F., Nix, D.A., Piel, J., Peterson, B., Ronshaugen, M., Sementchenko, V., et al. (2006). Comprehensive identification of *Drosophila* dorsal-ventral genes using a whole-genome tiling array, *Proceedings of the National Academy of Sciences USA*, **103**, 12763–12768.

Biernacki, P. and Waldorf, D. (1981). Snowball sampling, *Sociological Methods and Research*, **10**, 141–163.

Biggerstaff, M., Cauchernez, S., Reed, C., Gambhir, M., and Finelli, L. (2014). Estimates of the reproduction number for seasonal, pandemic, and zoonotic influenza: a systematic review of the literature, *BMC Infectious Diseases*, **14**, 480–499.

Biggs, N., Lloyd, E., and Wilson, R. (1998). *Graph Theory*, Oxford: Oxford University Press.

Billingsley, P. (1979). *Probability and Measure*, New York: Wiley.

Bishop, C.M. (2006) *Pattern Recognition and Machine Learning*, New York: Springer.

Bishop, Y.M.M., Fienberg, S.E., and Holland, P.W. (1975). *Discrete Multivariate Analysis*, Cambridge, MA: MIT Press.

Blagus, N., Šubelj, L., Weiss, G., and Bajec, M. (2015). Sampling promotes community structure in social and information networks, *Physica A*, **432**, 206–215.

Blei, D.M., Ng, A., and Jordan, M.I. (2003). Latent Dirichlet allocation, *Journal of Machine Learning Research*, **3**, 993–1022.

Blondel, V.D., Guillaume, J.-L., Lambiotte, R., and Lefebvre, E. (2008). Fast unfolding of communities in large networks, *Journal of Statistical Mechanics: Theory and Experiment*, P10008 (12 pages).

Blum, M., Luby, M., and Rubinfeld, R. (1993). Self-testing/correcting with applications to numerical problems, *Journal of the ACM*, **47**, 549–595.

Bollobás, B. (1980). A probabilistic proof of an asymptotic formula for the number of labelled regular graphs, *The European Journal of Combinatorics*, **1**, 311–316.

Bollobás, B. (1981). Degree sequences of random graphs, *Discrete Mathematics*, **33**, 1–19.

Bollobás, B. (1984). The evolution of random graphs, *Transactions of the American Mathematics Society*, **286**, 257–274.

Bollobás, B. (1985). *Random Graphs*, New York: Academic Press.

Bollobás, B. (2001). *Random Graphs*, 2nd ed., Cambridge: Cambridge University Press.

Bollobás, B. and Riordan, O. (2002). Mathematical results on scale-free random graphs. In *Handbook of Graphs and Networks* (S. Bornholdt and H.G. Schuster, eds.), New York: Wiley, pp. 1–34.

Bollobás, B. and Riordan, O. (2006). *Percolation*, Cambridge: Cambridge University Press.

Bollobás, B. and Riordan, O. (2012). A simple branching process approach to the phase transition in $G_{n,p}$, *The Electronic Journal of Combinatorics*. **19**, 8 pages.

Bollobás, B., Kozma, R., and Miklós, D. (eds.) (2008). *Handbook of Large-Scale Random Networks*, Bolyai Society Mathematical Studies, **18**, New York: Springer.

Bollobás, B., Riordan, O., Spencer, J., and Tusnády, G. (2001). The degree sequence of a scale-free random graph process, *Random Structures and Algorithms*, **18**, 279–290.

Bonacich, P. (2004). The invasion of the physicists, *Social Networks*, **26**, 285–288.

Bonetta, L. (2010). Protein–protein interactions: interactome under construction, *Nature*, **468**, 851–854.

Booth, L., Bruck, J., Franceschetti, M., and Meester, R. (2003). Covering algorithms, continuum percolation and the geometry of wireless networks, *The Annals of Applied Probability*, **13**, 722–741.

Borel, E. (1942). Sur l'empoi du thérème de Bernoulli pour faciliter le calcul d'une infinité de coefficients. Application au problème de l'attende à un guichet, *Comptes Rendus de l'Académie des Sciences, Paris*, **214**, 452–456.

Borgatti, S.P., Everett, M.G., and Freeman, L.C. (1999). *UCINET 5 for Windows: Software for Social Network Analysis*, Natick, MA: Analytic Technologies.

Borgs, C., Chayes, J.T., Cohn, H., and Holden, N. (2018). Sparse exchangeable graphs and their limits via graphon processes, *Journal of Machine Learning Research*, **18**, 1–71.

Borgs, C., Chayes, J.T., and Lovász, L. (2010). Moments of two-variable functions and the uniqueness of graph limits, *Geometric and Functional Analysis*, **19**, 1597–1619.

Borgs, C., Chayes, J.T., Lovász, L., Sós, V.T., and Vesztergombi, K. (2006). Counting graph homomorphisms. In: *Topics in Discrete Mathematics* (M. Kalzar, J. Kratochvil, M. Loebl, J. Matoušek, R. Thomas, and P. Valtr, eds.), New York: Springer, pp. 315–371.

Borgs, C., Chayes, J.T., Lovász, L., Sós, V.T., and Vesztergombi, K. (2008). Convergent sequences of dense graphs I: Subgraph frequencies, metric properties, and testing, *Advances in Mathematics*, **219**, 1801–1851.

Borgs, C., Chayes, J.T., Lovász, L., Sós, V.T., and Vesztergombi, K. (2012). Convergent sequences of dense graphs II: Multiway cuts and statistical physics, *Annals of Mathematics*, **176**, 151–219.

Borgs, C., Chayes, J., Lovász, L., Sós, V.T., Szegedy, B., and Vesztergambi, K. (2016). Graph limits and parameter testing, *STOC '06*, Seattle, WA.

Bramson, M. and Durrett, R. (1999). *Perplexing Problems in Probability: Festschrift in Honor of Harry Kesten*, Boston, MA: Birkhäuser.

Brandes, U., Delling, D., Gaertler, M., Görke, R., Hoefer, M., Nikoloski, Z., and Wagner, D. (2008). On modularity clustering, *IEEE Transactions on Knowledge Data Engineering*, **20**, 172–188.

Breiger, R.L., Boorman, S.A., and Arabie, P. (1975). An algorithm for clustering relational data with applications to social network analysis and comparison with multidimensional scaling, *Journal of Mathematical Psychology*, **12**, 328–383.

Breiman, L. (1996). Heuristics of instability and stabilization in model selection, *The Annals of Statistics*, **24**, 2350–2383.

Brin, S. and Page, L. (1998). The anatomy of a large-scale hypertextual Web search engine, *Computer Networks and IDNS Systems*, **30**, 107–117.

Britton, T., Juher, D., and Saldaña, J. (2016). A network epidemic model with preventative rewiring: comparative analysis of the initial phase, *Bulletin of Mathematical Biology*, **78**, 2427–2454.

Broadbent, S.R. and Hammersley, J.M. (1957). Percolation processes. 1. Crystals and mazes, *Proceedings of the Cambridge Philosophical Society*, **53**, 629–641.

Broder, A.Z., Glassman, S.C., and Manasse, M.S. (1997). Syntactic clustering of the web, *Proceedings of the 6th International World Wide Web Conference*.

Broido, A.D. and Clauset, A. (2018). Scale-free networks are rare, unpublished manuscript, available on arXiv (`arxiv.org/pdf/1801.03400.pdf`).

Bron, C. and Kerbosch, J. (1973). Algorithm 457: finding all cliques of an undirected graph, *Communications of the ACM*, **16**, 575–577.

Brooks, S., Gelman, A., Jones, G., and Meng, X.-L. (eds.) (2011). *Handbook of Markov Chain Monte Carlo*, Boca Raton, FL: CRC Press.

Brown, L.D. (1986). *Fundamentals of Statistical Exponential Families, With Applications in Statistical Decision Theory*, IMS Lecture Notes – Monograph Series, **9**, Hayward, CA: Institute of Mathematical Statistics.

Burton, R.M. and Keane, M. (1989). Density and uniqueness in percolation, *Communications in Mathematical Physics*, **121**, 501–505.

Cai, D., Ackerman, N., and Freer, C. (2015). *An Iterative Step-Function Estimator for Graphons*, MIT Technical Report, Department of Mathematics.

Cai, T.T. and Li, X. (2015). Robust and computationally feasible community detection in the presence of arbitrary outlier nodes, *The Annals of Statistics*, **43**, 1027–1059.

Cai, W., Wu, J., and Chung, A.C.S. (2006). Shape-based image segmentation using normalized cuts. In: *Proceedings of the 2006 IEEE International Conference on Image Processing*, 8–11 October, Atlanta, GA.

Callaway, D.S., Newman, M.E.J., Strogatz, S.H., and Watts, D.J. (2000). Network robustness and fragility: percolation on random graphs, *Physical Review Letters*, **85**, 5468–5471.

Carballido-Gamio, J., Belongie, S.J., and Majumdar, S. (2004). Normalized cuts for spinal MRI segmentation. *IEEE Transactions in Medical Imaging*, **23**, 36–44.

Carlin, B.P. and Louis, T.A. (2000). *Bayes and Empirical Bayes for Data Analysis*, 2nd ed., New York: Chapman & Hall/CRC.

Caron, F. and Fox, E.B. (2017). Sparse graphs using exchangeable random measures, *Journal of the Royal Statistical Society, Series B*, **79**, 1295–1366.

Carson, R. (1962). *Silent Spring*. Boston, MA: Houghton-Mifflin.

Cartwright, D. and Harary, F. (1958). Structural balance: a generalization of Heider's theory, *Psychological Review*, **63**, 277–293.

Casella, G. and Berger, R.L. (1990). *Statistical Inference*, Pacific Grove, CA: Wadsworth & Brooks/Cole.

Casey, M.J. and Vigna, P. (2015). The revolutionary power of digital currency, *The Wall Street Journal*, Review Section, 24–25 January, pp. C1–C2.

Castle, R. (2013). *Deadly Heat*, New York: Hyperion.

Cayley, A. (1889). A theorem on trees, *Quarterly Journal of Pure and Applied Mathematics*, **23**, 276–378.

Celisse, A., Daudin, J.-J., and Pierre, L. (2012). Consistency of maximum-likelihood and variational estimators in the stochastic block model, *Electronic Journal of Statistics*, **6**, 1847–1899.

Chan, S.H. and Airoldi, E.M. (2014). A consistent histogram estimator for exchangeable graph models, *Journal of Machine Learning Research Workshop and Conference Proceedings*, **32**, 208–216.

Chatterjee, S. (2015). Matrix estimation by universal singular value threshholding, *The Annals of Statistics*, **43**, 177–214.

Chatterjee, S. and Diaconis, P. (2013). Estimating and understanding exponential random graph models, *The Annals of Statistics*, **41**, 2428–2461.

Chaudhuri, A. (2014). *Network and Adaptive Sampling*, London: Taylor & Francis/CRC Press.

Chaudhuri, K., Chung, F., and Tsiatas, A. (2012). Spectral clustering of graphs with general degrees in the extended planted partition model, *Journal of Machine Learning Research*, **2012**, 1–23.

Chayes, J.T., Chayes, L., and Fröhlich, J. (1985). The low-temperature behavior of disordered magnets, *Communications in Mathematical Physics*, **100**, 399–437.

Cheeger, J.L. (1970). A lower bound for the smallest eigenvalue of the Laplacian, *Problems in Analysis* (Papers dedicated to Salomon Bochner), Princeton, NJ: Princeton University Press, pp. 195–199.

Chen, Y. and Xu, J. (2016). Statistical-computational tradeoffs in planted problems and submatrix localization with a growing number of clusters and submatrices, *Journal of Machine Learning*, **17**, 1–57.

Chernoff, H. (1952). A measure of asymptotic efficiency for tests of a hypothesis based on the sum of observations, *The Annals of Mathematical Statistics*, **23**, 493–507.

Choi, D. (2017). Co-clustering of nonsmooth graphons, *The Annals of Statistics*, **45**, 1488–1515.

Choi, D.S., Wolfe, P.J., and Airoldi, E.M. (2012). Stochastic blockmodels with a growing number of classes, *Biometrika*, **99**, 273–284.

Christakis, N.A. and Fowler, J.H. (2010). Social network sensors for early detection of contagious diseases, *PLoS ONE*, **5**, e12948.

Christensen, R. (1997). *Log-linear Models and Logistic Regression*, New York: Springer.

Chung, F. (1997). *Spectral Graph Theory*, 2nd ed., Regional Conference Series in Mathematics, No. 92, Providence, RI: American Mathematical Society.

Chung, F. (2007). Four Cheeger-type inequalities for graph partitioning algorithms, *ICCM 2007, Volume II, 1–4*, pp. 751–772.

Chung, F. (2008). A whirlwind tour of random graphs, unpublished manuscript (www.math.ucsd.edu/~fan/wp/randomg.pdf).

Chung, F. and Lu, L. (2002a). Connected components in random graphs with given expected degree sequences, *Annals of Combinatorics*, **6**, 125–145.

Chung, F. and Lu, L. (2002b). The average distances in random graphs with given expected degrees, *Proceedings of the National Academy of Sciences USA*, **99**, 15879–15882.

Chung, F. and Lu, L. (2003). Spectra of random graphs with given expected degrees, *Proceedings of the National Academy of Sciences USA*, **100**, 6313–6318.

Chung, F. and Lu, L. (2006). *Complex Graphs and Networks*, CMBS Lecture Series, No. 107, Providence, RI: American Mathematical Society.

Chung, F., Lu, L., Dewey, T.G., and Galas, D.J. (2003). Duplication models for biological networks, *Journal of Computational Biology*, **10**, 677–687.

Clauset, A., Newman, M.E.J., and Moore, C. (2004). Finding community structure in very large networks, *Physical Review E*, **70**, 066111.

Cohen, R., Erez, K., ben-Avraham, D., and Havlin, S. (2000). Resilience of the Internet to random breakdowns, *Physics Review Letters*, **85**, 4626–4628.

Coleman, J.S. (1958). Relational analysis: the study of social organizations with survey methods, *Human Organization*, **17**, 28–36.

Comets, F. and Janžura, M. (1998). A central limit theorem for conditionally centered random fields with and application to Markov fields, *Journal of Applied Probability*, **35**, 608–621.

Coore, D.N. (1999). *Botanical computing: a developmental approach to generating interconnect topologies on an amorphous computer*, Ph.D. dissertation, Department of Electrical Engineering and Computer Science, Massachusetts Institute of Technology.

Corander, J., Dahmström, K., and Dahmström, P. (1998). *Maximum Likelihood Estimation for Markov Graphs*, Unpublished Research Report 1998:8, Department of Satistics, University of Stockholm, Sweden.

Corander, J., Dahmström, K., and Dahmström, P. (2002). Maximum likelihood estimation for exponential random graph models. In: *Contributions to Social Network Analysis, Information Theory, and Other Topics in Statistics: A Festschrift in Honor of Ove Frank on the Occasion of His 65th Birthday*, Stockholm, Universtet Stockholm, pp. 1–17.

Cowels, M.K. and Carlin, B.P. (1996). Markov chain Monte Carlo convergence diagnostics: a comparative review, *Journal of the American Statistical Association*, **91**, 883–904.

Craig, B., Phelan, T., Siedlarek, J.-P., and Steinberg, J. (2020). Improving epidemic modeling with networks, *Economic Commentary*, Federal Reserve Bank of Cleveland, 2020-23. DOI: 10.26509/frbc-ec-202023.

Cramér, H. (1946). *Mathematical Methods of Statistics*, Princeton, NJ: Princeton University Press.

Crane, H. (2018). *Probabilistic Foundations of Statistical Network Analysis*, Boca Raton, FL: Chapman & Hall/CRC.

Crane, H. and Dempsey, W. (2016). Edge exchangeable models for network data, *Journal of the American Statistical Association*, **113**, 1311–1326.

Crane, H. and Towsner, S.P. (2016). The structure of combinatorial Markov processes, unpublished technical report, Rutgers University (arXiv:1603.05954).

Crawford, F.W., Wu, J., and Heimer, R. (2018). Hidden population size estimation from respondent-driven sampling: a network approach, *Journal of the American Statistical Association*, **113**, 755–766.

Cvetković, D. and Davidović, T. (2009). Multiprocessor interconnection networks. In: *Applications of Graph Spectra* (D. Cvetković and I. Gutman, eds.), *Zbornik radova*, **13**, Matematicki institut SANU, pp. 33–63.

Dall'Amico, L., Couillet, R., and Tremblay, N. (2020). Community detection in sparse time-evolving graphs with a dynamical Bethe-Hessian. In: *Proceedings of the 34th Conference on Neural Information Processing Systems (NeurIPS 2020)*, Vancouver, Canada (papers.nips.cc/paper/2020/file/54391c872fe1c8b4f98095c5d6ec7e c7-Paper.pdf).

Dall'Asta, L., Barrat, A., Barthélemy, M., and Vespignani, A. (2006). Integrated European project – DELIS 2006, *Vulnerability of Weighted Networks* (www.delis.upb.de/ paper/DELIS-TR-0340.pdf).

Darmois, G. (1935). Sur les lois de probabilité à estimation exhaustive, *Comptes Rendus de l'Académie des Sciences, Paris*, **200**, 1265–1266.

Darroch, J.N. and Ratcliff, D. (1972). Generalized iterative scaling for log-linear models, *Annals of Mathematical Statistics*, **43**, 1470–1480.

Daudin, J.-J. (2011). A review of statistical models for clustering networks with an application to a PPI network, *Journal de la Société Française de Statistique*, **152**, 111–125.

Daudin, J.-J., Picard, F., and Robin, S. (2008). A mixture model for random graphs, *Statistical Computing*, **18**, 173–183.

Davidson, E.H. (2009). Network design principles from the sea urchin embryo, *Current Opinions in Genetic Development*, **19**, 535–540.

Davidson, E.H. (2010). Emerging properties of animal gene regulatory networks, *Nature*, **468**, 911–920.

Davis, C. and Kahan, W.M. (1970). The rotation of eigenvectors by a perturbation, III, *SIAM Journal of Numerical Analysis*, **7**, 1–46.

Dawid, A.P. (1979). Conditional independence in statistical theory (with discussion), *Journal of the Royal Statistical Society, Series B*, **41**, 1–31.

Dawson, C. and Boston, W. (2018). Foreign auto makers look at more U.S. content, *The Wall Street Journal*, 6–7 October, p. B3.

Decelle, A., Krzakala, F., Moore, C., and Zdeborová, L. (2013). Asymptotic analysis of the stochastic block model for modular networks and its algorithmic applications, *Physical Review E*, **84**, 066106.

De Finetti, B. (1937). La prevision: ses lois logiques, ses sources subjectives, *Annals of the Institute Henri Poincaré*, **7**, 1–68. Reprinted as Foresight: its logical laws, its subjective sources. In: *Studies in Subjective Probability* (H.E. Kyburg, Jr. and H.E, Smokler, eds.), New York: Wiley, pp. 93–158.

De Gennes, P.-G. (1976a). On a relation between percolation theory and the elasticity of gels, *Journal de Physique Lettres*, **37**, 1.

De Gennes, P.-G. (1976b). La percolation: un concept unificateur, *La Recherche*, **72**, 919–927.

De la Concha, A., Martinez-Jaramillo, S., and Carmona, C. (2017). Multiplex financial networks: revealing the level of interconnectedness in the banking system, unpublished report (www.suomenpankki.fi/globalassets/de-la-concha_martinez-jaramillo_christian_final.pdf).

Delameter, P.L., Street, E.J., Leslie, T.F., Yang, T., and Jacobsen, K.H. (2019). Complexity of the basic reproduction number (R_0), *Emerging Infectious Diseases*, **25**, 1–4.

Delling, D. and Werneck, R.F. (2013). Faster customization of road networks. In: *12th Symposium on Experimental Algorithms*, *LNCS*, **7933**, 30–42, New York: Springer.

Dempster, A.P., Laird, N.M., and Rubin, D.B. (1977). Maximum likelihood from incomplete data via the EM algorithm (with discussion), *Journal of the Royal Statistical Society, Series B*, **39**, 1–38.

Dennis, C. and Gallagher, R. (eds.) (2001). *The Human Genome*, London: Nature/Palgrave.

De Solla Price, D.J. (1965). Networks of scientific papers, *Science*, **149**, 510–515.

De Solla Price, D. (1976). A general theory of bibliometric and cumulative advantage processes, *Journal of the American Society for Information Science*, **27**, 292–306.

Diaconis, P. (2009). The Markov chain Monte Carlo revolution, *Bulletin of the American Mathematical Society*, **46**, 179–205.

Diaconis, P. and Freedman, D. (1981). On the statistics of vision: the Julesz conjecture, *Journal of Mathematical Psychology*, **24**, 112–138.

Diaconis, P. and Janson, S. (2007). Graph limits and exchangeable random graphs, unpublished report, available on arXiv (arXiv:0712.2749).

Diesner, J., Frantz, T.L., and Carley, K.M. (2005). Communication networks from the Enron email corpus: "It's always about the people. Enron is no different." *Computational and Mathematical Organization Theory*, **11**, 201–228.

Donath, W.E. and Hoffman, A.J. (1973). Lower bounds for the partitioning of graphs, *Journal of Research and Development*, **17**, 420–425.

Donnat, C. and Holmes, S. (2018). Tracking network dynamics: a survey using graph distances, *The Annals of Applied Statistics*, **12**, 971–1012.

Doob, J.L. (1990). *Stochastic Processes*, revised edition, New York: Wiley.

Doreian, P., Batagelj, V., and Ferligoj, A. (2005). *Generalized Blockmodeling*, Cambridge: Cambridge University Press.

Doreian, P. and Stokman, F.N. (1997). *Evolution of Social Networks* (P. Doreian and F.N. Stokman, eds.), London: Routledge/Taylor & Francis.

Dorff, C. and Ward, M.D. (2012). Networks, dyads, and the social relations model, paper presented at the *2011 Annual Meetings of the American Political Science Association*, Seattle, WA.

Dorogovstev, S.N. and Mendes, J.F.F. (2002). Evolution of networks, *Advances in Physics*, **51**, 1079–1187.

Dorogovstev, S.N., Mendes, J.F.F., and Samukhin, A.N. (2000). Structure of growing networks with preferential linking, *Physics Review Letters*, **64**, 4633–4636.

Duchin, M. (2018). Geometry v. gerrymandering, *Scientific American*, **Nov**, 49–53.

Ducruet, C. and Lugo, I. (2013). Structure and dynamics of transportation networks: models, methods, and applications. In: *SAGE Handbook of Transport Studies* (J.P. Rodrigue, T.E. Notteboom, and J. Shaw, eds.), London: SAGE Publications, pp. 347–364.

Durrett, R. (1985). Some general results concerning the critical exponents of percolation processes, *Zeitschrift für Wahrscheinlichkeitstheorie und verwandte Gebeite*, **69**, 421–437.

Durrett, R. (2007). *Random Graph Dynamics*, Cambridge: Cambridge University Press.

Durrett, R. (2012). *Essentials of Stochastic Processes*, New York: Springer.

Easley, D. and Kleinberg, J. (2010). *Networks, Crowds, and Markets: Reasoning About a Highly Connected World*, Cambridge: Cambridge University Press.

Efron, B. (1975). Defining the curvature of a statistical problem (with applications to second-order efficiency) (with discussion), *The Annals of Statistics*, **3**, 1189–1242.

Efron, B. (1978). The geometry of exponential families, *The Annals of Statistics*, **6**, 362–376.

Elliott, M.L., Golub, B., and Jackson, M.O. (2014). Financial networks and contagion, *American Economic Review*, **104**, 3115–3153.

Elsässer, R., Královič, R., and Monien, B. (2003). Sparse topologies with small spectrum size, *Theoretical Computer Science*, **307**, 549–565.

Elton, C.S. (1927). *Animal Ecology*, New York: Macmillan.

En'ko, P.D. (1889). On the course of epidemics of some infectious diseases (in Russian), *Vrach*, **10**, 1008–1010, 1039–1042, 1061–1063. Translated by K. Dietz and appeared in *American Journal of Epidemiology*, **18**, 749–755.

Enright, J. and Kao, R.R. (2018). Epidemics on dynamic networks, *Epidemics*, **24**, 88–97.

Erdős, P. and Gallai, T. (1960). Graphs with points of prescribed degrees (in Hungarian). *Mat. Lapok*, **11**, 264–274.

Erdős, P. and Rényi, A. (1959). On random graphs, *Publications in Mathematics*, **6**, 290–297.

Erdős, P. and Rényi, A. (1960). On the evolution of random graphs, *Publications of the Mathematical Institute of the Hungarian Academy of Sciences*, **5**, 17–60.

Everett, M.G. and Borgatti, S.P. (1994). Regular equivalence: general theory, *Journal of Mathematical Sociology*, **19**, 29–52.

Everitt, B.S. and Hand, D.J. (1984). *Finite Mixture Distributions*, London: Chapman & Hall.

Fanusie, Y.J. amd Entz, A. (2017). *Terror Finance Briefing Book*, FDD's Center on Sanctions and Illicit Finance, Foundation for Defense of Democracies (www.defenddemocracy.org/media-hit/yaya-j-fanusie-terror-finance-briefing-book/).

Faulkner, R.R. (1983). *Music on Demand: Composers and Careers in the Hollywood Film Industry*, Piscataway, NJ: Transaction Books.

Faust, K. (1988). Comparison of methods for positional analysis: structural and general equivalences, *Social Networks*, **10**, 313–341.

Faust, K. (2006). Comparing social networks: size, density, and local structure, *Metodološki zvezki* (*Advances in Methodology and Statistics*), **3**, 185–216.

Feller, W. (1964). *An Introduction to Probability Theory and Its Applications*, Vol. 1, 2nd ed., New York: Wiley.

Fiedler, M. (1973). Algebraic connectivity of graphs, *Czechoslovak Mathematics Journal*, **23** (98), 298–305.

Fienberg, S.E., Meyer, M.M., and Wasserman, S.S. (1985). Statistical analysis of multiple sociometric relations, *Journal of the American Statistical Association*, **80**, 51–67.

Fienberg, S.E., Petrović, S., and Rinaldo, A. (2011). Algebraic statistics for p_1 random graph models: Markov bases and their uses. In: *Looking Back: Proceedings of a Conference in Honor of Paul W. Holland* (N.J. Doran and S. Sinharay, eds.), Lecture Notes in Statistics, **202**, 21–38, New York: Springer.

Fienberg, S.E. and Wasserman, S. (1981a). Categorical data analysis of single sociometric relations. In: *Sociological Methodology 1981* (S. Leinhardt, ed.), San Francisco, CA: Jossey-Bass, pp. 110–155.

Fienberg, S.E. and Wasserman, S.S. (1981b). Comment: an exponential family of probability distributions for directed graphs, *Journal of the American Statistical Association*, **80**, 51–67.

Fifield, B., Higgins, M., Imai, K., and Tarr, A. (2020). A new automated redistricting simulator using Markov chain Monte Carlo, *Journal of Computational and Graphical Statistics*, **29**, 715–728.

Fisher, M.E. (1998). Renormalization group theory: its basis and formulation in statistical physics, *Review of Modern Physics*, **70**, 653–681.

Fisher, R.A. (1934). Two new properties of mathematical likelihood, *Proceedings of the Royal Society, Series A*, **144**, 285–307.

Fishkind, D.E., Sussman, D.L., Tang, M., Vogelstein, J.T., and Priebe, C.E. (2012). Consistent adjacency-spectral partitioning for the stochastic block model when the model parameters are unknown, unpublished manuscript.

Flory, P.J. (1941). Molecular size distribution in three-dimensional gelation I–III, *Journal of the American Chemical Society*, **63**, 3083–3100.

Forsberg, W.L. and Pagano, M. (2008). A likelihood-based method for real-time estimation of the serial interval and reproductive number of an epidemic, *Statistics in Medicine*, **27**, 2999–3016.

Fortunato, S. (2010). Community detection in graphs, *Physics Reports*, **486**, 75–174.

Fortunato, S. and Barthélemy, M. (2007). Resolution limit in community detection, *Proceedings of the National Academy of Science of the USA*, **104**, 36–41.

Frank, O. (1991). Statistical analysis of change in networks, *Statistica Neerlandica*, **45**, 283–293.

Frank, O. and Snijders, T. (1994). Estimating the size of hidden populations using snowball sampling, *Journal of Official Statistics*, **10**, 53–67.

Frank, O. and Strauss, D. (1986). Markov graphs, *Journal of the American Statistical Association*, **81**, 832–842.

Freedman, D.A. (1971). *Markov Chains*, San Francisco, CA: Holden-Day.

Freeman, L.C. (2004). *The Development of Social Network Analysis: A Study in the Sociology of Science*, Vancouver: Empirical Press.

Freeman, L.C. (2011). The development of social network analysis – with an emphasis on recent events. In: *The SAGE Handbook of Social Network Analysis* (J. Scott and P. Carrington, eds.), Los Angeles, CA: SAGE Publications, pp. 26–39.

Friedman, R. and Hughes, A.L. (2003). The temporal distribution of gene duplication events in a set of highly conserved human gene families, *Molecular and Biological Evolution*, **20**, 154–161.

Frieze, A. and Kannan, R. (1999). Quick approximation to matrices and applications, *Combinatorica*, **19**, 175–220.

Fritsch, R. and Fritsch, G. (2000). *The Four Color Theorem: History, Topological Foundations, and Idea of Proof*, New York: Springer.

Frost, W.H. (1976). Some conceptions of epidemics in general, *American Journal of Epidemiology*, **103**, 141–151.

Fu, W., Song, L., and Xing, E.P. (2009). Dynamic mixed membership blockmodel for evolving networks. In: *Proceedings of the 26th International Conference on Machine Learning (ICML)*, Madison, WI: Omnipress, pp. 329–336.

Gamerman, D. and Lopes, H.F. (2006). *Markov Chain Monte Carlo: Stochastic Simulation for Bayesian Inference*, 2nd ed., Boca Raton, FL: Chapman & Hall/CRC Press.

Gandolfi, A., Keane, M., and Russo, L. (1988). On the uniqueness of the infinite occupied cluster in dependent two-dimensional site percolation, *The Annals of Probability*, **16**, 1147–1157.

Ganin, A.A., Kitsak, M., Marchese, D., Keisler, J.M., Seager, T., and Linkov, I. (2017). Resilience and efficiency in transportation networks, *Science Advances*, **3** (`advances.sciencemag.org/content/3/12/e1701079`).

Gavish, M. and Nadler, B. (2013). Normalized cuts are approximately inverse exit times, *SIAM Journal of Matrix Analysis*, **34**, 757–772.

Gawlinski, E.T. and Stanley, H.E. (1981). Continuum percolation in two dimensions: Monte Carlo tests of scaling and universality for non-interacting discs, *Journal of Physics A: Math. Gen.* **14**, L291–L299.

Gehrke, J., Ginsparg, P., and Kleinberg, J. (2003). Overview of the 2003 KDD Cup, *SIGKDD Explorations*, **5**, 149–151.

Gelfand, A.E. and Smith, A.F.M. (1990). Sampling-based approaches to calculating marginal densities, *Journal of the American Statistical Association*, **85**, 398–409.

Gelman, A. (1996). Inference and monitoring convergence. In: *Markov Chain Monte Carlo in Practice* (W.R. Gilks, S. Richardson, and D. Spiegelhalter, eds.), New York: Chapman & Hall, pp. 131–143.

Gelman, A. and Hill, J. (2007). *Data Analysis Using Regression and Multilevel/Hierarchical Models*, New York: Cambridge University Press.

Gelman, A. and Romano, G.A. (2010). "How many zombies do you know?" Using indirect survey methods to measure alien attacks and outbreaks of the undead (`arxiv.org/pdf/1003.6087.pdf`).

Gelman, A.E., Carlin, J.B., Stern, H.S., and Rubin, D.B. (1995). *Bayesian Data Analysis*, London: Chapman & Hall.

Geman, S. and Geman, D. (1984). Stochastic relaxation, Gibbs distributions, and the Bayesian restoration of images, *IEEE Transactions on Pattern Analysis and Machine Intelligence*, **6**, 721–741.

Geyer, C.J. (1991). Markov chain Monte Carlo maximum likelihood. In: *Computing Science and Statistics: Proceedings of the 23rd Symposium on the Interface* (E.M. Keramides, ed.), Fairfax Station, VA: Interface Foundation pp. 156–163.

Geyer, C.J. (1992). Practical Markov chain Monte Carlo, *Statistical Science*, **7**, 473–483.

Geyer, C.J. and Thompson, E.A. (1992). Constrained Monte Carlo maximum likelihood for dependent data (with discussion), *Journal of the Royal Statistical Society, Series B*, **54**, 657–699.

Gilbert, A.C. and Levchenko, K. (2004). Compressing network graphs. In: *Proceedings of the LinkKDD Workshop at the 10th ACM Conference on Knowledge Discover and Data Mining*, 22–25 August, Seattle, WA.

Gilbert, E.N. (1959). Random graphs, *Annals of Mathematical Statistics*, **30**, 1141–1144.

Gilbert, E.N. (1961). Random plane networks, *Journal of the Society for Industrial and Applied Mathematics*, **9**, 533–543.

Gile, K.J. (2011). Improved inference for respondent-driven sampling data with application to HIV prevalence estimation, *Journal of the American Statistical Association*, **106**, 135–146.

Gile, K.J. and Handcock, M.S. (2010). Respondent-driven sampling: an assessment of current methodology, *Sociological Methodology*, **40**, 285–327.

Gilks, W.R., Richardson, S., and Spiegelhalter, D.J. (1996). *Markov Chain Monte Carlo in Practice*, London: Chapman & Hall.

Gilks, W.R. and Roberts, G.O. (1996). Strategies for improving MCMC. In: *Markov Chain Monte Carlo in Practice* (W.R. Gilks, S. Richardson, and D. Spiegelhalter, eds.), New York: Chapman & Hall, pp. 89–114.

Gitterman, M. and Halpern, V.H. (2004). *Phase Transitions: A Brief Account With Modern Applications*, Singapore: World Scientific.

Girvan, M. and Newman, M.E.J. (2002). Community structure in social and biological networks, *Proceedings of the National Academy of Science of the USA*, **99**, 7821–7826.

Glasserman, P. and Young, H.P. (2015). Contagion in financial networks, Office of Financial Research Working Paper 15-21, Columbia University.

Gleiser, P. and Danon, L. (2003). Community structure in jazz, *Advances in Complex Systems*, **6**, 565–573.

Goel, S. and Salganik, M.J. (2009). Respondent-driven sampling as Markov chain Monte Carlo, *Statistics in Medicine*, **28**, 2202–2229.

Goel, S. and Salganik, M.J. (2010). Assessing respondent-driven sampling, *Proceedings of the National Academy of Sciences*, **107**, 6743–6747.

Goh, K.-I., Cusick, M.E., Valle, D., Childs, B., Vidal, M., and Barabási, A.-L. (2007). The human disease network, *Proceedings of the National Academy of Sciences of the United States of America*, **104**, 8685–8690.

Golden, K.M. (2009). Climate change and the mathematics of transport in sea ice, *Notices of the AMS*, **56**, 562–584.

Goldenberg, A., Zheng, A.X., Fienberg, S.E., and Airoldi, E.M. (2009). A survey of network models, *Foundations and Trends in Machine Learning*, **2**, 129–233.

Goldreich, O. (2008). *Computational Complexity: A Conceptual Perspective*, Cambridge: Cambridge University Press

Goldreich, O. (2017). *Introduction to Property Testing*, Cambridge: Cambridge University Press.

Goldreich, O., Goldwasser, S., and Ron, D. (1998). Property testing and its connection to learning and approximation, *Journal of the ACM*, **45**, 653–750.

Goldstein, E., Paur, K., Fraser, C., Kenah, E., Wallings, J., and Lipsitch, M. (2009). Reproductive numbers, epidemic spread and control in a community of households, *Mathematical Biosciences*, **221**, 11–25.

Goodman, L.A. (1961). Snowball sampling, *Annals of Mathematical Statistics*, **32**, 147–170.

Goodreau, S.M., Kitts, J.A., and Morris, M. (2009). Birds of a feather, or friend of a friend? Using exponential random graph models to investigate adolescent social networks, *Demography*, **46**, 103–125.

Gopalan, P.K. and Blei, D.M. (2013). Efficient discovery of overlapping communities in massive networks, *Proceedings of the National Academy of Sciences*, **110**, 14534–14539.

Grant, D.M., Pugmire, R.J., Fletcher, T.H., and Kerstein, A.R. (1989). Chemical model of coal devolatilization using percolation lattice statistics, *Energy Fuels*, **3**, 175–186.

Green, S., Serban, M., Scholl, R., Jones, N., Brigandt I., and Bechtelm W. (2018). Network analyses in systems biology: new strategies for dealing with biological complexity, *Synthese*, **195**, 1751–1777.

Greenwood, M. (1931). On the statistical measure of infectiousness, *Journal of Hygiene Cambridge*, **31**, 336–351.

Grimmett, G. (1989). *Percolation*, New York: Springer.

Grimmett, G. (1999). *Percolation*, 2nd ed., New York: Springer.

Grimmett, G. (2006). Uniqueness and multiplicity of infinite clusters. In: *Dynamics and Stochastics: Festschrift in Honor of Michael Keane*, IMS Lecture Notes-Monograph Series, pp. 24–36.

Grimmett, G. (2010). *Probability on Graphs: Random Processes on Graphs and Lattices*, Cambridge: Cambridge University Press.

Grimmett, G.R. and Newman, C.M. (1990). Percolation in $\infty + 1$ dimensions. In: *Disorder in Physical Systems: A Volume in Honour of John M. Hammersley* (G.R. Grimmett and D.J.A. Welsh, eds.), Oxford: Oxford University Press, pp. 167–190.

Grimmett, G.R. and Welsch, D. (1990). *Disorder in Physical Systems*, Oxford: Clarendon Press.

Grothendieck, A. (1953). Résumé de la théorie métrique des produits tensoriels topologiques, *Bol. Soc. Mat. Sao Paulo*, **8**, 1–79.

Guattery, S. and Miller, G.L. (1998). On the quality of spectral separators, *SIAM Journal on Matrix Analysis and Applications*, **19**, 701–719.

Gupta, P. and Kumar, P.R. (1998). Critical power for asymptotic connectivity in wireless networks. In: *Stochastic Analysis, Control, Optimization, and Applications: A Volume in Honor of W.H. Fleming* (W.M. McAneany, G. Yin, and Q. Zhang, eds.), Boston, MA: Birkhäuser.

Gupta, P. and Kumar, P.R. (2000). The capacity of wireless networks, *IEEE Transactions on Information Theory*, **46**, 388–404.

Haag, J. (1924). Sur un problème général de probabilités et ses diverses applications. In: *Proceedings of the International Congress of Mathematics*, Toronto, Canada, pp. 659–674.

Haan, S.W. and Zwanzig, R. (1977). Series expansions in continuum percolation model, *Journal of Physics*, **A10**, 1547–1555.

Haberman, S.J. (1974). *The Analysis of Frequency Data*, Chicago, IL: University of Chicago Press.

Haberman, S.J. (1981). Tests for independence in two-way contingency tables based on canonical correlation and on linear-by-linear interaction, *Annals of Statistics*, **9**, 1178–1186.

Hagen, L. and Kahng, A.B. (1992). New spectral methods for ratio cut partitioning and clustering, *IEEE Transactions on Computer-Aided Design*, **11**, 1074–1085.

Häggström, O. and Jonasson, J. (2006). Uniqueness and non-uniqueness in percolation theory, *Probability Surveys*, **3**, 289–344.

Häggström, O. and Peres, Y. (1999). Monotonicity of uniqueness for percolation on Cayley graphs: all infinite clusters are born simultaneously, *Probability and Related Fields*, **113**, 273–285.

Häggström, O., Peres, Y., and Schonmann, R.H. (1999). Percolation on transitive graphs as a coalescent process: relentless merging followed by simultaneous uniqueness. In: *Perplexing Problems in Probability*, Festschrift in Honor of Harry Kesten, New York: Springer, pp. 69–90.

Haldane, A.G. (2009). Rethinking the financial network, speech given at the Financial Student Association, Amsterdam, 28 April (www.bankofengland.co.uk/archive/Documents/historicpubs/speeches/2009/speech386.pdf).

Hall, P. (1985). On continuum percolation, *The Annals of Probability*, **13**, 1250–1266.

Hammersley, J.M. (1957). Percolation processes: lower bounds for the critical probability, *The Annals of Mathematical Statistics*, **28**, 790–795.

Hammersley, J.M. (1959). Bornes supérieures de la probabilité critique dans un processus de filtration, *Le Calcul de Probabilités et ses Applications*, Centre National de la Recherche Scientifique, Paris, pp. 17–37.

Hammersley, J.M. (1972). A few seedlings of research. In: *Proceedings of the Sixth Berkeley Symposium on Mathematical Statistics and Probability, Volume I: Theory of Statistics*, Berkeley, CA: University of California Press, pp. 345–394.

Hammersley, J.M. (1983). Origins of percolation theory, *Annals of the Israel Physical Society*, **5**, 47–57; *Journal of the Royal Statistical Society, Series B*, **16**, 23–38.

Hammersley, J.M., Feuerverger, A., Izenman, A., and Makani, K. (1969). A negative finding for the three-dimensional dimer problem, *Journal of Mathematical Physics*, **10**, 443–446.

Hammersley, J.M. and Morton, K.W. (1954). Poor man's Monte Carlo, *Journal of the Royal Statistical Society, Series B*, **16**, 23–38.

Handcock, M.S. (2003). Assessing degeneracy in statistical models of social networks, unpublished Working Paper 39, Center for Statistics and the Social Sciences, University of Washington, Seattle, WA.

Handcock, M.S., Raftery, A.E., and Tantrum, J.M. (2007). Model-based clustering for social networks, *Journal of the Royal Statistical Society, Series A*, **170**, 301–354.

Hara, T. and Slade, G. (1990). Mean-field critical behaviour for percolation in high dimensions, *Communications in Mathematical Physics*, **128**, 333–391.

Harris, K.M., Halpern, C.T., Whitsel, E.A., Hussey, J.M., Killeya-Jones, L.A., Tabor, J., and Dean, S.C. (2019). Cohort profile: the National Longitudinal Study of Adolescent to Adult Health, *International Journal of Epidemiology*, **48**, 1415–1415k.

Harris, K.M. and Udry, J.R. (2018). *National Longitudinal Study of Adolescent to Adult Health (AddHealth), 1994–2008*. Carolina Population Center, University of North Carolina–Chapel Hill, Inter-university Consortium for Political and Social Research.

Harris, T.E. (1960). A lower bound for the critical probability in a certain percolation process, *Proceedings of the Cambridge Philosophical Society*, **56**, 13–20.

Hastings, W.K. (1970). Monte Carlo sampling methods using Markov chains and their applications, *Biometrika*, **57**, 97–109.

Hawkins, D.M. (2014). The CUSUM and the EWMA head-to-head, *Quality Engineering*, **26**, 215–222.

Hayashi, J.-i. and Miura, K. (2004). Pyrolysis of Victorian brown coal. In: *Advances in the Science of Victorian Brown Coal*, (Li, C.-Z., ed.), Oxford: Elsevier, pp. 134–222.

He, X., Zha, H., Ding, C.H.Q., and Simon, H.D. (2002). Web document clustering using hyperlink structures, *Computational Statistics and Data Analysis*, **41**, 19–45.

Heath, D.L. and Sudderth, W.D. (1976). De Finetti's theorem for exchangeable random variables, *American Statistician*, **30**, 188–189.

Heckathorn, D.D. (1997). Respondent-driven sampling: a new approach to the study of hidden populations, *Social Forces*, **44**, 174–199.

Heckathorn, D.D. and Jeffri, J. (2001). Finding the beat: using respondent-driven sampling to study jazz musicians, *Poetics*, **28**, 307–329.

Herlau, T., Schmidt, M.N., and Mørup, M. (2013). The infinite degree corrected stochastic block model (`arXiv:1311.2520v2`).

Hermann, F. and Pfaffelhuber, P. (2016). Large-scale behavior of the partial duplication random graph, *Latin American Journal of Probability and Mathematical Statistics*, **13**, 687–710.

Herrmann, H.J., Stauffer, D., and Landau, D.P. (1983). Computer simulation of a model for irreversible gelation. percolation structures and processes, *Journal of Physics, A: General Mathematics*, **16**, 1221–1239.

Hiley, B.J. and Sykes, M.F. (1961). Probability of initial ring closure in the restricted random walk model of a macromolecule, *Journal of Chemical Physics*, **34**, 1531–1537.

Hill, S.B., Agarwal, D.K., Bell, R., and Volinsky, C. (2006). Building an effective representation for dynamic networks, *Journal of Computational and Graphical Statistics*, **15**, 584–608.

Hirsch, J.E. (2005). An index to quantify an individual's scientific research output, *Proceedings of the National Academy of Sciences*, **102**, 16569–16572.

Ho, Q., Song, L., and Xing, E.P. (2011). Evolving cluster mixed-membership blockmodel for time-varying networks, *Journal of Machine Learning Research*, **15**, 342–350.

Hoff, P.D. (2005). Bilinear mixed-effects models for dyadic data, *Journal of the American Statistical Association*, **100**, 286–295.

Hoff, P.D., Raftery, A.E., and Handcock, M.S. (2002). Latent space approaches to social network analysis, *Journal of the American Statistical Association*, **97**, 1090–1098.

Hoffman, B. (2006). *Inside Terrorism*, revised and enlarged edition. Columbia Studies in Terrorism and Irregular Warfare, New York: Columbia University Press.

Holland, P.W., Laskey, K.B., and Leinhardt, S. (1983). Stochastic blockmodels: some first steps, *Social Networks*, **5**, 109–137.

Holland, P.W. and Leinhardt, S. (1977). A dynamic model for social networks, *Journal of Mathematical Sociology*, **5**, 5–20.

Holland, P.W. and Leinhardt, S. (1981). An exponential family of probability distributions for directed graphs, *Journal of the American Statistical Association*, **76**, 33–50.

Holme, P. and Masuda, N. (2015). The basic reproduction number as a predictor for epidemic outbreaks in temporal networks, *PLoS ONE*, **10**, 0120567. Available at doi.org/10.1371/journal.pone.0120567.

Hoover, D.N. (1979). Relations on probability spaces and arrays of random variables, unpublished manuscript, Institute for Advanced Study, Princeton, NJ (stat.berkeley.edu/~aldous/Research/hoover.pdf).

Horvitz, D.G. and Thompson, D.J. (1952). A generalization of sampling without replacement from a finite population, *Journal of the American Statistical Association*, **47**, 663–685.

Hübler, C., Kriegel, H.-P., Borgwardt, K., and Ghahramani, Z. (2008). Metropolis algorithms for representative subgraph sampling, *ICDM 2008*, pp. 283–292.

Huerta, A.M., Salgado, H., Thieffry, D., and Collado-Vides, J. (1998). RegulonDB: a database on transcriptional regulation in *Escherichia coli, Nucleic Acids Research*, **26**, 55–59.

Huitsing, G. and Veenstra, R. (2012). Bullying in classrooms: participant roles from a social network perspective, *Aggressive Behavior*, **38**, 494–509.

Hummel, R.M., Hunter, D.R., and Handcock, M.S. (2012). Improving simulation-based algorithms for fitting ERGMs, *Journal of Computational and Graphical Statistics*, **21**, 920–939.

Hunter, D.R. and Handcock, M.S. (2006). Inference in curved exponential family models for networks, *Journal of Computational and Graphical Statistics*, **15**, 565–583.

Ipsen, M. and Mikhailov, A.S. (2002). Evolutionary reconstruction of networks, *Physical Review E*, **66**, 6–9.

Izenman, A.J. (1991). Recent developments in nonparametric density estimation, *Journal of the American Statistical Association*, **86**, 205–224.

Izenman, A.J. (2012a). Spectral embedding methods for manifold learning. In: *Manifold Learning Theory and Applications* (Y. Ma and Y. Fu, eds.), Baton Rouge, LA: CRC Press, pp. 1–36.

Izenman, A.J. (2012b). Introduction to manifold learning, *Computational Statistics*, **5**, 439–446.

Izenman, A.J. (2013). *Modern Multivariate Statistical Techniques: Regression, Classification, and Manifold Learning*, New York: Springer.

Izenman, A.J. (2021). Random matrix theory and its applications, *Statistical Science*, **36**, 421–442.

Jaakkola, T.S. (2000). Tutorial on variational approximation methods. In: *Advances in Mean Field Methods* (M. Opper and D. Saad, eds.), Cambridge, MA: MIT Press, pp. 129–159.

Jackson, M.O. (2008). *Social and Economic Networks*, Princeton, NJ: Princeton University Press.

Jacquez, J.A. (1987). A note on chain-binomial models of epidemic spread: what is wrong with the Reed–Frost formulation, *Mathematical Biosciences*, **87**, 73–82.

James, G.M., Sabatti, C., Zhou, N.M., and Zhu, J. (2010). Sparse regulatory networks, *The Annals of Applied Statistics*, **4**, 663–686.

Janson, S. (2013). *Graphons, Cut Norm and Distance, Couplings and Rearrangements*, NYJM Monographs, Vol. 4 (nyjm.albany.edu/m/2013/4v.pdf).

Janson, S. (2016). Graphons and cut metric on σ-finite measure spaces, unpublished report (arXiv:1608.01833).

Janson, S. and Luczak, M.J. (2008). A new approach to the giant component problem, *Random Structures and Algorithms*, **34**, 197–216.

Janson, S., Luczak, T., and Rucinski, A. (2000). *Random Graphs*, New York: Wiley.

Jeong, H., Mason, S.P., Barabási, A.-L., and Oltvai, Z.N. (2001). Lethality and centrality in protein networks, *Nature*, **411**, 41–42.

Ji, P. and Jin, J. (2016). Coauthorship and citation networks for statisticians, *The Annals of Applied Statistics*, **10**, 1779–1812.

Jones, G.L. and Hobert, J.P. (2001). Honest exploration of intractable probability distributions via Markov chain Monte Carlo, *Statistical Science*, **16**, 312–334.

Jordan, M.I., Ghahramani, Z., Jaakkola, T.S., and Saul, L.K. (1999). An introduction to variational methods for graphical models, *Machine Learning*, **37**, 183–233.

Joseph, A. and Yu, B. (2016). Impact of regularization on spectral clustering, *The Annals of Statistics*, **44**, 1765–1791.

Jurman, G., Visintainer, R., and Furlanello, C. (2011). An introduction to spectral distances in networks (extended version), *Frontiers in Artificial Intelligence and its Applications*, **226**, 227–234.

Jurman, G., Visintainer, R., Filosi, M., Riccadonna, S., and Furlanello, C. (2015). The HIM global metric and kernel for network comparison and classification, In *Proceedings of the 2015 IEEE International Conference on Data Science and Advanced Analytics*, **7**, 46109.

Kadanoff, L.P. (1966). Scaling laws for Ising models near T_c, *Physics*, **2**, 263–272.

Kahng, A.B., Lienig, J., Markov, I.L., and Hu, J. (2011). *VLSI Physical Design – From Graph Partitioning to Timing Closure*, New York: Springer.

Kallenberg, O. (2005). *Probabilistic Symmetries and Invariance Principles*, Probability and Its Applications, New York: Springer.

Karlin, S. (1966). *A First Course in Stochastic Processes*, New York: Academic Press.

Karlin, S. and Taylor, H.E. (2012). *A First Course in Stochastic Processes, Second Edition*, New York: Academic Press.

Karrer, B. and Newman, M. (2011a). Stochastic blockmodels and community structure in networks, *Physics Reviews E* **83**, 016107.

Karrer, B. and Newman, M.E.J. (2011b). Competing epidemics on complex networks, *Physical Review E*, **84**, 016107.

Kass, R.E. (1989). The geometry of asymptotic inference, *Statistical Science*, **4**, 188–219.

Kass, R.E. and Vos, P.W. (1997). *Geometrical Foundations of Asymptotic Inference*, New York: Wiley.

Katz, V. (1998). *A History of Mathematics: An Introduction*, Harlow: Addison-Wesley.

Kelmans, A.K. (1976). Comparison of graphs by their number of spanning trees, *Discrete Mathematics*, **16**, 241–261.

Kelmans, A.K. (1997). Transformations of a graph increasing its Laplacian polynomial and number of spanning trees, *European Journal of Combinatorics*, **18**, 35–48.

Kempe, D., Kleinberg, J., and Kumar, A. (2002). Connectivity and inference problems for temporal networks, *Journal of Computer and System Sciences*, **64**, 820–842.

Kendall, M.G. and Stuart, A. (1963). *The Advanced Theory of Statistics: I. Distribution Theory*, London: Charles Griffin.

Kermack, W.O. and McKendrick, A.G. (1927). Contributions to the mathematical theory of epidemics, *Proceedings of the Royal Society of London, Series A*, **115**, 700–721.

Kernighan, B.W. and Lin, S. (1970). An efficient heuristic procedure for partitioning graphs, *The Bell System Technical Journal*, **49**, 291–307.

Kernighan, B.W. and Pike, R. (1984). *The UNIX Programming Environment*, Englewood Cliffs, NJ: Prentice-Hall

Kernighan, B.W. and Ritchie, D.M. (1978). *The C Programming Language*, Englewood Cliffs, NJ: Prentice-Hall

Kertész, J. and Vicsek, T. (1982). Monte Carlo renormalisation group study of the percolation problem of discs with a distribution of radii, *Zeitschrift für Physik B, Condensed Matter*, **45**, 345–350.

Kessler, M.M. (1963a). Bibliographic coupling between scientific papers, *American Documentation*, **14**, 10–25.

Kessler, M.M. (1963b). An experimental study of bibliographic coupling between technical papers, *IEEE Transactions on Information Theory*, **9**, 49.

Kesten, H. (1980). The critical probability of bond percolation on the square lattice equals 1/2, *Communications in Mathematical Physics*, **74**, 41–59.

Kesten, H. (1981). Analyticity properties and power estimates of functions in percolation theory, *Journal of Statistical Physics*, **25**, 717–756.

Kesten, H. (1982). *Percolation Theory for Mathematicians*, Boston, MA: Birkhäuser.

Kesten, H. (1987). Percolation theory and first-passage percolation, *The Annals of Probability*, **15**, 1231–1271.

Kesten, H. (2002). Some highlights of percolation, *ICM*, **1**, 345–362.

Khintchine, A.Y. (1932). Sur les classes d'evénements équivalents, *Mat. Sbornik*, **39**, 40–43.

Kingman, J.F.C. (1978). Uses of exchangeability, The 1977 Wald Memorial Lectures, *The Annals of Probability*, **6**, 183–197.

Kiss, I., Berthouze, L., Taylor, T., and Simon, P. (2012). Modeling approaches for simple dynamic networks and applications to disease transmission models, *Proceedings of the Royal Society, Series A*, **468**, 1332–1355.

Klarreich, E. (2018). Scant evidence of power laws found in real-world networks, *Quanta Magazine*, 15 February (www.quantamagazine.org/scant-evidence-of-power-laws-found-in-real-world-networks-20180215). Reprinted in *The Atlantic*, 20 February (www.theatlantic.com/science/archive/2018/02/power-laws-networks/553562).

Kleinberg, J., Kumar, S.R., Raghavan, P., Rajagopalan, S., and Tomkins, A. (1999). The web as a graph: measurements, models, and methods, *Proceedings of the International Conference on Combinatorics and Computing*, 1–18.

Kleinfeld, J.S. (2002). The small world problem, *Society*, **39**, 61–66.

Kleinrock, L. (1964). *Communication Nets: Stochastic Message Flow and Delay*, New York: McGraw-Hill. (Reprinted, 1972, by Dover Publications.)

Klopp, O., Tsybakov, A., and Verzelen, N. (2017). Oracle inequalities for network models and sparse graphon estimation, *The Annals of Statistics*, **45**, 316–354.

Klosik, D.F. and Bornholdt, S. (2014). The citation wake of publications detects Nobel Laureates' papers, *PLoS ONE*, **9**, e113184. DOI: 10.1371/journal.pone.0113184.

Kolaczyk, E.D. (2009). *Statistical Analysis of Network Data: Methods and Models*, New York: Springer.

Kolaczyk, E.D. (2017). *Topics at the Frontier of Statistics and Network Analysis: (Re)Visiting the Foundations*, Cambridge: Cambridge University Press.

Kolaczyk, E.D. and Csárdi, G. (2014). *Statistical Analysis of Network Data with R*, New York: Springer.

Komlós, J., Shokoufandeh, A., Simonovits, M., and Szemerédi, E. (2002). The regularity lemma and its applications in graph theory. In: *Theoretical Aspects of Computer Science* (G.B. Khosrovshahi et al., eds.), LNCS 2292, New York: Springer, pp. 84–112.

Kondor, D., Pósfai, M., Csabai, I., and Vattay, G. (2014). Do the rich get richer? An empirical analysis of the Bitcoin transaction network, *PLoS ONE*, **9**(2), e86197.

Koonin, E.V., Wold, Y.I., and Karev, G.P. (eds.) (2006). *Power Laws, Scale-Free Networks, and Genome Biology*, New York: Springer Science+Business Media.

Koopman, B.O. (1936). On distributions admitting a sufficient statistic, *Transactions of the American Mathematical Society*, **39**, 399–409.

Kovanen, L., Karsai, M., Kaski, K., Kertész, K., and Saramäki, J. (2011). Temporal motifs in time-dependent networks, *Journal of Statistical Mechanics: Theory and Experiment*, **2011**. DOI: 10.1088/1742-5468/2011/11/P11005.

Krapivsky, P.L. and Redner, S. (2001). Organization of growing random networks, *Physical Review E*, **63**, 066123.1–066123.14.

Krapivsky, P.L. and Redner, S. (2005). Network growth by copying, *Physical Review E*, **71**, 036118.1–036118.7.

Krapivsky, P.L., Redner, S., and Leyvraz, F. (2000). Connectivity of growing random networks, *Physical Review Letters*, **85**, 4629–4632.

Krapivsky, P.L., Rodgers, G.J., and Redner, S. (2001). Degree distributions of growing networks, *Physics Review Letters*, **86**, 5401–5404.

Krivelevich, M. and Sudakov, B. (2013). The phase transition in random graphs – a simple proof, *Random Structures and Algorithms*, **43**, 131–138.

Krivitsky, P.N., Handcock, M.S., Raftery, A.E., and Hoff, P.D. (2009). Representing degree distributions, clustering, and homophily in social networks with latent cluster random effects models, *Social Networks*, **31**, 204–213.

Kruskal, J.B. (1956). On the shortest spanning subtree of a graph and the traveling salesman problem, *Proceedings of the American Mathematical Society*, **7**, 48–50.

Kumar, R., Raghavan, P., Rajagopalan, S., Sivakumar, D., Tomkins, A., and Upfal, E. (2000). Stochastic models for the web graph, *Proceedings of the 42nd Annual IEEE Symposium on the Foundations of Computer Science*, **41**, 57–65.

Kumar, S., Spezzano, F., Subrahmanian, V.S., and Faloutsos, C. (2016). Edge weight prediction in weighted signed networks, *IEEE International Conference on Data Mining (ICDM)*.

Kunegis, J., Blattner, M., and Moser, C. (2013). Preferential attachment in online networks: measurement and explanations, *WebSci'13*, 1–5 May, Paris, France.

Lamperti, J. (1977). *Stochastic Processes: A Survey of the Mathematical Theory*, New York: Springer.

Lancichinetti, A. and Fortunato, S. (2011). Limits of modularity maximization in community detection, *Physical Review E*, **84**, 066122.

Lancichinetti, A., Fortunato, S., and Kertész, J. (2009). Detecting the overlapping and hierarchical community structure in complex networks (arXiv:0802.1218v2).

Langville, A.N. and Meyer, C.D. (2005). Deeper inside PageRank, *Internet Mathematics*, **1**, 35–380.

Latouche, P., Birmelé, E., and Ambroise, C. (2011). Overlapping stochastic block models with application to the French political blogosphere, *Annals of Applied Statistics*, **5**, 309–336.

Latouche, P., Birmelé, E., and Ambroise, C. (2011). Overlapping clustering methods for networks. In: *Handbook of Mixed Membership Models and Their Applications* (E.M. Airoldi, D. Blei, E.A. Erosheva, and S.E. Fienberg, eds.), London: Taylor & Francis, Chapter 25.

Latouche, P. and Robin, S. (2015). Variational Bayes model averaging for graphon functions and motif frequencies inference in W-graph models, *Statistics and Computing*, **26**, 1173–1185.

Lauritzen, S.L. (1996). *Graphical Models*, Oxford: Clarendon Press.

Lazega, E. and van Duijn, M. (1997). Position in formal structure, personal characteristics, and choices of advisors in a law firm: a logistics regression model for dyadic network data, *Social Networks*, **19**, 375–397.

Le Bel, F. (2017). Structuring music by means of audio clustering and graph search algorithms, unpublished manuscript.

Lee, C. and Wilkinson, D.J. (2019). A review of stochastic block models and extensions for graph clustering, *Applied Network Science*, **4**, 122–172.

Lee, S., Kim, P., and Jeong, H. (2006). Statistical properties of sampled networks, *Physical Review E*, **73**, 016102.

Lee, T.I., Rinaldi, N.J., Robert, F., Odom, D.T., Bar-Joseph, Z., Gerber, G.K., et al. (2002). Transcriptional regulatory networks in *Saccharomyces cerevisiae*, *Science*, **298**, 799–804.

Lefevre, C. and Picard, P. (1989). On the formulation of discrete-time epidemic models, *Mathematical Biosciences*, **95**, 27–35.

Lehmann, E.L. (1959). *Testing Statistical Hypotheses*, New York: Wiley.

Lehmann, E.L. (1983). *Theory of Point Estimation*, New York: Wiley.

Lehmann, E.L. (1986). *Testing Statistical Hypotheses*, 2nd ed., New York: Wiley.

Leskovec, J. and Faloutsos, C. (2006). Sampling from large graphs. In: *Proceedings of the 12th ACM SIGKDD International Conference on Knowledge Discovery and Data Mining*, ACM, pp. 631–636.

Leskovec, J., Kleinberg, J., and Faloutsos, C. (2005). Graphs over time: densification laws, shrinking diameters, and possible explanations, *ACM SIGKDD International Conference on Knowledge Discovery and Data Mining*.

Leskovec, J., Kleinberg, J., and Faloutsos, C. (2007). Graph evolution: densification and shrinking diameters, *ACM Transactions on Knowledge Discovery From Data (ACM TKDD)*, **1**.

Leskovec, J., Lang, K., Dasgupta, A., and Mahoney, M. (2009). Community structure in large networks: natural cluster sizes and the absence of large well-defined clusters, *Internet Mathematics*, **6**, 29–123.

Li, J., Blakeley, D., and Smith?, R.J. (2011). The failure of R_0, *Computational and Mathematical Methods in Medicine*, **2011**, article ID 527610. (Note: the question mark in Smith? is part of the name.)

Light, S., Kraulis, P., and Elofsson, A. (2005). Preferential attachment in the evolution of metabolic networks, *BMC Genomics*, **6**, 159–169.

Liu, P., Guarrasi, V., and Sanyüce, A.E. (2015). Temporal network motifs: models, limitations, evaluation, *Journal of LaTeX Class Files*, **14**, 1–19 (`arxiv.org/pdf/2005.11817.pdf`).

Lloyd, C.J. (1999). *Statistical Analysis of Categorical Data*, New York: Wiley.

Lloyd, J.R., Orbanz, P., Ghahramani, Z., and Roy, D.M. (2012). Random function priors for exchangeable arrays with applications to graphs and relational data, *Advances in Neural Information Processing Systems*, **25**, 998–1006.

Locatelli, I., Trachsel, B., and Rousson, V. (2021). Estimating the basic reproduction number for COVID-19 in Western Europe, *PLoS ONE*, e0248731.

Loève, M. (1963). *Probability Theory*, 3rd ed., Princeton, NJ: D. Van Nostrand.

Lohr, S.L. (1999). *Sampling: Design and Analysis*, Belmont, CA: Duxbury Press.

Lorrain, F. and White, H.C. (1971). Structural equivalence of individuals in social networks, *Journal of Mathematical Sociology*, **1**, 49–80.

Lovász, L. (1967). Operations with structures, *Acta Mathematica Hungarica*, **18**, 321–328.

Lovász, L. (1993). Random walks on graphs: a survey, *Combinatorics*, **2**, 1–46.

Lovász, L. (2012). *Large Networks and Graph Limits*, Providence, RI: American Mathematical Society.

Lovász, L. and Szegedy, B. (2006). Limits of dense graph sequences, *Journal of Combinatorial Theory B*, **96**, 933–957.

Lovász, L. and Szegedy, B. (2007). Szemerédi's lemma for the analyst, *Geometric Function Analysis*, **17**, 252–270.

Lovász, L. and Szegedy, B. (2010). Testing properties of graphs and functions, *Israel Journal of Mathematics*, **178**, 113–156.

Luczak, T. (1990). Component behavior near the critical point of the random graph process, *Random Structure Algorithms*, **1**, 287–310.

Lusher, D., Kosinen, J., and Robins, G. (2013). *Exponential Random Graph Models for Social Networks: Theory, Methods, and Applications*, Cambridge: Cambridge University Press.

Lusseau, D., Schneider, K., Boisseau, O.J., Hasse, P., Slooten, E., and Dawson, S.M. (2003). The bottlenose dolphin community of Doubtful Sound features a large proportion of long-lasting associations, *Behavioral Ecology and Sociobiology*, **54**, 396–405.

Lyons, R. and Peres, Y. (2017). *Probability on Trees and Networks*, Cambridge: Cambridge University Press.

MacKay, D.J.C. (2003). *Information Theory, Inference, and Learning Algorithms*, Cambridge: Cambridge University Press.

MacQueen, J. (1967). Some methods for classification and analysis of multivariate observations, *Proceedings of the Fifth Berkeley Symposium on Mathematical Statistics and Probability*, **1**, 281–297.

Madhamshettiwar, P.B., Maetschke, S.R., Davis, M.J., Reverter, A., and Ragan, M.A. (2012). Gene regulatory network inference: evaluation and application to ovarian cancer allows the prioritization of drug targets, *Genome Medicine*, **4**, 41–55.

Madsen, L.T. (1979). *The Geometry of Statistical Models – A Generalization of Curvature*, Technical Report, Danish Medical Research Council, Statistical Research Unit Report 79-1.

Maduro, M.F. (2006). Endomesoderm specification in *Caenorhabditis elegans* and other nematodes, *Bioessays*, **28**, 1010–1022.

Maiya, A.S. and Berger-Wolf, T.Y. (2010). Sampling community structure, *International World Wide Web Conference*, 26–30 April Raleigh, NC.

Marion, B. (1994). Turing machines and computational complexity, *The American Mathematical Monthly*, **101**, 61–65.

McCann, K.S. (2011). *Food Webs*, Monographs in Population Biology, Princeton, NJ: Princeton University Press.

McComb, W.D. (2004). *Renormalization Methods: A Guide for Beginners*, Oxford: Clarendon Press.

McCullagh, P. and Nelder, J.A. (1989). *Generalized Linear Models*, 2nd ed., New York: Chapman & Hall/CRC.

McCulloh, I. and Carley, K.M. (2008). Social network change detection, *SSRN Electronic Journal*. Available at dx.doi.org/10.2139/ssrn.2726799.

McCulloh, I. and Carley, K.M. (2011). Detecting change in longitudinal social networks, *Journal of Social Structure*, **12**, 1–37.

McLachlan, G.J. and Basford, K.E. (1988). *Mixture Models: Inference and Applications to Clustering*, New York: Marcel Dekker.

Meester, R. and Roy, R. (1996). *Continuum Percolation*, Cambridge: Cambridge University Press.

Meilă, M. and Shi, J. (2001). A random walks view of spectral segmentation. In: *Proceedings of the International Workshop on AI and Statistics (AISTATS)*, *Proceedings of Machine Learning Research*, R3:203–208.

Menshikov, M.V. (1987). Coincidence of critical points in percolation problems, *Soviet Mathematics – Doklady*, **33**, 856–859. Quantitative estimates and strong inequalities for critical points of a graph and its subgraph, *Theory of Probability and Its Applications*, **32**, 544–547.

Menshikov, M.V., Molchanov, S.A., and Sidorenko, A.F. (1986). Percolation theory and some applications, *Itogi Nauki i Techniki* (Series of Probability Theory, Mathematical Statistics, Theoretical Cybernetics), **24**, 53–110.

Metropolis, N., Rosenbluth, A.W., Rosenbluth, M.N., Teller, A.H., and Teller, E. (1953). Equation of state calculations by fast computing machines, *The Journal of Chemical Physics*, **21**, 1087–1092.

Meyer, I. and Wilson, P.A. (2009). Sampling lesbian, gay, and bisexual populations, *Journal of Counseling Psychology*, **56**, 23–31.

Meyers, L.A. (2007). Contact network epidemiology: bond percolation applied to infectious disease prediction and control, *Bulletin of the American Mathematical Society*, **44**, 63–86.

Miller, B.A., Arcolano, N., and Bliss, N.T. (2013). Efficient anomoly detection in dynamic, attributed graphs, *Proceedings of the IEEE International Conference on Intelligence and Security Informatics*, 179–184.

Miller, B.A., Beard, M.S., and Bliss, N.T. (2011). Matched filtering for subgraph detection in dynamic networks, *IEEE Statistical Signal Processing Workshop*, 509–512.

Miller, E. (2016). *Patterns of Islamic State-Related Terrorism, 2002–2015*, START Background Report, College Park, MD (`www.start.umd.edu/pubs/START_IslamicStateTerrorismPatterns_BackgroundReport_Aug2016.pdf`).

Miller, J.C. (2009). Percolation and epidemics in random clustered networks, *Physical Review E*, **80**, 020901.

Miller, J.C. (2018). A primer on the use of probability generating functions in infectious disease modeling, *Infectious Disease Modelling*, **3**, 192–248.

Milo, R., Shen-Orr, S., Itzkovitz, S., Kashtan, N., Chklovskii, D., and Alon, U. (2002). Network motifs: simple building blocks of complex networks, *Science*, **298**, 824–827.

Molloy, M. and Reed, B. (1995). A critical point for random graphs with a given degree sequence, *Random Structures and Algorithms*, **6**, 161–179.

Molloy, M. and Reed, B. (2000). The size of the giant component of a random graph with a given degree sequence, *Combinatorics, Probability, and Computing*, **7**, 295–305.

Moon, J.W. and Moser, L. (1965). On cliques in graphs, *Israel Journal of Mathematics*, **3**, 23–28.

Moreno, J.L. (1932). *Application of the Group Method to Classification*, New York: National Committee on Prisons and Prison Labor.

Moreno, J.L. (1934). *Who Shall Survive?*, Washington, DC: Nervous and Mental Disease Publishing.

Moreno, J.L. and Jennings, H.H. (1938). Statistics of social configurations, *Sociometry*, **1**, 342–374.

Morrison, A.D. and Wilhelm, Jr., W.J. (2007). Investment banking: past, present, and future, *Journal of Applied Corporate Finance*, **19**, 8–20.

Moyer, D., Gutman, B., Prasa, G., Ver Steeg, G., and Thompson, P. (2015). Mixed membership stochastic blockmodels for the human connectome. Available at www-scf.usc.edu/~moyerd/pubs/mixed-membership-stochastic.

Mukherjee, K., Hasan, M.M., Boucher, C., and Kahveci, T. (2018). Counting motifs in dynamic networks, *BMC Systems Biology*, **12**, 6–17.

Mullen, E.K., Daley, M., Backx, A.G., and James, G. (2014). Gene co-citation networks associated with worker sterility in honey bees, *BMC Systems Biology*, **8**, 38–50.

Nachmias, A. and Peres, Y. (2007). Component sizes of the random graph outside the scaling window, *ALEA Latin American Journal of Probability and Mathematical Statistics*, **3**, 133–142.

Nachmias, A. and Peres, Y. (2008). The critical random graph, with martingales, *Israel Journal of Mathematics*, **176**, 29–41.

Nadakuditi, R.R. and Newman, M.E.J. (2013). Spectra of random graphs with arbitrary expected degrees, *Physical Review E*, **87**, 012803.1–012803.12.

Nadler, B. and Galun, M. (2006). Fundamental limitations of spectral clustering, *Proceedings of the 19th Annual Conference on Neural Information Processing Systems (NIPS)*, **19**, 1017–1024.

Nakamoto, S. (2008). Bitcoin: a peer-to-peer electronic cash system (`bitcoin.org/bitcoin.pdf`).

National Security Agency (2010). Revealing social networks of spammers, *The Next Wave: The National Security Agency's Review of Emerging Technologies*, **18**, 26–34.

Neal, P. (2003). SIR epidemics on a Bernoulli random graph, *Journal of Applied Probability*, **23**, 779–782.

Neal, R.M. (1993). *Probabilistic Inference Using Markov Chain Monte Carlo Methods*, Technical Report CRG-TR-93-1, Department of Computer Science, University of Toronto.

Newman, M.E.J. (2001a). Scientific collaboration networks. I. Network construction and fundamental results, *Physical Review E*, **64**, 016131.

Newman, M.E.J. (2001b). Scientific collaboration networks. II. Shortest paths, weighted networks, and centrality, *Physical Review E*, **64**, 016132.

Newman, M.E.J. (2003). The structure and function of complex networks, *SIAM Review*, **45**, 167–256.

Newman, M.E.J. (2004a). Detecting community structure in networks, *The European Physical Journal B*, **38**, 321–330.

Newman, M.E.J. (2004b). Fast algorithms for detecting community structure in networks, *Physical Review E*, **69**, 066133.

Newman, M.E.J. (2005). Power laws, Pareto distributions, and Zipf's law, *Contemporary Physics*, **46**, 323–351.

Newman, M.E.J. (2006). Modularity and community structure in networks, *Proceedings of the National Academy of Sciences*, **103**, 8577–8582.

Newman, M.E.J. (2009). Random graphs with clustering, *Physical Review Letters*, **103**, 058701.

Newman, M.E.J. (2010). *Networks: An Introduction*, Oxford: Oxford University Press.

Newman, M.E.J. (2013a). Spectral methods for network community detection and graph partitioning, *Physical Review E*, **88**, 04822.

Newman, M.E.J. (2013b). Community detection and graph partitioning, *Europhysics Letters*, **103**, 28003.1–28003.5.

Newman, M.E.J. and Girvan, M. (2004). Finding and evaluating community structure in networks, *Physical Review E*, **69**, 026113.1–026113.16.

Newman, C.M. and Schulman, L.S. (1981a). Infinite clusters in percolation models, *Journal of Statistical Physics*, **26**, 613–628.

Newman, C.M. and Schulman, L.S. (1981b). Number and density of persolating clusters, *Journal of Physics, Series A*, **14**, 1735–1743.

Newman, M.E.J., Barabási, A.-L., and Watts, D.J. (2006). *The Structure and Dynamics of Networks*, Princeton, NJ: Princeton University Press.

Newman, M.E.J., Strogatz, S.H., and Watts, D.J. (2001). Random graphs with arbitrary degree distributions and their applications, *Physical Review E*, **64**, 026118.1–026118.17.

Ng, A.Y., Jordan, M.I., and Weiss, Y. (2002). On spectral clustering: analysis and an algorithm. *NIPS 2001*.

Nguyen, G.H., Lee, J.B., Rossi, R.A., Ahmed, N.K., Koh, E., and Kim, S. (2018). Dynamic network embeddings: from random walks to temporal random walks. In: *Proceedings of the IEEE International Conference on Big Data*, Seattle, WA, pp. 1085–1092. Available at `johnboaz.github.io/files/BigData2018.pdf`.

Nienhuis, B., Riedel, E.K., and Schick, M. (1980). Variational renormalization-group approach to the q-state Potts model in two dimensions, *Journal of Physics A*, **13**, 189–192.

Nowicki, K. and Snijders, T.A.B. (2001). Estimation and prediction for stochastic blockstructures, *Journal of the American Statistical Association*, **96**, 1077–1087.

O'Connor, D. and Stephens, C.R. (eds.) (2001). RG2000: renormalization group theory in the new millennium, II. Held in Taxco, Mexico, January 1999. *Physics Reports*, **348**, 1–162.

Odell, P.L. and Feiveson, A.H. (1966). A numerical procedure to generate a sample covariance matrix, *Journal of the American Statistical Association*, **61**, 199–203.

Ogburn, E.L. and VanderWeele, T.J. (2017). Vaccines, contagion, and social networks, *The Annals of Applied Statistics*, **11**, 919–1023.

Ohno, S. (1970). *Evolution by Gene Duplication*, New York: Springer.

Olhede, S.C. and Wolfe, P.J. (2014). Network histograms and universality of blockmodel approximation, *Proceedings of the National Academy of Sciences, U.S.A*, **111**, 14722–14727.

Olston, C. and Najork, N. (2010). Web crawling, *Foundations and Trends in Information Retrieval*, **4**, 175–246.

Orbanz, P. and Roy, D.M. (2015). Bayesian models of graphs, arrays, and other exchangeable random structures, *IEEE Transactions on Pattern Analysis and Machine Intelligence*, **37**, 437–461.

Ormerod, P. and Wand, M.P. (2010). Explaining variational approximations, *The American Statistician*, **64**, 140–153.

Padgett, J.F. (1994). Marriage and elite structure in Renaissance Florence, 1282–1500, paper delivered to the Social Science History Association. Available at `home.uchicago.edu/jpadgett/papers/unpublished/maelite.pdf`.

Padgett, J.F. and Ansell, C.K. (1993). Robust action and the rise of the Medici, 1400–1434. *American Journal of Sociology*, **98**, 1259–1319.

Padole, V.B. and Chaudhari, D.S. (2012). Detection of brain tumor in MRI images using mean shift algorithm and normalized cut method, *International Journal of Engineering and Advanced Technology*, **1**, 53–56.

Palla, G., Derényi, I., Farkas, I., and Vicsek, T. (2005). Uncovering the overlapping community structure of complex networks in nature and society, *Nature*, **435**, 814–818.

Palmer, E.M. (1985). *Graphical Evolution: An Introduction to the Theory of Random Graphs*, New York: Wiley-Interscience.

Parfitt, D.E. and Shen, M.M. (2014). From blastocyst to gastrula: gene regulatory networks of embryonic stem cells and early mouse embryogenesis, *Philosophical Transactions of the Royal Society B*, **369**, 1657–1669.

Parzen, E. (2015). *Stochastic Processes*, New York: Dover Publications. (Originally published in 1962 by Holden-Day, San Francisco, CA.)

Pascual, M. and Dunne, J.A. (2006). *Ecological Networks: Linking Structure to Dynamics in Food Webs*, Oxford: Oxford University Press.

Pastor-Satorras, R. and Vespignani, A. (2004). *Evolution and Structure of the Internet*, Cambridge: Cambridge University Press.

Pearl, J. (2000). *Causality: Models, Reasoning, and Inference*, Cambridge: Cambridge University Press.

Penrose, M.D. (1997). The longest edge of the random minimal spanning tree, *Annals of Applied Probability*, **7**, 340–361.

Peskun, P.H. (1973). Optimum Monte Carlo sampling using Markov chains, *Biometrika*, **60**, 607–612.

Peterson, C. and Anderson, J.R. (1987), A mean field theory learning algorithm for neural networks, *Complex Systems*, **1**, 995–1019.

Petrović, S., Rinaldo, A., and Fienberg, S.E. (2010). Algebraic statistics for a directed random graph model with reciprocation. In: *Algebraic Methods in Stataistics and Probability, Volume II, Contemporary Mathematics*, **516**, Providence, RI: American Mathematical Society, pp. 261–283.

Petru, L. and Wiedermann, J. (2007). A model of an amorphous computer and its communication protocol. In: *SOFSEM 2007: Theory and Practice of Computer Science: 33rd Conference on Current Trends in Theory and Practice* (J. van Leeuwen, G.F. Italiano, W. van der Hoek, and C. Meinel, eds.), Lecture Notes on Computer Science 4362, New York: Springer, pp. 446–455.

Pike, G.E. and Seager, C.H. (1974). Percolation and conductivity: a computer study I, *Physical Review B*, **10**, 1421–1434.

Pitman, E.J.G. (1936). Sufficient statistics and intrinsic accuracy, *Proceedings of the Cambridge Philosophical Society*, **32**, 567–579.

Pitman, J. (1999). Coalescent random forests, *Journal of Combinatorial Theory, Series A*, **85**, 165–193.

Plischke, M. and Bergersen, B. (2006). *Equilibrium Statistical Physics*, 3rd ed., Singapore: World Scientific.

Pozsar, Z., Adrian, T., Ashcroft, A., and Boesky, H. (2012). *Shadow Banking*, Federal Reserve Bank of New York Staff Report No. 458 (`www.newyorkfed.org/medialibrary/media/research/staff_reports/sr458.pdf`).

Press, S.J. (1989). *Bayesian Statistics: Principles, Models, and Applications*, New York: Wiley.

Preston, R. (1994). *The Hot Zone: A Terrifying True Story of the Origins of the Ebola Virus*, Mass Market Paperback. Republished in 1999 by Anchor Books (Knopf Doubleday Publishing Group).

Prim, R.C. (1957). Shortest connection networks and some generalizations, *Bell System Technical Journal*, **36**, 1389–1401.

Pringle, D.J., Miner, J.E., Eicken, H., and Golden, K.M. (2009). Pore space percolation in sea ice single crystals, *Journal of Geophysical Research*, **114**, C12017.

Qin, T. (2015). Statistical justifications for computationally tractable network data analysis, unpublished Technical Report No. 1179, Department of Statistics, University of Wisconsin, Madison.

Qin, T. and Rohe, K. (2013). Regularized spectral clustering under the degree-corrected stochastic blockmodel. In: *Advances in Neural Information Processing Systems (NIPS)* (C.J.C. Burges, L. Bottou, M. Welling, Z. Ghahramani, and K.Q. Weinberger, eds.), **26**, 3120–2128.

Radicchi, F., Fortunato, S., Markines, B., and Vespignani, A. (2009). Diffusion of scientific credits and the ranking of scientists, *Physical Review E*, **80**, 056103.

Rahimi, A. and Recht, B. (2004). Clustering with normalized cuts is clustering with a hyperplane. In: *Workshop on Statistical Learning in Computer Vision*, Prague, Czech Republic.

Ramey, D.B. (1982). *A nonparametric test of bimodality with applications to cluster analysis*, Ph.D. dissertation, Department of Statistics, Yale University.

Rao, V.S., Srinivas, K., Sujini, G.N., and Kumar, G.N.S. (2014). Protein–protein interaction detection: methods and analysis, *International Journal of Proteomics* (dx.doi.org/ 10.1155/2014/147648).

Rao, V.S.H. and Durvasula, R. (eds.) (2013). *Dynamic Models of Infectious Diseases*, **2** *(Non-Vector-Borne Diseases)*, New York: Springer.

Reeds, J. (1975). Discussion of Efron's paper, *The Annals of Statistics*, **3**, 1234–1238.

Ribeiro, B. and Towsley, D. (2010). Estimating and sampling graphs with multidimensional random walks. In: *Proceedings of the 10th ACM SIGCOMM Conference on Internet Measurement, IMC 2010*, ACM, New York, pp. 390–403.

Ribeiro, B., Wang, P., and Towsley, D. (2010). On estimating degree distributions of directed graphs through sampling, UMass Technical Report UM-CS-2010-046.

Rinaldo, A., Fienberg, S.E., and Zhou, Y. (2009). On the geometry of discrete exponential families with application to exponential random graph models, *Electronic Journal of Statistics*, **3**, 446–484.

Rinaldo, A., Petrović, S., and Fienberg, S.E. (2010). On the existence of the MLE for directed random graph network model with reciprocation, unpublished manuscript (arXiv:1010.0745v1).

Rippel, E., Bar-Gill, A., and Shimkin, N. (2005). Fast graph-search algorithms for general-aviation flight trajectory generation, *Journal of Guidance, Control, and Dynamics*, **28**, 801–811.

Robert, C.P. and Casella, G. (2004). *Monte Carlo Statistical Methods, 2nd ed.*, New York: Springer.

Robert, C.P. and Casella, G. (2011). A short history of Markov chain Monte Carlo: subjective recollections from incomplete data, *Statistical Science*, **26**, 102–115.

Roberts, F.D.K. (1967). A Monte Carlo solution of a two-dimensional unstructured cluster problem, *Biometrika*, **54**, 625–628.

Roberts, F.D.K. and Storey, S.H. (1968). A three-dimensional cluster problem, *Biometrika*, **55**, 258–260.

Rodriguez, G. and Goldman, N. (1995). An assessment of estimation procedures for multilevel models with binary responses, *Journal of the Royal Statistical Society, Series A (Statistics in Society)*, **158**, 73–89.

Rohe, K. and Yu, B. (2012). Co-clustering for directed graphs: the stochastic co-blockmodel and a spectral algorithm, unpublished manuscipt (arXiv:1204.2296).

Rohe, K., Chatterjee, S., and Yu, B. (2011). Spectral clustering and the high-dimensional stochastic blockmodel, *The Annals of Statistics*, **39**, 1878–1915.

Rolland, T., Tasan, M., Charloteaux, B., Hao, T., Roth, F.P., and Vidal, M. (2014). A proteome-scale map of the human interactome network, *Cell*, **159**, 1212–1226.

Ron, D. (2008). Property testing: a learning theory perspective, *Foundations and Trends in Machine Learning*, **1**, 307–402 (www.wisdom.weizmann.ac.il/~oded/pt-intro.html).

Ron, D. (2010). Algorithmic and analysis techniques in property testing, *Foundations and Trends in Theoretical Computer Science*, **5**, 73–205.

Ross, S.M. (1996). *Stochastic Processes, 2nd ed.*, New York: Wiley.

Rossetti, G. and Cazabet, R. (2018). Community discovery in dynamic networks: a survey, *ACM Computing Surveys*, **51**, 1–37.

Rubinfeld, R. and Sudan, M. (1996). Robust characterization of polynomials with application to program testing, *SIAM Journal on Computing*, **25**, 252–271.

Russo, L. (1978). A note on percolation, *Zeitschrift für Wahrscheinlichkeitstheorie und Verwandte Gebiete*, **43**, 39–48.

Sailer, L.D. (1978). Structural equivalence: meaning and definition, computation and application, *Social Networks*, **1**, 73–90.

Salehi, M., Rabiee, H.R., and Rajabi, A. (2012). Sampling from complex networks with high community structures, *Chaos*, **22**, 023126.1–023126.12.

Salganik, M.J. (2006). Variance estimation, design effects, and sample size calculations for respondent-driven sampling, *Journal of Urban Health*, **83**, 98–112.

Salganik, M.J. and Heckathorn, D.D. (2004). Sampling and estimation in hidden populations using respondent-driven sampling, *Sociological Methodology*, **34**, 193–239.

Salter-Townshend, M. and Murphy, T.B. (2012). Variational Bayesian inference for the latent position cluster model for network data, *Computational Statistics and Data Analysis*, **57**, 661–671.

Salter-Townshend, M., White, A., Gollini, I., and Murphy, T.B. (2012). Review of statistical network analysis: models, algorithms, and software, *Statistical Analysis and Data Mining*, **5**, 243–264.

Sampson, S.F. (1968). *A novitiate in a period of change: an experimental and case study of social relationships*, unpublished Ph.D. dissertation, Department of Sociology, Cornell University.

Sarkar, P. and Bickel, P.J. (2013). Role of normalization in spectral clustering for stochastic blockmodels, *The Annals of Statistics*, **43**, 962–990.

Särndal, C.-E., Swensson, B., and Wretman, J. (1992). *Model Assisted Survey Sampling*, New York: Springer.

Saul, Z.M. and Filkov, V. (2007). Exploring biological network structure using exponential random graph models, *Bioinformatics Advance Access*, Oxford: Oxford University Press.

Schneider, C.M., Moreira, A.A., Andrade, J.S., Havlin, S., and Herrmann, H.J. (2011). Mitigation of malicious attacks on networks, *Proceedings of the National Academy of Sciences*, **108**, 3838–3841.

Schonmann, R.H. (1999). Stability of infinite clusters in supercritical percolation, *Probability Theory and Related Fields*, **113**, 287–300.

Schulman, L.S. and Seiden, P.E. (1982). Percolation analysis of stochastic models of galactic evolution, *Journal of Statistical Physics*, **27**, 83–118.

Schulman, L.S. and Seiden, P.E. (1986). Percolation and galaxies, *Science*, **233**, 425–431.

Schweikert, D.G. and Kernighan, B.W. (1972). A proper model for the partitioning of electrical circuits, *Proceedings of the 9th Design Automation Workshop*, pp. 57–62.

Schweitzer, F., Fagiolo, G., Sornette, D., Vega-Redondo, F., Vespignani, A. and White, D.R. (2009). Economic networks: the new challenges, *Science*, **325**, 422–425.

Scott, J. (2000). *Social Network Analysis*, 2nd ed., London: Sage.

Seiden, P.E. and Schulman, L.S. (1990). Percolation model of galactic structure, *Advances in Physics*, **39**, 1–54.

Semaan, S., Lauby, J., and Liebman, J. (2002). Street and network sampling in evaluation studies of HIV risk-reduction interventions, *AIDS Review*, **4**, 213–223.

Seymore, P.D. and Welsh, D.J.A. (1978). Percolation probabilities on the square lattice, *Annals of Discrete Mathematics*, **3**, 227–245.

Shafiei, M. and Chipman, H. (2010). Mixed-membership stochastic block-models for transactional networks, *Proceedings of the International Conference on Data Mining*, 1019–1024.

Shen-Orr, S.S., Milo, R., Mangan, S., and Alon, U. (2002). Network motifs in the transcriptional regulation network of *Eschericha coli*, *Nature Genetics*, **31**, 64–68.

Sherwani, N. (1993). *Algorithms for VLSI Physical Design Automation*, New York: Springer.

Shi, J. and Malik, J. (2000). Normalized cuts and image segmentation, *IEEE Transactions on Pattern Analysis and Machine Intelligence*, **22**, 888–905.

Shilts, R. (1987). *And the Band Played On: Politics, People, and the AIDs Epidemic*. New York: St. Martin's Press.

Sidow, A. (1996). Gen(om)e duplications in the evolution of early vertebrates, *Current Opinion in Genetics & Development*, **6**, 715–722.

Simmons, G.F. (1963). *Introduction to Topology and Modern Analysis*, New York: McGraw-Hill.

Simon, H.A. (1955). On a class of skew distribution functions, *Biometrika*, **42**, 425–440.

Sinclair, A. (1992). Improved bounds for mixing rates of Markov chains and multicommodity flow, *Combinatorics, Probability, and Computing*, **1**, 351–370.

Skarding, J., Gabrys, B., and Musial, K. (2021). Foundations and modelling of dynamic networks using Dynamic Graph Neural Networks: a survey, *IEEEAccess*, **9** (arXiv:2005.07496v2).

Slade, G. (1994). Self-avoiding walks, *The Mathematical Intelligencer*, **16**, 29–35.

Small, H.G. (1973). Cocitation in scientific literature – new measure of relationship between 2 documents, *Journal of the American Society for Information Science*, **24**, 265–269.

Smirnov, S. and Werner, W. (2001). Critical exponents for two-dimensional percolation, *Mathematical Research Letters*, **8**, 729–744.

Smith, A.L., Asta, D.M., and Calder, C.A. (2019). The geometry of continuous latent space models for network data, *Statistical Science*, **34**, 428–453.

Snijders, T.A.B. (2002). Markov chain Monte Carlo estimation of exponential random graph models, *Journal of Social Structure*, **3**, 1–40.

Snijders, T.A.B. and Nowicki, K. (1997). Estimation and prediction for stochastic blockmodels for graphs with latent block structure, *Journal of Classification*, **14**, 75–100.

Snijders, T.A.B. and Nowicki, K. (2007). *Manual for BLOCKS Version 1.8*, available at www.stats.ox.ac.uk/ snijders/Blocks18_man.pdf.

Spiegelhalter, D.J., Thomas, A., and Best, N.G. (1999). *Win-BUGS Version 1.2*, Cambridge: MRC Biostatistics Unit.

Spielman, D.A. and Teng, S.-H. (1996). Spectral partitioning works: planar graphs and finite element meshes, *Proceedings of the 37th Annual Symposium on Foundations of Computer Science*, 96–106.

Sporns, O. (2011). *Networks of the Brain*, Cambridge, MA: MIT Press.

Sporns, O. (2013). Structure and function of complex brain networks, *Dialogues in Clinical Neuroscience*, **15**, 247–262.

Spreen, M. (1992). Rare populations, hidden populations, and link-tracing designs: what and why? *Bulletin Methodologie Sociologique*, **36**, 34–58.

Spreen, M. and Coumans, M. (2003). A note on network sampling in drug abuse research, *Connections*, **25**, 27–35.

Stam, C.J. (2007). Graph theoretical analysis of complex networks in the brain, *Nonlinear Biomedical Physics*, **1**, 3–22.

Stauffer, D. and Aharony, A. (1992). *Introduction to Percolation Theory*, 2nd ed., London: Taylor & Francis.

Stauffer, D., Coniglio, A., and Adam, M. (1982). Gelation and critical phenomena. In: *Advances in Polymer Science*, **44**, New York: Springer.

Stephens, M. (2000). Dealing with label-switching in mixture models, *Journal of the Royal Statistical Society, Series B*, **62**, 795–809.

Stockmayer, W.H. (1943). Theory of molecular size distribution and gel formation in branched polymers II. General cross linking, *Journal of Chemical Physics*, **12**, 125–132.

Strauss, D. and Ikeda, M. (1990). Pseudolikelihood estimation for social networks, *Journal of the American Statistical Association*, **85**, 204–212.

Stumpf, M. and Wiuf, C. (2005). Sampling properties of random graphs: the degree distribution, *Physics Review E*, **72**, 036118.

Stumpf, M., Wiuf, C., and May, R. (2005). Subnets of scale-free networks are not scale-free: sampling properties of networks, *Proceedings of the National Academy of Sciences*, **102**, 4221–4224.

Sussman, D., Tang, M., Fishkind, D.E., and Priebe, C.E. (2012). A consistent adjacency spectral embedding for stochastic blockmodel graphs, *Journal of the American Statistical Association*, **107**, 1119–1128.

Szemerédi, E. (1978). Regular partitions of graphs, *Problèmes combinatoires et théorie des graphes* (Colloques Internationaux du CNRS, Université de Orsay, Orsay, 1976), **260**, 399–401.

Tanner, J.C. (1953). A problem of interference between two queues, *Biometrika*, **40**, 58–69.

Taylor, C. and Griffiths, P. (2005). Sampling issues in drug epidemiology. In: *Epidemiology of Drug Abuse* (Z. Sloboda, ed.), New York: Springer, pp. 79–98.

Thieffry, D., Huerta, A.M., Perez-Rueda, E., and Collado-Vides, J. (1998). From specific gene regulation to genomic networks: a global analysis of transcriptional regulation in *Escherichia coli*, *Bioessays*, **20**, 433–440.

Thompson, R.M. and Townsend, C.R. (1999). The effect of seasonal variation on the community structure and food-web attributes of two streams: implications for food-web science, *Oikos*, **87**, 75-88.

Thompson, S.K. (2012). *Sampling*, 3rd ed., New York: Wiley.

Tierney, L. (1994). Markov chains for exploring posterior distributions, *The Annals of Statistics*, **22**, 1701–1728.

Titterington, D.M., Smith, A.F.M., and Makov, U.E. (1985). *Statistical Analysis of Finite Mixture Distributions*, New York: Wiley.

Traag, V.A. (2021). leidenalg documentation. Available at `readthedocs/projects/ leidenalg/downloads/pdf/latest/`.

Ugander, J., Karrer, B., Backstrom, L., and Marlow, C. (2011). The anatomy of the Facebook social graph (`arXiv:1111.4503`).

Ulanowicz, R.E., Bondavilli, C., and Egnotovich, M.S. (1998). Network analysis of trophic dynamics in south Florida ecosystem, FY 97: the Florida Bay ecosystem, *Annual Report of the U.S. Geological Service Biological Resources Division*, pp. 98–123.

Van den Berg, J. and Keane, M. (1984). On the continuity of the percolation probability function, *Contemporary Mathematics*, **26**, 61–65.

Van den Berg, J. and Kesten, H. (1985). Inequalities with applications to percolation and reliability, *Journal of Applied Probability*, **22**, 556–569.

Van den Heuvel, M.P. and Sporns, O. (2013). Network hubs in the human brain, *Trends in Cognitive Sciences*, **17**, 683–696.

Van den Heuvel, M.P., Mandl, R., and Pol, H.H. (2008). Normalized cut group clustering of resting-state fMRI data, *PLoS ONE*, **3**, e2001.

Van den Heuvel, M.P. and Sporns, O. (2013). Network hubs in the human brain, *Cell, Trends in Cognitive Sciences*, Special Issue: The Connectome – Feature Review, **17**, 683–696.

Van der Hofstad, R, (2017a). *Random Graphs and Complex Networks*, Vol. 1, Cambridge Series in Statistical and Probabilistic Mathematics, Cambridge: Cambridge University Press.

Van der Hofstad, R, (2017b). *Random Graphs and Complex Networks*, Vol. 2, Cambridge: Cambridge University Press.

Vanderzande, C. (1998). *Lattice Models of Polymers*, Cambridge: Cambridge University Press.

Van Driessche, R. and Roose, D. (1995). An improved spectral bisection algorithm and its application to dynamic load balancing, *Parallel Computing*, **21**, 29–48.

Van Duijn, M.A.J. (1995). Estimation of a random effects model for directed graphs. In: *Symposium Statistische Software* (T.A.B. Snijders, M. Verbeek, B. Engel, J.C. Houwelingen, A. Keen, and G.J. Stemerdink, eds.), Interuniversitair Centrum ProGAMMA, pp. 113–132.

Van Duijn, M.A.J., Snijders, T.A.B., and Zijlstra, B.J.H. (2004). p_2: a random effects model with covariates for directed graphs, *Statistica Neerlandica*, **58**, 234–254.

Van Wijk, B.C.M., Stam, C.J., and Daffertshofer, A. (2010). Comparing brain networks of different sizes and connectivity density using graph theory, *PLoS ONE*, **5**, e13701.

Veenstra, R., Lindenberg, S., Zijlstra, B.J.H., de Winter, A.F., Verhulst, F.C., and Ormel, J. (2007). The dyadic nature of bullying and victimization: testing a dual-perspective theory, *Child Development*, **78**, 1843–1854.

Vehlow, C., Beck, F., Auwärter, P., and Weiskopf, D. (2015). Visualizing the evolution of communities in dynamic graphs, *Computer Graphics Forum*, **34**, 277–288.

Verdery, A.M., Mouw, T., Bauldry, S., and Mucha, P.J. (2015). Network structure and biased variance estimation in respondent-driven sampling, *PLoS ONE*, **11**, e0145296. Erratum: *PLoS ONE*, **11** (2016) e0148006.

Vicsek, T. and Kertész, J. (1981). Monte Carlo renormalization-group approach to percolation on a continuum: test of universality, *Journal of Physics A: Mathematical and General*, **14**, L31–L37.

Vieira, V. da F., Xavier, C.R., Ebecken, N.F.F., and Evsukoff, A.G. (2014). Performance evaluation of modularity based community detection algorithms in large scale networks, *Mathematical Problems in Engineering*, **2014**, art. 502809.

Volz, E. and Heckathorn, D.D. (2008). Probability-based estimation theory for respondent-driven sampling, *Journal of Official Statistics*, **24**, 79.

Von Bahr, B. and Martin-Löf, A. (1980). Threshold limit theorems for some epidemic processes, *Advances in Applied Probability*, **12**, 319–349.

Von Luxburg, U. (2007). A tutorial on spectral clustering, *Statistics and Computing*, **17**, 395–416.

Von Luxburg, U., Belkin, M., and Bousquet, O. (2008). Consistency of spectral clustering, *The Annals of Statistics*, **36**, 555–586.

Wagner, A. (1994). Evolution of gene networks by gene duplications: a mathematical model and its implications on genome organization, *Proceedings of the National Academy of Sciences of the USA*, **91**, 4387–4391.

Wagner, D. and Wagner, F. (1993). Between min cut and graph bisection. In: *Proceedings of the 18th International Symposium on Mathematical Foundations of Computer Science (MFCS)*, London: Springer, pp. 744–750.

Wainwright, M.J. and Jordan, M.I. (2008). Graphical models, exponential families, and variational inference, *Foundations and Trends in Machine Learning*, **1**, 1–305.

Wakita, K. and Tsurumi, T. (2007). Finding community structure in mega-scale social networks (extended abstract), *Proceedings of the 16th International Conference on World Wide Web*, New York, NY, pp. 1275–1276.

Walker, D., Xie, H., Yan, K.K., and Maslov, S. (2007). Ranking scientific publications using a model of network traffic, *Journal of Statistical Mechanics: Theory and Experiment*, p. 06010.

Wang, X.F. and Chen, G. (2003). Complex networks: small-world, scale-free, and beyond, *IEEE Circuits and Systems Magazine*, **3**, 6–20.

Wang, Y.J. and Wong, G.Y. (1987). Stochastic blockmodels for directed graphs, *Journal of the American Statistical Association*, **82**, 8–19.

Wasserman, S. (1980). Analyzing social networks as stochastic processes, *Journal of the American Statistical Association*, **75**, 280–294.

Wasserman, S, and Faust, K. (1994). *Social Network Analysis: Methods and Applications*, Cambridge: Cambridge University Press.

Wasserman, S. and Patterson, P. (1996). Logit models and logistic regressions for social networks: I. An introduction to Markov random graphs and p^*, *Psychometrika*, **60**, 401–426.

Watson, H.W. and Galton, F. (1874). On the probability of the extinction of families, *Journal of the Anthropological Institute of Great Britain and Ireland*, **4**, 138–144.

Watts, D.J. (1999). *Small Worlds*, Princeton, NJ: Princeton University Press.

Watts, D.J. (2004). *Six Degrees: The Science of a Connected Age*, New York: W.W. Norton & Co.

Watts, D.J. and Strogatz, S.H. (1998), Collective dynamics of 'small-world' networks, *Nature*, **393**, 440-442.

Wei, Y.C. and Cheng, C.K. (1991). Ratio cut partitioning for hierarchical designs, *IEEE Transactions on Computer-Aided Design*, **10**, 911–921.

Weirman, J.C. (1981). Bond percolation on honeycomb and triangular lattices, *Advances in Applied Probability*, **13**, 298–313.

Wejnert, C. (2009). An empirical test of respondent-driven sampling: point estimates, variance, degree measures, and out-of-equilibrium data, *Sociological Methodology*, **39**, 73–116.

Wejnert, C. and Heckathorn, D.D. (2007). Web-based network sampling: efficiency and efficacy of respondent-driven sampling for online research, *Sociological Methods and Research*, **37**, 105–134.

West, J.D., Bergstrom, T.C., and Bergstrom, C.T. (2010). *Collage and Research Libraries*, **71**, 236–244.

Westerby, J.D. (2012). *Dynamic Network Theory: How Social Networks Influence Goal Pursuit*, Washington, DC: American Psychological Association.

Whitaker, A. (2018). The Eureka moment that made Bitcoin possible, *The Wall Street Journal*, 26–27 May, p. C18.

White, D. and Reitz, K. (1983). Graph and semigroup homomorphisms, *Social Networks*, **5**, 193–234.

White, H.C., Boorman, S.A., and Breiger, R.L. (1976). Social structure from multiple networks: 1. blockmodels of roles and positions, *American Journal of Sociology*, **81**, 730–780.

Williams, R. (2011). Zachary Karate Club. Available at `sites.google.com/site/ucinetsoftware/datasets/zacharykarateclub`.

Wilson, K.G. (1983). The renormalization group and critical phenomena, *Reviews of Modern Physics*, **55**, 583–600. (This was also the Nobel Lecture, 8 December 1982.)

Wilson, R. (2003). *Four Colours Suffice*, London: Penguin Books.

Winship, C. and Mandel, M. (1983). Roles and positions: a critique and extension of the blockmodeling approach, *Sociological Methodology*, **14**, 314–344.

Wolfe, P.J. and Olhede, S.C. (2014). Nonparametric graphon estimation (`arXiv:1309.5936`).

Wolfe, K.H. and Shields, D.C. (1997). Molecular evidence for an ancient duplication of the entire yeast genome, *Nature*, **387**, 708–713.

Woodall, W.H., Zhao, M.J., Paynabar, K., Sparks, R., and Wilson, J.D. (2017). An overview and perspective on social network monitoring, *IISE Transactions*, **49**, 354–365.

Wormald, N.C. (1978). *Some problems in the enumeration of labelled graphs*, Ph.D. dissertation, University of Newcastle.

Xie, J., Kelley, S., and Szymanski, B.K. (2013). Overlapping community detection in networks: the state of the art and comparative study, *ACM Computing Surveys*, **45**, 1–37.

Xing, E.P., Fu, W., and Song, L. (2010). A state-space mixed membership blockmodel for dynamic network tomography, *The Annals of Applied Statistics*, **4**, 535–566.

Xu, K.S. and Hero III, A.O. (2014). Dynamic stochastic blockmodels for time-evolving social networks, *IEEE Journal of Selected Topics in Signal Processing*, **8**, 552–562.

Yang, J. and Leskovec, J. (2012). Community-affiliation graph model for overlapping network community detection. In: *Proceedings of the IEEE International Conference in Data Mining*, pp. 1170–1175.

Yang, J. and Leskovec, J. (2014). Overlapping communities explain core-periphery organization of networks, *Proceedings of the IEEE*, **102**, 1892–1902.

Yao, D. and Durrett, R. (2020). Epidemics on evolving graphs (arXiv2003.08534).

Yellen, J.L. (2013). Interconnectness and systemic risk: lessons from the financial crisis and policy implications, Remarks at the American Economic Association/American Finance Association Joint Luncheon, 4th January 2013.

Yin, H., Benson, A.R., Leskovec, J., and Gleich, D.F. (2017). Local higher-order graph clustering, *Proceedings of the 23rd ACM SIGKDD International Conference on Knowledge Discovery and Data Mining*.

Zachary, W.W. (1977). An information flow model for conflict and fission in small groups, *Journal of Anthropological Research*, **33**, 452–473.

Zallen, R. and Scher, H. (1971). Percolation on a continuum and the localization–delocalization transition in amorphous semiconductors, *Physical Review B*, **4**, 4471–4479.

Zhang, X., Nadakuditi, R.R., and Newman, M.E.J. (2014). Spectra of random graphs with community structure and arbitrary degrees, *Physical Review E*, **89**, 042816.

Zhang, Y., Kolaczyk, E.D., and Spencer, B.D. (2015). Estimating network degree distributions under sampling: an inverse problem, with applications to monitoring social media networks, *The Annals of Applied Statistics*, **9**, 166–199.

Zhao, Y., Levina, E., and Zhu, J. (2012). Consistency of community detection in networks under degree-corrected staochastic block models, *Annals of Statistics*, **40**, 2266–2292.

Zheng, T., Salganik, M.J., and Gelman, A. (2006). How many people do you know in prison? Using overdispersion in count data to estimate social structure in networks, *Journal of the American Statistical Association*, **101**, 409–423.

Zhou, S. (2004). *Parameterising and modelling the Internet topology*, unpublished Ph.D dissertation, Department of Electronic Engineering, Queen Mary College, University of London.

Zhou, Z. (2016). Systemic bank panics in financial networks, unpublished manuscript (zhenzhouecon.com/assets/jmp_draft.pdf).

Zijlstra, B.J.H., van Duijn, M.A.J., and Snijders, T.A.B. (2009). MCMC estimation for the p_2 network regression model with crossed random effects, *British Journal of Mathematical and Statistical Psychology*, **62**, 143–166.

Zijlstra, B.J.H., Veenstra, R., and van Duijn, M. (2008). A multilevel p_2 model with covariates for the analysis of binary bully–victim network data in multiple classrooms. In: *Modeling Dyadic and Interdependent Data in the Developmental and Behavioral*

Sciences (N.A. Card, J.P. Selig, and T.D. Little, eds.). New York: Routledge, pp. 369–389.

Zuev, S.A. and Sidorenko, A.F. (1985a). Continuous models of percolation theory I, *Theoretical and Mathematical Physics*, **62**, 51–58.

Zuev, S.A. and Sidorenko, A.F. (1985b). Continuous models of percolation theory II, *Theoretical and Mathematical Physics*, **62**, 171–177.

Index of Examples

Subject Index